Computing Essentials

Making IT work for you

INTRODUCTORY 2025

The O'Leary Series

Computing Concepts
- *Computing Essentials 2019*
- *Computing Essentials 2021*
- *Computing Essentials 2023*
- *Computing Essentials 2025*

Microsoft Office Applications
- *Microsoft® Windows 7: A Case Approach*
- *Microsoft® Office 2013: A Case Approach*
- *Microsoft® Office Word 2013: A Case Approach* Introductory Edition
- *Microsoft® Office Excel 2013: A Case Approach* Introductory Edition
- *Microsoft® Office Access 2013: A Case Approach* Introductory Edition
- *Microsoft® Office PowerPoint 2013: A Case Approach* Introductory Edition

Computing Essentials

Making  work for you

INTRODUCTORY 2025

Daniel A. O'Leary
Professor
City College of San Francisco

Timothy J. O'Leary
Professor Emeritus
Arizona State University

Linda I. O'Leary

COMPUTING ESSENTIALS 2025

Published by McGraw Hill LLC, 1325 Avenue of the Americas, New York, NY 10019. Copyright ©2024 by McGraw Hill LLC. All rights reserved. Printed in the United States of America. No part of this publication may be reproduced or distributed in any form or by any means, or stored in a database or retrieval system, without the prior written consent of McGraw Hill LLC, including, but not limited to, in any network or other electronic storage or transmission, or broadcast for distance learning.

Some ancillaries, including electronic and print components, may not be available to customers outside the United States.

This book is printed on acid-free paper.

1 2 3 4 5 6 7 8 9 LWI 29 28 27 26 25 24

ISBN 978-1-266-81687-1
MHID 1-266-81687-9

Cover Image: *metamorworks/Shutterstock*

All credits appearing on page or at the end of the book are considered to be an extension of the copyright page.

The Internet addresses listed in the text were accurate at the time of publication. The inclusion of a website does not indicate an endorsement by the authors or McGraw Hill LLC, and McGraw Hill LLC does not guarantee the accuracy of the information presented at these sites.

mheducation.com/highered

• Dedication

Maggie - A trusted friend and faithful companion.

Brief Contents

1. Information Technology, the Internet, and You 1
2. The Internet, the Web, and Electronic Commerce 23
3. Application Software 54
4. System Software 82
5. The System Unit 106
6. Input and Output 132
7. Secondary Storage 162
8. Communications and Networks 184
9. Privacy, Security, and Ethics 212
10. Information Systems 241
11. Databases 261
12. Systems Analysis and Design 284
13. Programming and Languages 305

The Evolution of the Computer Age 333

The Computer Buyer's Guide 344

Glossary 347

Index 370

Contents

INFORMATION TECHNOLOGY, THE INTERNET, AND YOU 1

Introduction 2
Information Systems 3
People 5
Software 6
    System Software 7
    Application Software 7

Making  work for you: Free Antivirus Program 8

Hardware 9
    Types of Computers 9
    Cell Phones 10
    Personal Computer Hardware 10

Data 12
Connectivity and the Mobile Internet 13
Careers in IT 14

A Look to the Future: Using and Understanding Information Technology 15

> Visual Summary 16
> Key Terms 19
> Multiple Choice 20
> Matching 21
> Open-Ended 21
> Discussion 22

THE INTERNET, THE WEB, AND ELECTRONIC COMMERCE 23

Introduction 24
The Internet and the Web 25
Internet Access 26
    Providers 26

Making IT work for you: Online Entertainment 27

    Browsers 29
Web Utilities 30

    Filters 30
    File Transfer Utilities 31
    Internet Security Suites 32
Communication 33
    Social Networking 33
    Blogs, Microblogs, Podcasts, and Wikis 34
    Messaging 35
    E-mail 36
Search Tools 38
    Search Engines 38
    Content Evaluation 38
Electronic Commerce 39
    Security 40
Cloud Computing 41
The Internet of Things 43
Careers in IT 44

A Look to the Future: Home Smart Home 45

> Visual Summary 46
> Key Terms 50
> Multiple Choice 51
> Matching 52
> Open-Ended 52
> Discussion 53

APPLICATION SOFTWARE 54

Introduction 55
Application Software 56
    App Stores 56
    User Interface 57
    Common Features 58
Mobile Apps 59
    Apps 59
General-Purpose Applications 60
    Word Processors 60
    Presentation Software 63
    Spreadsheets 64
    Database Management Systems 66

Specialized Applications 67
    Graphics Programs 67
    Video Game Design Software 68
    Web Authoring Programs 69
    Other Specialized Applications 70

Software Suites 70
    Office Suites 70
    Cloud Computing 70
    Specialized and Utility Suites 71

Careers in IT 71

Making work for you: Cloud Office Suites 72

A Look to the Future: The New Workplace Realities 74

> Visual Summary 75
> Key Terms 78
> Multiple Choice 79
> Matching 80
> Open-Ended 80
> Discussion 81

4 SYSTEM SOFTWARE 82

Introduction 83
System Software 84
Operating Systems 85
    Functions 85
    Features 85
    Categories 86

Mobile Operating Systems 88
Desktop Operating Systems 88
    Windows 88
    macOS 89
    UNIX and Linux 89
    Virtualization 90

Utilities 91

Making work for you: Virtual Assistant 92

    Operating System Utilities 94
    Utility Suites 97

Careers in IT 97

A Look to the Future: Making Better Computers by Making Them More Human 98

> Visual Summary 99
> Key Terms 102
> Multiple Choice 103
> Matching 104
> Open-Ended 104
> Discussion 105

5 THE SYSTEM UNIT 106

Introduction 107
System Unit 108
    Smartphones 108
    Tablets 108
    Laptops 109
    Desktops 109
    Wearable Computers 109

Making work for you: Gaming 110

    Components 112

System Board 112
Microprocessor 114
    Microprocessor Chips 114
    Specialty Processors 115

Memory 115
    RAM 115
    ROM 116
    Flash Memory 116

Expansion Cards and Slots 117
Bus Lines 118
    Expansion Buses 118

Ports 119
    Standard Ports 119
    Specialized Ports 119
    Cables 119

Power Supply 120
Electronic Data and Instructions 121
    Numeric Representation 121
    Character Encoding 122

Careers in IT 123

A Look to the Future: Brain–Computer Interfaces 124

Visual Summary 125
Key Terms 128
Multiple Choice 129
Matching 130
Open-Ended 130
Discussion 131

INPUT AND OUTPUT 132

Introduction 133
What Is Input? 134
Keyboard Entry 134
    Keyboards 134
Pointing Devices 135
    Touch Screens 135
    Mice 136
    Game Controllers 136
Scanning Devices 137
    Optical Scanners 137
    Card Readers 138
    Bar Code Readers 138
    RFID Readers 139
    Character and Mark Recognition Devices 139
Image-Capturing Devices 139
    Digital Cameras 139
    Webcams 140
Audio-Input Devices 140
    Voice Recognition Systems 140
What Is Output? 141
Monitors 141
    Features 141
    Flat-Panel Monitors 142
    E-book Readers 143
    Other Monitors 143
Printers 144
    Features 144
    Inkjet Printers 145
    Laser Printers 145
    3D Printers 145
    Other Printers 145
Audio-Output Devices 146
Combination Input and Output Devices 146

Headsets 146
Multifunctional Devices 146
Virtual Reality Head-Mounted Displays and Controllers 147
Drones 147
Robots 147

Making IT work for you: Headphones 148

Ergonomics 150
    Portable Computers 151
Careers in IT 152

A Look to the Future: The Internet of Things 153

Visual Summary 154
Key Terms 158
Multiple Choice 159
Matching 160
Open-Ended 160
Discussion 161

SECONDARY STORAGE 162

Introduction 163
Storage 164
Solid-State Storage 165
    Solid-State Drives 165
    Flash Memory Cards 165
    USB Flash Drives 165
Hard Disks 166
    Internal Storage 167
    External Hard Disks 167
    Network Drives 167
    Performance Enhancements 167
Optical Discs 168
Cloud Storage 170
Mass Storage Devices 171

Making IT work for you: Cloud Storage 172

    Enterprise Storage System 174
    Storage Area Network 175
Careers in IT 175

A Look to the Future: Next-Generation Storage 176

Visual Summary 177
Key Terms 180
Multiple Choice 181
Matching 182
Open-Ended 182
Discussion 183

A Look to the Future: Telepresence Lets You Be There without Actually Being There 204

Visual Summary 205
Key Terms 208
Multiple Choice 209
Matching 210
Open-Ended 210
Discussion 211

COMMUNICATIONS AND NETWORKS 184

Introduction 185
Communications 186
- Connectivity 186
- The Wireless Revolution 186
- Communication Systems 186

Communication Channels 187
- Wireless Connections 188
- Physical Connections 189

Connection Devices 189
- Modems 190
- Connection Service 190

Data Transmission 191

Making  work for you: The Mobile Office 192

- Bandwidth 194
- Protocols 194

Networks 195
- Terms 195

Network Types 197
- Local Area Networks 197
- Home Networks 198
- Wireless LAN 198
- Personal Area Networks 198
- Metropolitan Area Networks 198
- Wide Area Networks 199

Network Architecture 199
- Topologies 199
- Strategies 200

Organizational Networks 201
- Internet Technologies 201
- Network Security 201

Careers in IT 203

PRIVACY, SECURITY, AND ETHICS 212

Introduction 213
People 214
Privacy 214
- Big Data 214
- Private Networks 216
- The Internet and the Web 216
- Online Identity 219
- Major Laws on Privacy 220

Security 220
- Cybercrime 220
- Social Engineering 222
- Malicious Software 222
- Malicious Hardware 222
- Measures to Protect Computer Security 223

Ethics 227

Making work for you: Security and Technology 228

- Cyberbullying 229
- Copyright and Digital Rights Management 229
- Plagiarism 230

Careers in IT 231

A Look to the Future: End of Anonymity 232

Visual Summary 233
Key Terms 236
Multiple Choice 237
Matching 238
Open-Ended 238
Discussion 239

10
INFORMATION SYSTEMS 241

Introduction 242
Organizational Information Flow 243
    Functions 243
    Management Levels 244
    Information Flow 245
Computer-Based Information Systems 246
Transaction Processing Systems 247
Management Information Systems 248
Decision Support Systems 249
Executive Support Systems 250
Other Information Systems 252
    Expert Systems 253
Careers in IT 253

A Look to the Future: ChatGPT: Changing the Workplace 254

> Visual Summary 255
> Key Terms 257
> Multiple Choice 258
> Matching 259
> Open-Ended 259
> Discussion 260

11
DATABASES 261

Introduction 262
Data 263
Data Organization 263
    Key Field 264
    Batch versus Real-Time Processing 264
Databases 266
    Need for Databases 266
    Database Management 266
DBMS Structure 268
    Hierarchical Database 268
    Network Database 269
    Relational Database 269
    Multidimensional Database 270
    Object-Oriented Database 271
Types of Databases 272
    Individual 272
    Company 272
    Distributed 272
    Commercial 272
Database Uses and Issues 274
    Strategic Uses 274
    Security 274
Careers in IT 275

A Look to the Future: The Future of Crime Databases 276

> Visual Summary 277
> Key Terms 280
> Multiple Choice 281
> Matching 282
> Open-Ended 282
> Discussion 283

12
SYSTEMS ANALYSIS AND DESIGN 284

Introduction 285
Systems Analysis and Design 286
Phase 1: Preliminary Investigation 287
    Defining the Problem 287
    Suggesting Alternative Systems 288
    Preparing a Short Report 288
Phase 2: Systems Analysis 289
    Gathering Data 289
    Analyzing the Data 289
    Documenting Systems Analysis 291
Phase 3: Systems Design 291
    Designing Alternative Systems 292
    Selecting the Best System 292
    Writing the Systems Design Report 292
Phase 4: Systems Development 293
    Acquiring Software 293
    Acquiring Hardware 293
    Testing the New System 294
Phase 5: Systems Implementation 294
    Types of Conversion 294

Training 295

Phase 6: Systems Maintenance 295

Prototyping and Rapid Applications Development 296

Prototyping 296

Rapid Applications Development 296

Careers in IT 296

A Look to the Future: The Challenge of Keeping Pace 297

> Visual Summary 298
> Key Terms 302
> Multiple Choice 303
> Matching 304
> Open-Ended 304
> Discussion 304

PROGRAMMING AND LANGUAGES 305

Introduction 306

Programs and Programming 307

What Is a Program? 307

What Is Programming? 307

Step 1: Program Specification 308

Program Objectives 308

Desired Output 308

Input Data 309

Processing Requirements 309

Program Specifications Document 309

Step 2: Program Design 310

Top-Down Program Design 310

Pseudocode 311

Flowcharts 311

Logic Structures 311

Step 3: Program Code 314

The Good Program 314

Coding 314

Step 4: Program Test 315

Syntax Errors 316

Logic Errors 316

Testing Process 316

Step 5: Program Documentation 317

Step 6: Program Maintenance 318

Operations 319

Changing Needs 319

CASE and OOP 319

CASE Tools 320

Object-Oriented Software Development 320

Generations of Programming Languages 320

Machine Languages: The First Generation 321

Assembly Languages: The Second Generation 321

High-Level Procedural Languages: The Third Generation 321

Task-Oriented Languages: The Fourth Generation 322

Problem and Constraint Languages: The Fifth Generation 322

Careers in IT 323

A Look to the Future: Your Own Programmable Robot 324

> Visual Summary 325
> Key Terms 329
> Multiple Choice 330
> Matching 331
> Open-Ended 331
> Discussion 332

The Evolution of the Computer Age 333

The Computer Buyer's Guide 344

Glossary 347

Index 370

New to *Computing Essentials* 2025

To increase student motivation and engagement, a focus on smartphones has been added by increasing content and providing marginal tips offering practical advice for efficient smartphone use. While the coverage of other topics has not been reduced, this change offers a gateway to demonstrate the relevance of all types of computers to their lives. Additionally, every chapter's Making IT Work for You, Privacy, Ethics, and Community features have been carefully revaluated, enhanced, and/or replaced. More specific new coverage includes the following:

Chapter 2:
- Expanded coverage of Web 5.0
- Added coverage of AI and the emotional web
- Expanded coverage of Web 4.0
- Reorganized coverage of Web 1.0 to 3.0 to emphasize relationships between web generations
- Expanded coverage of the history of the ARPANET and the World Wide Web
- Expanded coverage of https and Internet security
- Expanded coverage of security risks with file downloads
- Added coverage of cryptocurrencies and blockchain
- Added coverage of search engines and AI technologies
- Expanded coverage of artificial intelligence and its role in deep fakes

Chapter 3:
- Added coverage of features, including grammar checkers
- Expanded coverage of the download and install process for desktop versus mobile apps
- Expanded coverage of mobile productivity tools
- Expanded coverage of desktop publishing software
- Added coverage of 3D modeling programs
- Expanded coverage of grammar checkers to include AI-based grammar checkers
- Added gaming coverage to include free-to-play

Chapter 4:
- Expanded coverage of latest mobile OSes
- Added coverage of Virtual Desktop Infrastructure
- Expanded coverage of the limitations of Chromebook operating systems
- Added coverage of macOS 14 Sonoma

Chapter 5:
- Expanded coverage of microchip multicore processing
- Added coverage of SIMs and eSIMs
- Added coverage of mobile AI specialty microprocessors

Chapter 6:
- Expanded coverage of wireless mouse technologies
- Added coverage of 1D versus 2D Bar Codes
- Added coverage of QR codes and MaxiCode
- Expanded coverage of RFID abilities
- Added coverage of monitor refresh rate
- Added coverage of gaming monitor features
- Added coverage of drone delivery services
- Added coverage of headphone special features
- Added coverage of eyestrain ergonomics

Chapter 7:
- Updated USB cybersecurity risks
- Expanded coverage of secondary storage performance enhancements
- Expanded coverage of drive compression

Chapter 8:
- Expanded coverage of satellite communications
- Expanded coverage of Virtual Private Networks
- Added coverage of public Wi-Fi security risks and protection
- Added coverage of artificial intelligence and network security

Chapter 9:
- Expanded coverage of major laws and online privacy
- Expanded coverage of virus security risks
- Added coverage of multifactor authentication
- Expanded coverage of dual authentication
- Added coverage of WPA3 security
- Added coverage of artificial intelligence and deep fakes
- Expanded coverage of cyberbullying

Preface

The 20th century brought us the dawn of the digital information age and unprecedented changes in information technology. In fact, the rate of change is clearly increasing. As we begin the 21st century, computer literacy is undoubtedly becoming a prerequisite in whatever career you choose.

The goal of *Computing Essentials* is to provide you with the basis for understanding the concepts necessary for success. *Computing Essentials* also endeavors to instill an appreciation for the effect of information technology on people, privacy, ethics, and our environment and to give you a basis for building the necessary skill set to succeed in the 21st century.

Times are changing, technology is changing, and this text is changing too. As students of today, you are different from those of yesterday. You put much effort toward the things that interest you and the things that are relevant to you. Your efforts directed at learning application programs and exploring the web seem, at times, limitless. On the other hand, it is sometimes difficult to engage in other equally important topics such as personal privacy and technological advances.

At the beginning of each chapter, we carefully lay out why and how the chapter's content is relevant to your life today and critical to your future. Within each chapter, we present practical tips related to key concepts through the demonstration of interesting applications that are relevant to your lives. Topics presented focus first on outputs rather than processes. Then, we discuss the concepts and processes.

Motivation and relevance are the keys. This text has several features specifically designed to engage and demonstrate the relevance of technology in your lives. These elements are combined with a thorough coverage of the concepts and sound pedagogical devices.

Visual Learning

VISUAL CHAPTER OPENERS

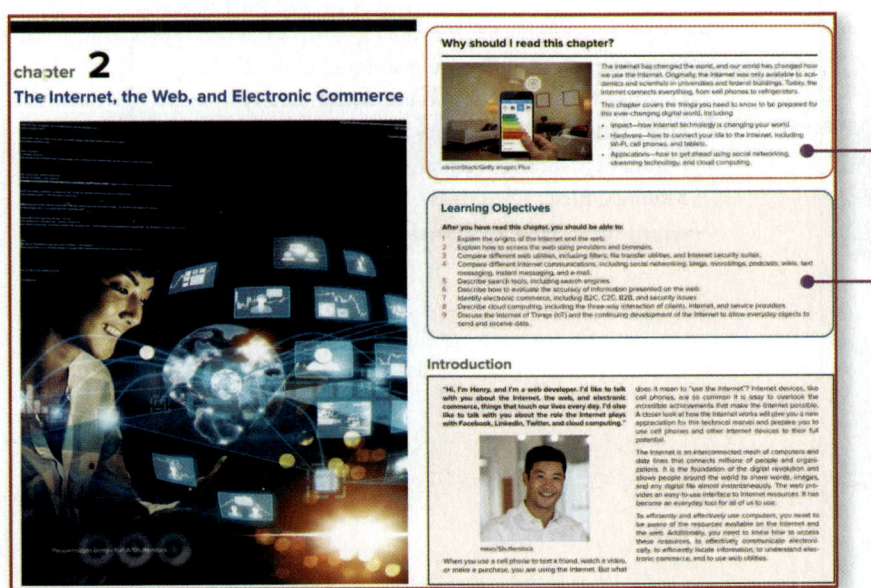

Each chapter begins with a Why Should I Read This Chapter? feature that presents a visually engaging and concise presentation of the chapter's relevance to the reader's current and future life in the digital world. Then a list of chapter learning objectives is presented providing a brief introduction to what will be covered in the chapter.

VISUAL SUMMARIES

Visual summaries appear at the end of every chapter and summarize major concepts covered throughout the chapter. Like the chapter openers, these summaries use graphics to reinforce key concepts in an engaging and meaningful way.

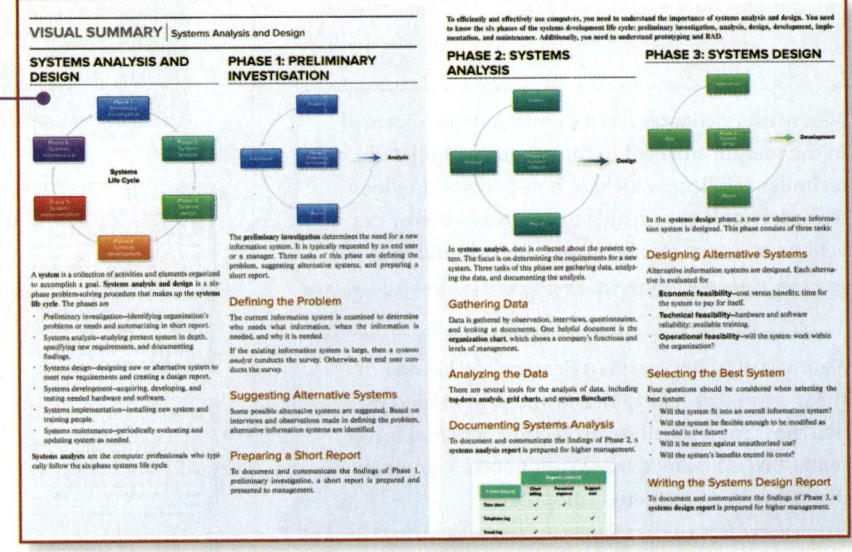

Unique Content

MAKING IT WORK FOR YOU

Special-interest topics are presented in the Making IT Work for You section found within nearly every chapter. These topics include Online Entertainment, Gaming, Virtual Assistants, and the Mobile Office.

PRIVACY, ETHICS, AND COMMUNITY

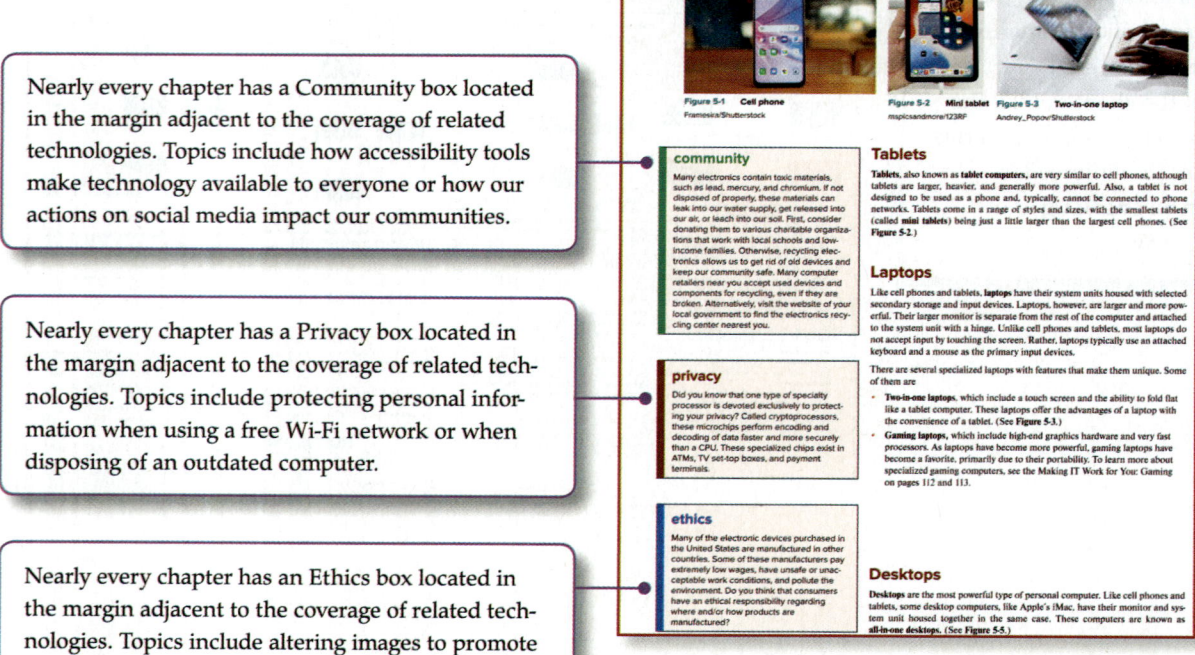

Nearly every chapter has a Community box located in the margin adjacent to the coverage of related technologies. Topics include how accessibility tools make technology available to everyone or how our actions on social media impact our communities.

Nearly every chapter has a Privacy box located in the margin adjacent to the coverage of related technologies. Topics include protecting personal information when using a free Wi-Fi network or when disposing of an outdated computer.

Nearly every chapter has an Ethics box located in the margin adjacent to the coverage of related technologies. Topics include altering images to promote a particular message and how the technology we use affects labor practices around the world.

Unique End-of-Chapter Discussion Materials

MAKING IT WORK FOR YOU

Making IT Work for You discussion questions are carefully integrated with the chapter's Making IT Work for You topics. The questions facilitate in-class discussion or written assignments focusing on applying specific technologies into a student's day-to-day life. They are designed to expand a student's awareness of technology applications.

PRIVACY

Privacy discussion questions are carefully integrated with the chapter's marginal Privacy box. The questions facilitate in-class discussion or written assignments focusing on critical privacy issues. They are designed to develop a student's ability to think critically and communicate effectively.

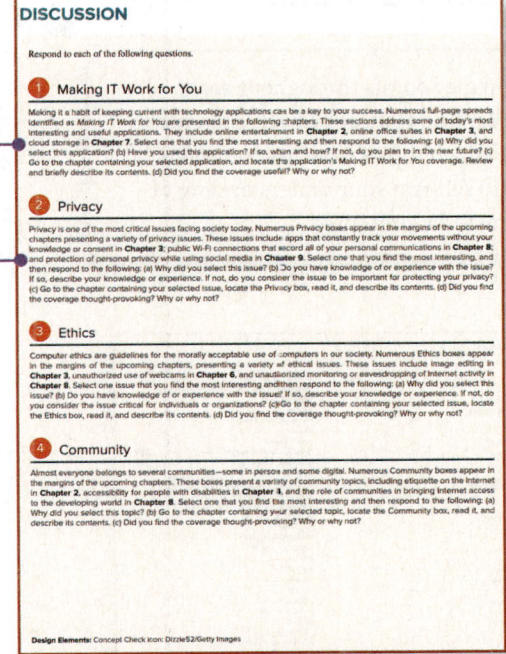

ETHICS

Ethics discussion questions are carefully integrated with the chapter's marginal Ethics boxes. The questions facilitate in-class discussion or written assignments focusing on ethical issues relating to technology. They are designed to develop a student's ability to think critically and communicate effectively.

COMMUNITY

Community discussion questions are carefully integrated with the chapter's marginal Community boxes. The questions facilitate in-class discussion or written assignments focusing on the impact of technology on communities. They are designed to develop a student's ability to think critically and communicate effectively.

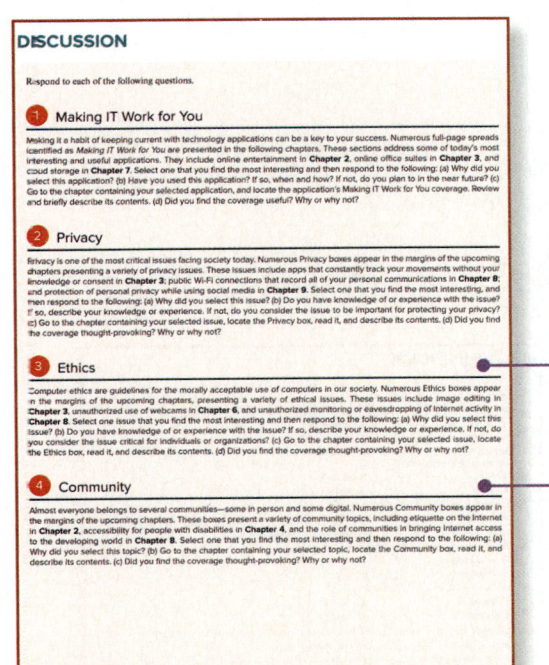

Reinforcing Key Concepts

CONCEPT CHECKS

Located at points throughout each chapter, the Concept Check cues you to note which topics have been covered and to self-test your understanding of the material presented.

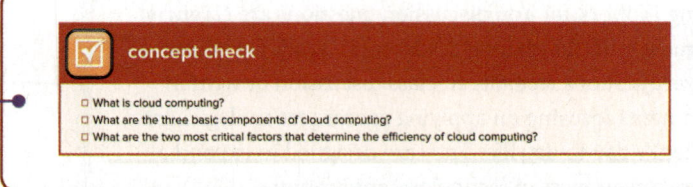

KEY TERMS

Throughout the text, the most important terms are presented in bold and are defined within the text. You will also find a list of key terms at the end of each chapter and in the glossary at the end of the book.

CHAPTER REVIEW

Following the Visual Summary, the chapter review includes material designed to review and reinforce chapter content. It includes a key terms list that reiterates the terms presented in the chapter, multiple-choice questions to help test your understanding of information presented in the chapter, matching exercises to test your recall of terminology presented in the chapter, and open-ended questions or statements to help review your understanding of the key concepts presented in the chapter.

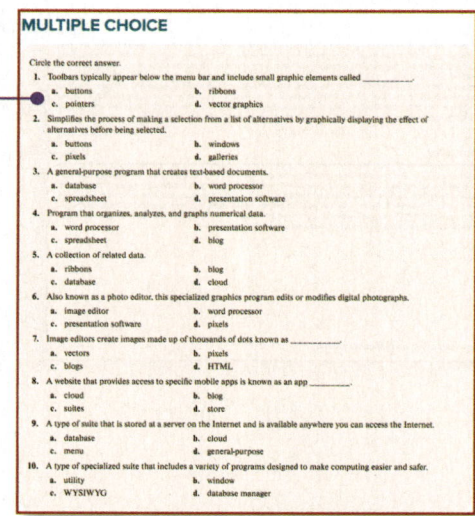

The Future of Information Technology

CAREERS IN IT

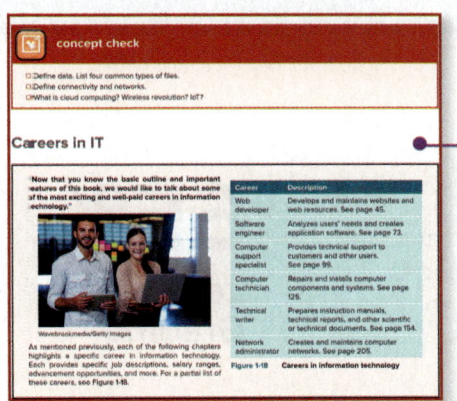

Some of the fastest-growing career opportunities are in information technology. Each chapter highlights one of the most promising careers in IT by presenting job titles, responsibilities, educational requirements, and salary ranges. Among the careers covered are webmaster, software engineer, and database administrator. You will learn how the material you are studying relates directly to a potential career path.

A LOOK TO THE FUTURE

Each chapter concludes with a brief discussion of a recent technological advancement related to the chapter material, reinforcing the importance of staying informed.

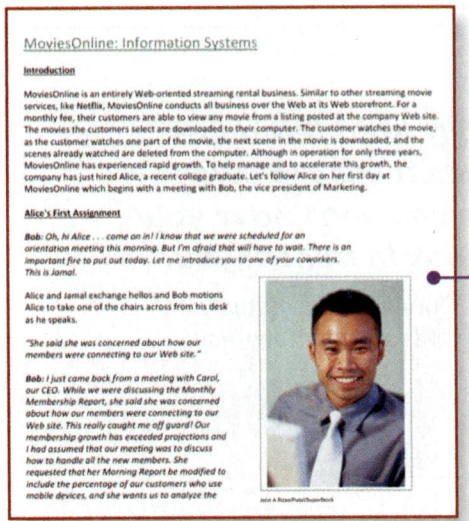

Found in Connect for *Computing Essentials* 2025, Using IT at MoviesOnline—A Case Study of a fictitious organization provides an up-close look at what you might expect to find on the job in the real world. You will follow Alice, a recent college graduate hired as a marketing analyst, as she navigates her way through accounting, marketing, production, human resources, and research, gathering and processing data to help manage and accelerate the growth of the three-year-old company.

A complete course platform

Connect enables you to build deeper connections with your students through cohesive digital content and tools, creating engaging learning experiences. We are committed to providing you with the right resources and tools to support all your students along their personal learning journeys.

65%
Less Time Grading

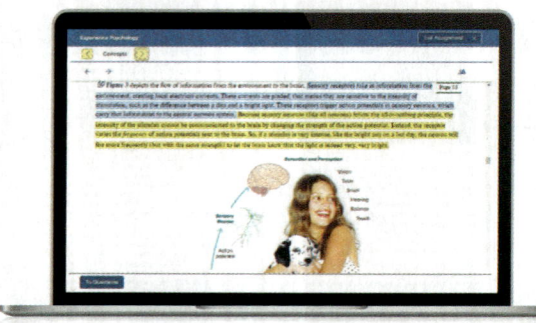

Laptop: Getty Images; Woman/dog: George Doyle/Getty Images

Every learner is unique

In Connect, instructors can assign an adaptive reading experience with SmartBook® 2.0. Rooted in advanced learning science principles, SmartBook® 2.0 delivers each student a personalized experience, focusing students on their learning gaps, ensuring that the time they spend studying is time well spent.
mheducation.com/highered/connect/smartbook

Study anytime, anywhere

Encourage your students to download the free ReadAnywhere® app so they can access their online eBook, SmartBook® 2.0, or Adaptive Learning Assignments when it's convenient, even when they're offline. And since the app automatically syncs with their Connect account, all of their work is available every time they open it. Find out more at **mheducation.com/readanywhere**

> *"I really liked this app—it made it easy to study when you don't have your textbook in front of you."*
>
> Jordan Cunningham, a student at *Eastern Washington University*

Effective tools for efficient studying

Connect is designed to help students be more productive with simple, flexible, intuitive tools that maximize study time and meet students' individual learning needs. Get learning that works for everyone with Connect.

Education for all

McGraw Hill works directly with Accessibility Services departments and faculty to meet the learning needs of all students. Please contact your Accessibility Services Office, and ask them to email **accessibility@mheducation.com**, or visit **mheducation.com/about/accessibility** for more information.

Affordable solutions, added value

Make technology work for you with LMS integration for single sign-on access, mobile access to the digital textbook, and reports to quickly show you how each of your students is doing. And with our Inclusive Access program, you can provide all these tools at the lowest available market price to your students. Ask your McGraw Hill representative for more information.

Solutions for your challenges

A product isn't a solution. Real solutions are affordable, reliable, and come with training and ongoing support when you need it and how you want it. Visit **supportateverystep.com** for videos and resources both you and your students can use throughout the term.

Updated and relevant content

Our new Evergreen delivery model provides the most current and relevant content for your course, hassle-free. Content, tools, and technology updates are delivered directly to your existing McGraw Hill Connect® course. Engage students and freshen up assignments with up-to-date coverage of select topics and assessments, all without having to switch editions or build a new course.

Support Materials in Connect

The Instructor's Manual offers lecture outlines with teaching notes and figure references. It provides definitions of key terms and solutions to the end-of-chapter material, including multiple-choice and open-ended questions.

The PowerPoint slides are designed to provide instructors with a comprehensive resource for lecture use. The slides include a review of key terms and topics, as well as artwork taken from the text to further explain concepts covered in each chapter.

The testbank contains over 2,200 questions categorized by level of learning (definition, concept, and application). This is the same learning scheme that is introduced in the text to provide a valuable testing and reinforcement tool.

SIMNET ONLINE TRAINING AND ASSESSMENT FOR OFFICE APPLICATIONS

SIMnet™ Online provides a way for you to test students' software skills in a simulated environment. SIMnet provides flexibility for you in your applications course by offering:

- Pretesting options
- Posttesting options
- Course placement testing
- Diagnostic capabilities to reinforce skills
- Web delivery of tests
- Learning verification reports

For more information on skills assessment software, please contact your local sales representative, or visit us at www.simnet-keepitsimple.com.

Acknowledgments

A special thank-you goes to the professors who took time out of their busy schedules to provide us with the feedback necessary to develop the 2025 edition of this text. The following professors offered valuable suggestions on revising the text:

Sharon LaVoy
Gateway Community College

Janel Veeser
Northeast Wisconsin Technical College

Harold Waterman
Anne Arundel Community College

Gigi Delk
Tyler Junior College

Marcie Yordy
Hillsborough Community College

Diego Tibaquira
Miami Dade College

Albena Belal
Gwinnett Technical College

Joseph Levens
Farmingdale State College

John Doyle
Indiana University Southeast

Pengtao Li
California State University

Becky McAfee
Hillsborough Community College

Rodney Koch
State University of New York at Cortland

Chen Ye
Purdue University

Mohammed El-Soussi
Santa Barbara City College

Lucy Mena-Quinn
Mildred Elley Business School

About the Authors

The O'Learys live in the American Southwest and spend much of their time engaging instructors and students in conversation about learning. In fact, they have been talking about learning for over 30 years. Something in those early conversations convinced them to write a book, to bring their interest in the learning process to the printed page.

The O'Learys form a unique team blending youth and experience. Dan has taught at the University of California at Santa Cruz, developed energy-related labs at NASA, and worked as a database administrator and as a consultant in information systems; he is currently a professor at the City College of San Francisco. Tim has taught courses at Stark Technical College in Canton, Ohio, and at Rochester Institute of Technology in upstate New York, and is currently a professor emeritus at Arizona State University. Linda offered her expertise at ASU for several years as an academic advisor. She also presented and developed materials for major corporations such as Motorola, Intel, Honeywell, and AT&T, as well as various community colleges in the Phoenix area.

Timothy O'Leary

Tim, Linda, and Dan have talked to and taught numerous students, all of them with a desire to learn something about computers and applications that make their lives easier, more interesting, and more productive.

Each new edition of an O'Leary text, supplement, or learning aid has benefited from these students and their instructors who daily stand in front of them (or over their shoulders).

Design Elements: Concept Check icons: Dizzle52/Getty Images

Computing Essentials

Making  work for you

INTRODUCTORY 2025

chapter 1
Information Technology, the Internet, and You

ideadesign/Shutterstock

Why should I read this chapter?

Matt Bird/Getty Images

The future of computers and digital technology promises exciting challenges and opportunities. Powerful software and hardware systems are changing the way people and organizations interact in their daily life and on the Internet.

This chapter introduces you to the skills and concepts you need to be prepared for this ever-changing digital world, including:

- Information systems—how the critical parts of technology interact.
- Efficiency and effectiveness—how to maximize the use of technology.
- Privacy, ethics, and community—how to integrate technology with people.
- Software, hardware, and data—understand the technology used in information systems.
- Connectivity and cloud computing—how the Internet, the web, and the wireless revolution are changing how we communicate and interact.

Learning Objectives

After you have read this chapter, you should be able to:

1. Explain the parts of an information system: people, procedures, software, hardware, data, and the Internet.
2. Distinguish between system software and application software.
3. Differentiate between the three kinds of system software programs.
4. Define and compare general-purpose, specialized, and mobile applications.
5. Identify the four types of computers and the five types of personal computers.
6. Describe the different types of computer hardware, including the system unit, input, output, storage, and communication devices.
7. Define data and describe document, worksheet, database, and presentation files.
8. Explain computer connectivity, the wireless revolution, the Internet, cloud computing, and IoT.

Introduction

"Welcome to *Computing Essentials*. I'm Katie, and this is Alan, we work in information technology. On the following pages, we'll be discussing some of the most exciting new developments in computer technology, including smartphones, tablets, and cloud computing. Let me begin this chapter by giving you an overview of the book and showing you some of its special features."

Wavebreakmedia/Getty Images

The purpose of this book is to help you become a highly efficient and effective computer user. This includes how to use (1) apps and application software; (2) all types of computer hardware, including mobile devices like smartphones, tablets, and laptops; and (3) the Internet. Becoming a highly efficient and effective computer user also requires a full understanding of the potential impact of technology on privacy and the environment as well as the role of personal and organizational ethics.

To effectively and efficiently use computers, you need to know the parts of an information system: people, procedures, software, hardware, data, and the Internet. You also need to understand the wireless revolution, the mobile Internet, and the web and to recognize the role of information technology in your personal and professional life.

Information Systems

When you think of a personal computer, perhaps you think of just the equipment itself. That is, you think of the screen or the keyboard. Yet there is more to it than that. The way to think about a personal computer is as part of an information system. An **information system** has several parts: *people, procedures, software, hardware, data,* and the *Internet.* (See **Figure 1-1**.)

- **People:** It is easy to overlook people as one of the parts of an information system. Yet this is what personal computers are all about—making **people**, **end users** like you, more productive.

    Later in this book, you will learn more about people and information systems in:
    - **Chapter 10** Information Systems
    - The introduction section at the beginning of each chapter
    - The Careers in IT section at the end of each chapter

- **Procedures:** The rules or guidelines for people to follow when using software, hardware, and data are **procedures**. These procedures are typically documented in manuals written by computer specialists. Software and hardware manufacturers provide manuals with their products. These manuals are provided in either printed or electronic form.

    Later in this book, you will learn more about procedures and information systems in:
    - **Chapter 2** The Internet, the Web, and Electronic Commerce
    - **Chapter 9** Privacy, Security, and Ethics
    - The Tips found throughout the book.
    - The Making IT Work for You section found within each chapter

- **Software:** A **program** consists of the step-by-step instructions that tell the computer how to do its work. **Software** is another name for a program or programs. The purpose of software is to convert **data** (unprocessed facts) into **information** (processed facts). For example, a payroll program would instruct the computer to take the number of hours you worked in a week (data) and multiply it by your pay rate (data) to determine how much you are paid for the week (information).

    Later in this book, you will learn more about software and information systems in:
    - **Chapter 3** Application Software
    - **Chapter 4** System Software
    - **Chapter 12** Systems Analysis and Design

- **Hardware:** The equipment that processes the data to create information is called **hardware**. It includes smartphones, tablets, keyboards, mice, displays, system units, and other devices. Hardware is controlled by software.

    Later in this book, you will learn more about hardware and information systems in:
    - **Chapter 5** The System Unit
    - **Chapter 6** Input and Output
    - **Chapter 7** Secondary Storage

- **Data:** The raw, unprocessed facts, including text, numbers, images, and sounds, are called data. Processed data yields information. Using the previous example of a payroll program, the data (number of hours worked and pay rate) is processed (multiplied) to yield information (weekly pay).

    Later in this book, you will learn more about data and information systems in:
    - **Chapter 11** Databases
    - **Chapter 13** Programming and Languages

- **Internet:** Almost all information systems provide a way to connect to other people and computers, typically using the Internet. This connectivity greatly expands the capability and usefulness of information systems.

    Later in this book, you will learn more about Internet and information systems in:
    - **Chapter 2** The Internet, the Web, and Electronic Commerce
    - **Chapter 8** Communications and Networks

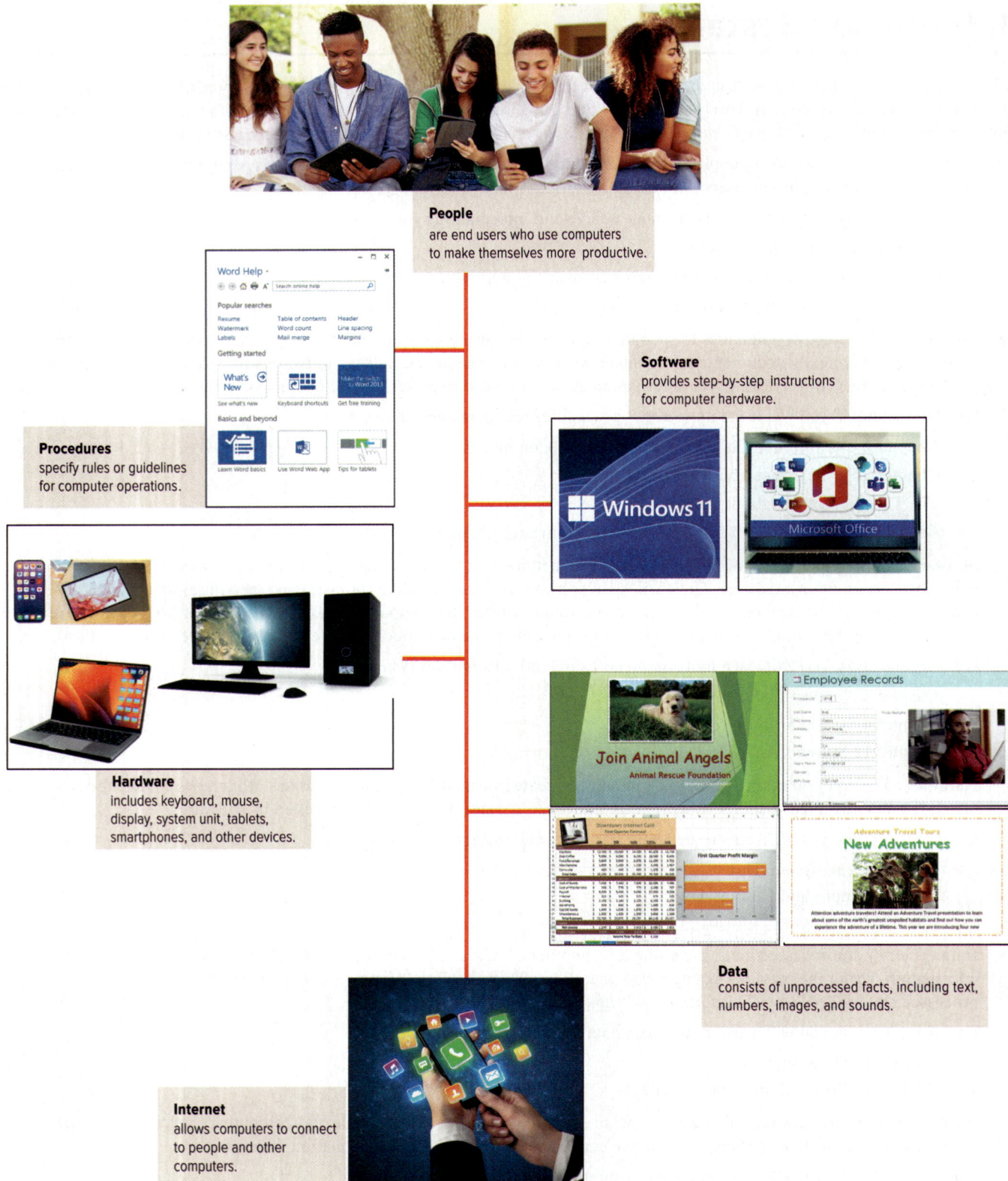

Figure 1-1 Parts of an information system

People: Shutterstock; **Procedures:** Microsoft Corporation; **Software:** Microsoft Corporation; monticello/Shutterstock; **Hardware:** (Smartphone): Kaspars Grinvalds/Shutterstock; (Tablet): Bloomberg/Getty Images; (Laptop): Quang NGUYEN DUC/Alamy Stock Photo; (Desktop) alxpin/Getty Images; **Data:** Microsoft Corporation; (Rescue puppy or dog): Zoom Pet Photography/Image Source/Getty Images; (Employee portrait): Sam Edwards/age fotostock; (Coffee): Stockbyte/Getty Images; (giraffe): Panksvatouny/Shutterstock; **Internet:** ra2 studio/Shutterstock

concept check

- What are the parts of an information system?
- What is a program?
- What is the difference between data and information?

People

People are surely the most important part of any information system. Our lives are touched every day by computers and information systems. Many times the contact is direct and obvious, such as when we create documents using a word processing program or when we connect to the Internet. (See **Figure 1-2**.) Other times, the contact is not as obvious.

Community

Every major technology has affected communities—but none in the unique ways that computers have. We have changed how we interact with our communities, both in the tools we use to communicate, such as social media posts, and in the ways we communicate, in emojis and podcasts. But technology has had a deeper impact on our communities than just the way we interact. It has forever changed how we find and identify our communities. Every day, people meet, discuss, and bond with others they have never met in person. The entire world feels a little smaller, with our communities extending around the globe.

Figure 1-2 People and computers
Prostock-studio/Shutterstock

Throughout this book you will find a variety of features designed to help you become an efficient and effective end user. These features include Making IT Work for You, Tips, Privacy, Community, Ethics, and Careers in IT.

- **Making IT Work for You.** Throughout this book you will find Making IT Work for You features that present numerous interesting and practical IT applications. For just a few of the Making IT Work for You topics, see **Figure 1-3**.

Application	Description
Free Antivirus Program	Protect your computer by installing and using a free antivirus program. See page 8.
Cloud Office Suites	Create and collaborate with others online to make better documents and presentations. See page 72.
Gaming	Delve into the world of video games and find the best video game hardware for you. See page 110.
Cloud Storage	Move your files online to synch files between devices or free up space on your digital devices. See page 172.
The Mobile Office	Get work done on the road; whether a business trip or your daily commute, these tools will help you make the most of your time. See page 192.

Figure 1-3 Making IT Work for You applications

tips

Are you getting the most out of your cell phone? Here are just a few of the tips to make your computing safer, more efficient, and more effective.

1. **Low battery.** Do you find that your cell phone's battery keeps its charge for less time than it used to? Here are some ways to make your battery last longer. See page 122.
2. **Cell phone cameras.** Capturing life's moments in a photo is easier and faster with a cell phone. But a few simple tips can make the process easier and your photos better. See page 69.
3. **Disaster planning.** Having a cell phone lost or stolen can be devastating. Follow these suggestions to make it easier to get your phone back or recover its data quickly. See page 225.
4. **Data usage.** Is your cell phone data plan costing you money? Are your cell phone apps using up your data plan without you knowing it? Take control of your data usage with the tips on page 169.
5. **Protecting your identity.** Identity theft is a growing problem and can be financially devastating if you are a victim. Some steps to protect your identity are on page 223.

Figure 1-4 Selected tips

- **Tips.** We all can benefit from a few tips or suggestions. Throughout this book you will find numerous tips to make your computing safer, more efficient, and more effective. These tips range from the basics of keeping your computer system running smoothly to how to protect your privacy while surfing the web. For a partial list of the tips presented in the following chapters, see **Figure 1-4**.
- **Privacy.** One of the most critical issues today is how to protect the privacy of our personal information. Throughout this book you will find Privacy boxes in the margin that present information about protecting our privacy.
- **Community.** Computers are changing the way we define and interact with our communities. In this chapter and the following ones, you will find Community boxes in the margins that present ways in which technology affects how we create and engage with our communities.
- **Ethics.** Most people agree that we should behave ethically. That is, we should follow a system of moral principles that direct our everyday lives. However, for any given circumstance, people often do not agree on the ethics of the situation. Throughout this book you will find numerous Ethics boxes posing a variety of different ethical/unethical situations for your consideration.
- **Careers in IT.** One of the most important decisions of your life is to decide upon your life's work or career. Perhaps you are planning to be a writer, an artist, or an engineer. Or you might become a professional in **information technology (IT)**. Each of the following chapters highlights a specific career in information technology. This feature provides job descriptions, projected employment demands, educational requirements, current salary ranges, and advancement opportunities.

concept check

- ☐ Which part of an information system is the most important?
- ☐ Describe the Making IT Work for You, Tips, and Privacy features.
- ☐ Describe the Environment, Ethics, and Careers in IT features.

Software

Software, as we mentioned, is another name for programs. Programs are the instructions that tell the computer how to process data into the form you want. In most cases, the words *software* and *programs* are interchangeable. There are two major kinds of software: *system software* and *application software*. You can think of application software as the kind you use. Think of system software as the kind the computer uses.

System Software

The user interacts primarily with application software. **System software** enables the application software to interact with the computer hardware. System software is "background" software that helps the computer manage its own internal resources.

System software is not a single program. Rather, it is a collection of programs, including the following:

- **Operating systems** are programs that coordinate computer resources, provide an interface between users and the computer, and run applications. Smartphones, tablets, and many other mobile devices use **embedded operating systems**, also known as **real-time operating systems (RTOS)**. Desktop computers use **stand-alone operating systems** like Windows 10 or macOS. (See **Figures 1-5** and **1-6**.) Networks use **network operating systems (NOS)**.

Figure 1-5 Windows 10
Microsoft Corporation

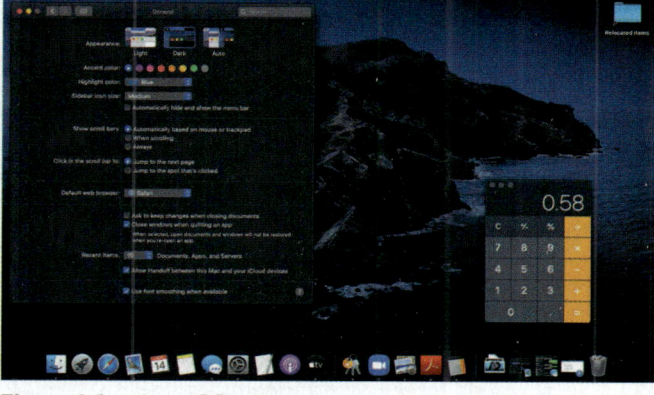

Figure 1-6 macOS
Apple, Inc.

- **Utilities** perform specific tasks related to managing computer resources. One of the most essential utility programs that every computer system should have is an antivirus program. These programs protect your computer system from **viruses** or malicious programs that are all too often deposited onto your computer from the Internet. These programs can damage software and hardware, as well as compromise the security and privacy of your personal data. If your computer does not have an antivirus program installed on it, you need to get one. To see how you can install a free antivirus program on your computer, see Making IT Work for You: Free Antivirus Program on page 8.

Application Software

Application software might be described as end-user software. Three types of application software are *general-purpose, specialized,* and *apps*.

General-purpose applications are widely used in nearly all career areas. They are the kinds of programs you have to know to be considered an efficient and effective end user. Some of the best known are presented in **Figure 1-7**.

Specialized applications include thousands of other programs that are more narrowly focused on specific disciplines and occupations. Two of the best known are graphics and web authoring programs.

Mobile apps, also known as **mobile applications** or simply **apps**, are small programs primarily designed for mobile devices such as smartphones and for tablets. There are over 5 million apps. The most popular mobile apps are for social networking, playing games, and downloading music and videos.

Type	Description
Word processors	Prepare written documents
Spreadsheets	Analyze and summarize numerical data
Database management systems	Organize and manage data and information
Presentation software	Communicate a message or persuade other people

Figure 1-7 General-purpose applications

Making  work for you

FREE ANTIVIRUS PROGRAM

Have you or someone you know had a slower computing experience due to a spyware infection? Even worse, perhaps a malicious piece of software stole crucial, personal information or caused a total system failure. Most of these problems can be averted by having an up-to-date antivirus program running in your computer's memory at all times. This exercise shows you how to download and install a free antivirus program if your computer does not yet have one. (Please note that the web is continually changing, and some of the specifics presented here may have changed.)

Getting Started First, make sure your computer does not have an antivirus or security suite running. If it does, be sure to completely uninstall that program, even if the subscription is expired. Now, follow these steps to install AVG, a popular, free antivirus program:

- Visit http://free.avg.com and click the *Download* button. You will be asked to click "save" to save the installation file to your computer.
- Run the installation file and follow the prompts.
- Select *Install Basic* to install the antivirus software. Once the program is installed, it will open automatically.

Using AVG Generally speaking, your antivirus program watches your system for malware and updates itself automatically. However, you can always download updates manually, set a schedule for full-system scans, and change basic settings for various components of the software.

- Click *Scan now* to run a full scan on your computer.
- Just to the right of that, click the button with the white cog to see the scan options, where you can set a schedule for automated scans.
- Click the *back arrow* to reach the main screen, where you can click various elements of the program to configure them. For example, clicking *Web* will allow you to turn on a feature that detects cookies that may be used to track your online activity.

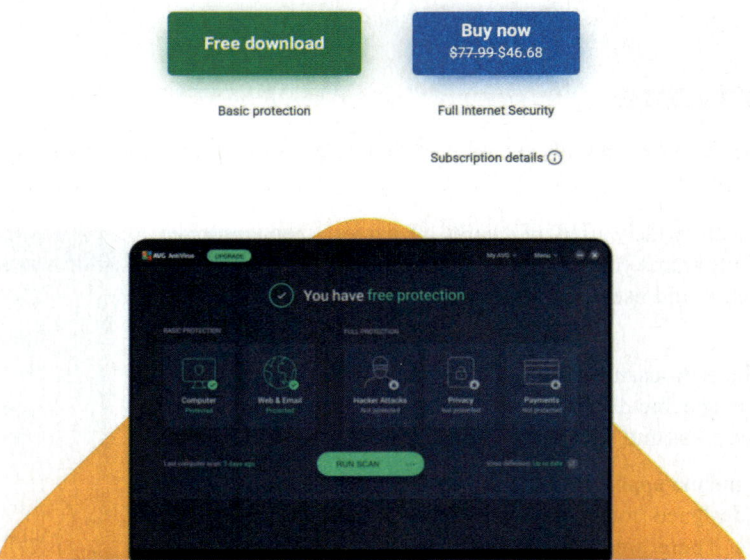

AVG Technologies

concept check

☐ Describe the two major kinds of software.
☐ Describe two types of system software programs.
☐ Define and compare general-purpose applications, specialized applications, and mobile apps.

Hardware

Computers are electronic devices that can follow instructions to accept input, process that input, and produce information. This book focuses principally on personal computers. However, it is almost certain that you will come in contact, at least indirectly, with other types of computers.

Types of Computers

There are four types of computers: supercomputers, mainframe computers, midrange computers, and personal computers.

- **Supercomputers** are the most powerful type of computer. These machines are special, high-capacity computers used by very large organizations. Supercomputers are typically used to process massive amounts of data. For example, they are used to analyze and predict worldwide weather patterns. IBM's Blue Gene supercomputer is one of the fastest computers in the world. (See **Figure 1-8**.)
- **Mainframe computers** occupy specially wired, air-conditioned rooms. Although not nearly as powerful as supercomputers, mainframe computers are capable of great processing speeds and data storage. For example, insurance companies use mainframes to process information about millions of policyholders.
- **Midrange computers,** also referred to as **servers**, are computers with processing capabilities less powerful than a mainframe computer yet more powerful than a personal computer. Originally used by medium-sized companies or departments of large companies to support their processing needs, today midrange computers are most widely used to support or serve end users for such specific needs as retrieving data from a database or supplying access to application software.

Figure 1-8 **Supercomputer**
DOE Photo/Alamy Stock Photo

Figure 1-9 Desktop
alxpin/Getty Images

Figure 1-10 Laptop
Kaspars Grinvalds/Shutterstock

Figure 1-11 Tablet
Bloomberg/Getty Images

- **Personal computers,** also known as **PCs**, are the least powerful, yet the most widely used and fastest-growing type of computer. There are five types of personal computers: *desktops, laptops, tablets, smartphones,* and *wearables.* **Desktop computers** are small enough to fit on top of or alongside a desk yet are too big to carry around. (See Figure 1-9.) **Laptop computers**, also known as **notebook computers,** are portable and lightweight and fit into most briefcases. (See Figure 1-10.) **Tablets**, also known as **tablet computers**, are smaller, lighter, and generally less powerful than laptops. Like a laptop, tablets have a flat screen but typically do not have a standard keyboard. (See Figure 1-11.) Instead, tablets typically use a virtual keyboard that appears on the screen and is touch sensitive.

Smartphones are the most widely used personal computer. Smartphones are cell phones with wireless connections to the Internet and processing capabilities. (See Figure 1-12.) Other mobile computers include **wearable devices** like Apple's Watch. (See Figure 1-13.)

Figure 1-12 Smartphone
Quang NGUYEN DUC/Alamy Stock Photo

Cell Phones

Many people are not aware that their cell phone is a computer, and this computer has many of the same components as desktops, laptops, and tablets. At one time, cell phones had very limited power and were used almost exclusively for making telephone calls. Now, almost all cell phones are powerful smartphones capable of connecting to the Internet and running any number of apps. In fact, nearly every cell phone purchased today is more powerful than the computers used to land the first person on the moon.

Today, over 99 percent of Americans under the age of 30 own a cell phone, and over 96 percent of those cell phones are smartphones. As a result, the two terms are becoming interchangeable. Reflecting this trend, we will use the terms *cell phone* and *smartphone* interchangeably.

Personal Computer Hardware

Hardware for a personal computer system consists of a variety of different devices. This physical equipment falls into four basic categories: system unit, input/output, secondary storage, and communication. Because we discuss hardware in detail later in this book, here we will present just a quick overview of the four basic categories.

- **System unit:** The **system unit** is a container that houses most of the electronic components that make up a computer system. Two important components of the system unit are *microprocessors* and *memory*. (See **Figure 1-14**.) The **microprocessor** controls and manipulates data to produce information. **Memory** is a holding area for data, instructions, and information. One type, **random-access memory (RAM)**, holds the program and data that are currently being processed. This type of memory is sometimes referred to as *temporary storage* because its contents will typically be lost if the electric power to the computer is disrupted.

- **Input/output: Input devices** translate data and programs that humans can understand into a form that the computer can process. The most common input devices are the **keyboard** and the **mouse**. **Output devices** translate the processed information from the computer into a form that humans can understand. The most common output device is the **display**, also known as a **monitor**.

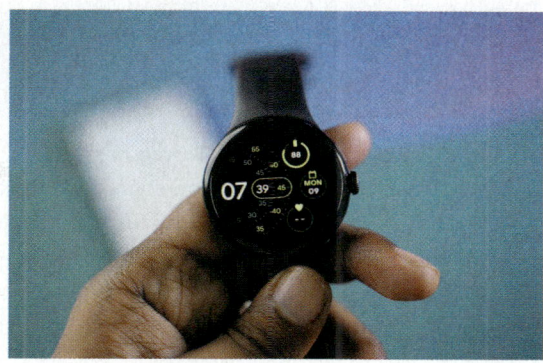

Figure 1-13   **Wearable**
Yasin Hasan/Shutterstock

Figure 1-14   **System unit**
(CPU): vetkit/Getty Images; (RAM): scanrail/123RF; (Microprocessor): Matveev Aleksandr/Shutterstock

- **Secondary storage:** Unlike memory, **secondary storage** holds data and programs even after electric power to the computer system has been turned off. The most important kinds of secondary media are *hard disks, solid-state storage,* and *optical discs.*

  Hard disks are typically used to store programs and very large data files. Using rigid metallic platters and read/write heads that move across the platters, data and information are stored using magnetic charges on the disk's surface. In contrast, **solid-state storage** does not have any moving parts, is more reliable, and requires less power. It saves data and information electronically similar to RAM except that it is not volatile. (See **Figure 1-15**.) **Optical discs** use laser technology to store data and programs. Three types of optical discs are **compact discs (CDs)**, **digital versatile** (or **video**) **discs (DVDs)**, and **Blu-ray discs (BD)**.

Figure 1-15   **Solid-state storage**
NMStudio789/Shutterstock

- **Communication:** At one time, it was uncommon for a personal computer system to communicate with other computer systems. Now, using **communication devices**, a personal computer routinely communicates with other computer systems located as near as the next office or as far away as halfway around the world, using the Internet. A **modem** is a widely used communication device that modifies audio, video, and other types of data into a form that can be transmitted across the Internet.

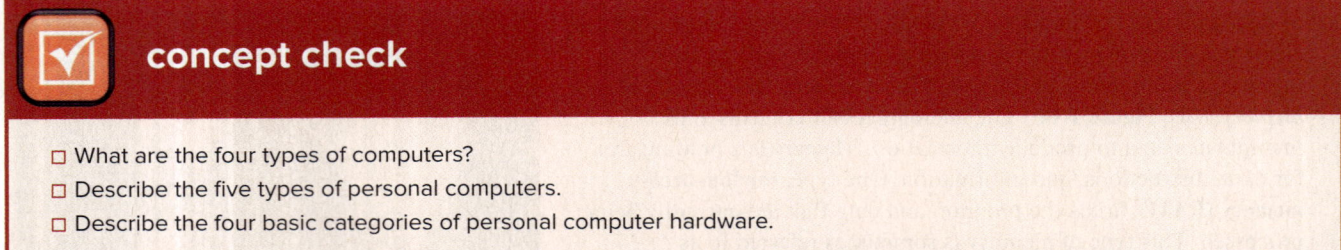

concept check

- What are the four types of computers?
- Describe the five types of personal computers.
- Describe the four basic categories of personal computer hardware.

Data

Data is raw, unprocessed facts, including text, numbers, images, and sounds. As we mentioned earlier, processed data becomes information. When stored electronically in files, data can be used directly as input for the system unit.

Four common types of files (see **Figure 1-16**) are

Figure 1-16 Four types of files: document, worksheet, database, and presentation
Presentation: Microsoft Corporation; Zoom Pet Photography/Image Source/Getty Images; **Database:** Microsoft Corporation; Sam Edwards/age fotostock; **Worksheet:** Stockbyte/Getty Images; **Document:** Microsoft Corporation; Panksvatouny/Shutterstock

- **Document files**, created by word processors to save documents such as memos, term papers, and letters.
- **Worksheet files**, created by electronic spreadsheets to analyze things like budgets and to predict sales.
- **Database files**, typically created by database management programs to contain highly structured and organized data. For example, an employee database file might contain all the workers' names, Social Security numbers, job titles, and other related pieces of information.
- **Presentation files**, created by presentation software to save presentation materials. For example, a file might contain audience handouts, speaker notes, and electronic slides.

Connectivity and the Mobile Internet

Connectivity is the capability of your personal computer to share information with other computers. Central to the concept of connectivity is the **network**. A network is a communications system connecting two or more computers. The largest network in the world is the **Internet**. It is like a giant highway that connects you to millions of other people and organizations located throughout the world. The **web** provides a multimedia interface to the numerous resources available on the Internet.

The Internet has driven the evolution of computers and their impact on our daily lives. The rate of technological change is accelerating at an ever-faster pace. Along with the Internet, three things that are driving the impact of technology on our lives are cloud computing, wireless communication, and the Internet of Things.

- **Cloud computing** uses the Internet and the web to shift many computer activities from a user's computer to computers on the Internet. Rather than relying solely on their computer, users can now use the Internet to connect to the cloud and access more powerful computers, software, and storage.
- **Wireless communication** has changed the way we communicate with one another. The rapid development and widespread use of wireless communication devices like tablets, cell phones, and wearable devices have led many experts to predict that wireless applications are just the beginning of the **wireless revolution**, a revolution that will dramatically affect the way we communicate and use computer technology.
- The **Internet of Things (IoT)** is the continuing development of the Internet that allows everyday objects embedded with electronic devices to send and receive data over the Internet. It promises to connect all types of devices, from computers to cell phones, to watches, to any number of everyday devices.

Wireless communication, cloud computing, and IoT are driving the mobile Internet. They promise to continue to dramatically affect the entire computer industry and how you and I will interact with computers and other devices. Each will be discussed in detail in the following chapters. For just a few of these mobile devices, see **Figure 1-17**.

Figure 1-17 Wireless communication devices
NurPhoto/Getty Images; Shutterstock Images, LLC; tinhkhuong/Shutterstock; Quang NGUYEN DUC/Alamy Stock Photo; Gabo_Arts/Shutterstock

concept check

- Define data. List four common types of files.
- Define connectivity and networks.
- What is cloud computing? Wireless revolution? IoT?

Careers in IT

"Now that you know the basic outline and important features of this book, we would like to talk about some of the most exciting and well-paid careers in information technology."

Wavebreakmedia/Getty Images

As mentioned previously, each of the following chapters highlights a specific career in information technology. Each provides specific job descriptions, salary ranges, advancement opportunities, and more. For a partial list of these careers, see Figure 1-18.

Career	Description
Web developer	Develops and maintains websites and web resources. See page 44.
Software engineer	Analyzes users' needs and creates application software. See page 71.
Computer support specialist	Provides technical support to customers and other users. See page 97.
Computer technician	Repairs and installs computer components and systems. See page 123.
Technical writer	Prepares instruction manuals, technical reports, and other scientific or technical documents. See page 152.
Network administrator	Creates and maintains computer networks. See page 203.

Figure 1-18   Careers in information technology

A LOOK TO THE FUTURE

Using and Understanding Information Technology

Matt Bird/Getty Images

The purpose of this book is to help you use and understand information technology. We want to help you become proficient and to provide you with a foundation of knowledge so that you can understand how technology is being used today and anticipate how technology will be used in the future. This will enable you to benefit from six important information technology developments.

The Internet and the Web

The Internet and the web are considered to be the two most important technologies for the 21st century. Understanding how to efficiently and effectively use the Internet to browse, communicate, and locate information is an essential skill. These issues are presented in **Chapter 2**, The Internet, the Web, and Electronic Commerce.

Powerful Software

The software that is now available can do an extraordinary number of tasks and help you in an endless number of ways. You can create professional-looking documents, analyze massive amounts of data, create dynamic multimedia web pages, and much more. Today's employers are expecting the people they hire to be able to effectively and efficiently use a variety of different types of software. General-purpose, specialized, and mobile applications are presented in **Chapter 3**. System software is presented in **Chapter 4**.

Powerful Hardware

Personal computers are now much more powerful than they used to be. Cell phones, tablets, and communication technologies such as wireless networks are dramatically changing the ways to connect to other computers, networks, and the Internet. However, despite the rapid change of specific equipment, their essential features remain unchanged. To become an efficient and effective end user, you should focus on these features. **Chapters 5 through 8** explain what you need to know about hardware. For those considering the purchase of a computer, an *appendix—The Computer Buyer's Guide*—is provided at the end of this book. This guide provides a very concise comparison of desktops, laptops, tablets, and cell phones.

Privacy, Security, and Ethics

What about people? Experts agree that we as a society must be careful about the potential of technology to negatively affect our lives. Specifically, we need to be aware of how technology can affect our personal privacy and our environment. Also, we need to understand the role and the importance of organizational and personal ethics. These critical issues are integrated in every chapter of this book as well as extensively covered in **Chapter 9**.

Organizations

Almost all organizations rely on the quality and flexibility of their information systems to stay competitive. As a member or employee of an organization, you will undoubtedly be involved in these information systems. In order to use, develop, modify, and maintain these systems, you need to understand the basic concepts of information systems and know how to safely, efficiently, and effectively use computers. These concepts are covered throughout this book.

Changing Times

Are the times changing any faster now than they ever have? Almost everyone thinks so. Whatever the answer, it is clear we live in a fast-paced age. *The Evolution of the Computer Age* section presented at the end of this book tracks the major developments since computers were first introduced.

After reading this book, you will be in a very favorable position compared with many other people in industry today. You will learn not only the basics of hardware, software, connectivity, the Internet, and the web, but also the most current technology. You will be able to use these tools to your advantage.

VISUAL SUMMARY | Information Technology, the Internet, and You

INFORMATION SYSTEMS

Shutterstock

The way to think about a personal computer is to realize that it is one part of an **information system.** There are several parts of an information system:

- **People** are an essential part of the system. The purpose of information systems is to make people, or **end users** like you, more productive.
- **Procedures** are rules or guidelines to follow when using software, hardware, and data. They are typically documented in manuals written by computer professionals.
- **Software (programs)** provides step-by-step instructions to control the computer to convert **data** into **information.**
- **Hardware** consists of the physical equipment. It is controlled by software and processes data to create information.
- **Data** consists of unprocessed facts, including text, numbers, images, and sound. **Information** is data that has been processed by the computer.
- The **Internet** allows computers to connect and share information.

To efficiently and effectively use the computer, you need to understand **information technology (IT)**, including software, hardware, data, and connectivity.

PEOPLE

Prostock-studio/Shutterstock

People are the most important part of an information system. This book contains several features to demonstrate how people just like you use computers. These features include the following:

- **Making IT Work for You** presents several interesting and practical applications. Topics include using online office suites and cloud storage.
- **Tips** offer a variety of suggestions on such practical matters as how to improve slow computer performance and how to protect your privacy while on the web.
- **Privacy** marginal boxes discuss threats to your personal privacy and suggest ways to protect yourself.
- **Community** boxes discuss the ways technology affects our communities. Computers are changing the way we find, create, and engage with our communities.
- **Ethics** boxes pose a variety of different ethical/unethical situations for your consideration.
- **Careers in IT** presents job descriptions, employment demands, educational requirements, salary ranges, and advancement opportunities.

To efficiently and effectively use computers, you need to understand the basic parts of an information system: people, procedures, software, hardware, data, and connectivity. You also need to understand the Internet and the web and to recognize the role of technology in your professional and personal life.

SOFTWARE

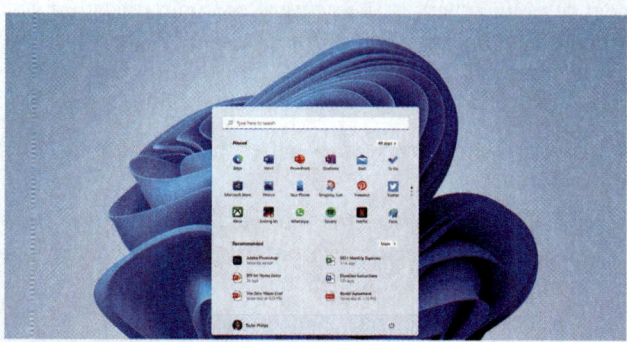

Microsoft Corporation

Software, or **programs**, consists of system and application software.

System Software

System software enables application software to interact with computer hardware.

- **Operating systems** coordinate resources, provide an interface, and run applications. Three types are **embedded (real-time, RTOS)**, **stand-alone**, and **network (NOS)**.
- **Utilities** perform specific tasks to manage computer resources.

Application Software

Application software includes general-purpose, specialized, and mobile applications.

- **General purpose**—widely used in nearly all career areas; programs include browsers, word processors, spreadsheets, database management systems, and presentation software.
- **Specialized**—focus more on specific disciplines and occupations; programs include graphics and web authoring.
- **Apps (mobile apps, mobile applications)**—designed for mobile devices; most popular are for text messaging, Internet browsing, and connecting to social networks.

HARDWARE

alxpin/Getty Images

Hardware consists of electronic devices that can follow instructions to accept input, process the input, and produce information.

Types of Computers

Supercomputer, **mainframe**, **midrange (server)**, and **personal computers (PCs)** are four types of computers. A personal computer can be a **desktop**, **laptop (notebook computer)**, **tablet**, **smartphone**, or a **wearable**.

Cell Phones

Today, almost all cell phones are smartphones; cell phone and smartphone are becoming interchangeable terms.

Personal Computer Hardware

There are four basic categories of hardware devices:

- **System unit** contains electronic circuitry, including **microprocessors** and **memory**. **Random-access memory (RAM)** holds the program and data currently being processed.
- **Input/output devices** are translators between humans and computers. **Input devices** include the **keyboard** and **mouse**. The most common output device is the computer **display (monitor)**.
- **Secondary storage** holds data and programs. Typical media include **hard disks**, **solid-state storage**, and **optical discs (CD, DVD, and Blu-ray)**.
- **Communication devices** allow personal computers to communicate with other computer systems. **Modems** modify audio, video, and other types of data for transmission across the Internet.

DATA

Data is the raw unprocessed facts about something. Common file types include

- **Document files** created by word processors.

Microsoft Corporation; Panksvatouny/Shutterstock

- **Worksheet files** created by spreadsheet programs.

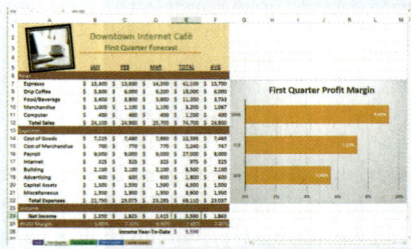

Microsoft Corporation; Stockbyte/Getty Images

- **Database files** created by database management programs.

Microsoft Corporation; Sam Edwards/age fotostock

- **Presentation files** created by presentation software programs.

Microsoft Corporation; Zoom Pet Photography/Image Source/Getty Images

CONNECTIVITY AND THE MOBILE INTERNET

Connectivity describes the ability of end users to use resources well beyond their desktops. Central to the concept of connectivity is the **network** or communication system connecting two or more computers. The **Internet** is the world's largest computer **network**. The **web** provides a multimedia interface to resources available on the Internet.

Along with the Internet, three other things are driving the impact of technology:

- **Cloud computing** uses the Internet and the web to shift many computer activities from a user's computer to computers on the Internet.
- **Wireless revolution** has changed the way we communicate and use computer technology. Wireless devices include tablets, cell phones, and watches.
- The **Internet of Things (IoT)** is the continuing development of the Internet that allows everyday objects embedded with electronic devices to send and receive data over the Internet.

CAREERS in IT

Career	Description
Web developer	Develops and maintains websites and web resources. See page 44.
Software engineer	Analyzes users' needs and creates application software. See page 71.
Computer support specialist	Provides technical support to customers and other users. See page 97.
Computer technician	Repairs and installs computer components and systems. See page 123.
Technical writer	Prepares instruction manuals, technical reports, and other scientific or technical documents. See page 152.
Network administrator	Creates and maintains computer networks. See page 203.

KEY TERMS

application software
apps
Blu-ray disc (BD)
cloud computing
communication device
compact disc (CD)
connectivity
data
database file
desktop computer
digital versatile disc (DVD)
digital video disc (DVD)
display
document file
embedded operating system
end user
general-purpose application
hard disk
hardware
information
information system
information technology (IT)
input device
Internet
IoT (Internet of Things)
keyboard
laptop computer
mainframe computer
memory
microprocessor
midrange computer
mobile app (application)
modem
monitor
mouse
network
network operating systems (NOS)
notebook computer
operating system
optical disc
output device
PC
people
personal computer
presentation file
procedures
program
random-access memory (RAM)
real-time operating system (RTOS)
secondary storage
server
smartphone
software
solid-state storage
specialized application
stand-alone operating system
supercomputer
system software
system unit
tablet
tablet computer
utility
virus
wearable device
web
wireless communication
wireless revolution
worksheet file

MULTIPLE CHOICE

Circle the correct answer.

1. This consists of the step-by-step instructions that tell the computer how to do its work.
   - **a.** program
   - **b.** browser
   - **c.** RAM
   - **d.** operating system

2. Another name for a program.
   - **a.** document
   - **b.** hardware
   - **c.** software
   - **d.** monitor

3. Enables the application software to interact with the computer hardware.
   - **a.** system software
   - **b.** utility software
   - **c.** RAM
   - **d.** output device

4. Type of computer that is small enough to fit on top of or alongside a desk yet is too big to carry around.
   - **a.** desktop
   - **b.** tablet
   - **c.** mainframe
   - **d.** supercomputer

5. A container that houses most of the electronic components that make up a computer system.
   - **a.** mainframe
   - **b.** system unit
   - **c.** hardware
   - **d.** utility

6. Devices that translate the processed information from the computer into a form that humans can understand.
   - **a.** program
   - **b.** input
   - **c.** document
   - **d.** output

7. Unlike hard disks, this type of storage does not have any moving parts, is more reliable, and requires less power.
   - **a.** network
   - **b.** cloud
   - **c.** general purpose
   - **d.** solid state

8. A device that modifies audio, video, and other types of data into a form that can be transmitted across the Internet.
   - **a.** monitor
   - **b.** modem
   - **c.** input
   - **d.** output

9. A type of file that might contain, for example, audience handouts, speaker notes, and electronic slides.
   - **a.** presentation
   - **b.** document
   - **c.** spreadsheet
   - **d.** database

10. A communications system connecting two or more computers.
    - **a.** modem
    - **b.** cloud computing
    - **c.** network
    - **d.** smartphone

MATCHING

Match each numbered item with the most closely related lettered item. Write your answers in the spaces provided.

a. cloud computing
b. document
c. general-purpose application
d. hardware
e. Internet
f. mainframe
g. memory
h. operating systems
i. secondary
j. wearable

_____ 1. RAM is a type of _____.
_____ 2. Examples include the keyboard, mouse, display, and system unit.
_____ 3. Although not as powerful as a supercomputer, this type of computer is capable of great processing speeds and data storage.
_____ 4. The Apple Watch is an example of this type of computer.
_____ 5. These programs coordinate computer resources, provide an interface, and run applications.
_____ 6. The type of file created by word processors—for example, memos, term papers, and letters.
_____ 7. Uses the Internet and the web to shift many computer activities from a user's computer to computers on the Internet.
_____ 8. The largest network in the world.
_____ 9. This category of applications includes the word processor.
_____ 10. Unlike memory, this type of storage holds data and programs even after electric power to the computer system has been turned off.

OPEN-ENDED

On a separate sheet of paper, respond to each question or statement.

1. Explain the parts of an information system. What part do people play in this system?
2. What is system software? What kinds of programs are included in system software?
3. Define and compare general-purpose applications, specialized applications, and apps. Describe some different types of general-purpose applications. Describe some types of specialized applications.
4. Describe the different types of computers. What is the most common type? What are the types of personal computers?
5. What is connectivity? What is a computer network? What are the Internet and the web? What are cloud computing, the wireless revolution, and IoT?

DISCUSSION

Respond to each of the following questions.

 ### Making IT Work for You

Making it a habit of keeping current with technology applications can be a key to your success. Numerous full-page spreads identified as *Making IT Work for You* are presented in the following chapters. These sections address some of today's most interesting and useful applications. They include online entertainment in **Chapter 2**, online office suites in **Chapter 3**, and cloud storage in **Chapter 7**. Select one that you find the most interesting and then respond to the following: (a) Why did you select this application? (b) Have you used this application? If so, when and how? If not, do you plan to in the near future? (c) Go to the chapter containing your selected application, and locate the application's Making IT Work for You coverage. Review and briefly describe its contents. (d) Did you find the coverage useful? Why or why not?

 ### Privacy

Privacy is one of the most critical issues facing society today. Numerous Privacy boxes appear in the margins of the upcoming chapters presenting a variety of privacy issues. These issues include apps that constantly track your movements without your knowledge or consent in **Chapter 3**; public Wi-Fi connections that record all of your personal communications in **Chapter 8**; and protection of personal privacy while using social media in **Chapter 9**. Select one that you find the most interesting, and then respond to the following: (a) Why did you select this issue? (b) Do you have knowledge of or experience with the issue? If so, describe your knowledge or experience. If not, do you consider the issue to be important for protecting your privacy? (c) Go to the chapter containing your selected issue, locate the Privacy box, read it, and describe its contents. (d) Did you find the coverage thought-provoking? Why or why not?

 ### Ethics

Computer ethics are guidelines for the morally acceptable use of computers in our society. Numerous Ethics boxes appear in the margins of the upcoming chapters, presenting a variety of ethical issues. These issues include image editing in **Chapter 3**, unauthorized use of webcams in **Chapter 6**, and unauthorized monitoring or eavesdropping of Internet activity in **Chapter 8**. Select one issue that you find the most interesting and then respond to the following: (a) Why did you select this issue? (b) Do you have knowledge of or experience with the issue? If so, describe your knowledge or experience. If not, do you consider the issue critical for individuals or organizations? (c) Go to the chapter containing your selected issue, locate the Ethics box, read it, and describe its contents. (d) Did you find the coverage thought-provoking? Why or why not?

 ### Community

Almost everyone belongs to several communities—some in person and some digital. Numerous Community boxes appear in the margins of the upcoming chapters. These boxes present a variety of community topics, including etiquette on the Internet in **Chapter 2**, accessibility for people with disabilities in **Chapter 4**, and the role of communities in bringing Internet access to the developing world in **Chapter 8**. Select one that you find the most interesting and then respond to the following: (a) Why did you select this topic? (b) Go to the chapter containing your selected topic, locate the Community box, read it, and describe its contents. (c) Did you find the coverage thought-provoking? Why or why not?

Design Elements: Concept Check icon: Dizzle52/Getty Images

chapter 2
The Internet, the Web, and Electronic Commerce

PeopleImages.comy—Yuri A/Shutterstock

Why should I read this chapter?

alexsl/iStock/Getty Images Plus

The Internet has changed the world, and our world has changed how we use the Internet. Originally, the Internet was only available to academics and scientists in universities and federal buildings. Today, the Internet connects everything, from cell phones to refrigerators.

This chapter covers the things you need to know to be prepared for this ever-changing digital world, including

- Impact—how Internet technology is changing your world.
- Hardware—how to connect your life to the Internet, including Wi-Fi, cell phones, and tablets.
- Applications—how to get ahead using social networking, streaming technology, and cloud computing.

Learning Objectives

After you have read this chapter, you should be able to:

1. Explain the origins of the Internet and the web.
2. Explain how to access the web using providers and browsers.
3. Compare different web utilities, including filters, file transfer utilities, and Internet security suites.
4. Compare different Internet communications, including social networking, blogs, microblogs, podcasts, wikis, text messaging, instant messaging, and e-mail.
5. Describe search tools, including search engines.
6. Describe how to evaluate the accuracy of information presented on the web.
7. Identify electronic commerce, including B2C, C2C, B2B, and security issues.
8. Describe cloud computing, including the three-way interaction of clients, Internet, and service providers.
9. Discuss the Internet of Things (IoT) and the continuing development of the Internet to allow everyday objects to send and receive data.

Introduction

"Hi, I'm Henry, and I'm a web developer. I'd like to talk with you about the Internet, the web, and electronic commerce, things that touch our lives every day. I'd also like to talk with you about the role the Internet plays with Facebook, LinkedIn, Twitter, and cloud computing."

mavo/Shutterstock

When you use a cell phone to text a friend, watch a video, or make a purchase, you are using the Internet. But what does it mean to "use the Internet"? Internet devices, like cell phones, are so common it is easy to overlook the incredible achievements that make the Internet possible. A closer look at how the Internet works will give you a new appreciation for this technical marvel and prepare you to use cell phones and other Internet devices to their full potential.

The Internet is an interconnected mesh of computers and data lines that connects millions of people and organizations. It is the foundation of the digital revolution and allows people around the world to share words, images, and any digital file almost instantaneously. The web provides an easy-to-use interface to Internet resources. It has become an everyday tool for all of us to use.

To efficiently and effectively use computers, you need to be aware of the resources available on the Internet and the web. Additionally, you need to know how to access these resources, to effectively communicate electronically, to efficiently locate information, to understand electronic commerce, and to use web utilities.

The Internet and the Web

In 1969, a computer network called **Advanced Research Project Agency Network (ARPANET)** was created. It connected computers at four universities: Stanford, UC Santa Barbara, UCLA, and the University of Utah. This network could only transmit text—no graphics, animations, or sound. Nevertheless, it quickly grew to span the globe, including the office of Tim Berners-Lee in Switzerland. In 1991, Tim introduced the **World Wide Web**, or **WWW**, or just the **Web**. This multimedia interface for the Internet would propel the Internet to become the largest and most widely used network in the world.

From its simple beginning in 1991, the web has matured through four generations, commonly called Web 1.0, 2.0, 3.0, and 4.0. Each generation is defined by a new technology that changes the relationship between content creator and audience.

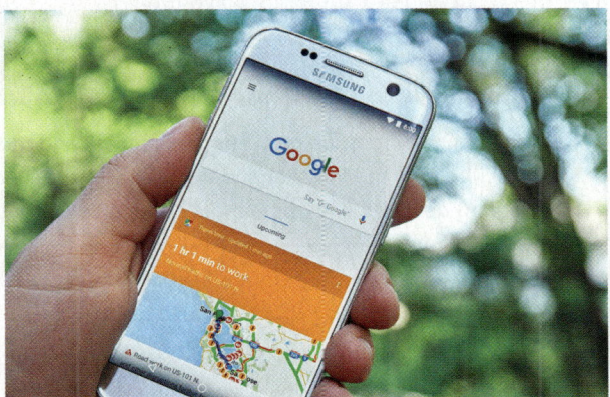

Figure 2-1 Web 3.0 application
dennizn/Shutterstock

- **Web 1.0** is defined by the use of search engines that connect everyone to web pages of interest. These static web pages can only be created or altered by computer programmers. These web pages were designed to appeal to the largest audience possible because they were costly and time consuming to change.

- **Web 2.0** is defined by the use of social media, websites that allow users to post changes to their social media pages without programming skills. This increased the size of the content creators but reduced the intended audience size of these pages to family and followers of the content creator.

- **Web 3.0** is defined by the innovation of the semantic web. The semantic web allows programs to identify relationships between data. For example, a program like Amazon's Alexa and Google Assistant can inform a user that a flight they are planning to take has been delayed (see **Figure 2-1**). This is accomplished by identifying the relationship between a flight scheduled on your calendar and flight delay information available on the Internet. The content creator is now a computer program, and the content audience is a single person—the user.

- **Web 4.0**, called the mobile web, is defined by the use of mobile tools that provide new sources of information for programs to provide users with information. This is an extension of Web 3.0, where programs draw relationships among data to assist individuals, but now the information available to these programs includes the information gathered by mobile devices. For example, your cell phone can provide your current location, or your smartwatch can report your heart rate. A program like Apple's Siri can use your current location, the location of your next calendar event, and local traffic reports on the Internet to alert you to leave early for an appointment because of a traffic jam between your current location and your next event.

- **Web 5.0** is still being developed, but researchers agree that it will be greatly impacted by the most recent innovation in Artificial Intelligence (AI). AI programs are designed to mimic human learning by adapting and learning from past experiences. The emotional web, where online systems respond to the emotional state of the user, will rely on AI systems to learn and tailor responses to the emotional state of the user.

It is easy to get the Internet and the web confused, but they are not the same thing. The Internet is the physical network. It is made up of wires, cables, satellites, and rules for exchanging information between computers connected to the network. Being connected to this network is often described as being **online**. The Internet connects millions of computers and resources throughout the world. The web is a multimedia interface to the resources available on the Internet. Every day over a billion users from nearly every country in the world use the Internet and the web. What are they doing? The most common uses are the following:

> ### ethics
> X, formerly Twitter, and other social media organizations ban users who post hateful or violent content. Some people feel that social media companies have an ethical responsibility to monitor and remove offensive or inaccurate content. Others say that censorship is a violation of an individual's right to free speech and that social media companies should never censor content. What do you think? Who should decide what information is shared on social media: the company or the users?

- **Communicating** is by far the most popular Internet activity. Every time someone receives a text, sends an e-mail, or shares photos or videos online, they are communicating on the Internet.

- **Shopping** is one of the fastest-growing Internet applications. From everyday items, like buying groceries, to special purchases, like buying a prom dress, the Internet provides a way to get the latest trends and the cheapest deals.

- **Searching** for information has never been more convenient. You can access some of the world's largest libraries directly from your home computer.

- **Education** or **e-learning** is another rapidly evolving web application. You can take driving school classes on your cell phone or earn college credit from your laptop. There are courses for every price range and interest.
- **Online entertainment** options are nearly endless. You can find the latest movies and news, listen your favorite songs, and play video games with friends around the world. To learn more about online entertainment, see Making IT Work for You: Online Entertainment on pages 27–28.

The first step to using the Internet and the web is to get connected, or to gain access to the Internet.

concept check

- What is the difference between the Internet and the web?
- Describe how the Internet and the web started. What are the four web generations?
- List and describe five of the most common uses of the Internet and the web.

tips

Are you getting the most out of your web browser? Here are a few suggestions to make you faster and more efficient.

1. **Bookmarks/Favorites Bar:** Most browsers have a bookmarks or favorites bar just below the address bar. Add your top 5 or 10 most-often-visited websites here. The next time you want to visit one of these sites, select it from the bookmarks/favorites list rather than entering the site's URL.
2. **Shortcuts:** Keyboard shortcuts are often faster than using a mouse. Use the following: F5 (refresh); Alt + left arrow (back); Ctrl + T (new tab); Ctrl + W (close tab); Ctrl + Enter (adds "www" and ".com" to any domain name you type in the address bar).
3. **Extensions/Add-Ons:** Many browsers, such as Chrome and Firefox, allow users to install small, third-party programs that extend, or add to, the capabilities of the browser. For example, the Chrome extension Grammarly uses AI learning tools to assist in the writing process.
4. **Configure Settings:** All browsers have a settings or options page that provides ways to secure and customize your Internet browsing. For example, you can deny websites from automatically opening up another browser window, called pop-ups (these annoying windows often contain advertisements). You can also set up your browser to always open the same websites on starting up your browser.

Internet Access

Your computer is a very powerful tool. However, it needs to be connected to the Internet to truly unleash its power. Once on the Internet, your computer becomes an extension of what seems like a giant computer—a computer that branches all over the world. When provided with a connection to the Internet, you can use a browser program to search the web.

Providers

The most common way to access the Internet is through an **Internet service provider (ISP)**. The providers are already connected to the Internet and provide a path or connection for individuals to access the Internet.

The most widely used commercial Internet service providers use telephone lines, cable, and/or wireless connections. Some of the best-known providers in the United States are AT&T, Comcast, T-Mobile, and Verizon.

As we will discuss in **Chapter 8**, users connect to ISPs using one of a variety of connection technologies including **digital subscriber line (DSL)**, **cable**, and **wireless modems**.

Making IT work for you

ONLINE ENTERTAINMENT

The amount of online entertainment can be overwhelming. Traditionally, film, television, and music studios have distributed their art through movie theaters, cable TV, and music CDs. Today, these entertainment industries have created online services to provide their movies, TV, and music. Some online entertainment never existed before the Internet, such as Social Media and podcasts. These forms of entertainment are incredibly popular—and unique to the Internet. With all these choices, how can you find the right online entertainment for you?

- **Online Video**—The Internet offers dozens of ways to view movies and TV on your Internet-connected devices. Here are four things to consider:
  - Is the service free or does it require a monthly subscription? Many big-name services charge a monthly fee, but there are also plenty of services that are free with ads.
  - Does the service include ads? Some services include ads even if you pay a monthly subscription fee.
  - What movies and shows are available? Each service has its own collection of available videos. Before subscribing, make sure you like what they have to offer.
  - Will the service work on your devices? Not every service works on every device. Make sure your devices are supported.

Paid subscription services, like Amazon Prime, offer popular film and TV shows, as well as original content.

True Images/Alamy Stock Photo

Some services are best known for their libraries of movies, full seasons of older TV shows, and original movies and shows exclusively available on their service. Services such as these include

- Amazon Prime (amazon.com/prime).
- Netflix (netflix.com).
- HBO Max (hbomax.com).

Some services offer a more TV-like experience, offering live TV channel content, including local news, national news, and sports coverage. Some of the most popular services that include live TV channels are

- Hulu (hulu.com).
- YouTube TV (tv.youtube.com).

- **Online Music**—You can access large music libraries online, with everything from contemporary hits to classical music. The things to consider when selecting an online music source are similar to choosing an online movie and TV service:
  - Is the service free, or does it require a monthly fee?
  - Does the service include ads?
  - What musicians and albums are available on the service?
  - Does the service work with your devices?

Popular online music services include

- Spotify (spotify.com).
- YouTube music (music.youtube.com).
- Tidal (tidal.com).

Streaming music services, such as Spotify, allow you to listen to vast libraries of music.

M4OS Photos/Alamy Stock Photo

Popular social media site TikTok constantly updates with new videos from users around the world.

M4OS Photos/Alamy Stock Photo

- **Social Media**—Social media websites offer a way to connect and interact with others about news, movies, and everyday life. Thoughts and ideas are posted instantly with sites like X, formerly Twitter, a popular platform for sharing short messages, videos, and photos. You can follow other X users to get automatic updates on the posts they make, and others can follow you to keep up to date on your posts.

Things to consider when choosing a social media app:

- Does the social media app work on your devices? Not all social media apps are available on all devices. For example, many social media apps only work on cell phones.
- Do your friends and family already belong to a social media site? While social media is a great way to meet new people, it is also a great way to keep up with friends and family. If you already know people who use a social media app, it could be a good way to try out social media.
- What types of interactions does the social media app support? Different social media sites support different types of interactions. For example, TikTok specializes in short videos, and Instagram encourages sharing photos. Choose an app that caters to what you are interested in sharing.

Popular social media sites include

- X (X.com).
- Facebook (facebook.com).
- Instagram (instagram.com).
- TikTok (tiktok.com).

Podcast apps, such as Google Podcasts, make it easy and convenient to subscribe and listen to podcasts.

Primakov/Shutterstock

- **Podcasts**—Podcasts are serialized audio programs available to listen to on demand. These programs can be nonfiction, such as news stories, lectures, topical discussions, and interviews, or they can be fiction, such as science fiction, drama, or suspense stories. You can use a podcast app to subscribe to a podcast and share podcasts with friends, receive alerts when a new podcast is available, and rate and comment on podcasts.

    Popular podcast apps include
    - Google Podcasts (podcasts.google.com).
    - Apple Podcasts (apple.com/apple-podcasts).
    - Spotify (spotify.com).

Browsers

Today it is common to access the Internet from a variety of devices like cell phones, tablets, and laptops. These devices use programs called **browsers** to provide access to web resources. These programs connect you to remote computers; open and transfer files; display text, images, and multimedia; and provide in one tool an uncomplicated interface to the Internet and web documents. Four well-known browsers are Apple Safari, Google Chrome, Microsoft Edge, and Mozilla Firefox. (See **Figure 2-2**.)

Figure 2-2   Browser
AntliiShutterstock

Browsers differ not only by designer but also by what device they work on. For example, Chrome is a web browser designed by Google that has a mobile version for cell phones and tablets and a desktop version for laptops and desktops. **Mobile browsers** are designed for the smaller touchscreens of mobile devices. They typically have larger buttons to select options and provide multitouch support for actions such as "pinch" or "stretch" to zoom in on web content. (See **Figure 2-3**.) **Desktop browsers** are designed for laptop and desktop computers, with smaller buttons and more on-screen options designed to take advantage of larger monitors and the precision of mouse input.

For browsers to connect to resources, the **location** or **address** of the resources must be specified. These addresses are called **uniform resource locators (URLs)**. All URLs have at least two basic parts.

Figure 2-3   Zoom web content
outline205/123RF

- The first part presents the protocol used to connect to the resource. As we will discuss in **Chapter 8**, **protocols** are rules for exchanging data between computers. The protocol *https* is used for web traffic and is one of the most widely used Internet protocols. The protocol https is a more secure version of the antiquated http protocol (the "s" in https stands for secure). You should avoid websites that use the insecure http protocol because data shared on the http protocol can be hacked and stolen.

Figure 2-4   Basic parts of a URL

- The second part presents the **domain name.** It indicates the specific address where the resource is located. In **Figure 2-4**, the domain is identified as www.google.com. The last part of the domain name following the dot (.) is the **top-level domain (TLD).** Also known as the **web suffix**, it typically identifies the type of organization. For example, *.com* indicates a commercial site. (See **Figure 2-5**.)

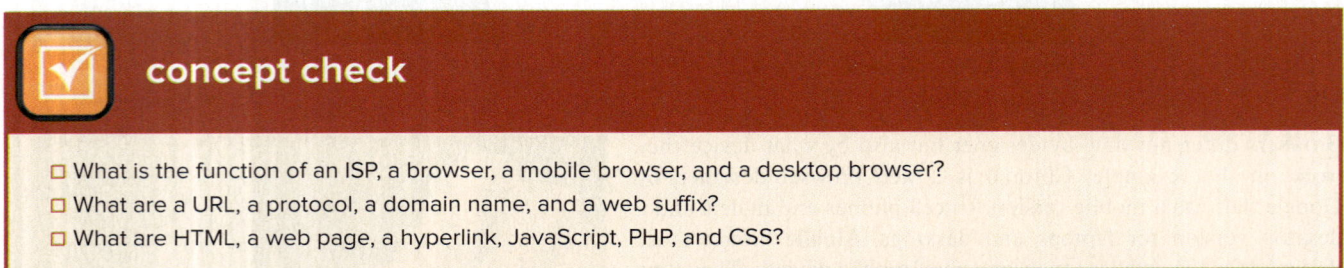

Domain	Type
.com	Commercial
.edu	Educational
.gov	Government
.mil	U.S. military
.net	Network
.org	Organization

Figure 2-5 Traditional top-level domains

Once the browser has connected to the website, a document file is sent back to your computer. This document typically contains **Hypertext Markup Language (HTML)**, a markup language for displaying web pages. The browser interprets the HTML formatting instructions and displays the document as a **web page.** For example, when your browser first connects to the Internet, it opens up to a web page specified in the browser settings. Web pages present information about the site along with references and **hyperlinks** or **links** that connect to other documents containing related information—text files, graphic images, audio, and video clips.

Various technologies are used to provide highly interactive and animated websites. These technologies include the following:

- **JavaScript** is a language often used within HTML documents to trigger interactive features, such as opening new browser windows and checking information entered in online forms. The Microsoft search engine Bing.com uses JavaScript to make its website more interactive and to assist its users by auto-filling search requests as they type content into the search box.
- **PHP**, like JavaScript, is a language often used within HTML documents to improve a website's interactivity. Unlike JavaScript, which typically executes on the user's computer, PHP executes on the website's computer.
- **Cascading style sheets (CSS)** are separate files referenced by, or lines inserted into, an HTML document that controls the appearance of a web page. CSS help ensure that related web pages have a consistent presentation or look. Netflix uses CSS to visually connect all its web pages.

concept check

- What is the function of an ISP, a browser, a mobile browser, and a desktop browser?
- What are a URL, a protocol, a domain name, and a web suffix?
- What are HTML, a web page, a hyperlink, JavaScript, PHP, and CSS?

Web Utilities

Utilities are programs that make computing easier. **Web utilities** are specialized utility programs that make the Internet and the web easier and safer to use. Some of these utilities are browser-related programs that either become part of your browser or are executed from your browser, while others work as separate stand-alone applications. Common uses for web utilities include filtering content and transferring files.

Filters

Filters block access to selected sites. Filter programs allow parents, as well as organizations, to block out selected sites and set time limits. (See **Figure 2-6**.) Additionally, these programs can monitor use and generate reports detailing the total time spent on the Internet and the time spent at individual websites. For a list of some of the best-known filters, see **Figure 2-7**.

Figure 2-6 Circle is a web filter
meetcircle.com

File Transfer Utilities

Using file transfer utility software, you can copy files to your computer from specially configured servers. This is called **downloading.** You also can use file transfer utility software to copy files from your computer to another computer on the Internet. This is called **uploading.** Three popular types of file transfer are web-based, BitTorrent, and FTP.

- **Web-based file transfer services** make use of a web browser to upload and download files. This eliminates the need for any custom software to be installed. Popular services include Microsoft's OneDrive (onedrive.com) and Google's Google Drive (drive.google.com).

- **BitTorrent** distributes file transfers across many different computers for more efficient downloads, unlike other transfer technologies whereby a file is copied from one computer on the Internet to another. A single file might be located on dozens of individual computers. When you download the file, each computer sends you a tiny piece of the larger file, making BitTorrent well suited for transferring very large files. Unfortunately, BitTorrent technology often has been used for distributing unauthorized copies of copyrighted music and video.

Filter	Site
Net Nanny	netnanny.com
Qustodio Parental Control	qustodio.com
Circle with Disney	meetcircle.com
Symantec Norton Family Premier	us.norton.com/norton-familypremier

Figure 2-7 Filters

- **File transfer protocol (FTP)** and **secure file transfer protocol (SFTP)** allow you to efficiently copy files to and from your computer across the Internet and are frequently used for uploading changes to a website hosted by an Internet service provider. FTP has been used for decades and still remains one of the most popular methods of file transfer.

It is important to be confident about any file you download from any source, including the file transfer utilities listed here. Opening any file has some risks of containing a virus. Do not open files from unknown sources and regularly maintain your device's antivirus and security tools.

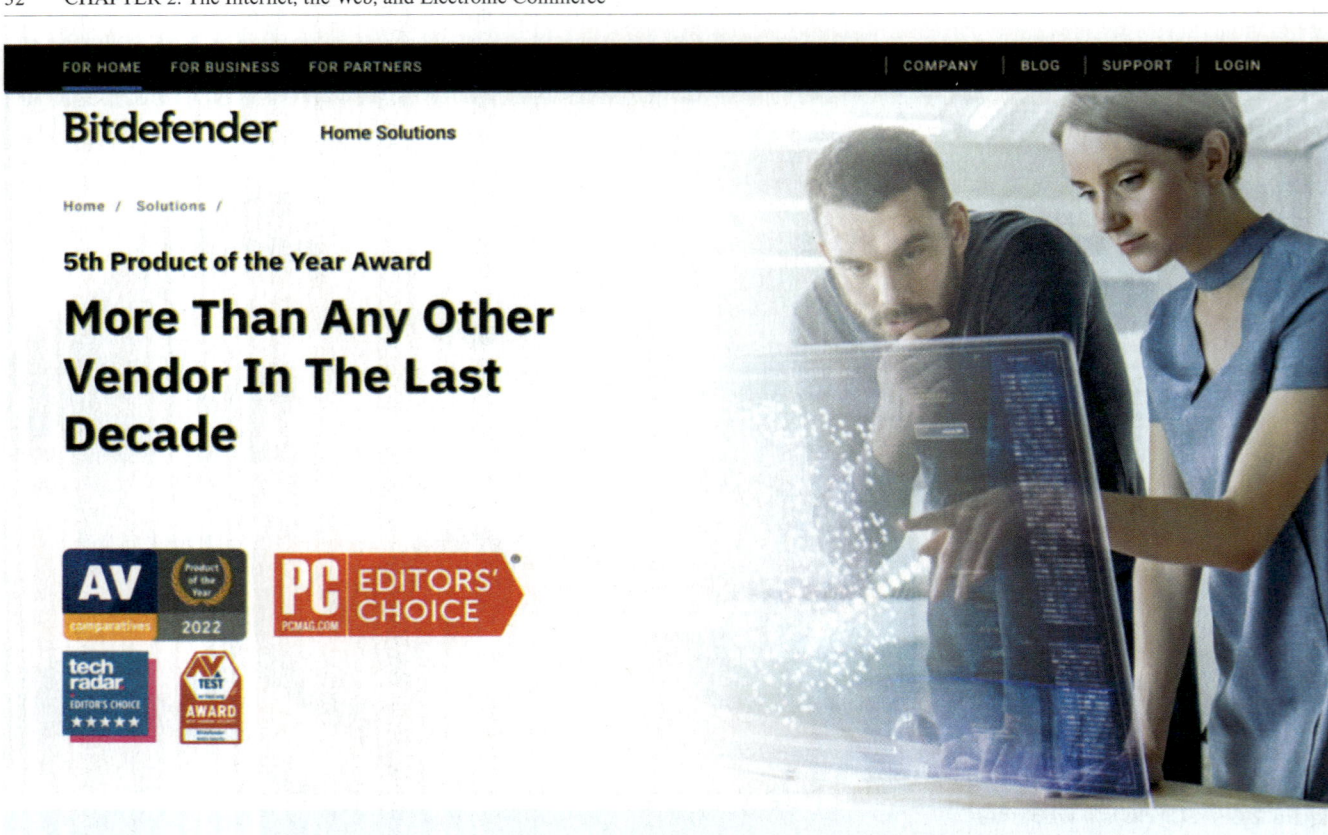

Figure 2-8 Security suite
Bitdefender

Internet Security Suites

Web Utility	Description
Filter	Blocks access to selected sites and sets time limits
File transfer	Upload and download files from servers
Internet security suite	Collection of utility programs for security and privacy

Figure 2-9 Web utilities

An **Internet security suite** is a collection of utility programs designed to maintain your security and privacy while you are on the web. These programs control spam, protect against computer viruses, provide filters, and much more. You could buy each program separately; however, the cost of the suite is typically much less. Two of the best-known Internet security suites are Symantec Norton Internet Security and Bitdefender Internet Security. These companies also offer apps for mobile devices: Bitdefender Mobile Security and Norton Mobile Security. (See **Figure 2-8**.)

For a summary of web utilities, see **Figure 2-9**.

 **concept check**

☐ What are web utilities? What are filters used for?
☐ What are file transfer utilities? Downloading? Uploading? Web-based file transfer services? FTP? SFTP?
☐ Define Internet security suites.

Communication

As previously mentioned, communication is the most popular Internet activity, and its impact cannot be overestimated. At a personal level, friends and family can stay in contact with one another even when separated by thousands of miles. At a business level, electronic communication has become a standard way to stay in touch with suppliers, employees, and customers. Some popular types of Internet communication are social networking, blogs, microblogs, podcasts, wikis, e-mail, and messaging.

Social Networking

Social networking is one of the fastest-growing and most significant Web 2.0 applications. Social networking sites focus on connecting people and organizations that share a common interest or activity. These sites typically provide a wide array of tools that facilitate meeting, communicating, and sharing. There are hundreds of social networking sites, but they share some common features:

- **Profiles** or **pages** allow you to share information about yourself or your business. Individuals create profiles while businesses create pages. Profiles and pages often include a photo that will be displayed along with the member name when they submit a post or send a message. Individuals typically use a photo of their face, while businesses often use a business logo. Other details that your profile or page might include are contact information and a short biography describing yourself or your business.
- **Groups** and **friends** are the other members on social media that you will communicate with. Friends are individuals you communicate with, while groups are communities of other members that share information and discuss specific topics. Groups often organize around a topic, event, or idea.
- **News feed** is the first page you see after logging into a social networking site. It typically consists of a collection of recent posts from friends, trending topics on the site, people's responses to your posts, and advertisements.
- **Share settings** on your social media account determine who can see your posts. The most common options include sharing with everyone, just your friends, or just a subset of your friends.

community

Every community has etiquette rules. On the Internet, these rules are called "netiquette," a combination of *Internet* and *etiquette*. When you enter a new community—online or in real life—it can be difficult to learn the etiquette rules. Fortunately, many social media sites have guidelines for new users to help them navigate the netiquette of social media. For example, did you know it was bad manners to type in all capital letters? It can make readers feel that you are screaming at them! Before communicating online, make sure you understand the netiquette of that community.

privacy

Have you ever seen one of those funny or not-so-funny embarrassing personal videos on the Internet? Unless you are careful, you could be starring in one of those videos. Without privacy settings, images and videos posted to these sites can be viewed and potentially reposted for all to see. If a social networking friend were to post an embarrassing video, would all your friends be able to see it? What about parents, teachers, or potential employers?

Figure 2-10   Instagram profile
Michael J Berlin/Alamy Stock Photo

Organization	Site
Facebook	facebook.com
LinkedIn	linkedin.com
Instagram	instagram.com
Pinterest	pinterest.com
Tumblr	www.tumblr.com
TikTok	tiktok.com
Snapchat	snapchat.com

Figure 2-11 Social networking sites

A social network often has an overall focus. For example, **LinkedIn** is a popular business-oriented social networking site. The profiles on LinkedIn share professional information, like resumes and job successes, and the corporate pages concentrate on sharing their corporate culture and promoting job openings. By contrast, the largest social networking site, **Facebook**, originally had a focus of connecting friends and family, but it has expanded to include news sites, entertainment sites, and a powerful way for businesses and organizations to connect with their audience.

While Facebook is the largest social network, there are many social networking sites, each with its own unique features and interactions. For example, Instagram is a social networking site that concentrates on sharing short videos, and Instagram puts an emphasis on sharing photos. (See **Figure 2-10**.) For a list of some of the most popular sites, see **Figure 2-11**.

All social media companies deal in storing and sharing information—from photos and messages to resumes and GPS data. When you entrust your data to a social media company, you risk them sharing your data with others. In 2023, Europe, Canada, and the United States restricted access to TikTok due to security concerns that the Chinese-owned social media app could spy on their government and citizens. You can learn more about the privacy concerns of social media companies in **Chapter 12**, as well as the privacy and ethics boxes located in each chapter.

Blogs, Microblogs, Podcasts, and Wikis

In addition to social networking sites, other Web 2.0 applications help ordinary people communicate across the web, including blogs, microblogs, podcasts, and wikis. These communication alternatives offer greater flexibility and security; however, they are often more complex to set up and maintain.

Many individuals create personal websites, called **blogs**, to keep in touch with friends and family. Blog postings are time-stamped and arranged with the newest item first. Often, readers of these sites are allowed to comment. Some blogs are like online diaries with personal information; others focus on information about a hobby or theme, such as knitting, electronic devices, or good books. Although most are written by individual bloggers, there are also group blogs with multiple contributors. Some businesses and newspapers also have started blogging as a quick publishing method. Several sites provide tools to create blogs. Two of the most widely used are Blogger and WordPress.

Much like a blog, a **microblog** allows an individual or company to share posts with an audience. However, microblogs are designed to be used with mobile devices and limit the size of posts. For example, **X, formerly Twitter**, one of the most popular microblogging sites, limits posts, also known as **posts**, to 280 characters. (See **Figure 2-12**.) **Instagram**, another popular microblogging site, is designed to share images and videos posts, with little to no written content.

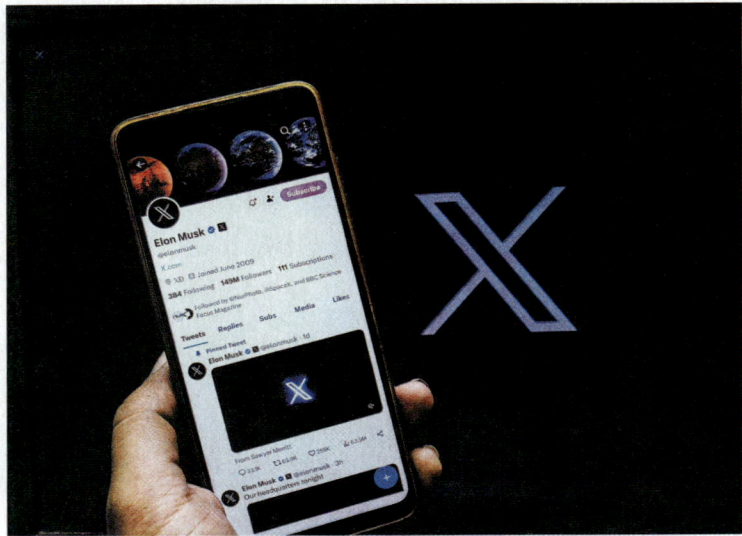

Figure 2-12 X
ZUMA Press, Inc./Alamy Stock Photo

Microblogging has become a significant media tool, used by politicians and celebrities to share moment-to-moment thoughts with their audience. For example, Taylor Swift has over 83 million followers on Twitter.

Podcasts are audio programs delivered over the Internet. Typically, a podcast is one audio program in a series of related podcasts. For example, the NPR podcast, Fresh Air, discusses contemporary art and issues. (See **Figure 2-13**.) Podcasts vary widely, from daily news to arts and comedy shows. For a list of popular podcasts, see **Figure 2-14**.

Figure 2-13 **A podcast**
Tada Images/Shutterstock

Category	Podcasts
Arts	Fresh Air
Comedy	The Joe Rogan Experience
News	The Daily
True Crime	Crime Junkie
Society & Culture	Stuff You Should Know

Figure 2-14 **Podcasts**

A **wiki** is a website specially designed to allow visitors to use their browser to add, edit, or delete the site's content. "Wiki" comes from the Hawaiian word for *fast,* which describes the simplicity of editing and publishing through wiki software. Wikis support collaborative writing in which there isn't a single expert author, but rather a community of interested people that builds knowledge over time. Perhaps the most famous example is **Wikipedia**, an online encyclopedia, written and edited by anyone who wants to contribute, that has millions of entries in over 20 languages.

Creating blogs and wikis are examples of web authoring. We will discuss web authoring software in detail in **Chapter 3**.

community

Our real life and our digital lives compete for our attention— sometimes with deadly results. In 2020, over 1.6 million car collisions were attributed to texting while driving. Engineers, programmers, and government officials are working hard to find innovative solutions to reduce these preventable tragedies. Does your cell phone have a "driving mode" or apps to help keep drivers focused on the road? What are the laws in your area regarding texting and driving?

concept check

☐ What is social networking? Profiles? Pages? Groups? Friends? News feeds? Share settings?
☐ What are blogs? Microblogs? X? Posts?
☐ What is a wiki? What is Wikipedia?

Messaging

Electronic messaging is a popular way to communicate quickly and efficiently with friends, family, and co-workers. This form of communication is particularly common on cell phones, where messages are short and informal, and instantaneous responses are the standard. The two most widely used forms of electronic messaging are text and instant messaging.

- **Text messaging**, also known as **texting** or **SMS (short message service)**, is the process of sending a short electronic message, typically fewer than 160 characters, using a wireless network to another person, who views the message on a mobile device such as a cell phone. Today, billions of people send text messages every day. It has become one of the most widely used ways to send very short messages from one individual to another. Text messaging was originally limited to characters, but you can now send images, video, and sound using a variation of SMS known as **MMS (multimedia messaging service)**. Although popular and convenient, there are downsides to using this technology in the wrong context. A great deal of attention has been directed toward texting while driving. A study by *Car and Driver* concluded that texting while driving had a greater negative impact on driver safety than being drunk. Several states have passed laws prohibiting texting while driving.

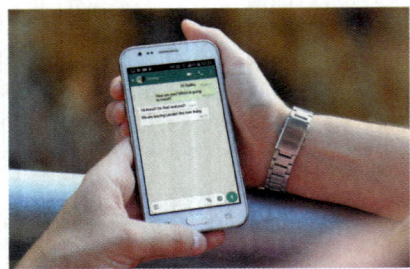

Figure 2-15 Instant messaging
Alejandro Guillermo Santiago Ruhl/Alamy Stock Photo

- **Instant messaging (IM)** allows two or more people to contact each other via direct, live communication. (See **Figure 2-15**.) To use instant messaging, you register with an instant messaging service and then specify a list of friends. Whenever you connect to the Internet, your IM service is notified. It then notifies your friends who are available to chat and notifies your friends who are online that you are available. You can then send messages directly back and forth. Most instant messaging programs also include videoconferencing features, file sharing, and remote assistance. Many businesses routinely use these instant messaging features. Three of the most popular instant messaging services are Facebook Messenger, WhatsApp, and Google Meet.

E-mail

Unlike electronic messaging, **e-mail** or **electronic mail** is used to communicate longer and more formal text. E-mail exchanges tend to take longer to write and are more carefully crafted than text message exchanges, which make them ideal for business communications and newsletters. Because e-mails are common to more formal communications, they do not include the casual use of abbreviations and loose use of punctuation more often found in text and instant messaging. A typical e-mail message has four basic elements: header, message, signature, and attachment. (See **Figure 2-16**.) The **header** appears first and typically includes the following information:

- **Addresses:** E-mail messages typically display the addresses of the intended recipient of the e-mail, anyone else who is to receive a copy of the e-mail, and the sender of the e-mail. The e-mail message in **Figure 2-16** is to dcoats@usc.edu, copied to aboyd@sdu.edu, and sent from cwillis@nyu.edu. E-mail addresses have two basic parts. (See **Figure 2-17**.) The first part is the user's name, and the second part is the domain name, which includes the top-level domain. In our example e-mail, *dcoats* is the recipient's user name. The server providing e-mail service for the recipient is usc.edu. The top-level domain indicates that the provider is an educational institution.

- **Subject:** A one-line description, used to present the topic of the message. Subject lines typically are displayed when a person checks his or her mailbox.

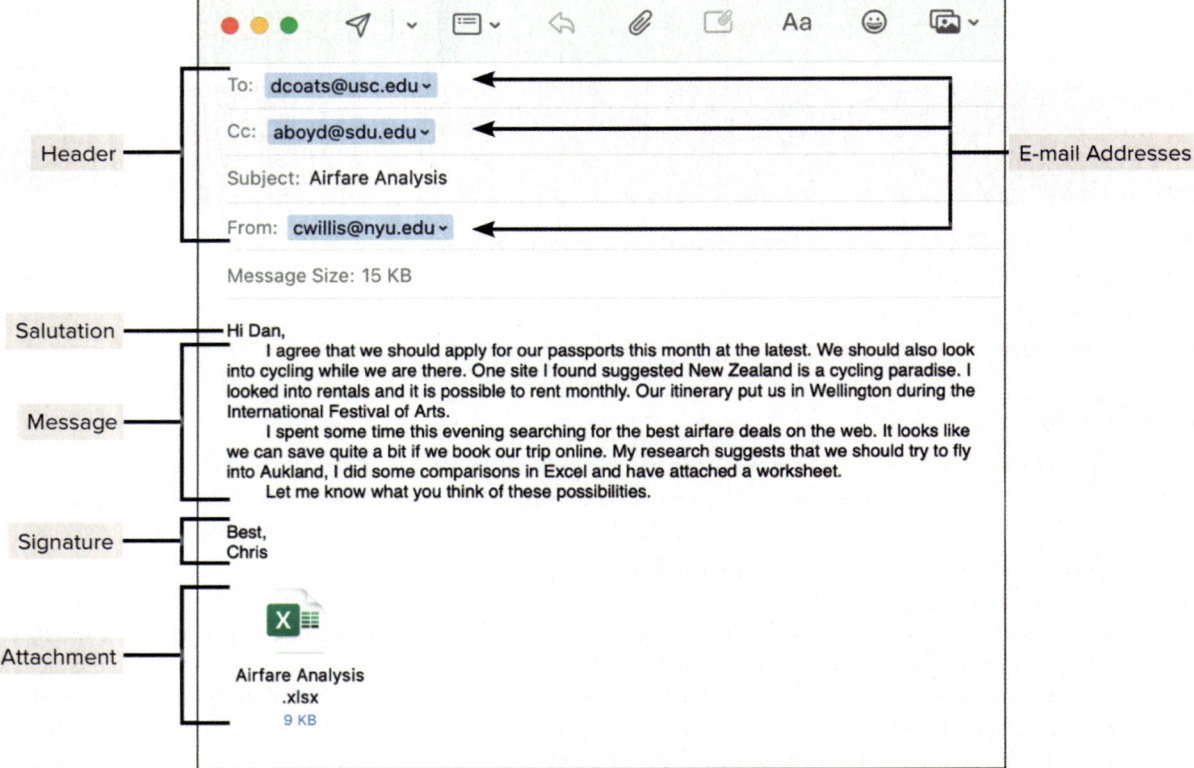

Figure 2-16 Basic elements of an e-mail message
Microsoft Corporation

The salutation begins the letter, with a short line identifying the letter or **message** comes next. Following the message, the **signature** identifies the sender. For many business communications, the signature is automatically inserted into the bottom of an e-mail and can be customized to include the sender's company name, address, and telephone number. Many e-mail programs allow you to include files such as documents and images as **attachments**. (See **Figure 2-16**.) Attachments are a common way to spread computer viruses. Do not open attachments from someone you do not know.

Figure 2-17 Two parts of an e-mail address

There are two basic types of e-mail systems: client-based and web-based.

- **Client-based e-mail systems** require a special program known as an **e-mail client** to be installed on your computer tablet, or cell phone. Before you can begin e-mailing, you need to run the e-mail client from your computer, which communicates with the e-mail service provider. Two of the most widely used e-mail clients are Apple's Mail and Microsoft's Outlook.
- **Web-based e-mail systems** do not require an e-mail program to be installed on your computer. Once your computer's browser connects to an e-mail service provider, a special program called a **webmail client** is run on the e-mail provider's computer and then you can begin e-mailing. This is known as **webmail**. Most Internet service providers offer webmail services. Three free webmail service providers are Google's Gmail, Microsoft's Outlook, and Yahoo!'s Yahoo! Mail.

For individual use, webmail is more widely used because it frees the user from installing and maintaining an e-mail client on every computer used to access e-mail. With webmail, you can access your e-mail from any computer anywhere that has Internet access.

E-mail can be a valuable asset in your personal and professional life. However, like many other valuable technologies, there are drawbacks too. Americans receive billions of unwanted and unsolicited e-mails every year. This unwelcome mail is called **spam**. Although spam is indeed a distraction and nuisance, it also can be dangerous. For example, computer **viruses** or destructive programs are often attached to unsolicited e-mail. Computer viruses will be discussed in **Chapter 4**.

In an attempt to control spam, antispam laws have been added to our legal system. For example, CAN-SPAM requires that every marketing-related e-mail provide an opt-out option. When the option is selected, the recipient's e-mail address is to be removed from future mailing lists. Failure to do so results in heavy fines. This approach, however, has had minimal impact since over 80 percent of all spam originates from servers outside the United States. A more effective approach to controlling spam has been the development and use of **spam blockers**, also known as **spam filters**. Most e-mail programs provide spam-blocking capabilities.

> **tips**
>
> Are you tired of sorting through an inbox full of spam? Here are a few spam reducing suggestions:
>
> 1. **Keep a low profile.** Many spammers collect e-mail addresses from personal web pages, social networking sites, and message boards. Be cautious when posting your address.
> 2. **Use caution when giving out your address.** Many companies collect and sell e-mail addresses to spammers. Be sure to read the privacy policy of a site before providing your address.
> 3. **Don't ever respond to spam.** Many are a trick to validate active e-mail addresses. These addresses are worth more to spammers, who then sell the addresses to other spammers.
> 4. **Use antispam and filter options.** Most e-mail programs and web-based e-mail services have antispam and filter options that can be configured. Use them.

 concept check

- What is text messaging? Texting? SMS? MMS? IM? Friends?
- Define e-mail header, address, subject, message, signature, and attachment.
- What are the two types of e-mail systems? What are viruses, spam, spam blockers, and spam filters?

Search Tools

The web can be an incredible resource, providing information on nearly any topic imaginable. With over 20 billion pages and more being added daily, the web is a massive collection of interrelated pages. With so much available information, locating the precise information you need can be difficult. Fortunately, a number of organizations called **search services** operate websites that can help you locate the information you need.

Search services maintain huge databases relating to information provided on the web and the Internet. The information stored in these databases includes addresses, content descriptions or classifications, and keywords appearing on web pages and other Internet informational resources. Special automated programs called **spiders** continually look for new information and update the search services' databases. Additionally, search services provide special programs called *search engines* that you can use to locate specific information on the web.

Search Engines

Search engines are specialized programs that assist you in locating information on the web and the Internet. To find information, you go to a search service's website and use its search engine. See **Figure 2-18** for a list of commonly used search engines.

Search Service	Site
Bing	www.bing.com
DuckDuckGo	www.duckduckgo.com
Google	www.google.com
Yahoo!	www.yahoo.com

Figure 2-18 Search engines

To use a search website, you enter a keyword or phrase reflecting the information you want. The search engine compares your entry against its database and returns the search result, or a list of sites on that topic. Each search result includes a link to the referenced web page (or other resource), along with a brief discussion of the information contained at that location. Many searches result in a large number of search results. For example, if you were to enter the keyword *music,* you would get billions of sites on that topic. Search engines order the search results according to those sites that most likely contain the information requested and present the list to you in that order, usually in groups of 10.

Because each search service maintains its own database, the search results returned by one search engine will not necessarily be the same results returned by another search engine. Therefore, when researching a topic, it is best to use more than one search engine.

Search engines now use special programs that mimic the thought process of the human brain. These programs are called artificial intelligence (AI). Google has released an AI search engine tool named Bard, and Microsoft's Bing now has a ChatGPT chat mode.

AI is used to improve search engine accuracy. AI programs can interpret complex questions, not just search keywords. Instead of returning a list of relevant sites, AI programs can report back the content in those sites in the context of the question you asked.

Content Evaluation

Search engines are excellent tools to locate information on the web. Be careful, however, how you use the information you find. Unlike most published material found in newspapers, journals, and textbooks, not all the information you find on the web has been subjected to strict guidelines to ensure accuracy. In fact, anyone can publish content on the web. Many sites allow anyone to post new material, sometimes anonymously and without evaluation. Some sites promote **fake news** or information that is inaccurate or biased. This can include manipulated videos known as **deep fakes**, where an individual's appearance or message is changed. Many of these sites are designed to look like legitimate news sites. Improvements in artificial intelligence programs and image manipulation have made deep fakes far easier to produce. This has made them more common and dangerous.

Before you believe a website's claims or share that site with friends and family, consider the following:

- **Authority.** Is the author an expert in the subject area? Is the site an official site for the information presented, or is the site an individual's personal website?

- **Accuracy.** Read beyond the headlines and consult other reputable sources on any surprising claims to separate the real news from the fake. Headlines can be misleading and fake news articles often make their statements seem accurate by referencing other fake news articles.
- **Objectivity.** Is the information factually reported, or does the author have a bias? Does the author appear to have a personal agenda aimed at convincing or changing the reader's opinion?
- **Currency.** Is the information up to date? Does the site specify the date when the site was updated? Are the site's links operational? If not, the site is most likely not being actively maintained.

> **community**
>
> Social media communities share news through likes, comments, and reposting articles. Unfortunately, many of these articles may be misleading or outright lies. When a user shares a news article, are they responsible for verifying that the article is true, or does that responsibility fall to the reader? Even if the articles are lies, social networking sites benefit from the clicks and views of these articles. Do social media companies have a responsibility to identify or eliminate fake news?

concept check

☐ What are search services, spiders, and search engines?
☐ What is fake news? Deep fakes?
☐ What are the four considerations for evaluating website content?

Electronic Commerce

Electronic commerce, also known as **e-commerce**, is the buying and selling of goods over the Internet. E-commerce has several advantages over retail stores:

- **Availability:** E-commerce sites never close
- **Accessibility:** E-commerce sites can be accessed from anywhere with an Internet connection.
- **Expense:** E-commerce sites do not have the expenses associated with owning or renting retail space.
- **Inventory:** E-commerce sites do not have to restock individual retail spaces; products are shipped directly from the warehouse to the consumer.

Of course, there are downsides to e-commerce. You cannot "try on" clothes before you buy them on an e-commerce site. Furthermore, when you buy from a retail store, you immediately receive your purchase—no need to wait for it to be delivered. Nevertheless, few observers suggest that e-commerce will replace brick-and-mortar businesses entirely. It is clear that both will coexist and that e-commerce will continue to grow.

Just like any other type of commerce, electronic commerce involves two parties: businesses and consumers. There are three basic types of electronic commerce:

- **Business-to-consumer (B2C)** commerce involves the sale of a product or service to the general public or end users. It is the fastest-growing type of e-commerce. Whether large or small, nearly every existing corporation in the United States provides some type of B2C support as another means to connect to customers. Because extensive investments are not required to create traditional retail outlets and to maintain large marketing and sales staffs, e-commerce allows start-up companies to compete with larger established firms. The three most widely used B2C applications are for online banking, financial trading, and shopping. Amazon.com is one of the most widely used B2C sites.
- **Consumer-to-consumer (C2C)** commerce involves individuals selling to individuals. C2C often takes the form of an electronic version of the classified ads or an auction. **Web auctions** are similar to traditional auctions except that buyers and sellers seldom, if ever, meet face to face. Sellers post descriptions of products at a website, and buyers submit bids electronically. Like traditional auctions, sometimes the bidding becomes highly competitive and enthusiastic. One of the most widely used auction sites is eBay.com. For a list of some of the most popular C2C sites, see **Figure 2-19**.

Organization	Site
eBay	ebay.com
Facebook Marketplace	facebook.com/marketplace
Etsy	etsy.com

Figure 2-19   C2C sites

- **Business-to-business (B2B)** commerce involves the sale of a product or service from one business to another. This is typically a manufacturer–supplier relationship. For example, a furniture manufacturer requires raw materials such as wood, paint, and varnish.

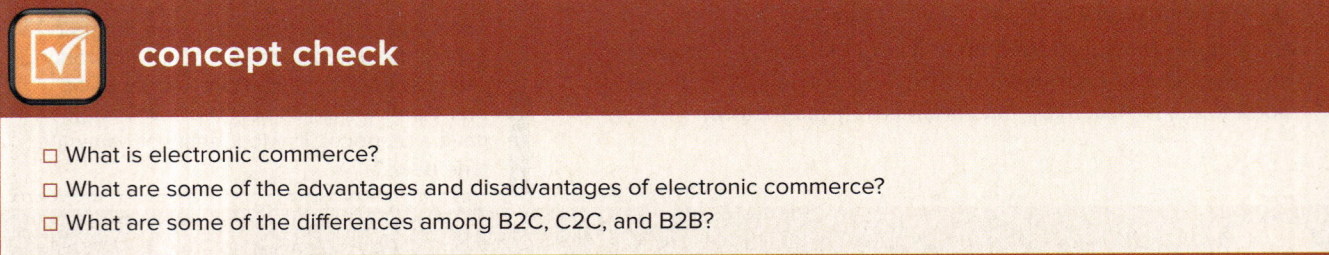

concept check

- What is electronic commerce?
- What are some of the advantages and disadvantages of electronic commerce?
- What are some of the differences among B2C, C2C, and B2B?

Security

The two greatest challenges for e-commerce are (1) developing fast, secure, and reliable payment methods for purchased goods and (2) providing convenient ways to submit required information such as mailing addresses and credit card information.

The two basic payment options are by credit card and by digital cash:

- Credit card purchases are faster and more convenient than check purchases. Credit card fraud, however, is a major concern for both buyers and sellers. We will discuss this and other privacy and security issues related to the Internet in **Chapter 9**.
- **Digital currency** is the Internet's equivalent to traditional cash. Buyers purchase digital currency from a third party (a bank that specializes in electronic currency) and use it to purchase goods. (See **Figure 2-20**.) Most digital currency is a digital version of traditional currency, such that a digital U.S. dollar has the same purchasing power as a traditional U.S. dollar. For example, the digital currency, Tether, is tied to the U.S. dollar. For a list of digital currency providers, see **Figure 2-21**. Some digital currencies, like **bitcoin**, have no traditional cash equivalent, and their transactions do not involve third-party banks. Such a currency is called a **cryptocurrency**. Cryptocurrencies use public ledgers, known as blockchains, to record all transactions of the cryptocurrency. Anyone can read the blockchain, and anyone can hold a copy of the blockchain, but it is nearly impossible to counterfeit the blockchain.

Organization	Site
Amazon	pay.amazon.com
Google	pay.google.com
Venmo	venmo.com
PayPal	www.paypal.com

Figure 2-20   Apple Pay offers digital currency
DenPhotos/Shutterstock

Figure 2-21   Digital currency providers

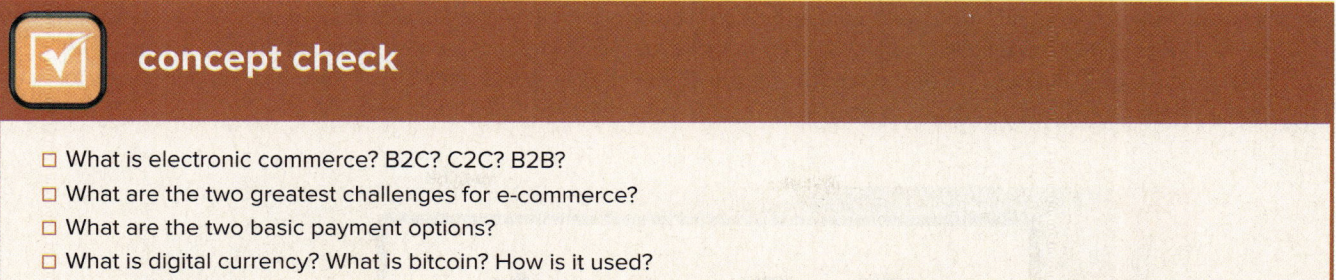

concept check

☐ What is electronic commerce? B2C? C2C? B2B?
☐ What are the two greatest challenges for e-commerce?
☐ What are the two basic payment options?
☐ What is digital currency? What is bitcoin? How is it used?

Cloud Computing

Typically, application programs are owned by individuals or organizations and stored on their computer system's hard disks. As discussed in **Chapter 1**, **cloud computing** uses the Internet and the web to shift many of these computer activities from the user's computer to other computers on the Internet.

Cloud computing frees users from owning, maintaining, and storing software and data. It provides access to these services from anywhere through an Internet connection. Several prominent firms are aggressively pursuing this new concept. These firms include Amazon, IBM, Intel, and Microsoft, to name just a few.

The basic components of cloud computing are clients, the network, and service providers. (See **Figure 2-22**.)

- Clients are corporations and end users who want access to data, programs, and storage. This access is to be available anywhere and anytime that a connection to the Internet is available. Clients can use their existing computers and applications to access the cloud. They do not need to buy, install, or maintain any new computers, application programs or data. Another name for the client side of the cloud is the **frontend**, which is the hardware and software that the user interacts with.
- The network is the connection between the clients and the providers. The most commonly used network is the Internet. Two of the most critical factors determining the efficiency of cloud computing are (1) the speed and reliability of the user's access to the Internet and (2) the Internet's capability to provide safe and reliable transmission of data and programs.

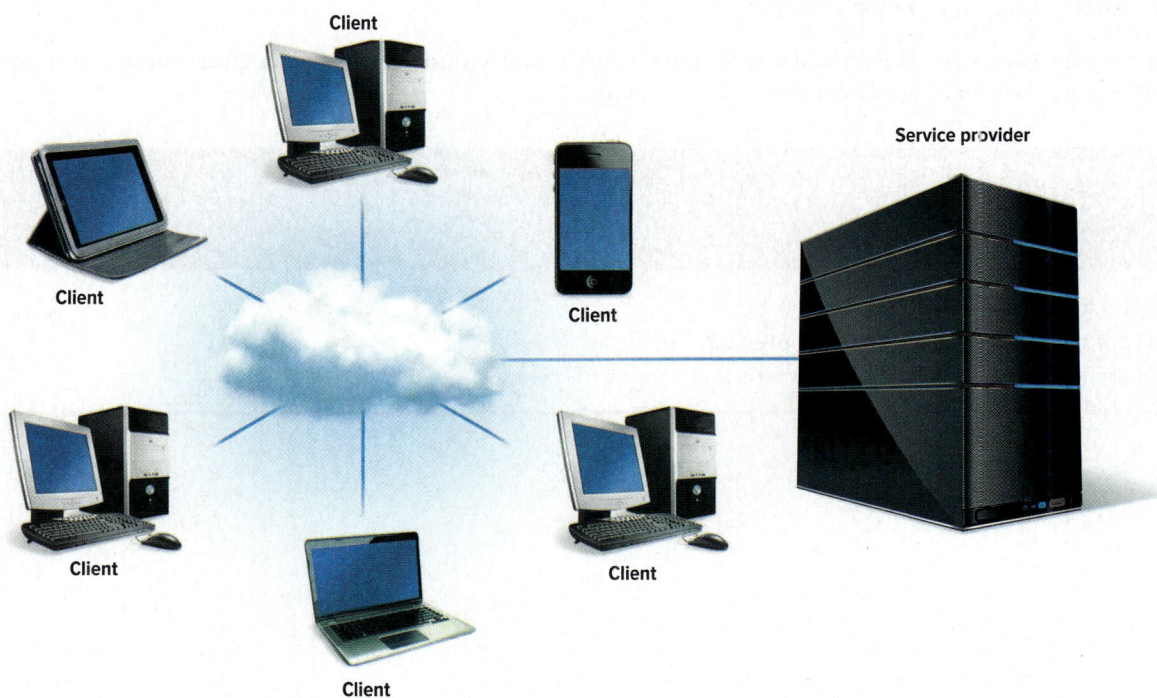

Figure 2-22 **Cloud computing**
Gravvi/Shutterstock

- Service providers are organizations with computers connected to the Internet that are willing to provide access to software, data, and storage. The **backend** is the name for the software and hardware that the user does not directly interact with. These providers may charge a fee or the services may be free. For example, Google Drive Apps provide free access to programs with capabilities similar to Microsoft's Word, Excel, and PowerPoint. (See **Figure 2-23**.)

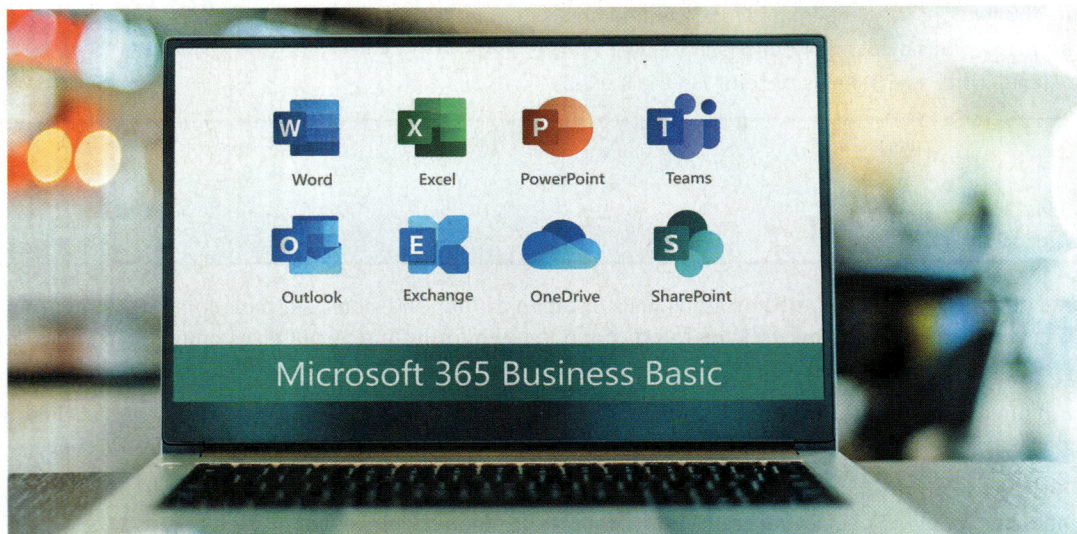

Figure 2-23   Web-based service (Microsoft Business Basics)
monticello/Shutterstock

Organizations that use cloud computing have one of three models to choose from:

- **Public Cloud** is the most common type of cloud computing. All cloud resources are shared between all individuals and organizations. Google Docs is a free public cloud service that provides pared-down versions of the features found in Microsoft Word, Excel, and PowerPoint. (See **Figure 2-23**.)
- **Private Cloud** is a cloud built exclusively for the use of one organization. Cloud resources will be shared between members of the organization. Only members of the organization use and access the cloud services.
- **Hybrid Cloud** is a combination of public and private cloud. Some applications are in the public cloud and those applications share resources with people outside the organization, whereas other applications are in the private cloud and run exclusively on the organization's computers.

In the following chapters, you will learn more about the services provided through cloud computing. You will also learn about security and privacy challenges associated with cloud computing.

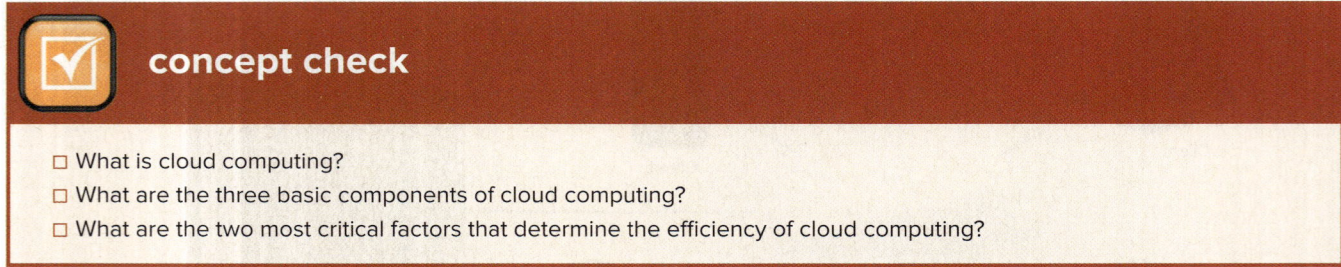

concept check

- ☐ What is cloud computing?
- ☐ What are the three basic components of cloud computing?
- ☐ What are the two most critical factors that determine the efficiency of cloud computing?

The Internet of Things

The Internet is becoming more and more a part of our lives. As discussed in **Chapter 1**, the **Internet of Things (IoT)** is the continuing development of the Internet that allows everyday objects embedded with electronic devices to send and receive data over the Internet. These everyday objects include cell phones, wearable devices, and even appliances and automobiles. For example, the Oura is a ring that monitors health data and sends that data to your cell phone or personal web page. (See **Figure 2-24**.) Google Nest is a collection of IoT devices for the home that allow you to answer your front door when you are not home, improve home security, and automate your home thermostat.

A Web 3.0 application can access the data on a smartwatch, combine that data with other data on the web, process the data, and send information back to another device. For example, Apple's Health App, a Web 3.0 application, can access your Apple Watch data, combine it with other related health data, analyze the data, and report back to you through your cell phone. These reports provide information about your health, including your heart rate, steps taken each day, and an estimate of daily calories burned. (See **Figure 2-25**.)

Figure 2-24 **Oura Ring**
Shutterstock-Pixelsquid/Shutterstock

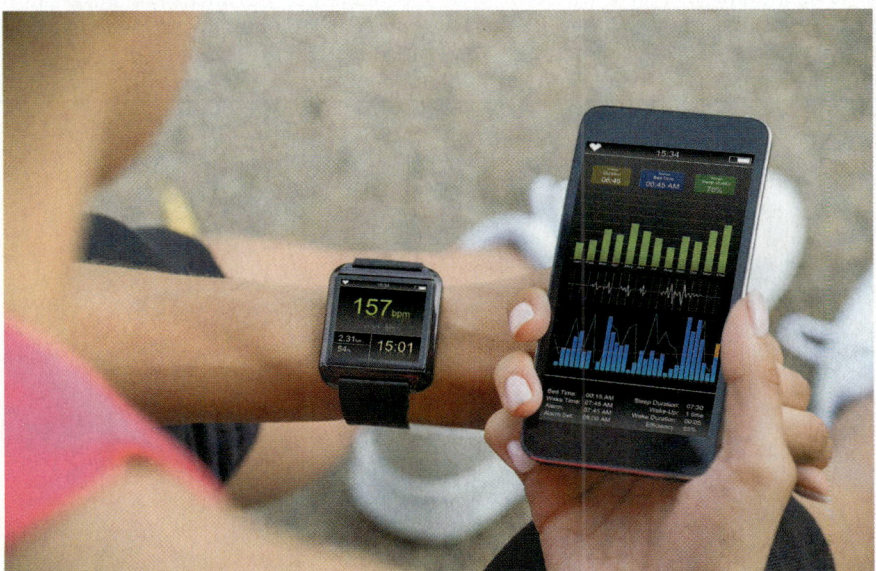

Figure 2-25 **Apple's Health App**
Andrey_Popov/Shutterstock

concept check

☐ What is the Internet of Things?
☐ What are Apple Watch and Apple's Health App? How do they work together?
☐ Discuss how Apple Watch and Apple's Health App are examples of using IoT.

Careers in IT

"Now that you've learned about the Internet, the web, and electronic commerce, I'd like to tell you about my career as a Web developer."

AnnaStills/iStock/Getty Images

Web developers develop and maintain websites and resources. The job may include backup of the company website, updating of resources, or development of new resources. Web developers are often involved in the design and development of the website. Some Web developers monitor traffic on the site and take steps to encourage users to visit the site. Web developers also may work with marketing personnel to increase site traffic and may be involved in the development of web promotions.

Employers look for candidates with a bachelor's or associate's degree in computer science or information systems and knowledge of common programming languages and web development software. Knowledge of HTML and PHP is considered essential. Those with experience using web authoring software and programs like Adobe Illustrator and Adobe Dreamweaver are often preferred. Good communication and organizational skills are vital in this position.

Web developers can expect to earn an annual salary of $59,000 to $82,000. In larger organizations, this position may be held by several employees, each specializing in a specific aspect of the company's web presence. With technological advances and increasing corporate emphasis on a web presence, experience in this field could lead to managerial opportunities.

A LOOK TO THE FUTURE

Home Smart Home

alexsl/iStock/Getty Images Plus

The technology in our homes has made dull tasks a little easier. Our parents' and grandparents' lives were changed by the introduction of the vacuum and the washing machine. Today, our lives are further impacted by new technologies in the home, and the biggest change is the use of the Internet. The introduction of Internet of Things (IoT) devices into our homes and the use of Web 5.0 technologies is doing more than making boring tasks easier. The future of technology in our homes is smart devices that monitor and react to us. These technologies will make our homes easier, healthier, and more sustainable.

Today, smart home systems like Google's Nest can adjust your thermostat to keep your home comfortable and energy bills low. Researchers at Cambridge Consultants are working on even smarter homes that will reduce waste, increase savings, and make for a more sustainable world. The smart home of the future will have smart trash cans that monitor and separate waste, recyclables, and compostable trash. Recyclables and compost can be repurposed within the home to create energy and reduce carbon emissions.

The smart home of the future will do more than clean up and recycle. It will monitor and adjust energy use based on your needs and the energy sources available. Interconnected IoT devices will work together to maximize their efficiency to make for a lower energy home. Devices monitor inside and outside temperature, news and weather reports, and fluctuating power availability from home solar panels and community wind turbines. These homes will also use sensors in the home to predict energy needs, such as increased hot water in the mornings when people shower and reduced air conditioning when you go away for a weekend.

Your smart home in the future will not just use less energy. It will use less space as well. Researchers at Samsung envision a home of smart walls that use adjustable surfaces and actuators to change the size of a room effectively, making empty rooms smaller and occupied rooms larger. When you are entertaining guests, the walls will reposition to create extra seating directly out of the wall. When the guests leave, the extra seating disappears back into the wall.

While these advances sound great, perhaps the greatest advantages of a smart home will be the ability to improve health and longevity. Currently, researchers are investigating the ways that IoT devices can help people make healthier choices. Smartwatches and other monitoring IoT wearables have been shown to increase people's activity level and improve health. The smart home of the future will monitor your health and adjust your home to make healthier choices, from furniture that reads your posture and guides you to improve how you sit, to refrigerators that monitor your food intake and promote healthier dining choices. IoT and Web 5.0 will allow smart homes to work seamlessly in the background to help you achieve your best you.

Of course, these advances include an increase in potential risks. A home that records every movement and activity could be used to spy on you. Companies that store and analyze your smart home data could use that data in ways you haven't agreed to, such as to sell products or to design and test new ideas. Finally, by putting your security and privacy in the hands of a smart home, you risk that smart home being hacked and opening your home to criminals.

What do you think? Are the conveniences of the smart home of the future worth the risks to your privacy and security? What smart home features exist right now that you would like to have? What smart home features that researchers are working on would be most valuable to you?

VISUAL SUMMARY | The Internet, the Web, and Electronic Commerce

INTERNET and WEB

dennizn/Shutterstock

Internet

Launched in 1969 with **ARPANET,** the **Internet** consists of the actual physical network.

Web

Introduced in 1991, the **web (World Wide Web, WWW)** provides a multimedia interface to Internet resources. Four generations: **Web 1.0** (links existing information, search engines), **Web 2.0** (creates dynamic content and social interaction, social media sites), **Web 3.0** (identifies relationships between data), **Web 4.0** (creates content from data from mobile devices), and **Web 5.0** (identifies and responds to human emotion).

Common Uses

The most common uses of the Internet and the web include

- Communication—the most popular Internet activity.
- Shopping—one of the fastest-growing Internet activities.
- Searching—access libraries and local, national, and international news.
- Education—**e-learning** or taking online courses.
- Online entertainment—movies, news, music, and video games.

INTERNET ACCESS

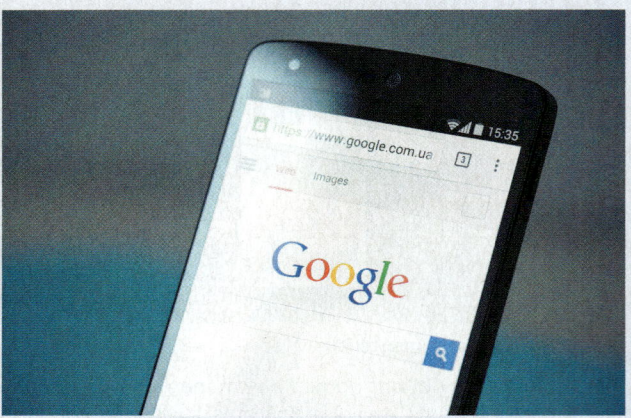

Antlii/Shutterstock

Once connected to the Internet, your computer seemingly becomes an extension of a giant computer that branches all over the world.

Providers

Internet service providers connected to the Internet provide a path to access the Internet. Connection technologies include **DSL, cable,** and **wireless modems.**

Browsers

Browsers (**mobile** and **desktop**) provide access to web resources. Some related terms are

- URLs—**locations** or **addresses** to web resources; two parts are **protocol** and **domain name; top-level domain (TLD)** or **web suffix** identifies type of organization.
- HTML—commands to display **web pages; hyperlinks (links)** are connections.

Technologies providing interactive, animated websites include **JavaScript** (executes on the user's computer to trigger interactive features); **PHP** (operates within HTML to trigger interactive features and check online forms); and **CSS** (controls the appearance of web pages).

To efficiently and effectively use computers, you need to be aware of resources available on the Internet and web, to be able to access these resources, to effectively communicate electronically, to efficiently locate information, to understand electronic commerce, and to use web utilities.

WEB UTILITIES

meetcircle.com

Web utilities are specialized utility programs that make using the Internet and the web easier and safer.

Filters

Filters are used by parents and organizations to block certain sites and to monitor use of the Internet and the web.

File Transfer Utilities

File transfer utilities copy files to **(downloading)** and from **(uploading)** your computer. There are three types:

- **File transfer protocol (FTP)** and **secure file transfer protocol (SFTP)** allow you to efficiently copy files across the Internet.
- **BitTorrent** distributes file transfers across many different computers.
- **Web-based file transfer services** make use of a web browser to upload and download files.

Internet Security Suite

An **Internet security suite** is a collection of utility programs designed to protect your privacy and security on the Internet.

Web Utility	Description
Filter	Blocks access to selected sites and sets time limits
File transfer	Upload and download files from servers
Internet security suite	Collection of utility programs for security and privacy

COMMUNICATION

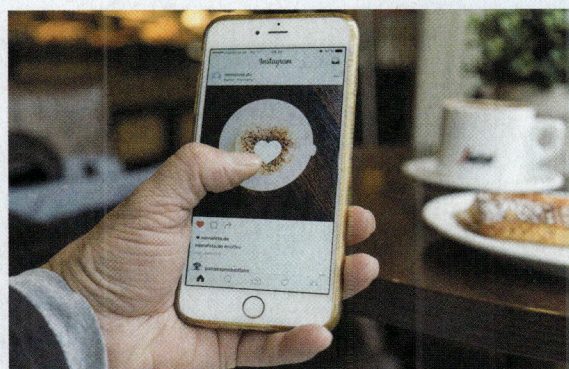

Michael J Berlin/Alamy Stock Photo

Social Networking

Social networking sites connect people and organizations that share a common interest or activity. Common features include **profiles**, **pages**, **groups**, **friends**, **news feeds**, and **share settings.** Two well-known sites are **LinkedIn** and **Facebook.**

Blogs, Microblogs, Podcasts, and Wikis

Other sites that help individuals communicate across the web are blogs, microblogs, podcasts, and wikis.

- **Blogs** are typically personal websites to keep in touch with friends and family. Some are like online diaries. Businesses, newspapers, and others also use blogs as a quick publishing method.
- **Microblogs** use short sentences. **X, formerly Twitter** allows 280 characters per **post. Instagram** is primarily used to share images and videos.
- **Podcasts** are audio programs delivered over the Internet. Typically, a podcast is one audio program in a series of related podcasts.
- A **wiki** is a website designed to allow visitors to use their browsers to add, edit, or delete the site's content. **Wikipedia** is one of the most popular wikis.

Category	Podcasts
Arts	99% Invisible
Comedy	The Joe Rogan Experience
News	The Daily
Science & Medicine	Radiolab
Society & Culture	This American Life

COMMUNICATION

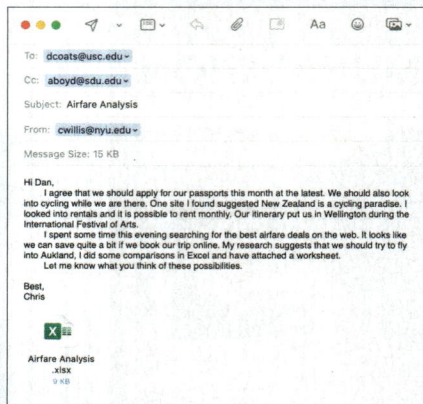

Microsoft Corporation

Messaging

Although e-mail is the most widely used, there are two other messaging systems:

- **Text messaging,** also known as **texting** and **SMS (short message service),** is a process of sending short electronic messages, typically fewer than 160 characters. Texting while driving is very dangerous and illegal in several states.
- **Instant messaging (IM)**—supports live communication between **friends.** Most instant messaging programs also include videoconferencing features, file sharing, and remote assistance.

E-mail

E-mail (electronic mail) is the transmission of electronic messages. There are two basic types of e-mail systems:

- **Client-based e-mail systems** use **e-mail clients** installed on your computer.
- **Web-based e-mail systems** use **webmail clients** located on the e-mail provider's computer. This is known as **webmail.**

A typical e-mail has three basic elements: **header** (including **address, subject,** and perhaps **attachment**), **message,** and **signature.**

Spam is unwanted and unsolicited e-mail that may include a computer **virus** or destructive programs often attached to unsolicited e-mail. **Spam blockers,** also known as **spam filters,** are programs that identify and eliminate spam.

SEARCH TOOLS

Search Service	Site
Bing	www.bing.com
DuckDuckGo	www.duckduckgo.com
Google	www.google.com
Yahoo!	www.yahoo.com

Search services maintain huge databases relating to website content. The information stored in these databases includes addresses, content descriptions or classifications, and keywords appearing on web pages and other Internet informational resources. **Spiders** are programs that update these databases.

Search Engines

Search engines are specialized programs to help locate information. To use, enter a keyword or phrase, and a list of search results is displayed.

Content Evaluation

Not all information you find on the web has been subjected to strict guidelines to ensure accuracy. Many sites allow anyone to post new material, sometimes anonymously and without critical evaluation. Some sites promote **fake news** (inaccurate or biased information) and/or **deep fake** videos. Many fake news sites are designed to look like legitimate news sites.

To evaluate the accuracy of information found on the web, consider the following:

- Authority. Is the author an expert? Is the site official or does it present one individual's or organization's opinion?
- Accuracy. Consult other reputable sources to verify surprising claims. Fake news articles often reference other fake news articles.
- Objectivity. Is the information factual or does the author have a bias? Does the author appear to have a personal agenda to convince or form a reader's opinion?
- Currency. Is the information up to date? Does the site specify when information is updated? Are the site's links operational?

ELECTRONIC COMMERCE

DenPhotos/Shutterstock

Electronic commerce, or **e-commerce,** is the buying and selling of goods over the Internet. Three basic types are

- **Business-to-consumer (B2C)** commerce, which involves sales from business to the general public.
- **Consumer-to-consumer (C2C)** commerce, which involves sales between individuals. **Web auctions** are similar to traditional auctions except buyers and sellers rarely, if ever, meet face to face.
- **Business-to-business (B2B)** commerce, which involves sales from one business to another, typically a manufacturer–supplier relationship.

Security

The two greatest challenges for e-commerce are the development of

- Safe, secure payment methods. Two types are credit cards and **digital currency. Bitcoins** exist only on the Internet.
- Convenient ways to provide required information.

CLOUD COMPUTING

Cloud computing shifts many computer activities from the user's computer to other computers on the Internet. The three basic components to cloud computing are

- Clients—corporations and end users. The **frontend** is the hardware and software that the clients directly interact with.
- Internet—provides connection between clients and providers.
- Service providers—organizations with computers connected to the Internet that are willing to provide access to software, data, and storage. The **backend** is the hardware and software that is not directly accessed by the user.

INTERNET of THINGS

Andrey_Popov/Shutterstock

The **Internet of Things (IoT)** is a continuing development of the Internet allowing everyday objects embedded with electronic devices to send and receive data over the Internet. Objects include cell phones, wearable devices, and even coffeemakers. For example:

- The Fitbit is an IoT device (bracelet) that monitors health data and sends that data to a cell phone or personal web page.
- Apple's Health App is a Web 3.0 application that can access Fitbit data, combine it with other related health data, analyze the data, and report back to you through a cell phone.

The Apple Health App reports provide information about users' health, including heart rate, steps taken each day, and an estimate of daily calories burned.

AnnaStills/iStock/Getty Images

CAREERS in IT

Web developers develop and maintain websites and web resources. A bachelor's or associate's degree in computer science or information systems and knowledge of common programming languages and web development software are required. Expected salary range is $59,000 to $82,000.

KEY TERMS

address
Advanced Research Project Agency Network (ARPANET)
attachment
bitcoin
backend
BitTorrent
blog
browser
business-to-business (B2B)
business-to-consumer (B2C)
cable
cascading style sheets (CSS)
client-based e-mail system
cloud computing
consumer-to-consumer (C2C)
cryptocurrency
deep fake
desktop browser
digital currency
domain name
downloading
digital subscriber line (DSL)
e-commerce
e-learning
electronic commerce
electronic mail
e-mail
e-mail client
Facebook
fake news
file transfer protocol (FTP)
filter
friend
frontend
group
header
hyperlink
Hypertext Markup Language (HTML)
Instagram
instant messaging (IM)
Internet
Internet of Things (IoT)
Internet security suite
Internet service provider (ISP)
JavaScript
link
LinkedIn
location
message
microblog
MMS (multimedia messaging service)
mobile browser
news feed
online
page
PHP
podcast
post
profile
protocol
search engine
search service
secure file transfer protocol (SFTP)
share settings
signature
SMS (short message service)
social networking
spam
spam blocker
spam filter
spider
subject
texting
text messaging
top-level domain (TLD)
uniform resource locator (URL)
uploading
virus
web
Web 1.0
Web 2.0
Web 3.0
Web 4.0
Web 5.0
web auction
web-based e-mail system
web-based file transfer service
webmail
webmail client
web developer
web page
web suffix
web utility
wiki
Wikipedia
wireless modem
World Wide Web
WWW
X

MULTIPLE CHOICE

Circle the correct answer.

1. When the Internet launched, it was a network called:
   - a. DSL
   - b. LAN
   - c. ARPANET
   - d. CSS

2. This Internet activity is associated with sending and receiving e-mails.
   - a. shopping
   - b. communicating
   - c. e-learning
   - d. online entertainment

3. The physical network that is the world's largest network is called:
   - a. the World Wide Web
   - b. the Internet
   - c. ARPANET
   - d. SFTP

4. This generation of the web that brought about social media.
   - a. Web 1.0
   - b. Web 2.0
   - c. Web 3.0
   - d. Web 4.0

5. An example of a microblogging site is:
   - a. Facebook
   - b. TikTok
   - c. X
   - d. Microsoft

6. The most common way to access the Internet is through a(n) _____.
   - a. cell phone
   - b. ISP
   - c. SFTP
   - d. TikTok

7. Transmission of electronic messages over the Internet.
   - a. Web 3.0
   - b. B2B
   - c. hyperlink
   - d. e-mail

8. Two popular instant messaging services are WhatsApp and Facebook _____.
   - a. Social
   - b. Meet
   - c. Messenger
   - d. ISP

9. A business-oriented social networking site.
   - a. TikTok
   - b. Instagram
   - c. LinkedIn
   - d. Facebook

10. Electronic commerce involving individuals selling to individuals.
    - a. B2C
    - b. C2C
    - c. B2B
    - d. I2I

MATCHING

Match each numbered item with the most closely related lettered item. Write your answers in the spaces provided.

a. HTML
b. service providers
c. Internet
d. protocols
e. X
f. digital currency
g. downloading
h. groups
i. web-based
j. hits

____ 1. The name for using a file transfer utility software to copy files to your computer from specially configured servers on the Internet.
____ 2. Using a keyword, a search engine returns a list of related sites known as _____.
____ 3. This is the Internet's equivalent to traditional cash.
____ 4. The continuing Internet development that allows objects to send and receive data over the Internet.
____ 5. On social media, communities of individuals who share a common interest typically create _____.
____ 6. The type of e-mail account that does not require an e-mail program to be installed on a user's computer.
____ 7. A very popular microblogging site.
____ 8. The network that connects computers all over the world.
____ 9. The rules for exchanging data between computers.
____ 10. Three basic components to cloud computing are clients, Internet, and _____.

OPEN-ENDED

On a separate sheet of paper, respond to each question or statement.

1. Discuss the Internet and web, including their origins, the four generations of the web, and the most common uses.
2. Describe how to access the Internet. What are providers? Define desktop and mobile browsers, and discuss URLs, HTML, CSS, JavaScript, PHP, and mobile browsers.
3. What are web utilities? Discuss filters, file transfer utilities, and Internet security suites.
4. Discuss Internet communications, including social networking, blogs, microblogs, podcasts, wikis, client-based and web-based e-mail, and text and instant messaging.
5. Define search tools, including search services. Discuss search engines and fake news. Describe how to evaluate the content of a website.
6. Describe electronic commerce, including business-to-consumer, consumer-to-consumer, business-to-business e-commerce, and security.
7. Discuss the Internet of Things (IoT). Describe how Apple Watch and Apple's Health App are examples of how an IoT device can interact with a Web 3.0 application.
8. What is cloud computing? Describe three basic components of cloud computing.

DISCUSSION

Respond to each of the following questions.

Making IT Work for You: ONLINE ENTERTAINMENT

Finding the right online video library can be challenging. In this discussion, you will use the website justwatch.com, a search site that can tell you where you can view TV shows and movies online. Review the Making IT Work for You: Online Entertainment on pages 27–28 and then respond to the following: (a) Choose 3 TV shows that you would enjoy and search for those titles on the website justwatch.com to learn what online video services offer those TV shows. Go online and investigate the price and features of the services that you have found. What service would you choose and why? (b) Ask a friend or family member what their favorite 3 TV shows are. Investigate the online video services that offer those shows. What service would you recommend for your friend or family member and why?

Privacy: SOCIAL NETWORKING

When a Facebook friend posts a picture, video, or text that includes you, who can view that post? Review the Privacy box on page 33, and respond to the following: (a) Who should be responsible for ensuring privacy on social networking sites? Defend your position. (b) Do you think that most people are aware of their privacy settings on Facebook? Have you ever checked your settings? Why or why not? (c) Investigate and then summarize the default security settings for a social networking website such as Facebook.

Ethics: MONITORING CONTENT

Review the Ethics box on page 25 and then respond to the following: (a) Do social media companies have an ethical responsibility to monitor content? Defend your position. (b) If you responded yes to (a), who should have the power to determine or to arbitrate what is and what is not objectionable? If you responded no to (a), can you think of any situations in which it would not be appropriate to block or remove content? Be specific and defend your position.

Community: FAKE NEWS

Review the Community box on page 35 and then respond to the following: (a) Do you or anyone you know share news through social media? What types of news do they share? (b) Can you identify the original source of this news? Have you ever seen news shared on social media that does not have a clearly identified or trustworthy source? (c) How do you identify a trusted source shared on social media? How would you correct someone who has shared fake news with you?

Design Elements: Concept Check icon: Dizzle52/Getty Images

chapter 3
Application Software

ra2 studio/Shutterstock

Why should I read this chapter?

Gorodenkoff/Shutterstock

The power and capability of application software is exploding. We can expect applications beyond our imagination, and to control these applications entirely with our voice, gestures, and thoughts.

This chapter covers the innovations in applications you can expect to see in the future and the most important applications found in businesses around the world, including:

- General-purpose applications—how to create documents, analyze data, make presentations, and organize information.
- Special-purpose applications—how to use programs for image editing, web page creation, and video game development, and how to locate and use mobile apps.
- Software suites—how to use suites and cloud-based applications.

Learning Objectives

After you have read this chapter, you should be able to:

1. Identify general-purpose applications.
2. Describe word processors, spreadsheets, presentation programs, and database management systems.
3. Describe specialized applications, such as graphics, web authoring, and video game development programs.
4. Describe mobile apps and app stores.
5. Identify software suites.
6. Describe office suites, cloud suites, specialized suites, and utility suites.

Introduction

"Hi, I'm Mia, and I'm a software engineer. I'd like to talk with you about application software and how to access these traditional programs using cloud computing."

SeventyFour/Shutterstock

You might think that only programmers at tech companies like Amazon and Google need to know how to use applications. However, applications do a lot more than that program webpages and databases. Every industry, at every level, uses applications to create reports, analyze data, schedule meetings, and communicate important information.

Think of the personal computer as an electronic tool. You may not consider yourself very good at typing, calculating, organizing, presenting, or managing information. However, a personal computer can help you do all these things and much more. All it takes is the right kinds of software.

To efficiently and effectively use computers, you need to understand the capabilities of general-purpose application software, which includes word processors, spreadsheets, presentation programs, and database management systems. You also need to know about integrated packages and software suites.

Application Software

As we discussed in **Chapter 1**, there are two kinds of software. **System software** works with end users, application software, and computer hardware to handle the majority of technical details. **Application software** can be described as end-user software and is used to accomplish a variety of tasks.

Application software can be divided into three categories. One of the most popular categories, **mobile apps**, consists of applications designed for cell phones and tablets. Another category, **general-purpose applications**, includes word processing programs, presentation software, spreadsheets, and database management systems. Finally, **specialized applications** include thousands of programs that are more narrowly focused on specific disciplines and occupations.

App Stores

An **app store** is where you can download applications to work with your mobile, laptop, and desktop devices. Some applications are free, but many cost a nominal fee. Three of the best-known app stores are Apple's App Store, Google Play, and the Microsoft Store (see **Figure 3-1**).

Furthermore, some applications require a two-step process of download and installation, whereas others combine the process into one step. Most mobile devices will download and install in one step from the app store. However, many applications for laptops or desktops require a two-step process of downloading the application from the app store and then installing the application by running the downloaded file.

Applications are written for a particular type of device. For example, an app designed for Apple's iPhone will not work on a Windows laptop. Therefore, most app stores focus on apps for a particular type of device. For example, the Microsoft Store focuses on apps for Windows laptops and desktops, while the Apple App Store has apps designed to work on Apple devices. For a list of some of the more widely used app stores, see **Figure 3-2**.

Figure 3-1 Apple's App Store
ymgerman/Shutterstock

Store	Focus	Site
Apple App Store	Apple devices	apps.apple.com
Microsoft Store	Windows devices	apps.microsoft.com/store
Google Play	Android devices	play.google.com
Amazon	Android devices	amazon.com/appstore

Figure 3-2 App stores

User Interface

A **user interface** is the portion of the application that allows you to control and to interact with the program. Depending on the application, you can use a mouse, a keyboard, a stylus, a finger, and/or your voice to communicate with the application. Most general-purpose applications use a mouse and a **graphical user interface (GUI)** that displays graphical elements called **icons** to represent familiar objects. The mouse controls a **pointer** on the screen that is used to select items such as icons. Another feature is the use of windows to display information. A **window** is simply a rectangular area that can contain a document, program, or message. (Do not confuse the term *window* with the various versions of Microsoft's Windows operating systems, which are programs.) More than one window can be opened and displayed on the computer screen at one time.

The standard GUI uses a system of menus, toolbars, and dialog boxes. (See **Figure 3-3**.)

- **Menus** present commands that are typically displayed in a **menu bar** at the top of the screen.
- **Toolbars** typically appear below the menu bar and include small graphic elements called **buttons** that provide shortcuts for quick access to commonly used commands.
- **Dialog boxes** provide additional information and request user input.

Many applications, and Microsoft applications in particular, use an interface known as the **Ribbon GUI**, which changes based on the needs of the user. This GUI uses a system of interrelated ribbons, tabs, and galleries. (See **Figure 3-4**.)

- **Ribbons** replace toolbars and menus by organizing commonly used commands into sets of related activities. These activities are displayed as tabs and appear in the first ribbon.
- **Tabs** divide the ribbon into major activity areas. Each tab is then organized into **groups** that contain related items. Some tabs, called **contextual tabs**, appear only when they are needed and anticipate the next operation to be performed by the user.
- **Galleries,** like dialog boxes, provide additional options and simplify choosing an option by showing the effect.

Figure 3-3   Traditional graphical user interface
Microsoft Corporation

Figure 3-4 Ribbon GUI
Microsoft Corporation

community

Fake Reviews

Shopping apps help customers make smart purchases by posting consumers' reviews of products. For example, Amazon allows its community of users to rate products from one to five stars. You can even search for products based on the product's average star rating. Unfortunately, these ratings can also be faked; there are entire companies that offer services to post fake reviews and falsely improve a product's ratings. Have you ever encountered product reviews that you suspect were fake? Do you trust online reviews? Should companies be held accountable for the reviews posted on their sites, or is it a case of "buyer beware"?

Common Features

Most applications provide a variety of features to make entering/presenting, editing, and formatting documents easy. Some of the most common features include

- Spelling and grammar checker—looks for misspelled words or grammatical errors.
- Alignment—centers, right-aligns, or left-aligns numbers and characters.
- Fonts and font sizes—specify the size and style of entered numbers and text.
- Character effects—provide a variety of different typefaces, such as bold or italics.
- Edit options—provide easy ways to edit text, such as cut, copy, and paste.
- Find and replace—searches a document for a specific word or phrase, with the option to replace the word or phrase with something else.

concept check

- What are app stores? What are they used for?
- List three categories of application software.
- What is a graphical user interface? What are windows, menus, toolbars, and dialog boxes?
- What is the Ribbon GUI? What are ribbons, tabs, and galleries?
- Discuss some of the most common features in application programs.

Mobile Apps

Mobile apps or **mobile applications** are add-on programs for a variety of mobile devices, including cell phones and tablets. Sometimes referred to simply as **apps**, mobile apps have been widely used for years. Many standard apps are included free with a mobile OS, including:

- e-mail apps to view and reply to work and personal e-mails.
- to-do list and calendar apps to organize and plan work and personal activities.
- alarm and timer apps to help you wake up or remember to do things.
- messaging and phone call apps to communicate with co-workers and friends.

With the introduction of cell phones, tablets, and wireless connections to the Internet, mobile capabilities are almost limitless.

Apps

The breadth and scope of available mobile applications for cell phones and other mobile devices are ever expanding. There are over 2.5 million mobile apps on the Google Play store alone. Some of the most widely used are for listening to music, viewing video, social networking, shopping, and game playing. See **Figure 3-5** for a list of some of the most widely used mobile apps.

Category	Popular Apps
Music	Spotify and YouTube Music
Video	Netflix and YouTube
Social Networking	TikTok and Instagram
Shopping	Amazon
Games	Crossy-road and Stardew Valley

Figure 3-5 Mobile apps

- Music apps stream music, organize playlists, and recommend new artists. The Spotify app offers free music with ads. For a monthly fee, apps such as Spotify and YouTube Music offer ad-free listening, offline listening, and other extra features.
- Video apps are a favorite way to watch movie trailers and video blogs. The YouTube app offers access to free online videos such as movie trailers and video blogs. The Netflix app provides access to more professional videos for a fee.
- Social networking apps let you use your mobile device to share vacation photos, check in at your favorite coffee shop, or invite friends to a party. The most widely used social media app is Facebook, but other apps, such as Instagram and TikTok, are gaining in popularity.
- Shopping apps make shopping online quick and easy. These apps will help you find products, read customer reviews, and make purchases.
- Gaming apps are an enjoyable way to pass the time on a cell phone. Some game apps cost money, but others work on a **free-to-play** model, where the initial game is free, but the more advanced features require an in-app purchase. Games vary from Crossy-road, a simple arcade-style game, to Stardew Valley, a complex **role-playing game (RPG)** where gamers create a farm, grow crops, and raise livestock. (See **Figure 3-6**.)

privacy

Although mobile apps on your cell phone are an amazing way to share your life with friends and family on social media, it is easy to forget that these apps are constantly recording your location, photos you take, and local businesses that you visit. Privacy advocates are concerned that this data might be used for unintended purposes. For example, employers could access and use this information when deciding who to hire. In fact, such cases are documented. Employers argue that they should consider every aspect of a future or current employee, including how an employee's media presence could reflect on the company. What do you think?

Figure 3-6   Fortnite game
Yurii_Dr/Alamy Stock Photo

Many apps are written for a particular type of mobile device and will not run on other types. For example, an app designed for Apple's iPhone may not work with Google's Android.

concept check

- What are mobile apps? What are they used for?
- Describe two types of game apps. What is an RPG?

General-Purpose Applications

As mentioned previously, general-purpose applications include word processors, spreadsheets, presentation software, and database management systems.

Word Processors

Word processors create text-based **documents** and are one of the most flexible and widely used software tools. All types of people and organizations use word processors to create memos, letters, and reports. Organizations create newsletters, manuals, and brochures to provide information to their customers. Students and researchers use word processors to create reports.

Word processors include many valuable tools to help you create clear and professional writing. These can include changing the appearance of a document, by altering the font or color of text, to making suggestions on how to write more clearly, with spelling and grammar suggestions. Microsoft Word is the most widely used word processor. Other popular word processors include Apple Pages and Google Docs. Google Docs includes grammar checkers that use AI to understand complex grammar structures.

Assume that you have accepted a job as an advertising coordinator for Adventure Travel Tours, a travel agency specializing in active adventure vacations. Your primary responsibilities are to create and coordinate the company's promotional materials, including flyers and travel reports. To see how you could use Microsoft Word as the advertising coordinator for the Adventure Travel Tours, see **Figures 3-7** and **3-8**.

Spell Checker
Correcting spelling and typing errors identified by the **spell checker** creates an error-free and professional-looking document.

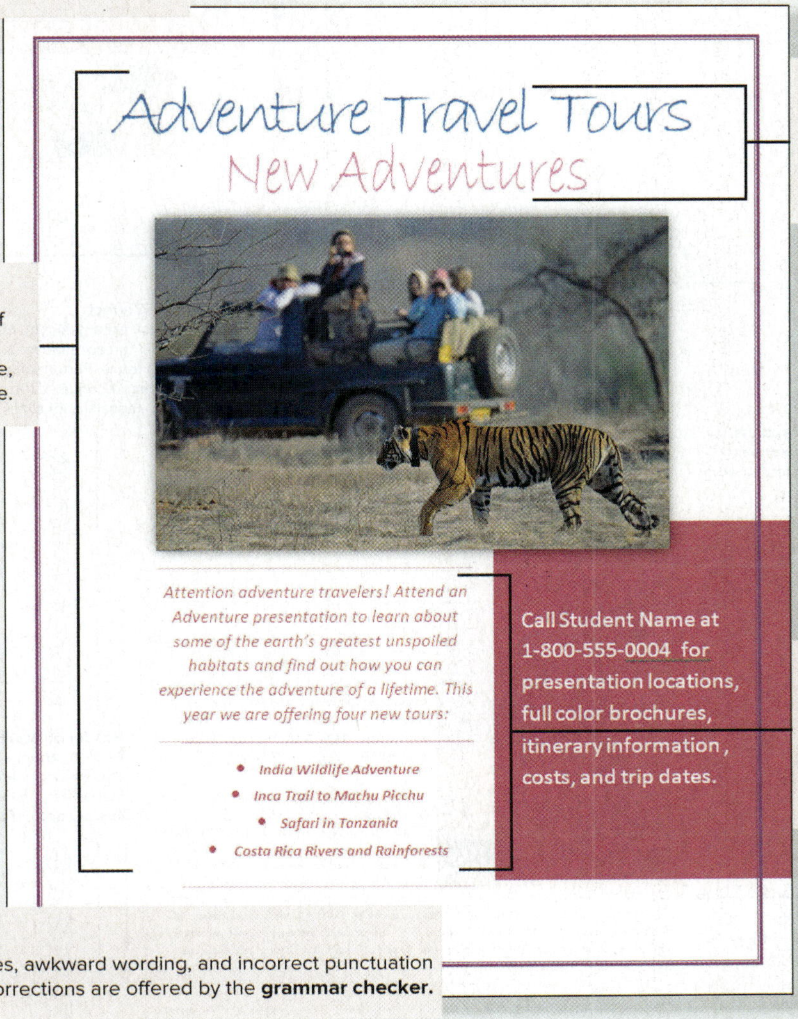

Fonts and Font Size
Using interesting **fonts** and a large **font size** in the flyer's title grabs the reader's attention.

Center-Aligning
Center-aligning all of the text in the flyer creates a comfortable, balanced appearance.

Character Effects
Adding **character effects** such as bold and color makes important information stand out and makes the flyer more visually interesting.

Grammar Checker
Incomplete sentences, awkward wording, and incorrect punctuation are identified, and corrections are offered by the **grammar checker.**

Figure 3-7 Flyer
Aditya "Dicky" Singh/Alamy Stock Photo

Creating a Flyer

You have been asked to create a promotional advertising flyer. After discussing the flyer's content and basic structure with your supervisor, you start to enter the flyer's text. As you enter the text, the spell checker and grammar checker catch some spelling and grammatical errors. Once the text has been entered, you proofread the text and then focus your attention on enhancing the visual aspects of the flyer. You add a photograph and experiment with different character and paragraph formats, including fonts, font sizes, colors, and alignments.

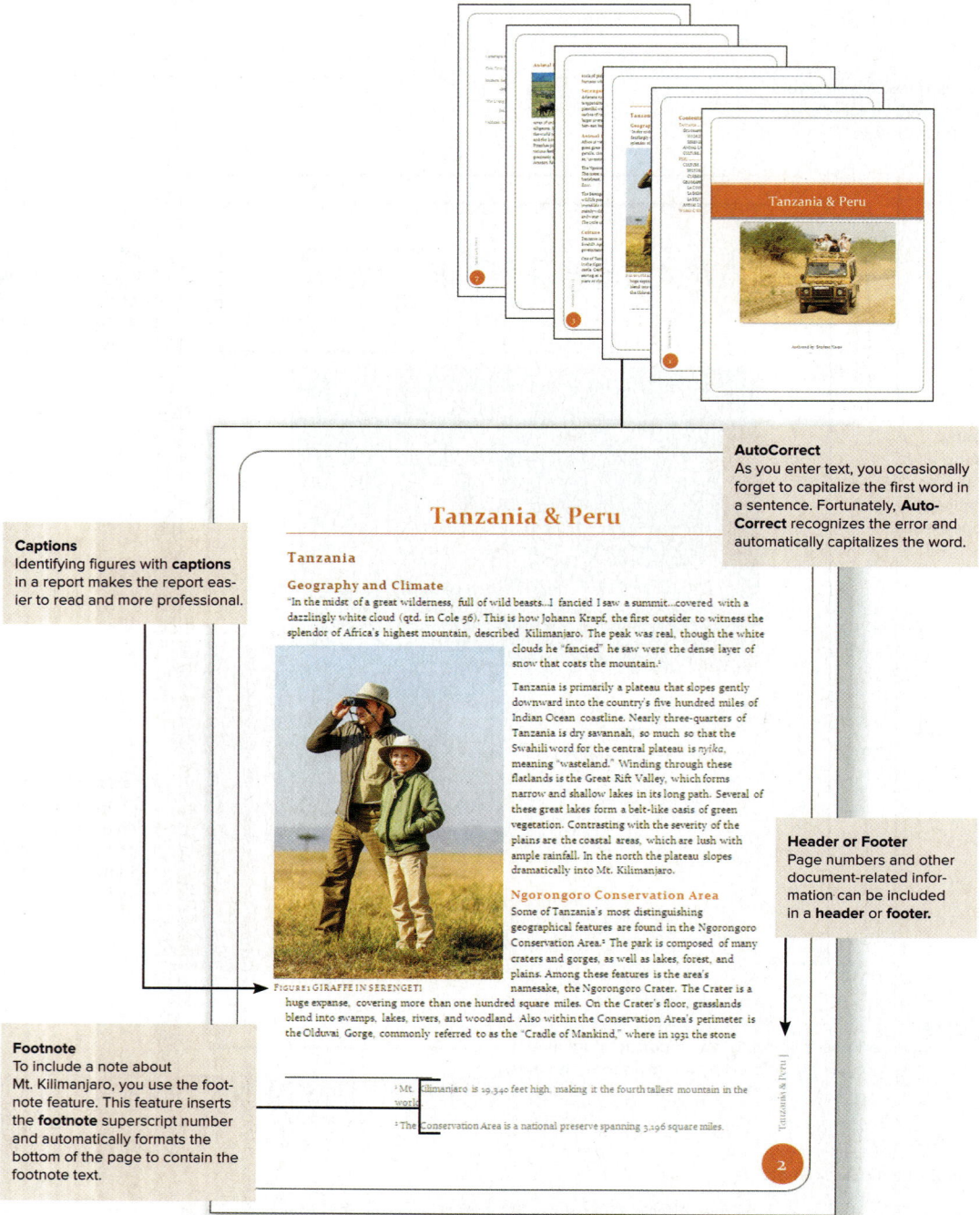

Figure 3-8 Report
imageBROKER.com GmbH & Co. KG/Alamy Stock Photo; BlueOrange Studio/Shutterstock

Creating a Report

Your next assignment is to create a report on Tanzania and Peru. After conducting your research, you start writing your paper. As you enter the text for the report, you notice that the AutoCorrect feature automatically corrects some grammar and punctuation errors. Your report includes several figures and tables. You use the captions feature to keep track of figure and table numbers, to enter the caption text, and to position the captions. You use the footnote feature to assist in adding notes to further explain or comment on information in the report.

Finally, you prepare the report for printing by adding header and footer information. •

Presentation Software

Presentation software are programs that combine a variety of visual objects to create attractive, visually interesting presentations. They are excellent tools to communicate a message and to persuade people.

People in a variety of settings and situations use presentation software programs to make their presentations. For example, marketing managers use presentation software to present proposed marketing strategies to their superiors. Salespeople use these programs to demonstrate products and encourage customers to make purchases. Students use presentation software to create high-quality class presentations.

Three of the most widely used presentation software programs are Microsoft PowerPoint, Apple Keynote, and Google Slides.

Assume that you have volunteered for the Animal Rescue Foundation, a local animal rescue agency. You have been asked to create a powerful and persuasive presentation to encourage other members from your community to volunteer. To see how you could use Microsoft PowerPoint, see **Figure 3-9**.

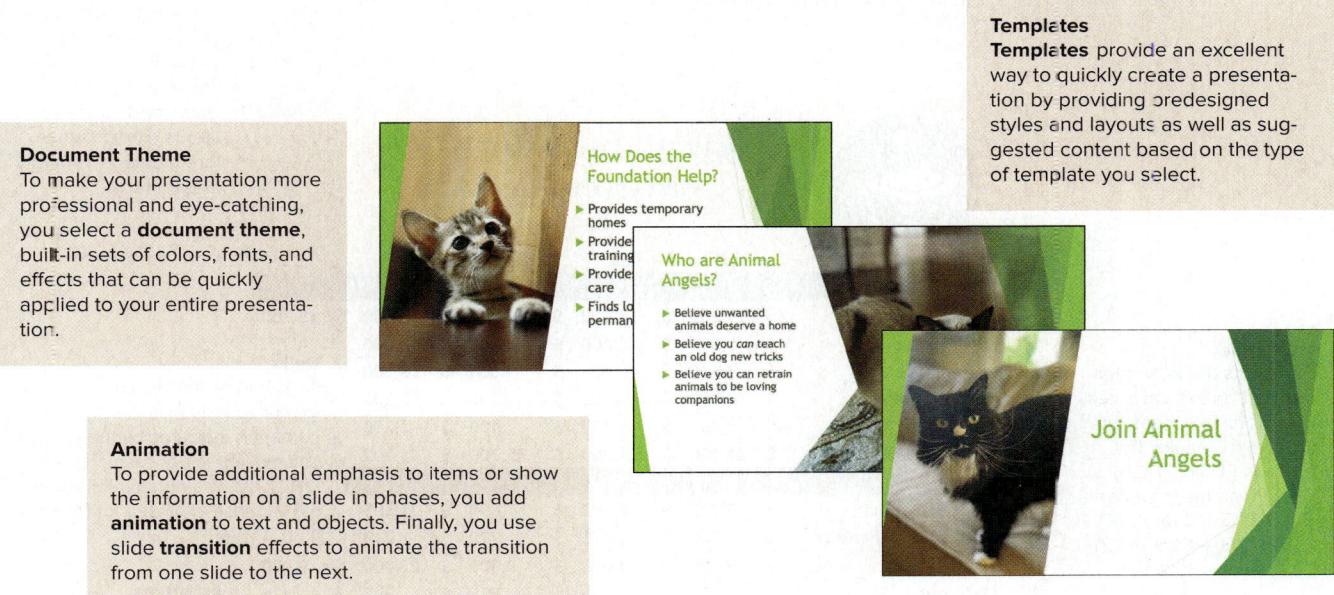

Figure 3-9 Presentation
Roylee_photosunday/Shutterstock; Animal Angels

Creating a Presentation

You have been asked to create a powerful and persuasive presentation for the director of the foundation designed to encourage other members from your community to volunteer. The first step is to meet with the director of the foundation to determine the content of the presentation. Then, using PowerPoint, you begin creating the presentation by selecting a presentation template and document theme. After entering the content, you add interest to the presentation by adding animation to selected objects and using slide transition effects. •

Spreadsheets

Spreadsheets organize, analyze, and graph numeric data such as budgets and financial reports. Once used exclusively by accountants, spreadsheets are widely used by nearly every profession. Marketing professionals analyze sales trends. Financial analysts evaluate and graph stock market trends. Students and teachers record grades and calculate grade point averages.

The most widely used spreadsheet program is Microsoft Excel. Other spreadsheet applications include Apple Numbers and Google Sheets.

Assume that you have just accepted a job as manager of the Downtown Internet Café. This café provides a variety of flavored coffees as well as Internet access. One of your responsibilities is to create a financial plan for the next year. To see how you could use Microsoft Excel as the manager for the Downtown Internet Café, see **Figures 3-10** and **3-11**.

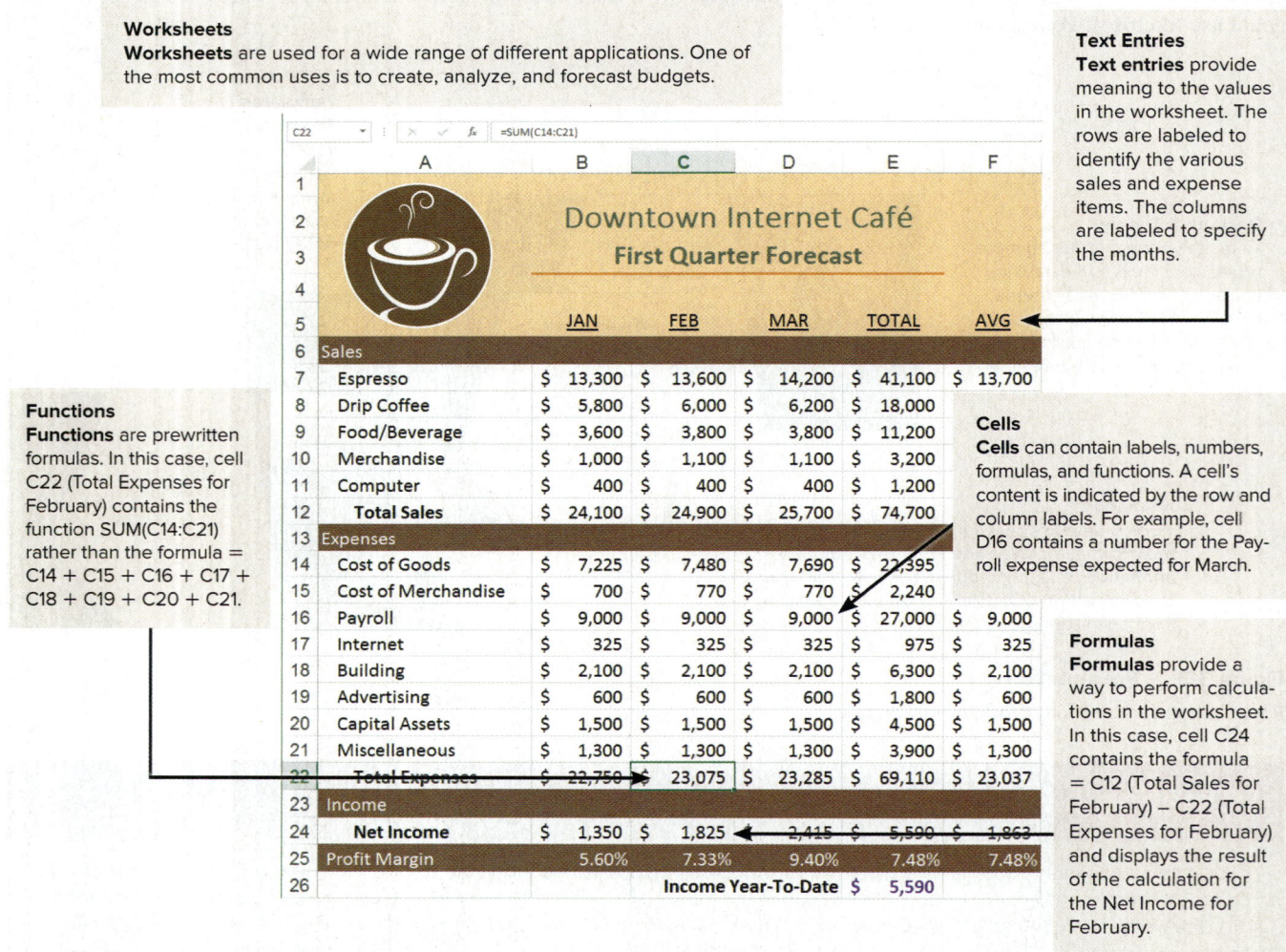

Figure 3-10 First-quarter forecast

Microsoft Corporation; Achmad fandhy akhbar/Getty Images

Creating a Sales Forecast

Your first project is to develop a first-quarter sales forecast for the café. You begin by studying sales data and talking with several managers. After obtaining sales and expense estimates, you are ready to create the first-quarter forecast. You start structuring the worksheet by inserting descriptive text entries for the row and column headings. Next, you insert numeric entries, including formulas and functions to perform calculations. To test the accuracy of the worksheet, you change the values in some cells and compare the recalculated spreadsheet results with hand calculations. •

CHAPTER 3: Application Software   65

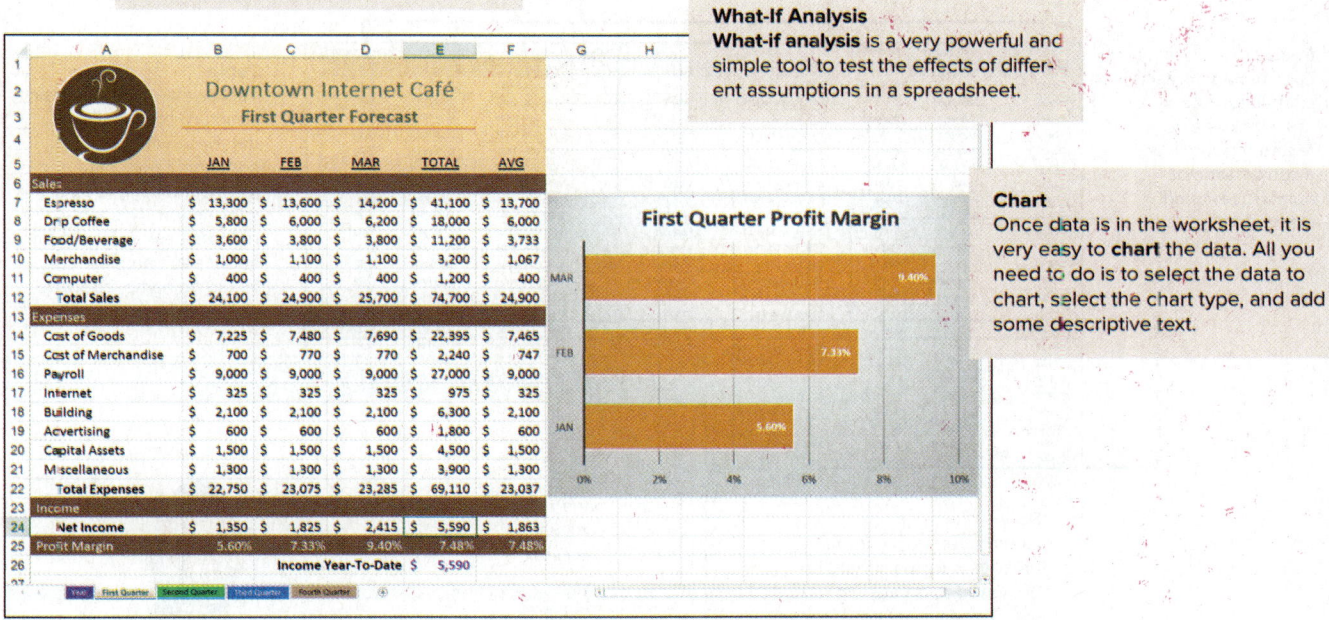

Workbook
The first worksheet in a **workbook** is often a summary of the following worksheets. In this case, the first worksheet presents the entire year's forecast. The subsequent worksheets provide the details.

Sheet Name
Each worksheet has a unique **sheet name.** To make the workbook easy to navigate, it is a good practice to always use simple yet descriptive names for each worksheet.

What-If Analysis
What-if analysis is a very powerful and simple tool to test the effects of different assumptions in a spreadsheet.

Chart
Once data is in the worksheet, it is very easy to **chart** the data. All you need to do is to select the data to chart, select the chart type, and add some descriptive text.

Figure 3-11   Annual forecast and analysis
Microsoft Corporation; Achmad fandhy akhbar/Getty Images

Analyzing Your Data

After presenting the First-Quarter Forecast to the owner, you revise the format and expand the workbook to include worksheets for each quarter and an annual forecast summary. You give each worksheet a descriptive sheet name. At the request of the owner, you perform a what-if analysis to test the effect of different estimates for payroll, and you use a chart to visualize the effect.

Database Management Systems

A **database** is a collection of related data. It is the electronic equivalent of a file cabinet. A **database management system (DBMS)** or **database manager** is a program that sets up, or structures, a database. It also provides tools to enter, edit, and retrieve data from the database. All kinds of individuals use databases, from hospital administrators recording patient information to police officers checking criminal histories. Colleges and universities use databases to keep records on their students, instructors, and courses. Organizations of all types maintain employee databases.

Three widely used database management systems designed for personal computers are Microsoft Access, Apple FileMaker, and Oracle Database Express Edition.

Assume that you have accepted a job as an employment administrator for the Lifestyle Fitness Club. To see how you could use Microsoft Access, see **Figure 3-12**.

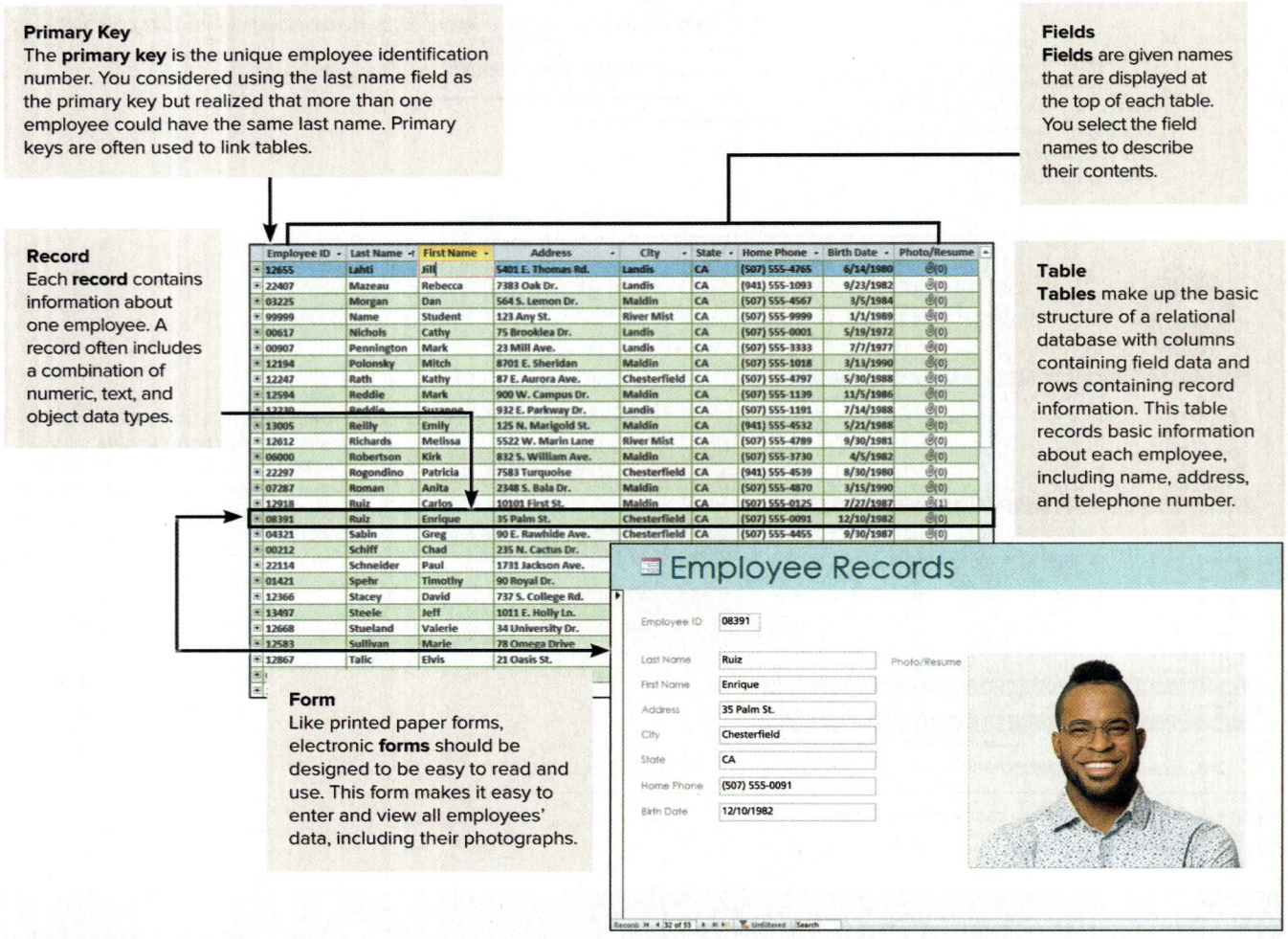

Primary Key
The **primary key** is the unique employee identification number. You considered using the last name field as the primary key but realized that more than one employee could have the same last name. Primary keys are often used to link tables.

Fields
Fields are given names that are displayed at the top of each table. You select the field names to describe their contents.

Record
Each **record** contains information about one employee. A record often includes a combination of numeric, text, and object data types.

Table
Tables make up the basic structure of a relational database with columns containing field data and rows containing record information. This table records basic information about each employee, including name, address, and telephone number.

Form
Like printed paper forms, electronic **forms** should be designed to be easy to read and use. This form makes it easy to enter and view all employees' data, including their photographs.

Figure 3-12   Database
Microsoft Corporation; fizkes/Shutterstock

Creating a Database

You have been asked to create an employee database to replace the club's manual system for recording employee data. Using Microsoft Access, you design the basic structure or organization of the new database system to include a table that will make entering data and using the database more efficient. You create the table structure by specifying the fields and primary key field. To make the process faster and more accurate, you create a form and enter the data for each employee as a record in the table. •

concept check

☐ What are word processors? What are they used for?
☐ What are presentation software programs? What are they used for?
☐ What are spreadsheets? What are they used for?
☐ What are database management systems? What are they used for?

Specialized Applications

While general-purpose applications are widely used in nearly every profession, specialized applications are widely used within specific professions. These programs include graphics programs and web authoring programs.

Graphics Programs

Graphics programs are widely used by professionals in the graphic arts profession. They use video editors, image editing programs, illustration programs, and desktop publishing programs.

- **Video editors** are used to edit videos to enhance quality and appearance of the videos you capture using your cell phone or other devices. You can readily add special effects, music tracks, titles, and on-screen graphics.

  Just a few years ago, video editors were used only by professionals with expensive specialized hardware and software. Now, there are several free or inexpensive editors designed to assist the amateur videographer. Three well-known video editors are Movavi Video Editor Plus for Android phones and tablets, Apple iMovie for Apple devices, and Adobe Premier Elements 2023 for a professional tool on laptops and desktops. (See **Figure 3-13**.)

Figure 3-13   Video editor
ifeelstock/Alamy Stock Photo

ethics

Image editing software has made it easy to alter any photo or video to correct for a variety of different imperfections. However, some professionals can use these programs to significantly manipulate the content or meaning of a photo or video. Such changes are often intended to influence the opinions or emotions of the viewer. Supporters argue that this type of editing is acceptable and is just another way to express an opinion or feeling from an editor. Critics note that this type of image and video manipulation is unethical because it intentionally misleads the viewer and often creates unobtainable or unhealthy definitions of beauty. What do you think?

Figure 3-14 Bitmap image

- **Image editors**, also known as **photo editors**, are specialized graphics programs for editing or modifying digital photographs. They are often used to touch up photographs to remove scratches and other imperfections. The photographs consist of thousands of dots, or **pixels**, that form images, often referred to as **bitmap** or **raster** images. One limitation of bitmap images, however, is that when they are expanded, the images can become pixelated, or jagged on the edges. For example, when the letter *A* in **Figure 3-14** is expanded, the borders of the letter appear jagged, as indicated by the expanded view.

  For cell phones and tablets, Snapseed is a popular image editor. Most laptops and desktops come with a free photo editor; the two most common are Microsoft Photo and Apple's Photos. Finally, for professional photographers, Adobe Photoshop is the gold standard in image editors.

- **Illustration programs**, also known as **drawing programs**, are used to create and edit vector images. While bitmap images use pixels to represent images, **vector images**, also known as **vector illustrations**, use geometric shapes or objects. These objects are created by connecting lines and curves, avoiding the pixelated or ragged edges created by bitmap images. (See **Figure 3-15**.) Because these objects can be defined by mathematical equations, they can be rapidly and easily resized, colored, textured, and manipulated. An image is a combination of several objects. Illustration programs are often used for graphic design, page layout, and creating sharp artistic images. Popular illustration programs include Procreate, Adobe Illustrator, and Autodesk Sketchbook.

Figure 3-15 Vector image

- **Desktop publishing programs**, or **page layout programs**, are graphic design tools used by artists to create online and print content. These tools focus on the professional display of graphics and text in a print or online format. This includes precise color reproduction on different types of paper and ink, templates for industry-standard magazine, book, and other publishing layouts, and compatibility with professional publishing tools.

  Popular desktop publishing programs include Adobe InDesign and Microsoft Publisher. These programs provide the capability to create text and graphics; however, typically graphic artists import these elements from other sources, including word processors, digital cameras, scanners, image editors, illustration programs, and image galleries.

- **3D Modeling programs** are programs that allow an artist to create and manipulate 3D objects drawn in the computer. These programs are used to create animated films and TV, video games, and magazine and book illustrations. Popular examples include AutoDesk's Maya and Adobe's Substance 3D Modeler.

concept check

- ☐ What are video editors?
- ☐ What are image editors? Bitmap images?
- ☐ What are illustration programs? Vector images?
- ☐ What are desktop publishing programs?

Video Game Design Software

Have you ever thought about designing a video game? The largest gaming titles, like *Call of Duty* or *Grand Theft Auto* take teams of artists, programmers, and managers. However, you can experiment and create some impressive games on your own with the right software. The first step is to visualize the game by thinking about the game's length and plot. The second step is to choose the right video game design software.

Video game design software will help you organize your thoughts and guide you through the game design process, including character development and environmental design. There are many choices from free software to very expensive software designed for professional game designers. The video game design software behind some of the biggest games include the Unreal Game Engine, the Unity development kit, and GameMaker Studio 2. (See **Figure 3-16**.)

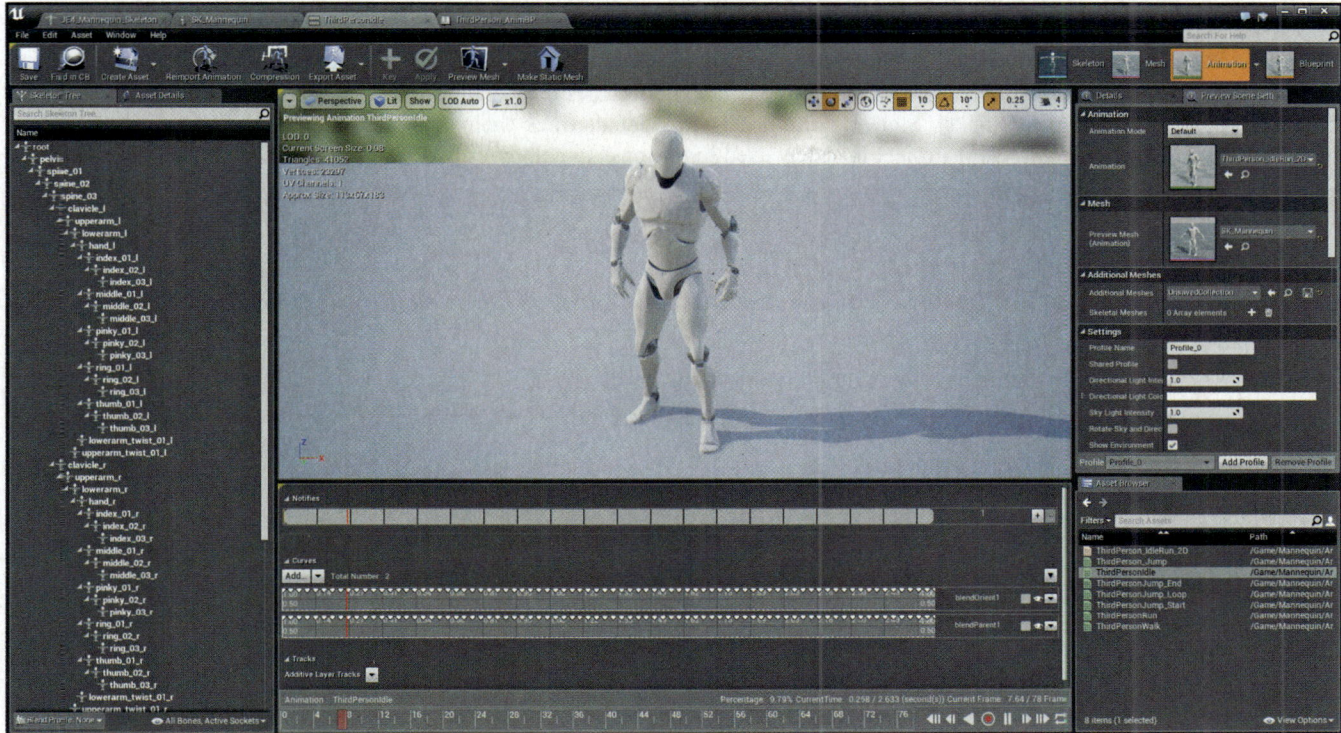

Figure 3-16 Video game design software
Cryengine

Web Authoring Programs

There are over a billion websites on the Internet, and more are being added every day. Corporations use the web to reach new customers and to promote their products. Individuals create online diaries or commentaries, called **blogs**. Creating a site is called **web authoring**.

Almost all websites consist of interrelated web pages. As we mentioned in **Chapter 2**, web pages are typically HTML (Hypertext Markup Language) and CSS (cascading style sheets) documents. With knowledge of HTML and a simple text editor, you can create web pages. Even without knowledge of HTML, you can create simple web pages using a word processing package like Microsoft Word.

More specialized and powerful programs, called **web authoring programs**, are typically used to create sophisticated commercial sites. Also known as **web page editors** and **HTML editors**, these programs provide support for website design and HTML coding. Some web authoring programs are **WYSIWYG (what you see is what you get) editors**, which means you can build a page without interacting directly with HTML code. WYSIWYG editors preview the page described by HTML code. Widely used web authoring programs include Adobe Dreamweaver and Froala 4.0.

> ### tips
>
> **Do you take photos with your cell phone? Almost everyone does. Here are some suggestions to keep in mind.**
>
> **1 Quickly launch the camera app.** Sometimes a photo opportunity comes along and you don't have time to log into your cell phone. Did you know that you can quickly access your cell phone's camera without logging in?
>
> - For Android cell phones: Double-click the power button.
> - For iOS cell phones: Tap the screen to wake up your phone, then swipe left.
>
> **2 Compose your shot.** Your cell phone camera can help you line up or balance your photos by displaying gridlines on the screen.
>
> - For Android cell phones: Open the camera app, open *settings*, and turn on the *Assistive Grid* option.
> - For iOS cell phones: Click on the *settings* icon from the home screen, then select *Camera*. Finally, turn on the *Grid* option.

Other Specialized Applications

There are numerous other specialized applications, including accounting, personal finance, and project management applications. Accounting applications such as Intuit QuickBooks help companies record and report their financial operations. Personal financial applications such as Quicken Starter Edition help individuals track their personal finances and investments. Project management software like Microsoft Project is widely used in business to help coordinate and plan complicated projects.

concept check

- What is video game design software?
- What are blogs? Web authoring? Web authoring programs? WYSIWYG?

Software Suites

A **software suite** is a collection of separate application programs bundled together and made available as a group. Four types of suites are office suites, cloud suites, specialized suites, and utility suites.

tips

Your cell phone's Internet browser has many of the same features as a desktop or laptop browser. To get the most out of your mobile browser, consider the following.

 Add Bookmarks. The ability to quickly save and load your favorite websites is a huge timesaver. To save a web page to your bookmarks:

- For Android Chrome: Go to the web page you want to add to the bookmarks, and tap the star in the address bar.
- For iOS Safari: Press and hold the icon of an open book. Then select *Add Bookmark*.

 Open a Bookmark. To open a web page, you have saved as a bookmark:

- For Android Chrome: Tap the more button (an icon of three vertical dots) and select *Bookmarks* to see your saved bookmarks. Simply tap on the bookmark you wish to open.
- For iOS Safari: Tap the icon of an open book. This will list your saved bookmarks. Simply tap the bookmark you wish to open.

Office Suites

Office suites, also known as **office software suites** and **productivity suites**, contain general-purpose application programs that are typically used in a business situation. Productivity suites commonly include a word processor, spreadsheet, database manager, and a presentation application. The best known is Microsoft Office. Another well-known productivity suite is Apple iWork.

Cloud Computing

Cloud suites or **online office suites** are stored at a server on the Internet and are available anywhere you can access the Internet. Documents created using cloud applications can also be stored online, making it easy to share and collaborate on documents with others. One downside to cloud applications is that you are dependent on the server providing the application to be available whenever you need it. For this reason, when using cloud applications, it is important to have backup copies of your documents on your computer and to have a desktop office application available to use. Popular cloud office suites include Google Workspace Microsoft 365 and Apple iWorks. (See **Figure 3-17**.) To learn more about cloud office suites, see Making IT Work for You: Cloud Office Suites on pages 72 and 73.

Figure 3-17 Cloud suite
monticello/Shutterstock

Specialized and Utility Suites

Two other types of suites that are more narrowly focused are specialized suites and utility suites:

- **Specialized suites** focus on specific applications. These include graphics suites like Adobe Creative Cloud, recording, audio editing, and mastering suites like Magix's Sound Forge Pro 15., financial planning suites like Moneytree Software's TOTAL Planning Suite, and many others.
- **Utility suites** include a variety of programs designed to make computing easier and safer. Two of the best known are iolo's System Mechanic Ultimate Defence and AVG TuneUp. (Utility suites will be discussed in detail in **Chapter 4**.)

 concept check

- ☐ What is a software suite? What are the advantages of purchasing a suite?
- ☐ What is the difference between a traditional office suite and a cloud or online suite?
- ☐ What is a specialized suite? What is a utility suite?

Careers in IT

"Now that you have learned about application software, I'd like to tell you about my career as a software engineer."

SeventyFour/Shutterstock

Software engineers analyze users' needs and create application software. Software engineers typically have experience in programming but focus on the design and development of programs using the principles of mathematics and engineering.

A bachelor's or an advanced specialized associate's degree in computer science or information systems and an extensive knowledge of computers and technology are required by most employers. Internships may provide students with the kinds of experience employers look for in a software engineer. Those with specific experience with web applications may have an advantage over other applicants. Employers typically look for software engineers with good communication and analytical skills.

Software engineers can expect to earn an annual salary in the range of $70,000 to $182,000. Starting salary is dependent on both experience and the type of software being developed. Experienced software engineers are candidates for many other advanced careers in IT.

Making IT work for you

CLOUD OFFICE SUITES

Do you work with a team on documents, presentations, or spreadsheets? Do your teammates use different computers and work in different locations? Do you sometimes need to work remotely—from home, a hotel, or in a coffee shop? Remote teamwork is growing in popularity, and cloud office suites can give you the tools to succeed.

If your company or school already uses a cloud office suite, such as Google Workspace, the best online tool may be the one you already have.

monticello/Shutterstock

Choosing a Cloud Office Suite The three biggest cloud office suites are Microsoft 365, Google Workspace, and Apple's iWork. Each is unique, with different prices, strengths, and weaknesses. Here are a few things to consider when choosing a cloud office suite for your team:

- **What does your team use now?** To use a cloud office suite, everyone in your team will need to sign up online and set up their devices. This can be time-consuming, but if you are already using one of these tools, the work may already be done. For example, if your team currently uses Google Docs, setting up a collaborative workspace could be as simple as sending an e-mail.

- **How experienced are your teammates?** A team knowledgeable in Microsoft Office can produce impressive, customized results quickly and professionally. However, the time and training necessary to become an expert in Microsoft Office may be too much for teammates with little or no computer experience. Other collaborative office suites, such as Apple iWorks and Google Workspace, are recognized as being easier for new users.

  When choosing a cloud office suite, consider the experience and knowledge of your teammates to make the best choice for your team.

Collaborative office suites, like Microsoft 365, often emphasize their ease of setup and tools for working with co-workers.

Jirapong Manustrong/Shutterstock

- Each cloud suite offers different online storage options and price points. To learn more about pricing and options, check out these websites:

iWorks	apple.com/iwork
Microsoft 365	microsoft.com/microsoft-365
Google Workspace	workspace.google.com

Collaborating with Your Group Online collaborative office suite tools share many important features:

- **Document versions**—When changes are made to a document by members of the team, the collaborative tools create a new version of the document. This feature allows team members to explore changes in the document without fear of losing important work.
- **Member changes**—when a team member changes a document, the collaborative tool tells the group about the change by highlighting the change and identifying who made it. Use this tool to track members' impact on the document and to quickly see what changes have been made.
- **Adding members to the team**—As the document progresses, you may want to get more opinions. Collaborative tools let you add new people to your team, but also limit what they can do. Some team members may be able to view and edit the document, whereas others may only be able to view it. This is a good tool to show someone your work without concern that they might accidentally edit something.

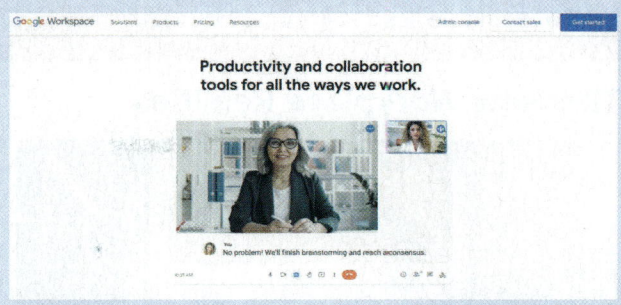

Most collaborative cloud office suites, like Google Workspace, offer the ability to see who made changes to a document and when and to undo those changes if necessary.

Google, INC.

Many office suites, like Apple's iWorks, are designed to work with mobile devices; and offline utility suites, such as Apple's word processor, Pages, has apps for computers, laptops, and cell phones.

Apple Inc.

Mobile Tools For many, working at a laptop or desktop computer is the easiest way to compose a document or presentation. However, when traveling on a crowded plane or walking down the hallway with a classmate, you can use cloud office suites on your tablet or phone. This is a great way to review a document, make a small comment, or quickly show a document to a teacher or friend. Apple, Google, and Microsoft all have apps to download to cell phones and tablets that allow you to view your online documents and make simple changes with interfaces designed for the smaller screens and touch interfaces of tablets and cell phones.

A LOOK TO THE FUTURE

The New Workplace Realities

Gorodenkoff/Shutterstock

Working from home has become increasingly popular. Many no longer commute to work but, instead, have a home office or work from the kitchen table. However, working from home is lacking in important features that office buildings often provide. Virtual reality (VR) and augmented reality (AR) technologies offer a future where you work from home but experience all the benefits of working at the office. A virtual office can offer a completely online office place, while augmented reality blends digital elements into your real-world environment.

Augmented reality glasses and complex camera arrays are looking to vastly change the way in which we videoconference. Imagine a meeting between yourself in L.A. and a co-worker in London. You both enter specialized rooms with multiple cameras that capture your position and appearance. You both wear augmented reality eyeglasses. In London, your co-worker sits down at a table. AR software records her position and appearance in London and sends that information to you in L.A. Your AR software projects the appearance and position of your co-worker onto the transparent AR glasses. Although you can see the room around you, the image of your co-worker is overlaid on your vision of the room to give the appearance that she is in the room as well.

As the meeting progresses, you present your ideas to your co-worker through charts and graphs. These charts float around the room in virtual screens. You decide to bring in a third member to the meeting. Your co-workers in the office outside the augmented reality meeting room are using virtual reality headsets, doing their work in virtual offices. You ask your manager to join you, and the manager joins the meeting by changing his virtual location to the meeting room. Because the VR headset covers your manager's face, he appears in the room as an avatar, a virtual representation of your manager.

Virtual reality office spaces have the potential to completely remove the need for office space. Workers could use VR software at home and complete their work with co-workers in a virtual office. Augmented reality programs could allow you to try on clothes without going to the store, overlaying a projection of an outfit to model the style and fit of a garment. Although not common yet, these tools are currently being developed by researchers and scientists.

Soon the glasses themselves may be unnecessary. Developments in flexible and translucent screens may one day be so compact that your AR glasses could be AR contact lenses. Scientists at UC Berkley are working on a power supply that is the size of a grain of rice and transmits power through ultrasonic pulses—allowing energy to pass through the body and be absorbed by the power supply. In this future, a computer screen will always be at your disposal, with hand gestures and head position monitored to create virtual keyboards that only you can see. Would you enjoy working at a virtual desk and attending virtual meetings, or do you think that virtual work would be less productive?

VISUAL SUMMARY | Application Software

APPLICATION SOFTWARE

Microsoft Corporation

The three categories of application software are **general purpose**, **specialized**, and **mobile**.

App Stores

An **app store** is typically a website that provides access to specific mobile apps that can be downloaded either for a nominal fee or free of charge. Two of the best-known stores are Apple's App Store and Google Play. Most of the best-known app stores specialize in applications for a particular line of mobile devices; other less well-known stores provide apps for a wide variety of mobile devices.

Store	Focus	Site
Apple App Store	Apple devices	apps.apple.com
Microsoft Store	Windows devices	apps.microsoft.com/store
Google Play	Android devices	play.google.com
Amazon	Android devices	amazon.com/appstore

User Interface

You control and interact with a program using a **user interface**. A **graphical user interface (GUI)** uses **icons** selected by a mouse-controlled **pointer**. A **window** contains a document, program, or message. Software programs with a traditional GUI have

- **Menus**—present commands listed on the **menu bar**.
- **Toolbars**—contain **buttons** for quick access to commonly used commands.
- **Dialog box**—provides additional information or requests user input.

APPLICATION SOFTWARE

Software programs with a **Ribbon GUI** have

- **Ribbons**—replace menus and toolbars.
- **Tabs**—divide ribbons into **groups**. **Contextual tabs** automatically appear when needed.
- **Galleries**—graphically display alternatives before they are selected.

Common Features

Common features include spell checkers, alignment, fonts and font sizes, character effects, and editing options.

MOBILE APPS

ymgerman/Shutterstock

Mobile apps (mobile applications, apps) are add-on programs for mobile devices. Traditional applications include address books, to-do lists, alarms, and message lists.

Apps

Popular apps include those for music, videos, social networking, shopping, and game playing.

- Pandora and Spotify provide popular music apps.
- YouTube and Netflix provide streaming video apps.
- Facebook and Instagram provide social networking apps.
- Amazon provides a shopping app.
- Minecraft and Final Fantasy are popular game-playing apps. Mindcraft involves exploring and reconstructing a world. Players may buy additional environments to explore. Final Fantasy is a **role-playing game (RPG)** where gamers can join a team of adventurers and battle across fantastic worlds.

To efficiently and effectively use computers, you need to understand the capabilities of general-purpose and specialized application software. Additionally, you need to know about mobile applications and software suites.

GENERAL-PURPOSE APPLICATIONS

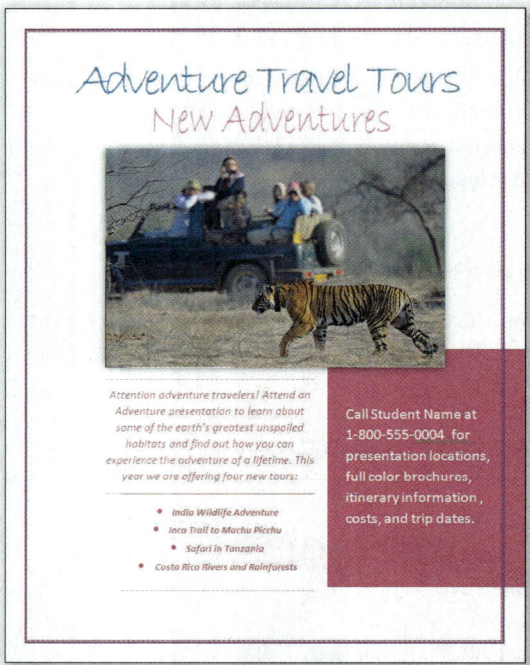

Aditya "Dicky" Singh/Alamy Stock Photo

General-purpose applications include word processors, presentation software, spreadsheets, and database management systems.

Word Processors

Word processors create text-based documents. Individuals and organizations use word processors to create memos, letters, and reports. Organizations also create newsletters, manuals, and brochures to provide information to their customers. Microsoft Word is the most widely used word processor. Others include Apple Pages and Google Workspace.

Presentation Software

Presentation software are programs that combine a variety of visual objects to create attractive, visually interesting presentations. They are excellent tools to communicate a message and to persuade people. People in a variety of settings and situations use presentation software programs to make their presentations more interesting and professional. Three of the most widely used presentation software programs are Microsoft PowerPoint, Apple Keynote, and Google Slides.

GENERAL-PURPOSE APPLICATIONS

Spreadsheets

Spreadsheets organize, analyze, and graph numeric data such as budgets and financial reports. They are widely used by nearly every profession. Microsoft Excel is the most widely used spreadsheet program. Others include Apple Numbers and Google Sheets.

Animal Angels

Database Management Systems

A **database** is a collection of related data. A **database management system (DBMS)** or **database manager** is a program that structures a database. It provides tools to enter, edit, and retrieve data from the database. Organizations use databases for many purposes, including maintaining employee records. Two widely used database management systems designed for personal computers are Microsoft Access and Apple FileMaker.

fizkes/Shutterstock

SPECIALIZED APPLICATIONS

Cryengine

Specialized applications are widely used within specific professions. They include graphics programs, video game design software, and web authoring programs.

Graphics Programs

Graphics programs are used by graphic arts professionals.

- **Video editors** edit videos to enhance quality and appearance.
- **Image editors (photo editors)** edit digital photographs consisting of thousands of dots, or **pixels**, that form **bitmap** or **raster** images.
- **Illustration programs (drawing programs)** create and edit vector images. **Vector images (vector illustrations)** use geometric shapes.
- **Desktop publishing programs (page layout programs)** mix text and graphics to create professional-quality publications.
- **3D Modeling programs** are programs that allow an artist to create and manipulate 3D objects drawn in the computer.

Video Game Design Software

Video game design software helps to organize thoughts and guide users through the game design process, including character development and environmental design.

Web Authoring Programs

Web authoring is the process of creating a website. Individuals create online diaries called **blogs**. **Web authoring programs (web page editors, HTML editors)** create sophisticated commercial websites. Some are **WYSIWYG (what you see is what you get)** editors.

SOFTWARE SUITES

monticello/Shutterstock

A **software suite** is a collection of individual application packages sold together.

- **Office suites (office software suites** or **productivity suites)** contain professional-grade application programs.
- **Cloud suites (online office suites)** are stored on servers and available through the Internet.
- **Specialized suites** focus on specific applications such as graphics.
- **Utility suites** include a variety of programs designed to make computing easier and safer.

CAREERS in IT

Software engineers analyze users' needs and create application software. A bachelor's or advanced specialized associate's degree in computer science or information systems and extensive knowledge of computers and technology are required. Expected salary range is $70,000 to $182,000.

KEY TERMS

- app
- application software
- app store
- bitmap
- blog
- button
- cloud suite
- contextual tab
- database
- database management system (DBMS)
- database manager
- desktop publishing program
- dialog box
- document
- drawing program
- free-to-play
- gallery
- general-purpose application
- graphical user interface (GUI)
- graphics program
- group
- HTML editor
- icon
- illustration program
- image editor
- menu
- menu bar
- mobile app
- mobile application
- office software suite
- office suite
- online office suite
- page layout program
- photo editor
- pixel
- pointer
- presentation software
- productivity suite
- raster
- ribbon
- Ribbon GUI
- role-playing game (RPG)
- software engineer
- software suite
- specialized application
- specialized suite
- spreadsheet
- system software
- tab
- toolbar
- user interface
- utility suite
- vector illustration
- vector image
- video editor
- video game design software
- web authoring
- web authoring program
- web page editor
- window
- word processor
- WYSIWYG (what you see is what you get) editor

MULTIPLE CHOICE

Circle the correct answer.

1. Toolbars typically appear below the menu bar and include small graphic elements called _____.
   - a. buttons
   - b. ribbons
   - c. pointers
   - d. vector graphics

2. Simplifies the process of making a selection from a list of alternatives by graphically displaying the effect of alternatives before being selected.
   - a. buttons
   - b. windows
   - c. pixels
   - d. galleries

3. A general-purpose program that creates text-based documents.
   - a. database
   - b. word processor
   - c. spreadsheet
   - d. presentation software

4. Program that organizes, analyzes, and graphs numerical data.
   - a. word processor
   - b. presentation software
   - c. spreadsheet
   - d. blog

5. A collection of related data.
   - a. ribbons
   - b. blog
   - c. database
   - d. cloud

6. Also known as a photo editor, this specialized graphics program edits or modifies digital photographs.
   - a. image editor
   - b. word processor
   - c. presentation software
   - d. pixels

7. Image editors create images made up of thousands of dots known as _____.
   - a. vectors
   - b. pixels
   - c. blogs
   - d. HTML

8. A website that provides access to specific mobile apps is known as an app _____.
   - a. cloud
   - b. blog
   - c. suites
   - d. store

9. A type of suite that is stored at a server on the Internet and is available anywhere you can access the Internet.
   - a. database
   - b. cloud
   - c. menu
   - d. general-purpose

10. A type of specialized suite that includes a variety of programs designed to make computing easier and safer.
    - a. utility
    - b. window
    - c. WYSIWYG
    - d. database manager

MATCHING

Match each numbered item with the most closely related lettered item. Write your answers in the spaces provided.

a. system
b. window
c. vector
d. spreadsheets
e. word processors
f. presentation software
g. desktop publishing
h. web authoring programs
i. cloud suite
j. blog

____ 1. Programs that organize, analyze, and graph numerical data such as budgets and financial reports.
____ 2. Also known as an online office suite.
____ 3. Programs that allow you to mix text and graphics to create publications of professional quality.
____ 4. A rectangular area that can contain a document, program, or message.
____ 5. Programs that create text-based documents.
____ 6. Programs that combine a variety of visual objects to create attractive, visually interesting presentations.
____ 7. Programs typically used to create sophisticated commercial websites.
____ 8. The type of image that consists of geometric shapes.
____ 9. An online diary or commentary.
____ 10. This type of software works with end users, application software, and computer hardware to handle the majority of technical details.

OPEN-ENDED

On a separate sheet of paper, respond to each question or statement.

1. Explain the difference between general-purpose and specialized applications. Also discuss the common features of application programs, including those with traditional and ribbon graphical user interfaces.
2. Discuss general-purpose applications, including word processors, spreadsheets, database management systems, and presentation software.
3. Discuss specialized applications, including graphics programs, video game design software, web authoring programs, and other professional specialized applications.
4. Describe mobile apps, including popular apps and app stores.
5. Describe software suites, including office suites, cloud suites, specialized suites, and utility suites.

DISCUSSION

Respond to each of the following questions.

Making IT Work for You: CLOUD OFFICE SUITES

Review the Making IT Work for You: Cloud Office Suites on pages 72–73 and then respond to the following: (a) Do you currently use a cloud office suite? If so, what types of documents do you typically create? If not, then list some possible benefits a cloud office suite could provide. (b) Do you collaborate with others on creating documents? What are some types of documents you create with others that can take advantage of cloud collaborative tools? (c) Using a search engine or other type of research tool, identify and list a few differences between the online general-purpose applications: Google Workspace and Microsoft's 365. Which one would work best for your needs? Why?

Privacy: CELL PHONE TRACKING

Review the Privacy box on page 60, and respond to the following: (a) Do you think that cell phone tracking is a violation of your privacy? If yes, what can be done? If no, explain your position. (b) Does a company that tracks your movements have the right to sell this information to other companies? Would your opinion change if the company sells your location information but does not reveal your identity? State and defend your position. (c) Does the government have the right to subpoena GPS information from an app maker? Why or why not? (d) Are there any circumstances in which it would be acceptable/justifiable for a company to reveal location data to the government or another company? If so, give some examples.

Ethics: IMAGE EDITING

Review the Ethics box on page 67. Using a search engine or other research tool, find examples of digital photo or video editing that have resulted in controversy, and then respond to the following: (a) Do you see any ethical issues related to altering photographs or videos? (b) What do you consider the boundary to be between acceptable editing and deceptive or misleading practices? (c) How does such editing affect courtrooms, where visual evidence is often presented? (d) Do you feel the old saying "seeing is believing" needs to be reconsidered for the digital age? Defend your answers.

Community: ONLINE REVIEWS

Review the Community box on page 58 and then respond to the following: (a) What products do you purchase that can be purchased online? Do online stores offer user reviews of this or similar products? Would you ever write a review for a product bought online? (b) How much importance do you put into a user review? Do you trust user reviews? If a review states that the reviewer received the product for free or at a discount to review the product, does this affect the value of the review? (c) What do you look for in a good review? What qualities of a review make you trust it? What qualities would make you suspicious of a review?

Design Elements: Concept Check icon: Dizzle52/Getty Images

chapter 4
System Software

ImageFlow/Shutterstock

CHAPTER 4: System Software

Why should I read this chapter?

Blue Planet Studio/Shutterstock

Your cell phone can receive automatic software updates over the Internet, but there are many software improvements that require you, the user, to take action. In the future, software may diagnose and repair problems much like your body's immune system protects your health. But for now, your computers are at risk from dangerous viruses and software failures.

This chapter covers the things you need to know to protect your computer and data today and to prepare you for tomorrow, including:

- Mobile operating systems—learn the key features of the operating systems that control tablets and cell phones.
- Desktop operating systems—discover how operating systems control and protect desktop and laptop computers.
- Utilities—protect your computer from viruses and perform important maintenance tasks.

Learning Objectives

After you have read this chapter, you should be able to:

1. Describe the differences between system software and application software.
2. Identify the four types of system software programs.
3. Explain the basic functions, features, and categories of operating systems.
4. Compare mobile operating systems, including iOS and Android.
5. Compare desktop operating systems, including Windows, macOS, UNIX, Linux, and virtualization.
6. Explain the purpose of utilities and utility suites.
7. Identify the six most essential utilities.

Introduction

"Hi, I'm Ray, and I'm a computer support specialist. I'd like to talk with you about system software, programs that do a lot of the work behind the scenes so that your electronic devices keep running smoothly. I'd also like to talk about the mobile operating systems that control cell phones and other small portable computers."

AnnaStills/iStock/Getty Images

When most people use a computer, they use a cell phone to send messages, play games, or view social media. We typically think of using a computer as using the applications that have become an important part of our everyday lives. Most of us agree that they are great . . . as long as they are working.

We usually do not think about the more mundane and behind-the-scenes computer activities: loading and running programs, coordinating networks that share resources, organizing files, protecting us from viruses, performing periodic maintenance to avoid problems, and controlling hardware devices so that they can communicate with one another. Typically, these activities go on behind the scenes without our help.

That is the way it should be, and the way it is, as long as everything is working perfectly. But what if new application programs are not compatible and will not run on our current computer system? What if we get a computer virus? What if our hard disk fails? What if we buy a new digital video camera and can't store and edit the images on our computer system? What if our computer starts to run slower and slower?

These issues may seem mundane, but they are critical. This chapter covers the vital activities that go on behind the scenes. A little knowledge about these activities can go a long way to making your computing life easier. To efficiently and effectively use computers, you need to understand the functionality of system software, including operating systems, utility programs, and device drivers.

System Software

End users use application software to accomplish specific tasks. For example, we use messaging applications to send text and photos with our cell phones. However, end users also use system software. **System software** works with end users, application software, and computer hardware to handle the majority of technical details. For example, system software controls where a word processing program is stored in memory, how commands are converted so that the system unit can process them, and where a completed document or file is saved. (See **Figure 4-1**.)

System software is not a single program. Rather, it is a collection or a system of programs that handle hundreds of technical details with little or no user intervention. System software consists of four types of programs:

- **Operating systems** coordinate computer resources, provide an interface between users and the computer, and run applications.
- **Utilities** perform specific tasks related to managing computer resources.
- **Device drivers** are specialized programs that allow particular input or output devices to communicate with the rest of the computer system.
- **Language translators** convert the programming instructions written by programmers into a language that computers understand and process.

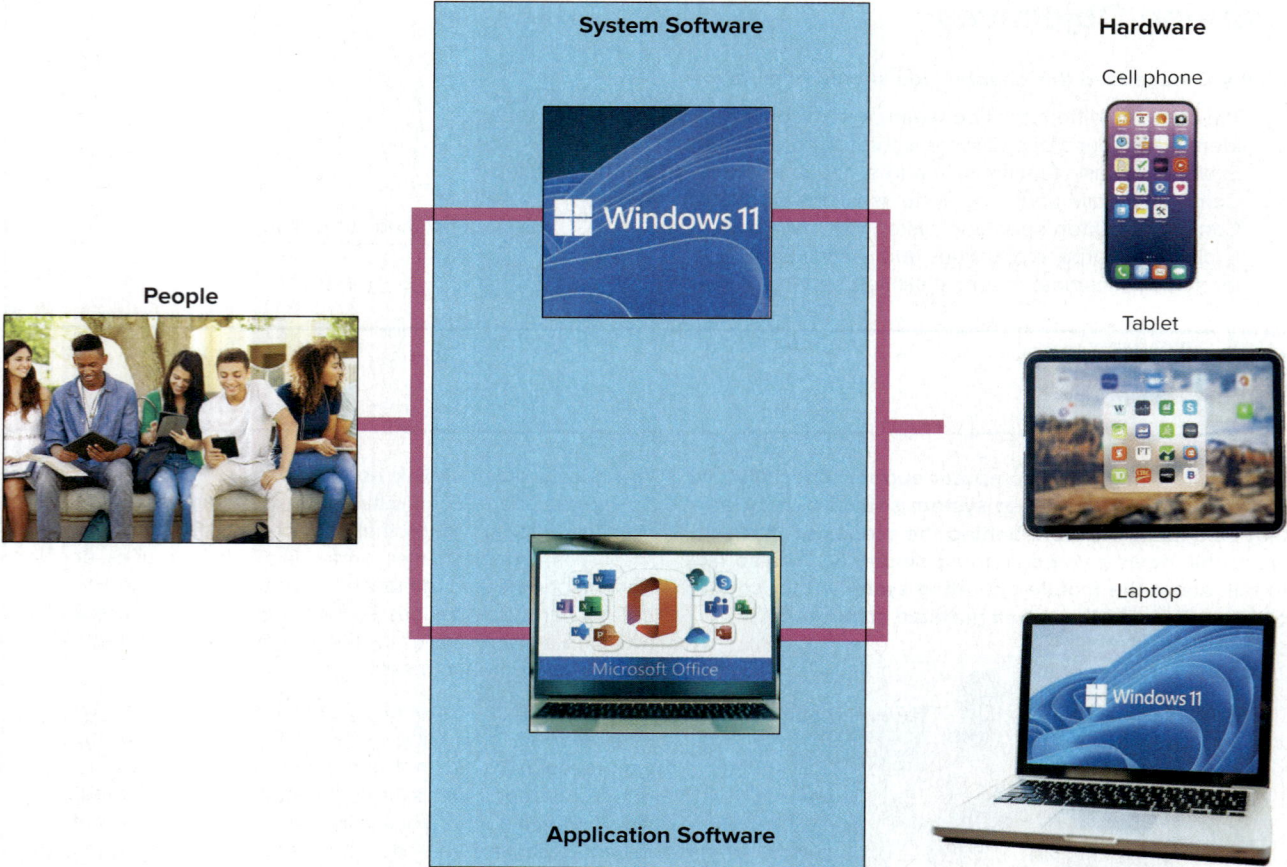

Figure 4-1 System software handles technical details

People: Shutterstock; **System Software:** Microsoft Corporation; **Application Software:** monticello/Shutterstock; **Hardware:** Kaspars Grinvalds/Shutterstock; **(Tablet):** Prime Stock Photo/Alamy Stock Photo; **(Laptop):** Rawf8/Alamy Stock Photo

Operating Systems

An **operating system** is a collection of programs that handle many of the technical details related to using a computer. In many ways, an operating system is the most important type of computer program. Without a functioning operating system, your computer would be useless.

Functions

Every computer has an operating system, and every operating system performs a variety of functions. These functions can be classified into three groups:

- **Managing resources:** Operating systems coordinate all the computer's resources, including memory, processing, storage, and devices such as printers and monitors. They also monitor system performance, schedule tasks, provide security, and start up the computer.
- **Providing user interface:** Operating systems allow users to interact with application programs and computer hardware through a **user interface**. Originally, operating systems used a character-based interface in which users communicated with the operating system through written commands such as "Copy A: report.txt C:". Today, most operating systems use a **graphical user interface (GUI)**. As we discussed in **Chapter 3**, a graphical user interface uses graphical elements such as icons and windows. More recently, many operating systems now include voice assist tools. Much like a graphical user interface offers users a visual way to interact with application programs and computer hardware, voice assist tools allow a user to directly issue voice commands.
- **Running applications:** Operating systems load and run applications such as word processors and spreadsheets. Most operating systems support **multitasking**, or the ability to switch between different applications stored in memory. With multitasking, you could have a web browser and a messaging app running at the same time and switch easily between the two applications. The program that you are currently working on is described as running in the **foreground**. The other program or programs are running in the **background**.

Features

Starting or restarting a computer is called **booting** the system. There are two ways to boot a computer: a warm boot and a cold boot. A **warm boot** occurs when the computer is already on and you restart it without turning off the power. A warm boot can be accomplished in several ways. For many computer systems, they can be restarted by simply pressing a sequence of keys. Starting a computer that has been turned off is called a **cold boot**.

You typically interact with the operating system through the graphical user interface. Most provide a place, called the **desktop**, that provides access to computer resources. (See **Figure 4-2**.) Some important features common to most operating systems and application programs include:

- **Icons**—graphic representations for a program, type of file, or function.
- **Pointer**—controlled by a mouse, trackpad, or touch screen, the pointer changes shape depending on its current function. For example, when shaped like an arrow, the pointer can be used to select items such as an icon.
- **Windows**—rectangular areas for displaying information and running programs.
- **Menus**—provide a list of options or commands that can be selected.

ethics

As our devices have gotten smarter, their operating systems have begun accepting voice commands. This allows a user to speak commands to the device. A device may be controlled by speaking a key phrase—such as "OK Google," "Hey Siri," or "Alexa"— followed by a command or question, such as "what time is it?" In order for the device to recognize when a user says the device's key phrase, it must always be recording and analyzing users' conversations. These recordings are often shared over the Internet with companies that use this information to improve the voice recognition software and search tools. Could such data be used unethically? Who should oversee this data, and what limits should be placed on its use?

community

Did you know that your operating system plays a key role in making sure that everybody has access to technology? All major operating systems include accessibility features that make our devices easier to use for people with disabilities. For people with low vision, there are tools to magnify portions of the screen or read written text out loud. For people with dyslexia, there are fonts that can make identifying letters and words easier. For the deaf or hard of hearing, audible alerts can be set to also offer visual cues, like flashing the screen when a beep or noise is played. These features are included free in your device's operating system, and you can configure them in your device's settings menus.

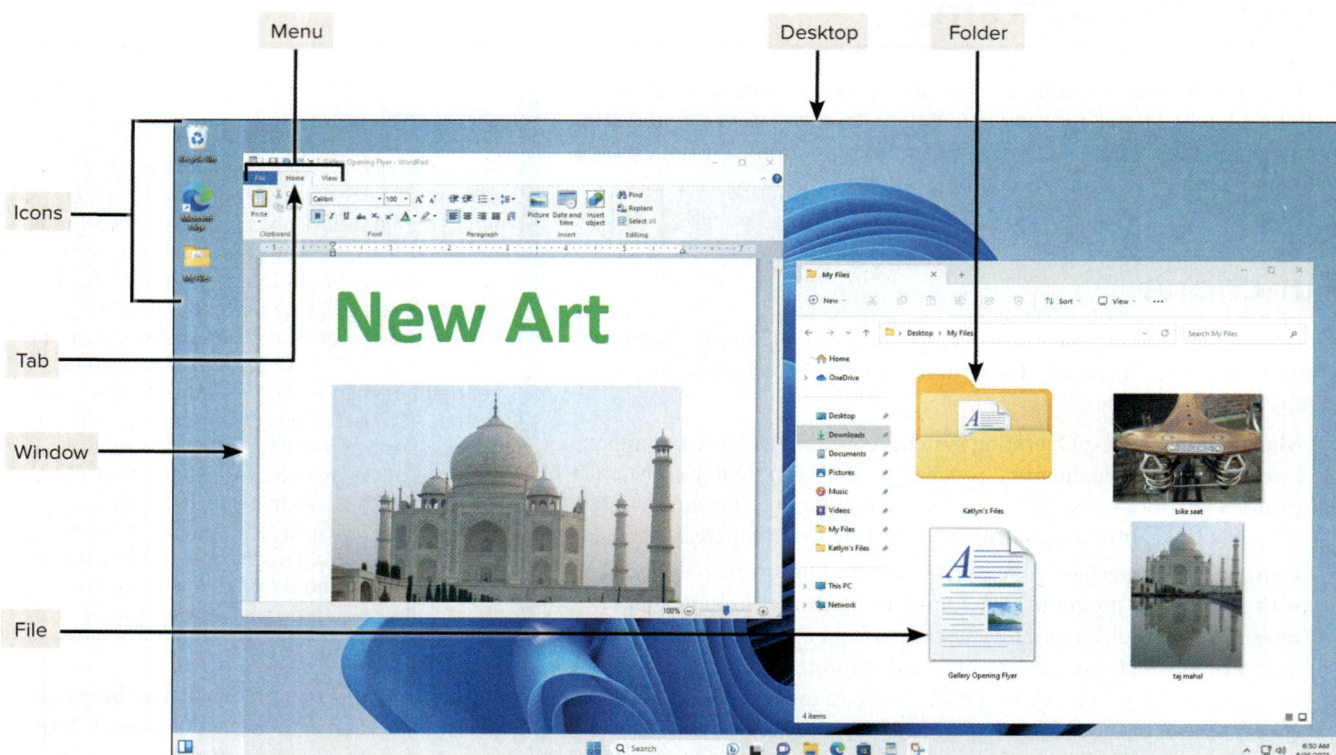

Figure 4-2 Desktop
Microsoft Corporation

- **Tabs**—divide menus into major activity areas such as format and page layout.
- **Dialog boxes**—typically provide information or request input.
- **Help**—provides online assistance for operating system functions and procedures.
- **Gesture control**—ability to control operations with finger movements, such as swiping, sliding, and pinching.

Most offices have filing cabinets that store important documents in folders. Similarly, most operating systems store data and programs in a system of files and folders. **Files** are used to store data and programs. Related files are stored within a **folder**, and, for organizational purposes, a folder can contain other folders, or subfolders. For example, you might organize your electronic files in the *Documents* folder on your hard disk. This folder could contain other folders, each named to indicate its contents. One might be "Computer Class" and could contain all the files you have created (or will create) for this course.

Figure 4-3 Embedded operating systems control smartwatches
charnsitr/Shutterstock

Categories

While there are hundreds of different operating systems, there are only three basic categories: embedded, stand-alone, and network.

- **Embedded operating systems** are used in cell phones and tablets, as well as video game systems and thousands of other small electronic devices. Also known as **real-time operating systems (RTOS)**, these operating systems are uniquely designed to work exclusively with (i.e., embedded into) a particular device's hardware. Typically designed for a specific application, embedded operating systems are essential in the evolution of IoT where many everyday devices are able to communicate with one another, as discussed in **Chapter 1**. For example, Watch OS was developed by Apple exclusively for the Apple Watch. (See **Figure 4-3**.)

- **Stand-alone operating systems** are used in a single desktop or laptop computer. (See **Figure 4-4**.) Also called **desktop operating systems**, these operating systems are located on the computer's hard disk. Often desktop computers and laptops are part of a network. In these cases, the desktop operating system works with the network to share and coordinate resources. Two of the most popular stand-alone operating systems are Apple's macOS and Microsoft's Windows.
- **Network operating systems (NOS)** are used to control and coordinate computers that are networked or linked together. These networks can range in size, from a few computers in an office to large networks that span an entire company or university. An NOS can be composed of several smaller networks and many types of computers.

Figure 4-4 Laptops use stand-alone operating systems

Africa Studio/Shutterstock

Network operating systems are typically located on one of the connected computers' hard disks. Called the **network server**, this computer coordinates all communication between the other computers. Popular network operating systems include Linux, Windows Server, and UNIX.

The operating system is often referred to as the **software environment** or **software platform**. A computer's platform dictates what applications will run on that device. Almost all application programs are designed to run with a specific platform. For example, Apple's iMovie software is designed to run with the macOS environment and will not run on an Android cell phone or Windows laptop. Many applications, however, have different versions, each designed to operate with a particular platform. For example, there are two versions of Microsoft Office, one designed to work with Windows and another designed to work with macOS.

tips

Do you often type on your cell phone's virtual keyboard? You can customize your keyboard to a different language or download a specialized keyboard at the app store.

1. For Android cell phones: Tap on the Settings icon from the home screen, then choose *Languages and Inputs*, then *Current Keyboard*, and finally *Keyboard*. From here you can select a different keyboard.
2. For iOS cell phones: Tap on the *Settings* icon from the home screen, then select *General* and *Keyboard*. From here, you can customize your keyboards or add new keyboards to your phone.

 concept check

☐ What is system software? What are the four kinds of system software programs?
☐ What is an operating system? Discuss operating system functions and features.
☐ Describe each of the three categories of operating systems.

Mobile Operating Systems

Figure 4-5 Apple's iPad and iPhone use iOS mobile operating system

19 STUDIO/Shutterstock; Den Rozhnovsky/Shutterstock

Mobile operating systems, also known as **mobile OS**, are a type of embedded operating system. Just like other computer systems, mobile computers—including cell phones, tablets, and wearable computers—require an operating system. These mobile operating systems are less complicated and more specialized for wireless communication.

While there are numerous mobile operating systems, two of the best known are Android and iOS.

- **Android** was developed by Google and is used as the platform for many popular cell phones, such as the Samsung Galaxy and the Google Pixel phones. Improvements include increased stability, security, and improved accessibility with larger fonts.
- **iOS 17**, announced in 2023, was developed by Apple and is used as the platform for Apple's mobile devices, the iPad and iPhone. (See **Figure 4-5**.) New features include artificial intelligence-enhanced spell checking and improved standby mode for always-on displays.

In the last chapter, we discussed that not all mobile applications will run on all cell phones. That is because an app is designed to run on a particular software platform or operating system. Before downloading an app, be sure that it is designed to run with the mobile operating system on your mobile device.

 concept check

- What is a mobile operating system?
- List the most widely used mobile operating systems.
- Which mobile operating system works with the iPhone? Which mobile operating system is owned by Google?

privacy

Most cell phone operating systems include a virtual assistant, like Apple's Siri or Google Assistant, that scans your personal information such as calendar events, e-mails, text messages, and GPS location to anticipate your needs and help organize your life. However, most users of these complex programs are unaware that they copy and transfer data from texts, e-mails, and GPS back to company servers. Should your operating system be gathering and sharing this data? Who should be in charge of what data is gathered and how it is stored?

Desktop Operating Systems

Every personal computer has an operating system controlling its operations. The most widely used desktop operating systems are Windows, macOS, UNIX, and Linux.

Windows

Microsoft's **Windows** is the most widely used personal computer operating system. Because its market share is so large, more application programs have been developed to run under Windows than any other operating system. Windows comes in a variety of different versions and is designed to run with a variety of different microprocessors.

The most recent version of Windows, **Windows 11**, was announced in 2021. (See **Figure 4-6**.) Windows 11 is Microsoft's desktop, laptop, and tablet operating system. This version of Windows includes several innovations, such as a simplified user interface and improved support for mobile devices, such as tablets and 2-in-1 laptops. One of the biggest changes with Windows 11 is its ability to run Android apps on a Windows device.

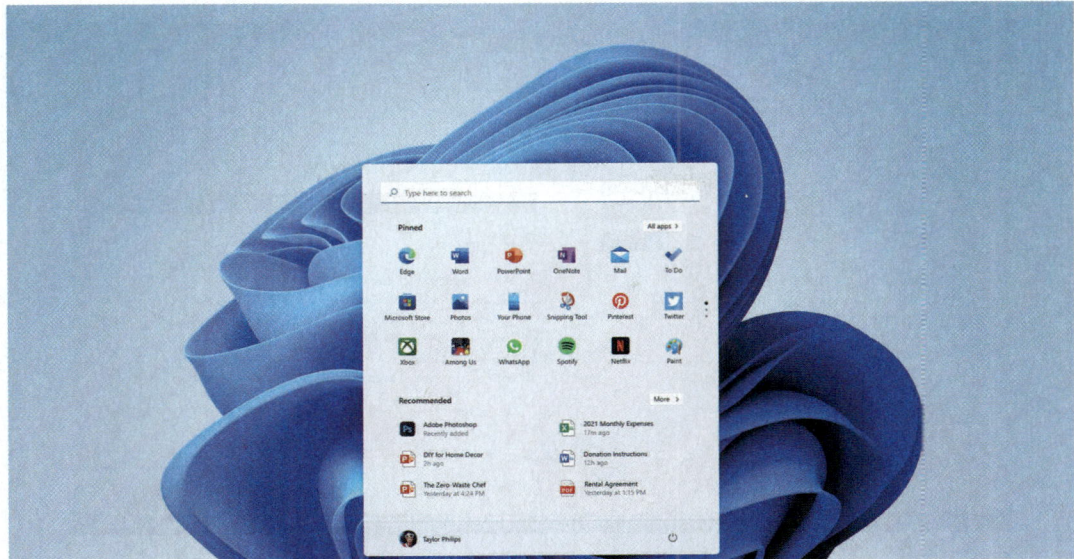

Figure 4-6 Windows 11
Microsoft Corporation

macOS

Apple has been the leader in the development of powerful and easy-to-use personal computer operating systems since its introduction of the Macintosh personal computer in 1984. Designed to run only with Apple computers, **macOS** is not as widely used as the Windows operating system. As a result, fewer application programs have been written for it. However, with increasing sales of Apple computers, the use of macOS has been rapidly increasing and is widely recognized as one of the most innovative operating systems. The popular iOS and iPadOS mobile operating systems are based on the macOS.

macOS is the most widely used Mac desktop operating system. **macOS 12 Sonoma,** announced in 2023, brings new innovations to Mac computers, including interactive widgets, improved video conferencing tools, and security enhancements for browsing the web. (See **Figure 4-7**.)

Figure 4-7   macOS Monterey
Apple, Inc.

UNIX and Linux

The **UNIX** operating system was originally designed in the late 1960s to run on minicomputers in network environments. Over the years, UNIX has evolved with numerous different versions. Now, it is widely used by servers on the web, mainframe computers, and very powerful personal computers. There are a large number of different versions of UNIX.

Linux is an operating system that extended one of the UNIX versions. It was originally developed by a graduate student at the University of Helsinki, Linus Torvalds, in 1991. He allowed free distribution of the operating system code and encouraged others to modify and further develop the code. Programs released in this way are called **open source**. Linux is a popular and powerful alternative to the Windows operating system. (See **Figure 4-8**.) Linux has been the basis of several other operating systems. For example, Google's **Chrome OS** and the Android mobile operating system are based on Linux.

Chrome OS integrates with web servers to run applications and to perform other traditional operating system functions. This capability has made Chrome OS a popular choice for inexpensive notebook computers that use cloud computing and cloud storage to do things that would normally require much more expensive hardware. However, this also means that popular software found on Windows and macOS may not be available for Chrome OS. Furthermore, the applications that are available for Chrome OS may not perform well without a fast Internet connection.

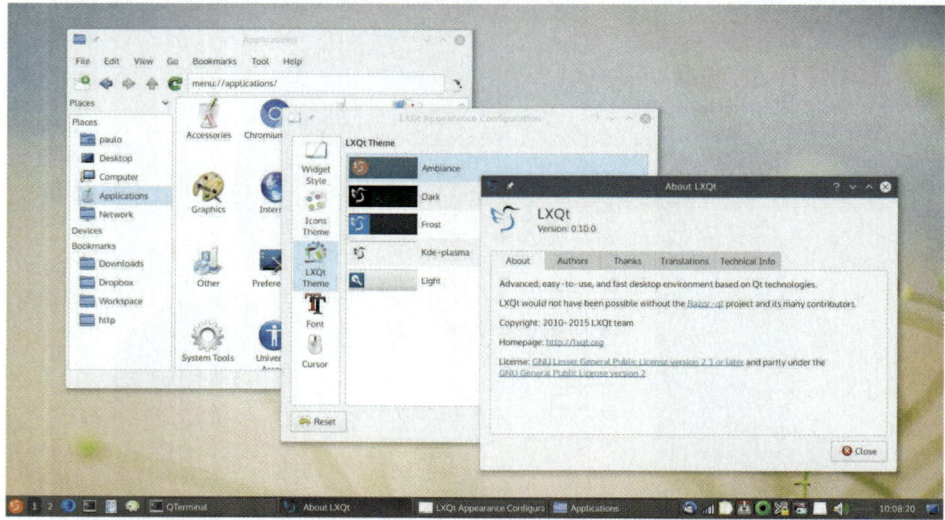

Figure 4-8 Linux

Linux

Virtualization

As we have discussed, application programs are designed to run with particular operating systems. What if you wanted to run two or more applications, each requiring a different operating system? One solution would be to install each of the operating systems on a different computer. There is, however, a way in which a single physical computer can support multiple operating systems that operate independently. This approach is called **virtualization**.

When a single physical computer runs a special program known as **virtualization software**, it operates as though it were two or more separate and independent computers, known as **virtual machines**. Each virtual machine appears to the user as a separate, independent computer with its own operating system. The operating system of the physical machine is known as the **host operating system**. The operating system for each virtual machine is known as the **guest operating system**. Users can readily switch between virtual computers and programs running on them. There are several programs that create and run virtual machines. Two such programs, Parallels and Oracle VM VirtualBox, allow a user on a Mac to run Windows programs in macOS. (See **Figure 4-9**.)

Figure 4-9 macOS running Windows 11 in a virtual machine.

Microsoft Corporation

You can also use virtual machines that are not running on your computer. A **Virtual Desktop Infrastructure (VDI)** is a collection of hardware and software that allows a user to remotely access virtual machines over the Internet. This is valuable for large organizations that want to offer the same desktop environment to many users without having to manage and update each user's computer.

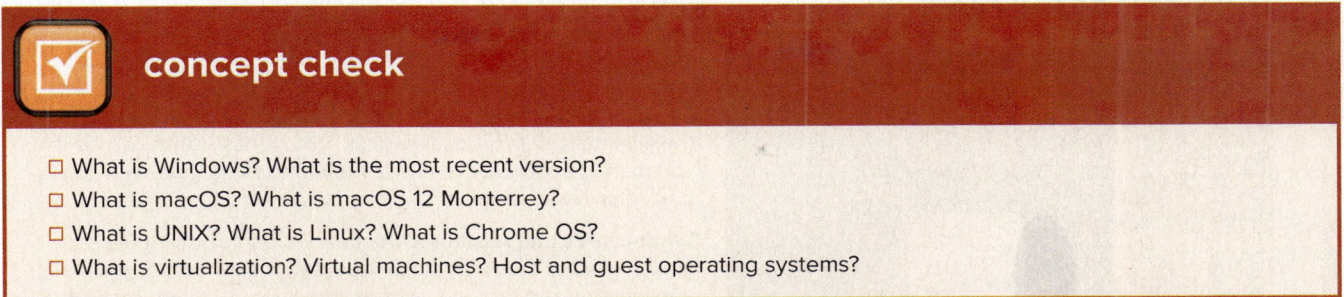

concept check

☐ What is Windows? What is the most recent version?

☐ What is macOS? What is macOS 12 Monterrey?

☐ What is UNIX? What is Linux? What is Chrome OS?

☐ What is virtualization? Virtual machines? Host and guest operating systems?

Utilities

Ideally, personal computers would continuously run without problems. However, that simply is not the case. All kinds of things can happen such as files or apps can go missing, storage space can fill up, and data can be lost. These events can make computing very frustrating. That's where utilities come in. **Utilities** are specialized programs designed to make computing easier. There are hundreds of different utility programs. The most essential are

- **Search programs** provide a quick and easy way to search or examine an entire computer system to help you find specific applications, data, or other files.
- **Storage management programs** help solve the problem of running out of storage space by providing lists of application programs, stored videos, and other program files so that you can eliminate unused applications or archive large files elsewhere.
- **Backup programs** make copies of files to be used in case the originals are lost or damaged. Windows 11 comes with a free backup program, the File History tool, and macOS has a backup feature named Time Machine.
- **Antivirus programs** guard your computer system against viruses or other damaging programs that can invade your computer system. Popular antivirus programs include Norton AntiVirus and Bitdefender's Antivirus Plus. Antivirus protection is an important part of your security online. You can also learn more about free antivirus programs in the Making IT Work for You section of **Chapter 1**.
- **Troubleshooting** or **diagnostic programs** recognize and correct problems, ideally before they become serious.
- **Virtual assistants** are utilities that accept commands through text or speech to allow intuitive interaction with your computer, cell phone, or tablet and coordinate personal data across multiple applications. Microsoft Windows 11 has the virtual assistant **Cortana**, and Apple's macOS has **Siri**. To learn more about virtual assistants, see the Making IT Work for You: Virtual Assistant on pages 92 and 93.

Most operating systems provide some utility programs. Even more powerful utility programs can be purchased separately or in utility suites.

Making IT work for you

VIRTUAL ASSISTANT

Google Assistant lets a user dictate a text message. Most intelligent assistants will coordinate with your phone's features to send texts and e-mails, or make calls.

Lukmanazis/Shutterstock

Your cell phone already organizes much of your life. It contains a list of your contacts, alarms, and timers you can set, and a calendar of upcoming events. However, coordinating all these details can be challenging. Virtual assistants can help you manage your day, review e-mails and texts, and even suggest local restaurants.

What Is a Virtual Assistant? As defined earlier, a virtual assistant is a utility that coordinates personal data across several applications and provides a more intuitive way to interact with your computer, phone, or tablet. Most operating systems offer a virtual assistant; Apple offers Siri, Amazon has Alexa, and Google has Google Assistant. The most common way to interact with these utilities is through voice commands.

Voice Assist One of the most important features of your virtual assistant is its ability to listen to spoken commands and respond verbally. This allows you to text, review e-mails, and even schedule events while keeping your hands free.

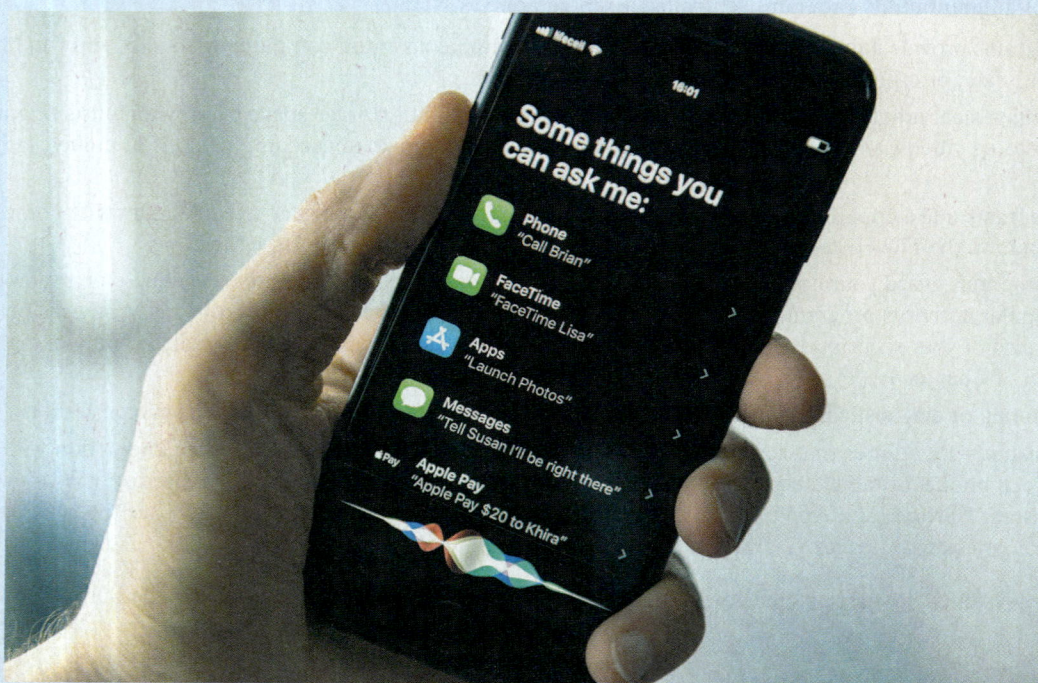

Apple's Siri accepts a voice command to make a phone call. Intelligent assistants interpret voice commands and can perform simple actions, such as setting alarms, sending e-mails or text messages, and alerting you to upcoming calendar events.

Anatolii Babii/Alamy Stock Photo

Initially, you must get your assistant's attention with a key phrase, such as Hey Siri, OK Google, or Alexa, depending on your device. After you alert your virtual assistant that you want help, you can ask it to manage your personal information or look up information on the Internet. Some common requests are:

- "What is the weather like today?"
- "Do I have any new e-mails?"
- "Add milk to the grocery list."
- "Text Mom that I am going to be late."

Making Plans Many virtual assistants review your e-mail and text messages for key words that indicate a calendar event. Virtual assistants scan texts and e-mails for indications of calendar events, such as a text that reads "Dinner tomorrow at 8 pm?" or an e-mail receipt for airline tickets.

Your virtual assistant then uses that information to add events to your calendar, such as "Dinner" at 8:00 pm tomorrow or "flight" with duration and departure and landing times—even correcting for time zone changes and including a link in the calendar event to quickly retrieve the original e-mail.

Connecting Different Data Points Virtual assistants can use multiple pieces of information to make suggestions to your schedule. For example, your virtual assistant uses your phone's GPS, traffic information on the Internet, and the details on your calendar to remind you when you need to leave for your next appointment if you want to arrive there on time.

Security, Settings, and Limitations Although many of these features are set up by default, you may have to change the security and setting of your device to allow your virtual assistant access to your e-mails, text messages, photos, microphone, and GPS location.

Microsoft Cortana monitors Windows apps for text that looks like calendar events. Most intelligent assistants can monitor texts and e-mails for potential calendar events.

ymgerman/Shutterstock

You may also find that the applications you use do not work well with your Assistant. For example, Siri only schedules events on the iOS calendar app, and Google Assistant will only search your e-mails for calendar events if you are using Google's e-mail service, Gmail. As a general rule, assistants only work well with the apps from the same developer—Siri works with Apple tools, Cortana works with Microsoft software, and Google Assistant works with Google applications.

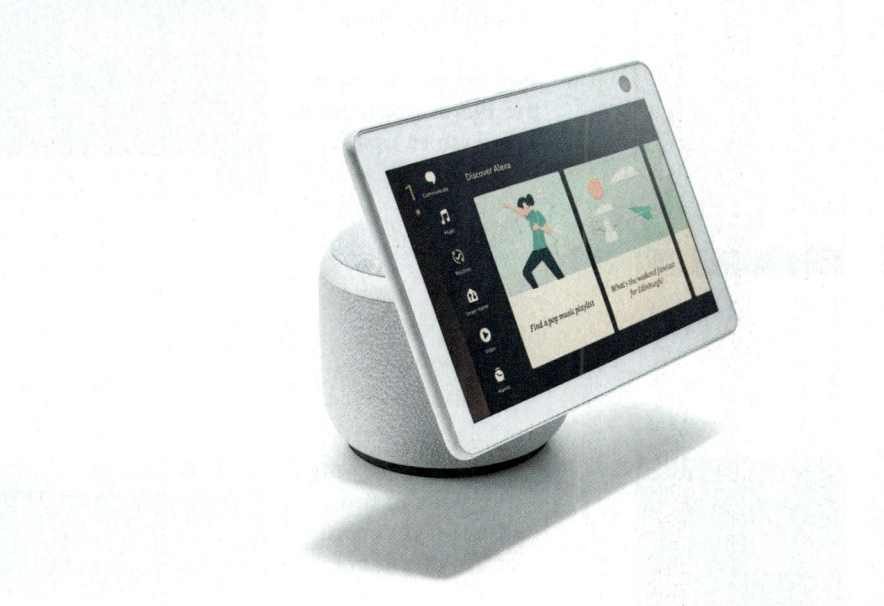

Amazon's Alexa integrates with home automation tools. You can change the control your personal assistant has—from the ability to read your e-mails and dim the house lights to simply alerting you to calendar events.

Photo by Phil Barker/Future Publishing via Getty Images

tips

Have you ever experienced problems after installing a new program or changing system settings? If so, the System Restore utility can help by reversing the changes and bringing your computer to a previous point in time. For Windows 11:

1. Go to the Windows 11 start screen and then type "recovery" in the search box.

2. Click on *Open System Restore* from the Advanced recovery tools list.

3. Follow the prompts, and choose a restore point.

4. Click the *Finish* button to start the process.

Operating System Utilities

Most operating systems are accompanied by several common utility programs, including a search program, a storage management program, and a backup program.

Search

Most cell phones have hundreds of apps and thousands of pictures and files. How can you find a particular app or file you need among all the others? Search programs can search your device for you to separate what you need from all the other apps and files. You will find a **search program** on mobile operating systems, such as Android and iOS, as well as laptops and desktops running Windows and macOS. (See **Figure 4-10**.)

Android

- On your Android phone, open the Google app.
- In the search box, type what you're looking for.

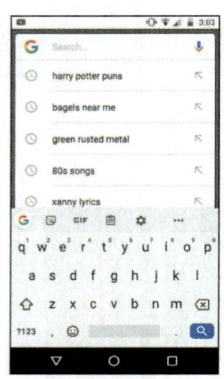

Windows

- In the lower left of the desktop, click in the text field with the magnifying glass.
- Type in the name of the file or application you wish to open.

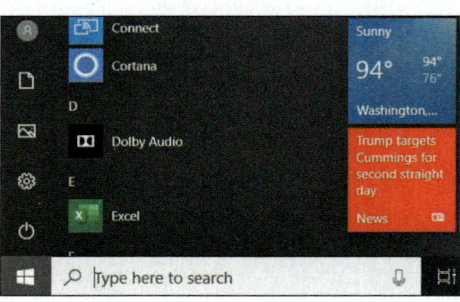

iOS

- Swipe down from the middle of the Home screen.
- Tap the Search field, then enter what you're looking for.

macOS

- Press the magnifying glass in the upper-right corner of the menu bar.
- Type in the name of the file or application you wish to open.

Figure 4-10 Search programs: Android, iOS, Windows, and macOS

Google INC.; Apple Inc.; Microsoft Corporation; Apple Inc.

CHAPTER 4: System Software 95

Storage Management

As you use your computer, your hard disk will fill with photos, videos, documents, and applications. You may find that as you run out of hard drive space, your computer slows down. Most operating systems include utilities to organize and view your hard drive usage to identify old or unused files. Using these tools, you can move or eliminate files and give your operating system the space it needs to run at peak efficiency.

Storage management programs can identify and remove unused files and applications. You will find a storage management program on mobile operating systems, such as Android and iOS, as well as laptops and desktops running Windows and macOS. (See **Figure 4-11**.)

Android

- Open your device's Settings app.
- Tap Storage.

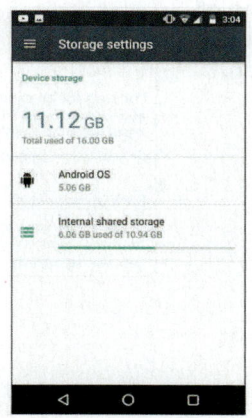

Windows

- In the lower left of the desktop, click in the text field with the magnifying glass.
- Type in the name of the file or application you wish to open.

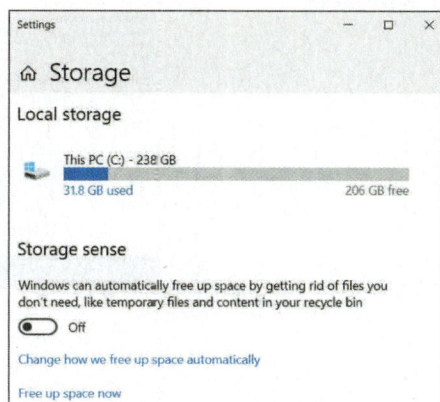

iOS

- Search for settings and select the settings icon.
- Select "General."
- Select iPad Storage or iPhone Storage (depending on device).

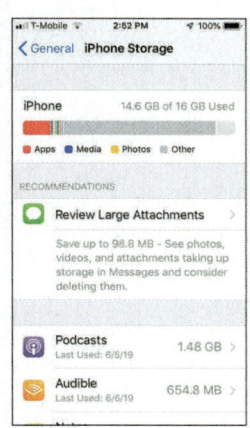

macOS

- Search for "About This Mac."
- Select "About This Mac."
- Click on Storage.

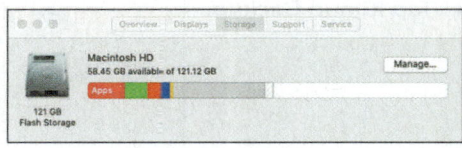

Figure 4-11 Storage management programs: Android, iOS, Windows, and macOS
Google INC.; Apple Inc.; Microsoft Corporation; Apple Inc.

Backup

It can be devastating to drop a cell phone or have a laptop stolen. Not only is the device itself expensive to replace, but the files and photos on the device may be lost forever. **Backup programs** should be run frequently to avoid the unexpected loss of files from your devices. All major operating systems offer backup utilities to save a copy of your device's data. (See **Figure 4-12**.)

Android

- Go to Settings.
- Select Personal.
- Finally, select Backup and reset.

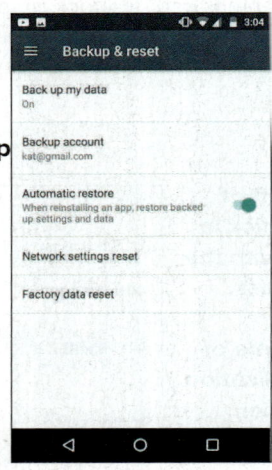

Windows

- In the lower left of the desktop, click in the text field with the magnifying glass.
- Type in the name of the file or application you wish to open.

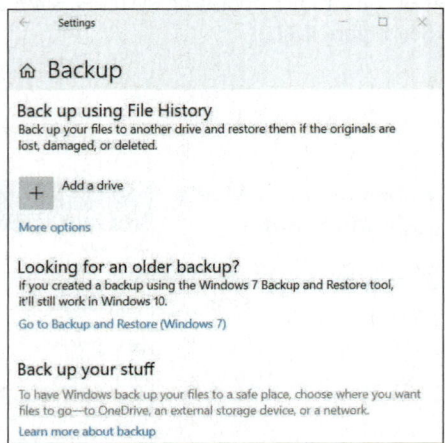

iOS

- Go to Settings > [your name] and tap iCloud.
- Tap iCloud Backup.

macOS

- Search for "System Preferences."
- Click on "Time Machine."

Figure 4-12 Backup programs: Android, iOS, Windows, and macOS

Google INC.; Apple Inc.; Microsoft Corporation; Apple Inc.

Utility Suites

Like application software suites, **utility suites** combine several programs into one package. Buying the package is less expensive than buying the programs separately. Some of the best-known utility suites are Bitdefender and Norton. (See **Figure 4-13**.) These suites provide a variety of utilities. Some programs will improve your computer's performance, while other programs will protect your system from dangerous computer **viruses**. You can "catch" a computer virus in many ways, including by opening attachments to e-mail messages and downloading software from the Internet. (We will discuss computer viruses in detail in **Chapter 9**.)

Figure 4-13 Norton's utility suite
Norton

concept check

☐ What are utility programs? Discuss six essential utilities.
☐ What are virtual assistants? What is Cortana? What is Siri?
☐ What is the difference between a utility and a utility suite?

Careers in IT

"Now that you know about system software, I'd like to tell you about my career as a computer support specialist."

AnnaStills/iStock/Getty Images

Computer support specialists provide technical support to customers and other users. They also may be called technical support specialists or help-desk technicians. Computer support specialists manage the everyday technical problems faced by computer users. They resolve common networking problems and may use troubleshooting programs to diagnose problems. Most computer support specialists are hired to work within a company and provide technical support for other employees and divisions. However, it is increasingly common for companies to contract with businesses that specialize in offering technical support.

Employers generally look for individuals with either an advanced associate's degree or a bachelor's degree to fill computer support specialist positions. Degrees in information technology or information systems may be preferred. However, because demand for qualified applicants is so high, those with practical experience and certification from a training program increasingly fill these positions. Employers seek individuals with customer service experience who demonstrate good analytical, communication, and people skills.

Computer support specialists can expect to earn an annual salary of $42,000 to $56,000. Opportunities for advancement are very good and may involve design and implementation of new systems.

A LOOK TO THE FUTURE

Making Better Computers by Making Them More Human

Blue Planet Studio/Shutterstock

Imagine that the next time your computer breaks, instead of getting it fixed, it heals itself. Many of our greatest technological breakthroughs are replicating the amazing complexity of nature. Scientists are continuing to look to nature to bring about the next wave of computing innovations. These include holding information in the way a brain holds information, with synapses and nerves to think, remember, and process. In searching for the next great improvement in computing, scientists are turning to human biological systems.

Today, most computers automatically perform standard maintenance operations such as defragging hard disks. Most search for viruses and eliminate them before they can become a problem. In the future, computers may not only fix viruses and other software issues but also identify and resolve hardware problems using autonomic computing.

Autonomic computing is a computing model that allows machines to run with little human intervention. It has the potential to revolutionize the way we interact with computers. The Cloud and Autonomic Computer Center is the leading research group devoted to developing self-healing computers. This center includes the National Science Foundation, leading universities (such as the University of Arizona and Texas Tech University), and industry leaders (such as Mitsubishi and Dell). Their objective is to free businesses and individuals from time-consuming computer maintenance by developing systems that are self-maintaining and virtually invisible to the user.

Autonomic processes in machines are modeled after human autonomic processes. For example, each of us has an autonomic system that automatically controls our breathing, our heart rate, and many other bodily functions. Scientists hope autonomic computing will behave in a similar manner and maintain computer systems without human intervention. Such computers would not have self-awareness but would be self-correcting.

These autonomic computing systems, however, are not artificial intelligence systems. Although autonomic systems automatically perform standard operations, they do not have human cognitive abilities or intelligence. These systems are limited to reacting to their own systems and have limited capability to learn from experience to correct errors. In order to adapt to conditions beyond this, scientists have turned to artificial intelligence.

Artificial intelligence can be found in computers around the world, from tiny cell phones to towering mainframes. These programs mimic the human brain. Researchers at MIT are working on neuromorphic computer chips. A neuromorphic chip is a computer chip that computes using an architecture like a brain, with neurons and synapses. They intend to use resistive computing, a process that uses analog voltages to store and compute data, instead of the digital method used today that is limited to binary values. This allows the chip to handle artificial intelligence tasks much more efficiently than today's hardware.

It is not only hardware that benefits from mimicking the human brain. Artificial intelligence is a field of programming where programs are designed to learn, adapt, and grow. Current technologies, such as OpenAI's ChatGPT and Google's Bard, provide a human-like discussion as these tools use knowledge from the Internet to form conversation answers to any question. In the future, these human-like conversations will guide us in everything from choosing what to make for dinner to making important medical decisions.

Given the potential for a self-maintaining server and a human-like computer chip, the possibility of a similar system designed for a personal computer seems less like a dream and more like a reality. What do you think? Will personal computers ever think and care for themselves?

VISUAL SUMMARY | System Software

SYSTEM SOFTWARE

monticello/Shutterstock

System software works with end users, application programs, and computer hardware to handle many details relating to computer operations.

Not a single program but a collection or system of programs, these programs handle hundreds of technical details with little or no user intervention.

Four kinds of systems programs are operating systems, utilities, device drivers, and language translators.

- **Operating systems** coordinate resources, provide an interface between users and the computer, and run programs.
- **Utilities** perform specific tasks related to managing computer resources.
- **Device drivers** allow particular input or output devices to communicate with the rest of the computer system.
- **Language translators** convert programming instructions written by programmers into a language that computers can understand and process.

Blue Planet Studio/Shutterstock

OPERATING SYSTEMS

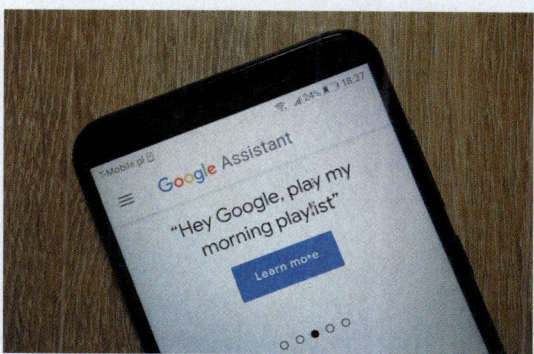
Piotr Swat/Shutterstock

Operating systems (software environments, software platforms) handle technical details.

Functions

Functions include managing resources, providing a **user interface** (**graphical user interface**, or **GUI**, provides visual interface; **voice assist tools** accept spoken commands), and running applications. **Multitasking** allows switching between different applications stored in memory; current programs run in the **foreground**; and other programs run in the **background**.

Features

Booting starts **(cold)** or restarts **(warm)** a computer system. The **desktop** provides access to computer resources. Common features include **icons**, **pointers**, **windows**, **menus**, **tabs**, **dialog boxes**, **help**, and **gesture control**. Data and programs are stored in a system of **files** and **folders**.

Categories

Three categories of operating systems are

- **Embedded**—also known as **real-time operating systems (RTOS)**; used with handheld computers; operating system stored within the device.
- **Stand-alone (desktop)**—controls a single computer; located on the hard disk.
- **Network (NOS)**—controls and coordinates networked computers; located on the **network server**.

Operating systems are often called **software environments** or **software platforms**.

To efficiently and effectively use computers, you need to understand the functionality of system software, including operating systems and utility programs.

MOBILE OPERATING SYSTEMS

19 STUDIO/Shutterstock

Mobile operating systems (mobile OS) are embedded in every cell phone and tablet. These systems are less complicated and more specialized for wireless communication than desktop operating systems.

Some of the best known are Google's Android and Apple's iOS.

- **Android** was developed by Google and is based on the Linux operating system. It is used as the operating system for some of the most popular cell phone lines, including the Samsung Galaxy S1 Ultra and Google Pixel 4a.
- **iOS** was developed by Apple and is based on macOS. It is used as the platform for Apple's mobile devices, the iPad and iPhone.

Not all mobile applications will run on all cell phones. That is because an app is designed to run on a particular software platform or operating system. Before downloading an app, be sure that it is designed to run with the mobile operating system on your mobile device.

DESKTOP OPERATING SYSTEMS

Microsoft Corporation

Windows

Windows is designed to run with many different microprocessors. The most recent version is **Windows 11**. Windows 11 was introduced in 2021. It includes the ability to run mobile Android apps; improved innovations for mobile computing; and integration of instant messaging, videoconferencing, and collaborative work tools.

macOS

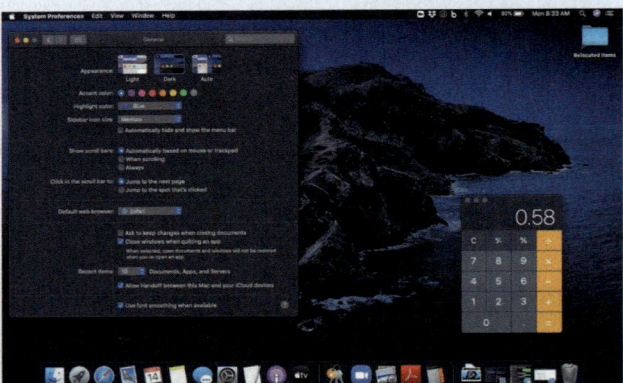

Apple, Inc.

macOS, an innovative, powerful, easy-to-use operating system, runs on Macintosh computers. **macOS** was introduced in 2021 and includes improved videoconferencing, notification customization, and the ability to use one mouse and keyboard across multiple devices.

DESKTOP OPERATING SYSTEMS

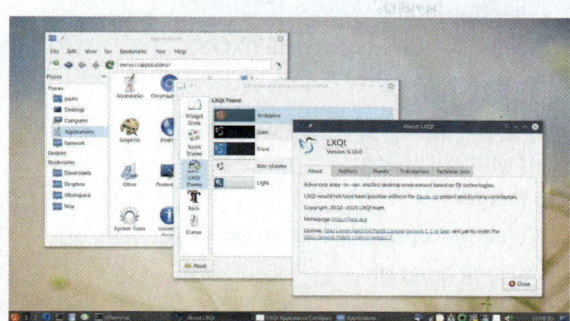
Linux

UNIX and Linux

UNIX was originally designed to run on minicomputers in network environments. Now, it is widely used by servers on the web, mainframe computers, and very powerful personal computers. There are many different versions of UNIX. One version, **Linux,** a popular and powerful alternative to the Windows operating system, is **open-source** software. Google's **Chrome OS** is based on Linux. It integrates with web servers to run applications and to perform other traditional operating system functions. Chrome OS is a popular choice for inexpensive notebook computers using cloud computing and cloud storage. One limitation of these computers is that their efficiency is dependent upon the speed of their Internet connection.

Virtualization

Microsoft Corporation

Virtualization allows a single physical computer to support multiple operating systems. Virtualization software (Parallels and VMware) allows the single physical computer to operate as two or more separate and independent computers known as **virtual machines.** Host operating systems run on the physical machine. **Guest operating systems** operate on virtual machines. **Virtual Desktop Infrastructure (VDI)** is a collection of hardware and software that allows a user to remotely access virtual machines over the Internet.

UTILITIES

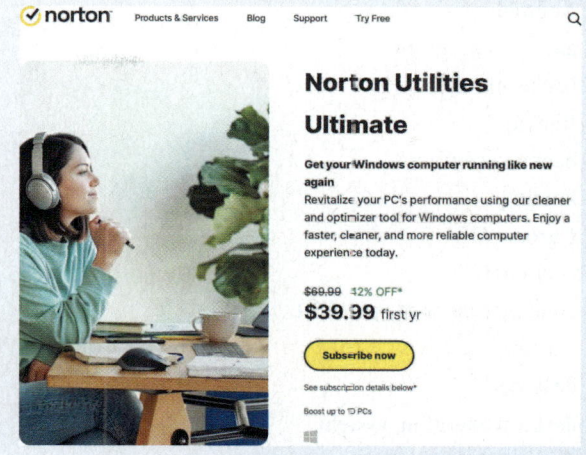
Norton

Operating System Utilities

Utilities are specialized programs designed to make computing easier. The most essential follow:

- **Search programs** provide a quick and easy way to search or examine an entire computer system to locate applications, data, or other files.
- **Storage management programs** help solve the problem of running out of storage space by providing lists of program files.
- **Backup programs** make copies of files to be used in case the originals are lost or damaged.
- **Antivirus programs** guard against viruses or other damaging programs.
- **Troubleshooting (diagnostic) programs** recognize and correct problems.
- **Virtual assistants** accept commands through text or speech and coordinate personal data across multiple applications.

Utility Suites

Utility suites combine several programs into one package. Computer **viruses** are dangerous programs.

CAREERS in IT

Computer support specialists provide technical support to customers and other users. Degrees in computer science or information systems are preferred plus good analytical and communication skills. Expected salary range is $42,000 to $56,000.

KEY TERMS

Android
antivirus program
background
Backup
backup program
booting
Chrome OS
cold boot
computer support specialist
Cortana
desktop
desktop operating system
device driver
diagnostic program
dialog box
embedded operating system
file
folder
foreground
gesture control
graphical user interface (GUI)
guest operating system
help
host operating system
icon
iOS
language translator
Linux
macOS
macOS 12 Monterey
menu
mobile operating system

mobile OS
multitasking
network operating system (NOS)
network server
open source
operating system
pointer
real-time operating system (RTOS)
search program
Siri
software environment
software platform
stand-alone operating system
storage management program
system software
tab
troubleshooting program
UNIX
user interface
utilities
utility suite
virtual assistant
Virtual Desktop Infrastructure (VDI)
virtualization
virtualization software
virtual machine
virus
voice assist tool
warm boot
window
Windows
Windows 11

MULTIPLE CHOICE

Circle the correct answer.

1. Programs that perform specific tasks related to managing computer resources.
   - a. troubleshooting
   - b. translators
   - c. virtual
   - d. utilities

2. Restarting a running computer without turning off the power.
   - a. cold boot
   - b. warm boot
   - c. multitasking
   - d. virtualization

3. Type of operating system that controls and coordinates networked computers.
   - a. iOS
   - b. mobile OS
   - c. NOS
   - d. Chrome OS

4. An operating system is often referred to as the software environment or software _____.
   - a. interface
   - b. driver
   - c. platform
   - d. suite

5. Switching among different applications.
   - a. multitasking
   - b. multicore
   - c. platforming
   - d. virtualization

6. A type of software that allows a single physical computer to operate as though it were two or more separate and independent computers.
   - a. virtualization
   - b. multitasking
   - c. embedded operating system
   - d. platform

7. Mobile operating system that is owned by Google and is widely used in many cell phones.
   - a. Android
   - b. macOS
   - c. Windows
   - d. iOS

8. Type of program that guards computer systems from viruses and other damaging programs.
   - a. embedded operating system
   - b. antivirus
   - c. background
   - d. storage management

9. What operating system is operating on each virtual machine?
   - a. embedded
   - b. UNIX
   - c. mobile
   - d. guest

10. Sonoma is the name of which type of operating system?
    - a. iOS
    - b. Windows
    - c. macOS
    - d. Chrome OS

MATCHING

Match each numbered item with the most closely related lettered item. Write your answers in the spaces provided.

a. Chrome
b. iOS
c. system
d. translators
e. icon
f. pointer
g. multitasking
h. antivirus program
i. search programs
j. utility suites

____ 1. This is an essential utility program that provides a quick and easy way to find a specific application or other file.
____ 2. Examples of this include Bitdefender and Norton.
____ 3. This is the ability to switch between different applications stored in memory.
____ 4. Graphic representation for a program, type of file, or function.
____ 5. Essential utility that protects against damaging programs that can invade your computer system.
____ 6. This operating system feature is controlled by a mouse and changes shape depending on its current function.
____ 7. This type of software works with users, application software, and computer hardware to handle the majority of technical details.
____ 8. The programs that convert programming instructions written by programmers into a language that computers understand and process are language _____.
____ 9. The operating system based on Linux, designed for notebook computers that use cloud computing and cloud storage.
____ 10. The mobile operating system developed by Apple.

OPEN-ENDED

On a separate sheet of paper, respond to each question or statement.

1. Describe system software. Discuss each of the four types of system programs.
2. Define operating systems. Describe the basic features and the three categories of operating systems.
3. What are mobile operating systems? Describe the leading mobile operating systems.
4. What are desktop operating systems? Compare Windows, macOS, Linux, and Chrome OS. Discuss virtualization.
5. Discuss utilities. What are the most essential utilities? What is a utility suite?

DISCUSSION

Respond to each of the following questions.

 ### Making IT Work for You: VIRTUAL ASSISTANTS

Review the Making IT Work for You: Virtual Assistants on pages 92 and 93. Then respond to the following: (a) Do you currently have a cell phone, tablet, or PC with an operating system that includes a virtual assistant? If you do, what are the operating system and the name of the virtual assistant? (If you do not, use one of the virtual assistants mentioned on pages 92 and 93 to respond to this question and the following questions.) Research your virtual assistant online and answer the following questions: (b) What are some of the basic commands that the assistant understands? (c) What information does the assistant have access to? Can it read your e-mail? Your calendar? (d) How does your OS assistant differ from other OS assistants? What can it do that others can't? What do you wish it could do?

 ### Privacy: VIRTUAL ASSISTANTS

Review the Privacy box on page 88, and respond to the following: (a) Have you ever sent a personal text or e-mail that you wouldn't want anyone else to see? Would you be uncomfortable with a company, like Apple or Google, seeing that text or e-mail? Would you be willing to let a company see that text or e-mail if it promised not to share the information? (b) What if the company profited from the information you shared with it? Would you deserve a share of that profit? (c) Search online for the privacy settings and privacy agreements of a popular virtual assistant, such as Siri or Google Assistant, and answer the following questions: What control do you have over the information the assistant sees? Who does it share that information with? What can the assistant do with that information?

 ### Ethics: CONVERSATION MONITORING

Review the Ethics box on page 85, and respond to the following: (a) Which devices in your home have a microphone? Are any of these devices connected to the Internet? (b) Do you or someone you know have a device that includes a voice assist tool (such as a cell phone or PC with a modern OS)? Which tool do you/they have? Does it have an option to always be listening to conversations? (c) If you visited someone's home and they told you there was a device in the house that recorded all conversations and shared those conversations over the Internet, would that concern you? Would it be ethical for you to ask them to turn the device off? Why or why not?

 ### Community: ACCESSIBILITY FEATURES

Review the Community box on page 85, and then respond to the following: (a) Do you or someone you know have a disability that makes using a computing device difficult? In what ways might people with disabilities find␣t difficult to use a cell phone or laptop? (b) Explore the accessibility settings for a computing device you have access to—a cell phone, laptop, or desktop. What settings does that device have to assist the visual or hearing impaired? (c) Try using one of these settings. What is it like to use the accessibility features? Are these features easy to find and use?

Design Elements: Concept Check icon: Dizzle52/Getty Images

chapter **5**

The System Unit

Blue Andy/Shutterstock

Why should I read this chapter?

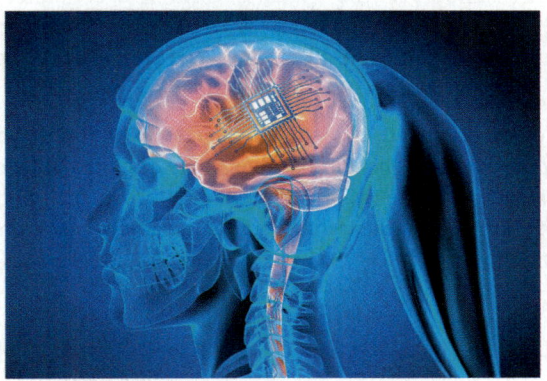

peterschreiber.media/iStock/Getty Images Plus

System units are getting smaller, faster, cheaper, and more powerful. These staggering improvements are resulting in microchips integrating with all aspects of our lives. For example, in the future, we will see everything from microchips embedded in the brain that improve mental capacity to sensors that can literally read your mind.

This chapter covers the things you need to know to be prepared for this ever-changing digital world, including:

- Types of computers—learn the strengths and weaknesses of cell phones, tablets, laptops, desktops, and wearable computers.
- Computer components—understand the impact microprocessors and memory have on the power of a computer.
- Peripherals and upgrades—expand your computer's abilities and speed.

Learning Objectives

After you have read this chapter, you should be able to:

1 Differentiate among the five basic types of system units.
2 Describe system boards, including sockets, slots, and bus lines.
3 Recognize different microprocessors, including microprocessor chips and specialty processors.
4 Compare different types of computer memory, including RAM, ROM, and flash memory.
5 Explain expansion cards and slots.
6 Describe bus lines, bus widths, and expansion buses.
7 Describe ports, including standard and specialized ports.
8 Identify power supplies for cell phones, tablets, laptops, and desktops.
9 Explain how a computer can represent numbers and encode characters electronically.

Introduction

"Hi, I'm Liz, and I'm a computer technician. I'd like to talk with you about the different types of system units for personal computers. I'd also like to talk about various electronic components that make your computer work."

wavebreakmedia/Shutterstock

There are many different types of computers, from cell phones to desktops, and they don't look very similar, so what makes them all computers? How are they similar?

While cell phones and desktops may look different on the outside, they all share a similar structure on the inside, the system unit. The system unit is what determines a computer's speed, capacity, and flexibility. In learning about the system unit, you will be prepared to make a smart choice on your next cell phone upgrade, tablet purchase, or even a laptop or desktop purchase. (The Computer Buyer's Guide at the end of this book provides additional information on purchasing the right desktop or laptop for your needs.)

If you open up any computer, you will see that it is basically a collection of electronic circuitry. Although there is no need to understand how all these components work, it is important to understand the principles. Armed with this knowledge, you will be able to confidently make sound purchasing and upgrading decisions.

To efficiently and effectively use computers, you need to understand the functionality of the basic components in the system unit, including the system board, microprocessor, memory, expansion slots and cards, bus lines, ports, cables, and power supply units.

System Unit

> **ethics**
>
> Many of the electronic devices purchased in the United States are manufactured in other countries. Some of these manufacturers pay extremely low wages, have unsafe or unacceptable work conditions, and pollute the environment. Do you think that consumers have an ethical responsibility regarding where and/or how products are manufactured?

The **system unit**, also known as the **system chassis**, is a container that houses most of the electronic components that make up a computer system. Some system units are located in a separate case, whereas others share a case with other parts of the computer system.

As we have previously discussed, a **personal computer** is the most widely used type of computer. It is the most affordable and is designed to be operated directly by an end user. The five most common types are cell phones, tablets, laptops, desktops, and wearable computers. Each has a unique type of system unit.

Smartphones

As we discussed in **Chapter 1**, almost everyone has a **cell phone**, and almost every cell phone is a **smartphone**. Smartphones are the most popular type of personal computer. They are effectively a thin slab that is almost all monitor, with the system unit, secondary storage, and all electronic components located behind the monitor. Designed to comfortably fit into the palm of one hand and accept finger touches on the monitor as the primary input, cell phones have become the indispensable handheld computer. (See **Figure 5-1**.)

Figure 5-1 Cell phone
Framesira/Shutterstock

Figure 5-2 Mini tablet
mspicsandmore/123RF

Figure 5-3 Two-in-one laptop
Andrey_Popov/Shutterstock

Figure 5-4 Ultrabook
Quang NGUYEN DUC/Alamy Stock Photo

Tablets

Tablets, also known as **tablet computers**, are very similar to cell phones, although tablets are larger, heavier, and generally more powerful. Also, a tablet is not designed to be used as a phone and, typically, cannot be connected to phone networks. Tablets come in a range of styles and sizes, with the smallest tablets (called **mini tablets**) being just a little larger than the largest cell phones. (See **Figure 5-2**.)

Laptops

Like cell phones and tablets, **laptops** have their system units housed with selected secondary storage and input devices. Laptops, however, are larger and more powerful. Their larger monitor is separate from the rest of the computer and attached to the system unit with a hinge. Unlike cell phones and tablets, most laptops do not accept input by touching the screen. Rather, laptops typically use an attached keyboard and a mouse as the primary input devices.

There are several specialized laptops with features that make them unique. Some of them are

- **Two-in-one laptops**, which include a touch screen and the ability to fold flat like a tablet computer. These laptops offer the advantages of a laptop with the convenience of a tablet. (See **Figure 5-3**.)
- **Gaming laptops**, which include high-end graphics hardware and very fast processors. As laptops have become more powerful, gaming laptops have become a favorite, primarily due to their portability. To learn more about specialized gaming computers, see the Making IT Work for You: Gaming on pages 110 and 111.
- **Ultrabooks**, also known as **ultraportables** or **mini notebooks**, are lighter and thinner with longer battery life than most laptops. They accomplish these advantages by leaving out components such as optical drives and using energy-efficient microprocessors. (See **Figure 5-4**.)

Figure 5-5 All-in-one desktop
Kaspars Grinvalds/Shutterstock

Desktops

Desktops are the most powerful type of personal computer. Like cell phones and tablets, some desktop computers, like Apple's iMac, have their monitor and system unit housed together in the same case. These computers are known as **all-in-one desktops**. (See **Figure 5-5**.)

However, most desktops have their system unit in a separate case. This case contains the system's electronic components and selected secondary storage devices. Input and output devices, such as a mouse, keyboard, and monitor, are located outside the system unit. This type of system unit is designed to be placed either horizontally or vertically. Desktop system units that are placed vertically are sometimes referred to as a **tower unit** or **tower computer**. (See **Figure 5-6**.)

Figure 5-6 Tower unit
Kateryna998/Shutterstock

Wearable Computers

Wearable computers, also known as **wearable devices**, are one of the first evolutionary steps to the Internet of Things (IoT), as discussed in **Chapter 2**. These devices contain an embedded computer on a chip that is typically much smaller and less powerful than a cell phone's. The most common wearable computers are smartwatches and activity trackers. (See **Figure 5-7**.)

- **Smartwatches** like Apple's Watch. This device acts as a watch, fitness monitor, and communication device. For example, the Apple Watch connects to an iPhone to display e-mails, text messages, and calendar reminders on the user's wrist.
- **Activity trackers** as developed by companies such as Garmin and Fitbit. These are often worn on the wrist, like a smartwatch, but are more narrowly focused on tracking health, exercise, and sleep data. They also connect wirelessly to desktops, laptops, and smartphones to record and share data.

Figure 5-7 Smartwatch
PixieMe/Shutterstock

Making IT work for you

GAMING

Would you like to spend less time bored and more time playing games with friends? Would you like to turn a simple walk into a sci-fi shoot out or an evening in front of the TV into an adventure through the Middle Ages? Video games can turn the mundane into the magical, and with so many game styles and gaming devices, there is a video game for every interest and budget.

Mobile Gaming You may already own one of the most popular gaming devices on the planet—a cell phone. As cell phones have gained powerful hardware, better screens, and faster Internet connections, they have become the most popular gaming device. Here are three things you should know about mobile gaming.

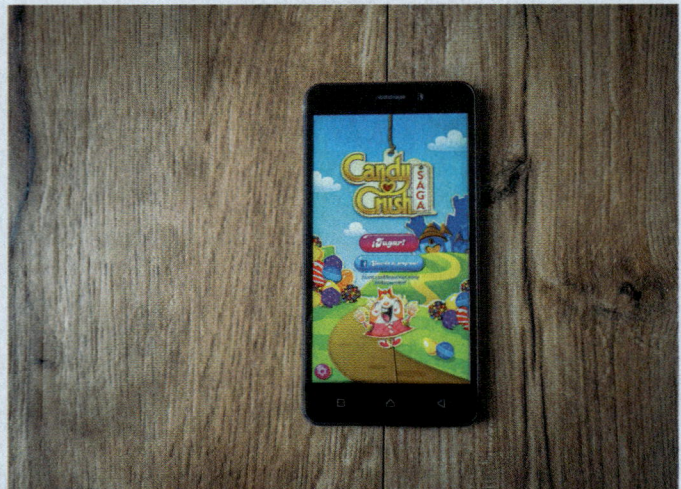

Minecraft and other mobile games are designed to take advantage of cell phone and tablet interfaces.

alexat25/Shutterstock

- **Casual gaming**—Mobile games are often designed to be "casual games," meaning the games are easy to pick up and play in short bursts. These games are ideal for short bursts of play when waiting for the bus or taking a quick study break. Popular examples of casual gaming are Candy Crush Saga and Monument Valley 2. Some casual gamers take it a step further with games that teach a skill or improve your memory, such as Lumosity or Elevate.

- **Augmented reality**—More sophisticated mobile games use the mobile device's GPS and camera to create "augmented reality" multiplayer games, where the game's graphics are a mixture of video images from the phone's camera and fantasy elements in the game realm. Examples of these games are Pokémon Go, Angry Birds AR: Isle of Pigs, and Google's Ingress Prime.

- **Hidden costs**—Just because a game is free, doesn't mean it isn't taking something of value. Many free games will take your time by requiring that you watch ads. Some games require high data usage that can rack up charges on your phone bill, and some may try to access personal information on your phone that you don't intend to share. Be aware of your data usage, read any fine print, and exercise caution when a game asks for credit card information or permission to access data on your phone.

Console Gaming The most popular video game hardware isn't a desktop or laptop computer. It is a specialized computer called a gaming console. Consoles have specialized hardware that produce high-quality graphics and sound and are designed to take advantage of TVs and home theaters. They provide wireless controllers for playing from the living room couch and HDMI ports to support standard TV resolutions and sound configurations. Also, because consoles are standardized and popular, they often have large libraries of games to choose from.

The three most popular consoles are Nintendo Switch, Microsoft Xbox Series X and S, and the Sony PlayStation 5. If you are looking for an immersive video game that goes beyond the casual gamer experience, consoles offer great performance at a reasonable price with excellent online features and gaming libraries. Here are a few things to consider when deciding what console to purchase:

- **Friends**—If you want to play online games with your friends, you will need to get the same console that your friends have. If you want to play games when friends come over to visit, Nintendo offers many of the most innovative cooperative games with an emphasis on party games meant to be played together simultaneously.

- **Input devices**—Each console has its own controllers, with different strengths and weaknesses. The Microsoft Xbox and Sony PlayStation have standard gamepads with joysticks and buttons ergonomically placed for playing games with both hands. The Nintendo Switch JoyCon controller has the features of a gamepad but also separates into two handheld motion controllers that can track arm gestures to play games. Finally, VR controllers, such as found with Meta's Quest 2 and Sony's PlayStation VR2 have buttons and joysticks like a gamepad, record arm gestures like a motion controller, and track hand and finger gestures to interact with the VR world.

- **Games**—Although many games are offered across all consoles, some are exclusive to only one console. Further, each console has types of games that they are best known for. Nintendo, for example, tends to attract more casual and family-friendly gamers, whereas the Xbox is famous for its intense online competition. Look for a console with a catalog of games that appeal to you.

Sony's PlayStation VR with PS VR headset. Each console has its own unique input/output devices.

agencies/Shutterstock

PC Gaming While mobile devices and console systems are the most common and popular gaming devices, gaming desktops and laptops offer unparalleled experiences that attract video games' biggest fans. Gaming PCs offer the highest degree of customizability, with specialized graphics cards and input devices. Although an entry-level gaming desktop isn't much more expensive than a typical desktop, the price can quickly escalate as you approach the cutting edge of graphics and sound technology. Here are some things to consider before buying a gaming PC:

A powerful gaming PC includes specialized hardware for running high-performance graphics.

Gorodenkoff/Shutterstock

- **Indy games**—Beyond offering the highest-quality video game experience, gaming PCs also have access to niche and independent (indy) video game developers, who cannot easily release games on consoles or mobile devices.

- **Upgrades**—Desktop PCs allow the computer hobbyist to upgrade hardware as technology improves. Gamers who are willing to open up their computer and replace parts can keep their gaming PC playing the newest games without having to buy a whole new computer.

- **Compatibility**—The fact that PC hardware is so customizable can make it difficult to be sure that the game you purchase is compatible with your PC configuration. PC gamers must know their system specifications when buying video games to be sure that the game will run on their system.

Components

Personal computers come in a variety of different sizes, shapes, and capabilities. Although they look different and each has its own unique features, they share surprisingly similar components, including system boards, microprocessors, and memory. (See **Figure 5-8**.)

Cell phone

Tablet

Wearable

Laptop

Desktop

Figure 5-8   **System unit components**
Cell phone: Andrew Berezovsky/Shutterstock; **Tablet:** lowepix/Alamy Stock Photo; **Wearable:** Chemari/Shutterstock; **Laptop:** Raw Group/Shutterstock; **Desktop:** Volodymyr Krasyuk/Shutterstock

concept check

- What is the system unit?
- Describe and compare the five most common types of personal computers.
- What is an ultrabook? What is an all-in-one? What is a tower unit? What is a mini tablet?

System Board

The **system board** is also known as the **mainboard** or **motherboard**. The system board controls communications for the entire computer system. All devices and components connect to the system board, including external devices like keyboards and monitors and internal components like hard drives and microprocessors. The system board acts as a data path and traffic monitor, allowing the various components to communicate efficiently with one another.

For cell phones, tablets, and wearable computers, the system board is located behind the screen. For laptops and desktops, the system board is typically located at the bottom of the system unit or along one side. It is a flat circuit board covered with a variety of different electronic components, including sockets, slots, and bus lines. (See **Figure 5-9**.)

Figure 5-9　Desktop system board
BonD80/Shutterstock

- **Sockets** provide a connection point for small specialized electronic parts called chips. **Chips** consist of tiny circuit boards etched onto squares of sandlike material called silicon. These circuit boards can be smaller than the tip of your finger. (See **Figure 5-10**.) A chip is also called a **silicon chip**, **semiconductor**, or **integrated circuit**. Chips typically are mounted onto **chip carriers**. (See **Figure 5-11**.) These carriers plug either directly into sockets on the system board or onto cards that are then plugged into slots on the system board. Sockets are used to connect the system board to a variety of different types of chips, including the microprocessor and memory chips.
- **Slots** provide a connection point for specialized cards or circuit boards. These cards provide expansion capability for a computer system. For example, a wireless networking card plugs into a slot on the system board to provide a connection to a local area network.
- Connecting lines called **bus lines** provide pathways that support communication among the various electronic components that are either located on the system board or attached to the system board.

Figure 5-10　Chip
Szasz-Fabian Jozsef/Shutterstock

community

Many electronics contain toxic materials, such as lead, mercury, and chromium. If not disposed of properly, these materials can leak into our water supply, get released into our air, or leach into our soil. First, consider donating them to various charitable organizations that work with local schools and low-income families. Otherwise, recycling electronics allows us to get rid of old devices and keep our community safe. Many computer retailers near you accept used devices and components for recycling, even if they are broken. Alternatively, visit the website of your local government to find the electronics recycling center nearest to you.

Figure 5-11　Chip mounted onto a chip carrier
ktsdesign/Shutterstock

Generally, the system board found on a desktop is larger than that found on a laptop, and much larger than one found on a tablet, cell phone, or wearable computer. Although these system boards vary in size, speed, power, and versatility, they nevertheless all perform the same function of communicating between the components of the personal computer.

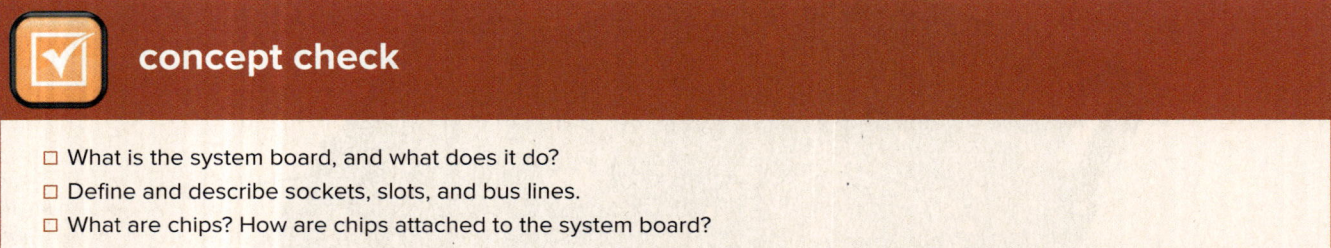

concept check

- What is the system board, and what does it do?
- Define and describe sockets, slots, and bus lines.
- What are chips? How are chips attached to the system board?

Microprocessor

In most personal computer systems, the **central processing unit (CPU)** or **processor** is contained on a single chip called the **microprocessor**. The microprocessor is the "brains" of the computer system. It has two basic components: the control unit and the arithmetic-logic unit.

- **Control unit:** The **control unit** tells the rest of the computer system how to carry out a program's instructions. It directs the movement of electronic signals between memory, which temporarily holds data, instructions, and processed information, and the arithmetic-logic unit. It also directs these control signals between the CPU and input and output devices.
- **Arithmetic-logic unit:** The **arithmetic-logic unit**, usually called the **ALU**, performs two types of operations: arithmetic and logical. **Arithmetic operations** are the fundamental math operations: addition, subtraction, multiplication, and division. **Logical operations** consist of comparisons such as whether one item is equal to (=), less than (<), or greater than (>) the other.

Microprocessor Chips

Processor	Manufacturer
A16 Bionic	Apple
Snapdragon 8 Gen 2	Qualcomm
Dimensity 92000	Vivo

Figure 5-12   Popular mobile microprocessors

Processor	Manufacturer
Ryzen 7840U	AMD
M2 Max	Apple
Meteor Lake	Intel

Figure 5-13   Popular desktop microprocessors

Unit	Speed
Microsecond	Millionth of a second
Nanosecond	Billionth of a second
Picosecond	Trillionth of a second
Femtosecond	Quadrillionth of a second

Figure 5-14   Processing speeds

Microprocessors are an important part of any computing device and are tailored to the needs of the device it serves. There are two major categories of microprocessors developed today: mobile and desktop. Mobile processors are used in cell phones and tablets and try to strike a balance between the processing power of the processor and the power the processor draws from the mobile device's battery. See **Figure 5-12** for a listing of popular mobile processors. Desktop processors are used in laptops and desktops. These processors are less concerned with the energy use of the processor and are more powerful than mobile processors. See **Figure 5-13** for a listing of popular desktop processors. Chip processing capacities are often expressed in word sizes. A **word** is the number of bits (such as 32 or 64) that can be accessed at one time by the CPU. A computer designed to process 64-bit words has greater processing capacity. Other factors affect a computer's processing capability, including how fast it can process data and instructions.

The processing speed of a microprocessor is typically represented by its **clock speed**, which is related to the number of times the CPU can fetch and process data or instructions in a second. Devices that don't need high processing power and want to use as little energy as possible, such as wearable devices, typically process data and instructions in millionths of a second, or microseconds. More powerful devices, such as cell phones, tablets, laptops, and desktops, are much faster and process data and instructions in billionths of a second, or nanoseconds. Supercomputers, by contrast, operate at speeds measured in picoseconds—1,000 times as fast as personal computers. In the near future, we can expect processor speeds to be 1,000 times faster than that, operating at speeds measured in femtoseconds. (See **Figure 5-14**.) Logically, the higher a microprocessor's clock speed, the faster the microprocessor. However, some processors can handle multiple instructions per cycle or tick of the clock; this means that clock speed comparisons can only be made between processors that work the same way.

At one time, personal computers were limited by microprocessors that could support a single CPU that controlled operations. These computers were limited to processing one program at a time. Now, many personal computers have **multicore processors** that can provide two or more separate and independent CPUs. For example, a quad-core processor could have one core computing a complex Excel spreadsheet, a second core creating a report using Word, a third core locating a record using Access, and a fourth core running a multimedia presentation—all at the same time. If a fifth program is processed, it will require that one of the cores be assigned to process two programs. This core will divide its computing power between the two programs, causing both programs to run slower than they would if they were running on individual cores. In this way, computers with more cores can run more programs faster.

For multicore processors to be used effectively, computers must understand how to divide tasks into parts that can be distributed across each core—an operation called **parallel processing**. Software developers use this technology for a wide range of applications from scientific programs to sophisticated computer games.

Specialty Processors

In addition to microprocessor chips, a variety of more specialized processing chips have been developed and are widely used. For example, many cars have more than 100 separate specialty processors to control nearly everything from fuel efficiency to satellite entertainment and tracking systems. **Coprocessors**, one of the more popular specialty processors, are specialty chips designed to improve specific computing operations. For example, to support fast processing of virtual environments, a standard feature in gaming computers is a **graphics coprocessor** known as a **GPU (graphics processing unit)**. These processors are designed to handle a variety of specialized tasks such as displaying 3D images and encrypting data. Most cell phones and tablets have specialty processors to efficiently show and store videos. Mobile devices often use many AI-enhanced tools, such as voice and image recognition, conversational virtual assistants, and security enhancements. These AI programs are powered through specialty AI chips.

privacy

Did you know that one type of specialty processor is devoted exclusively to protecting your privacy? Called cryptoprocessors, these microchips perform encoding and decoding of data faster and more securely than a CPU. These specialized chips exist in ATMs, TV set-top boxes, and payment terminals.

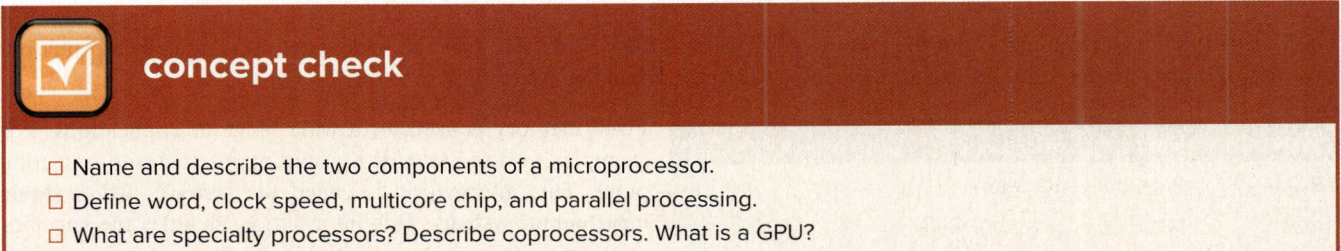

concept check

- Name and describe the two components of a microprocessor.
- Define word, clock speed, multicore chip, and parallel processing.
- What are specialty processors? Describe coprocessors. What is a GPU?

Memory

Memory is a holding area for data, instructions, and information. Like microprocessors, **memory** is contained on chips connected to the system board. There are three well-known types of memory chips: random-access memory (RAM), read-only memory (ROM), and flash memory.

RAM

Random-access memory (RAM) chips hold the program (sequence of instructions) and data that the CPU is presently processing. (See **Figure 5-15**.) RAM is called temporary or volatile storage because everything in most types of RAM is lost as soon as the computer is turned off. It is also lost if there is a power failure or other disruption of the electric current going to the computer.

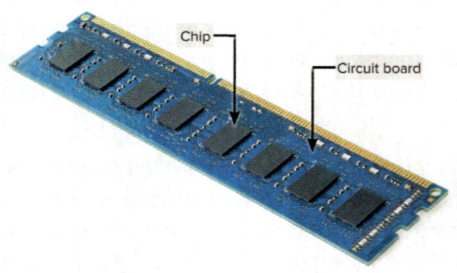

Figure 5-15 RAM chips mounted on circuit board

Olexandr Panchenko/Shutterstock

Cache (pronounced "cash") **memory** improves processing by acting as a temporary high-speed holding area between the memory and the CPU. The computer detects which information in RAM is most frequently used and then copies that information into the cache. When needed, the CPU can quickly access the information from the cache.

Having enough RAM is important! For example, to use Windows 11, Microsoft recommends that your computer have at least 4 GB, or 4 billion bytes, of RAM. Some applications, such as photo editing software, may require even more. Fortunately, additional RAM can be added to a computer system by inserting an expansion module called a **DIMM (dual in-line memory module)** into the system board. The capacity or amount of RAM is expressed in bytes. There are four commonly used units of measurement to describe memory capacity. (See **Figure 5-16**.)

Unit	Capacity
Megabyte (MB)	1 million bytes
Gigabyte (GB)	1 billion bytes
Terabyte (TB)	1 trillion bytes
Petabyte (PB)	1 quadrillion bytes

Figure 5-16 Memory capacity

Even if your computer does not have enough RAM to hold a program, it might be able to run the program using **virtual memory**. With virtual memory, large programs are divided into parts, and the parts are stored on a secondary device, usually a hard disk. Each part is then read into RAM only when needed. In this way, computer systems are able to run very large programs.

ROM

Read-only memory (ROM) chips have information stored in them by the manufacturer. Unlike RAM chips, ROM chips are not volatile and cannot be changed by the user. "Read only" means that the CPU can read, or retrieve, data and programs written on the ROM chip. However, the computer cannot write—encode or change—the information or instructions in ROM.

Not long ago, ROM chips were typically used to contain almost all the instructions for basic computer operations. For example, ROM instructions are needed to start a computer, to access memory, and to handle keyboard input. Recently, however, flash memory chips have replaced ROM chips for many applications.

Flash Memory

Flash memory offers a combination of the features of RAM and ROM. Like RAM, it can be updated to store new information. Like ROM, it does not lose that information when power to the computer system is turned off.

Type	Use
RAM	Programs and data
ROM	Fixed start-up instructions
Flash	Flexible start-up instructions

Figure 5-17 Memory

Flash memory is used for a wide range of applications. For example, it is used to store the start-up instructions for a computer. This information is called the system's **BIOS (basic input/output system)**. This information includes the specifics concerning the amount of RAM and the type of keyboard, mouse, and secondary storage devices connected to the system unit. If changes are made to the computer system, these changes are reflected in flash memory.

See **Figure 5-17** for a summary of the three types of memory.

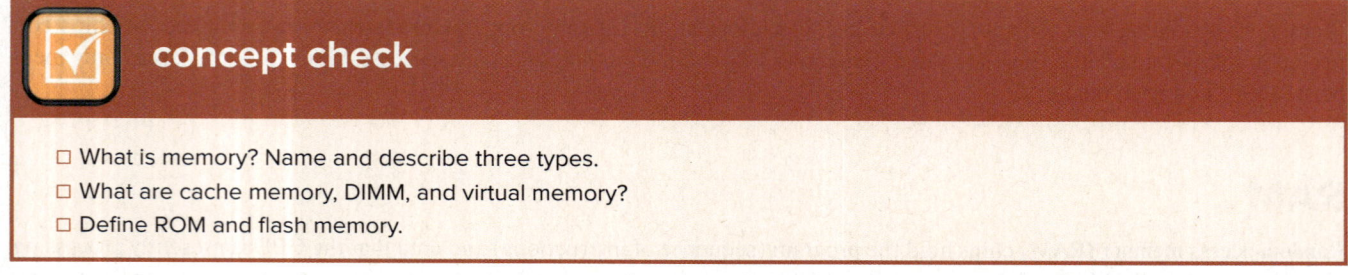

concept check

- What is memory? Name and describe three types.
- What are cache memory, DIMM, and virtual memory?
- Define ROM and flash memory.

Expansion Cards and Slots

To meet the size constraints of smaller mobile devices that need lots of storage, fingernail-size expansion cards known as **SD cards** have been developed. Mostly popular with high-end digital cameras, these cards plug into expansion slots located within many cell phones, tablets, and laptops. (See **Figure 5-18**.) Cell phones often have slots that use **SIM (Subscriber Identity Module) Cards**, a fingernail-sized card that uniquely identifies a customer to the cell phone company and allows the customer to access the cell phone network. These cards can appear very similar to SD cards, as they both store data and are a similar size. However, a SIM card stores much less data than an SD card and the primary purpose of a SIM card is to identify a customer account on a cell phone network, not the storage of large files. Modern cell phones do not use eSIMs Card, but use eSIMs that store the customer's identifying data on a specialized microchip on the cell phone's system board.

Figure 5-18 SD card

tkyszk/Shutterstock

As previously mentioned, many personal computers allow users to expand their systems by providing **expansion slots** on the system board. Users can insert optional devices known as **expansion cards** into these slots. (See **Figure 5-19**.) Ports on the cards allow cables to be connected from the expansion cards to devices outside the system unit. (See **Figure 5-20**.) There are a wide range of different types of expansion cards. Some of the most commonly used expansion cards are

Figure 5-19 Expansion cards fit into slots on the system board

Andrey Solovev/Alamy Stock Photo

- **Graphics cards,** which provide high-quality 3D graphics and animation for games and simulations. While many personal computer systems have a GPU connected directly to the system board, others connect through a graphics card. This card can contain one or more GPU chips and is standard for most gaming computers.
- **Network interface cards (NIC)**, also known as **network adapter cards**, are used to connect a computer to a network. (See **Figure 5-21**.) The network allows connected computers to share data, programs, and hardware. The network adapter card connects the system unit to the network via a network cable.
- **Wireless network cards** allow computers to be connected without cables. As we will discuss in **Chapter 8**, wireless networks in the home are widely used to share a common Internet connection. Each device on the network is equipped with a wireless network card that communicates with the other devices.

Figure 5-20 Expansion card with three ports

Volodymyr Krasyuk/Shutterstock

Figure 5-21 Network interface card

DeSerg/Shutterstock

Many computers include graphics and network capabilities embedded in the system board, making a graphics or network card unnecessary. However, graphics and network technologies are advancing quickly, and having an expansion card allows users to upgrade their computers to the newest technology without having to replace the system board.

concept check

☐ List and describe three commonly used expansion cards.
☐ What are SD cards? How are they used?
☐ What are expansion slots and cards? What are they used for?

Bus Lines

Figure 5-22 Bus is a pathway for bits
raigvi/Shutterstock

As mentioned earlier, a **bus line**—also known simply as a **bus**—connects the parts of the CPU to each other. Buses also link the CPU to various other components on the system board. (See **Figure 5-22**.) A bus is a pathway for bits representing data and instructions. The number of bits that can travel simultaneously down a bus is known as the **bus width**.

A bus is similar to a multilane highway that moves bits rather than cars from one location to another. The number of traffic lanes determines the bus width. A highway (bus line) with more traffic lanes (bus width) can move traffic (data and instructions) more efficiently. For example, a 64-bit bus can move twice as much information at a time as a 32-bit bus. Why should you even care about what a bus line is? Because as microprocessor chips have changed, so have bus lines. Bus design or bus architecture is an important factor relating to the speed and power for a particular computer. Additionally, many devices, such as expansion cards, will work with only one type of bus.

Every computer system has two basic categories of buses. One category, called **system buses**, connects the CPU to memory on the system board. The other category, called **expansion buses**, connects the CPU to other components on the system board, including expansion slots.

Expansion Buses

Computer systems typically have a combination of different types of expansion buses. The principal types are USB, FireWire, and PCIe.

- **Universal serial bus (USB)** is widely used today. External USB devices are connected from one to another or to a common point or hub and then onto the USB bus. The USB bus then connects to the PCI bus on the system board. The current USB standard is USB 3.4.
- **FireWire buses** are similar to USB buses but are more specialized. They are used primarily to connect audio and video equipment to the system board.
- **PCI Express (PCIe)** is widely used in many of today's most powerful computers. Unlike most other buses that share a single bus line or path with several devices, the PCIe bus provides a single dedicated path for each connected device.

concept check

☐ What is a bus, and what is bus width?
☐ What is the difference between a system and an expansion bus?
☐ Discuss three types of expansion buses.

Ports

A **port** is a socket for external devices to connect to the system unit. A cell phone typically uses a port to recharge its battery. (See **Figure 5-23**.) Some ports connect directly to the system board, while others connect to cards that are inserted into slots on the system board. Some ports are standard features of most computer systems, and others are more specialized. (See **Figure 5-24**.)

Standard Ports

Most desktop and laptop computers come with a standard set of ports for connecting **peripherals**, or external devices, such as a monitor and keyboard. The most common ports are:

- **Universal serial bus (USB) ports** can be used to connect several devices to the system unit and are widely used to connect keyboards, mice, printers, storage devices, and a variety of specialty devices. A single USB port can be used to connect many USB devices to the system unit. There are many types of USB ports.
    - **USB-A** is the port found on most laptops and desktops.
    - **USB-B** ports are found on peripheral devices, like digital cameras and cell phones.
    - **USB-C** is the newest USB port, found on high-end cell phones and laptops. USB-C is faster, smaller, and easier to use than previous USB types. Some USB-C ports also support **Thunderbolt 3**, a high-speed version of the USB-C port.
- **High-Definition Multimedia Interface (HDMI) ports** provide high-definition video and audio, making it possible to use a computer as a video jukebox or an HD video recorder.
- **Ethernet ports** are a high-speed networking port that has become a standard for many of today's computers. Ethernet allows you to connect multiple computers for sharing files, or to a DSL or cable modem for high-speed Internet access.

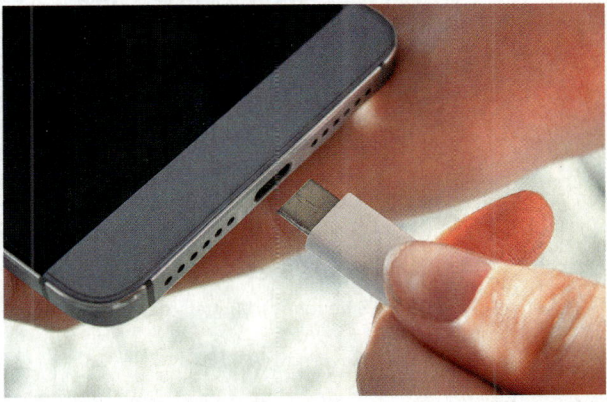

Figure 5-23   Cell phone port
OleJohny/Shutterstock

Figure 5-24   Ports
Dmitrii Anikin/Alamy Stock Photo

Specialized Ports

In addition to standard ports, there are numerous specialty ports. The most common include:

- **DisplayPorts (DP)** are audiovisual ports typically used to connect large monitors. They are popular on gaming computers with high-end graphics cards.
- **DVI (Digital Video Interface) ports** connect digital monitors to your computer. These ports can only send video signals and cannot send audio signals. They can be found mainly on desktops.
- **FireWire ports** provide high-speed connections to specialized FireWire devices such as camcorders and storage devices.

Cables

Cables are used to connect exterior devices to the system unit via the ports. One end of the cable is attached to the device, and the other end has a connector that is attached to a matching connector on the port. Standard cables include USB, HDMI, Thunderbolt, and Ethernet. (See **Figure 5-25**.)

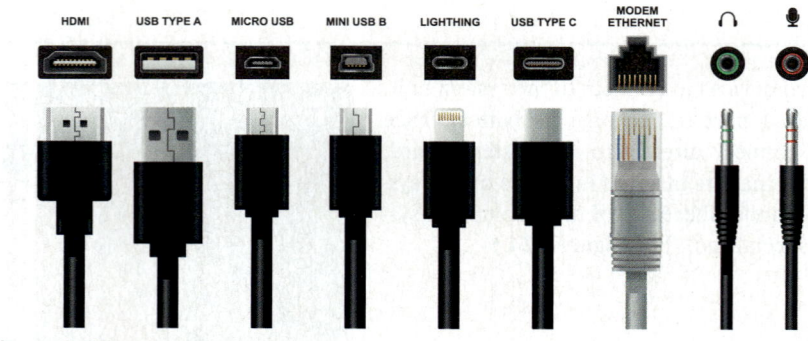

Figure 5-25 Standard cables

Siberian Art/Shutterstock

Power Supply

Computers require direct current (DC) to power their electronic components and to represent data and instructions. DC power can be provided indirectly by converting alternating current (AC) from standard wall outlets or directly from batteries.

- Most cell phones and tablets are powered directly from batteries that are recharged using an **AC adapter**. These adapters plug into standard wall outlets and convert AC to DC. With cell phones and tablets, a USB cable often connects the AC adapter to the mobile device. Some cell phones, however, can use a **wireless charging platform**, eliminating the cable. (See **Figure 5-26**.) Most tablets, mobile devices, and wearable computers can operate only using battery power. Their AC adapters or charging platforms are only used to recharge batteries.

Figure 5-26 Wireless charging platform

sashkin7/123RF

- Like cell phones and tablets, laptops typically use AC adapters that are located outside the system unit. (See **Figure 5-27**.) Unlike cell phones and tablets, these computers can be operated either using an AC adapter plugged into a wall outlet or using battery power. Their batteries typically provide sufficient power for up to eight hours before they need to be recharged.

- Desktop computers have a **power supply unit** located within the system unit. (See **Figure 5-28**.) This unit plugs into a standard wall outlet, converts AC to DC, and provides the power to drive all of the system unit components.

  Unlike cell phones, tablets, or laptops, desktop computers do not have batteries and cannot operate unless connected to a power source.

tips

Does your laptop seem to be losing its charge sooner than it used to? These batteries do lose power over time; however, you can take some steps to slow down the aging process.

1. **Balance adapter and battery use.** The best practice is to use the laptop on battery power for a little while without draining it completely (e.g., 20 percent charge), followed by charging it back to 100 percent. Modern batteries should not be drained to 0 percent each day.

2. **Calibrate it.** Your laptop's manufacturer will recommend that you calibrate, or reset, your battery every few months. Follow its guidelines on the web or in your instruction manual, as it will ensure that the battery meter in your operating system is accurate and that you are getting the expected charge time.

3. **Avoid excessive heat.** High temperatures can accelerate the deterioration of modern batteries. Therefore, avoid exposure to excessive heat and consider purchasing a laptop cooler or fan.

4. **Proper storage.** If you are not going to use your laptop for a few weeks, most manufacturers recommend that you remove the battery.

Figure 5-27 AC adapter
New Africa/Shutterstock

Figure 5-28 Power supply unit
hodim/Shutterstock

concept check

- What are ports? What do they do?
- Describe four standard ports and three specialized ports.
- What is a power supply unit? AC adapter? Charging platform?
- What is the difference between a SIM card and an SD Card?

Electronic Data and Instructions

Have you ever wondered why it is said that we live in a digital world? It's because computers cannot recognize information the same way you and I can. People follow instructions and process data using letters, numbers, and special characters. For example, if we wanted someone to add the numbers 3 and 5 together and record the answer, we might say "please add 3 and 5." The system unit, however, is electronic circuitry and cannot directly process such a request.

Our voices create **analog**, or continuous, signals that vary to represent different tones, pitches, and volume. Computers, however, can recognize only **digital** electronic signals. Before any processing can occur within the system unit, a conversion must occur from what we understand to what the system unit can electronically process.

Numeric Representation

What is the most fundamental statement you can make about electricity? It is simply this: It can be either on or off. Indeed, many forms of technology can make use of this two-state on/off, yes/no, present/absent arrangement. For instance, a light switch may be on or off, or an electric circuit open or closed. A specific location on a tape or disk may have a positive charge or a negative charge. This is the reason, then, that a two-state or binary system is used to represent data and instructions.

The decimal system that we are all familiar with has 10 digits (0, 1, 2, 3, 4, 5, 6, 7, 8, 9). The **binary system**, however, consists of only two digits—0 and 1. Each 0 or 1 is called a **bit**—short for binary digit. In the system unit, the 1 can be represented by a negative charge and the 0 by no electric charge. In order to represent numbers, letters, and special characters, bits are combined into groups of eight called **bytes**. Whenever you enter a number into a computer system, that number must be converted into a binary number before it can be processed.

tips

Is your cell phone struggling to make it through the day on a single charge? Before you replace the battery, consider the following suggestions.

1 **Monitor power usage.** Some apps are always running, even when you aren't using your cell phone. To monitor what apps are using the most power, check out the power usage features in your mobile OS and close the apps you are not using.

- For Android cell phones: Click on the *Settings* icon from the home screen, then choose *Device*, then *Battery* or *Settings*, then *Power*, and finally *Battery*.
- For iOS cell phones: Click on the *Settings* icon from the Home screen, then choose *Battery*.

2 **Reduce screen brightness.** One of the greatest drains on your cell phone's battery is the monitor. By reducing screen brightness, you can greatly extend your battery life.

- For Android cell phones: Swipe down from the top of the screen to display the notification shade and then adjust the brightness slider (an icon that looks like a sun) to lower your screen brightness.
- For iOS cell phones: Swipe down from the top-right corner of the screen to display the control center and then adjust the brightness bar (an icon that looks like a sun) to lower your screen brightness.

3 **Reduce background services.** When your battery gets low, you can set your cell phone to reduce its battery usage by performing background processes less frequently.

- For Android: Swipe down from the top of the screen to display the notification shade and tap the *Battery Saver* button.
- For iOS cell phones: Click on the Settings icon from the Home screen and choose *Battery* and tap the *Low Power Mode* button.

Any number can be expressed as a binary number. Binary numbers, however, are difficult for humans to work with because they require so many digits. Instead, binary numbers are often represented in a format more readable by humans. The **hexadecimal system**, or **hex**, uses 16 digits (0, 1, 2, 3, 4, 5, 6, 7, 8, 9, A, B, C, D, E, F) to represent binary numbers. Each hex digit represents four binary digits, and two hex digits are commonly used together to represent 1 byte (8 binary digits). (See **Figure 5-29**.) You may have already seen hex when selecting a color in a website design or drawing application, or when entering the password for access to a wireless network.

Character Encoding

As we've seen, computers must represent all numbers with the binary system internally. What about text? How can a computer provide representations of the nonnumeric characters we use to communicate, such as the sentence you are reading now? The answer is character encoding schemes or standards.

Character encoding standards assign a unique sequence of bits to each character. Historically, personal computers used the **ASCII (American Standard Code for Information Interchange)** to represent characters, while mainframe computers used **EBCDIC (Extended Binary Coded Decimal Interchange Code)**. These schemes were quite effective; however, they are limited. ASCII, for example, only uses 7 bits to represent each character, which means that only 128 total characters could be represented. This was fine for most characters in the English language but was not large enough to support other languages such as Chinese and Japanese. These languages have too many characters to be represented by the 7-bit ASCII code.

The explosion of the Internet and subsequent globalization of computing have led to a new character encoding called **Unicode**. The Unicode standard is the most widely used character encoding standard and is recognized by virtually every computer system. The first 128 characters are assigned the same sequence of bits as ASCII to maintain compatibility with older ASCII-formatted information. However, Unicode uses a variable number of bits to represent each character, which allows non-English characters and special characters to be represented. Unicode can be written in UTF-16 or UTF-8. UTF-16 is the older Unicode standard, and each character is a minimum of 16 bits. The newer standard, UTF-8, can have characters as small as 8 bits, making it more efficient than UTF-16.

Decimal	Binary	Hex
00	00000000	00
01	00000001	01
02	00000010	02
03	00000011	03
04	00000100	04
05	00000101	05
06	00000110	06
07	00000111	07
08	00001000	08
09	00001001	09
10	00001010	0A
11	00001011	0B
12	00001100	0C
13	00001101	0D
14	00001110	0E
15	00001111	0F

Figure 5-29 **Numeric representations**

concept check

☐ What is the difference between an analog and a digital signal?
☐ What are decimal and binary systems? How are they different?
☐ Compare EBCDIC, ASCII, and Unicode.

Careers in IT

"**Now that you know about system units, I'd like to tell you about my career as a computer technician.**"

wavebreakmedia/Shutterstock

Computer technicians repair and install computer components and systems. They may work on everything from personal computers and mainframe servers to printers. Some computer technicians are responsible for setting up and maintaining computer networks. Experienced computer technicians may work with computer engineers to diagnose problems and run routine maintenance on complex systems. Job growth is expected in this field as computer equipment becomes more complicated and technology expands.

Employers look for those with certification or an associate's degree in computer repair. Computer technicians also can expect to continue their education to keep up with technological changes. Good communication skills are important in this field.

Computer technicians can expect to earn an annual salary in the range of $43,000 to $53,000. Opportunities for advancement typically come in the form of work on more advanced computer systems. Some computer technicians move into customer service positions or go into sales.

A LOOK TO THE FUTURE

Brain–Computer Interfaces

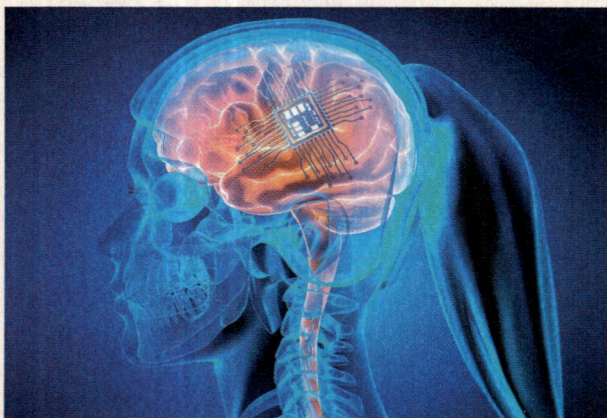

peterschreiber.media/iStock/Getty Images Plus

Can you imagine a future when we become half human and half machine? Today, we use computer implants to help people hear and to control their heartbeat. Computer chips are being implanted into people with Parkinson's disease to help them control their movements and to reduce other impacts of the condition. But these devices are simple compared to the future of brain–computer interfaces imagined by the researchers and entrepreneurs of Silicon Valley.

The human brain stores information in neurons using electrical impulses. Using neural imaging, specialized equipment can read the intensity and location of these impulses or thought patterns. This technology currently allows amputees to control robotic limbs. This case and almost all other applications, however, require brain surgery and cumbersome equipment. Researchers are currently trying to reduce the size and invasiveness of these technologies. Scientists at Florida International University are investigating devices so small that they can travel through the bloodstream to the brain. In the future, a simple shot may allow your computer to tell what you are thinking.

With improvements in the quality of neural imaging and neural hardware, we may soon be able to image all the brain's electrical impulses. For example, when you think of turning on the house lights, a computer can take an image of your brain thinking about turning on the house lights. Later, when you have that thought again, the computer can compare your brain's image to previous images and realize that when those areas of the brain are active, you want the house lights to turn on.

Not only may computers someday read your mind—they someday may write to it as well. Memories are stored in the electrical impulses of the brain, and by injecting tiny magnetic particles into the brain, scientists can stimulate neurons, essentially creating memories. Currently, our understanding of how the brain operates is too limited, and our ability to manipulate neurons is too coarse, to allow us to accurately transmit information. However, someday you may think, "I would like to be able to speak French," and through neural stimulation, a computer would generate the memories necessary for you to speak French.

Many ethicists are concerned by the potential use of such technologies to improve human abilities instead of treating medical conditions. For example, with tiny chips being able to store so much, people could use brain implants to improve their memory. This can lead to a variety of scenarios where the individual with the implants would have an advantage over those who don't have them. However, others disagree, arguing that the integration of technology and biology is to be expected and it is nothing more than the next step in human evolution. If such a technology became widespread and affordable, would you opt to receive a chip implant?

VISUAL SUMMARY | The System Unit

SYSTEM UNIT

Andrey_Popov/Shutterstock

System unit (system chassis) contains electronic components. The most common **personal computers** are **cell phone, tablet, laptop, desktop,** and **wearable computers**.

Smartphones

Smartphones, more commonly called **cell phones**, are the most popular personal computer. Almost all monitors with system unit, storage, and electronics located behind the monitor.

Tablet

A **tablet**, or **tablet computer**, is like a cell phone except larger, heavier, and more powerful. A **mini tablet** has a smaller screen with less functionality than traditional tablet.

Laptop

Laptops, compared to cell phones, are larger, are more powerful, have a separate monitor, and use a keyboard and mouse. Specialty laptops include **two-in-one laptops; gaming laptops** (high-end graphics and fast processors); **ultrabooks**, also known as **ultraportables;** and **mini notebooks**.

Desktop

A **desktop** is the most powerful personal computer; **tower unit (tower computer)** has vertical system unit; an **all-in-one desktop** combines system unit and monitor.

Wearable Computers

Wearable computers (wearable devices) contain an embedded computer on a chip. The most common wearable computers are **smartwatches** and **activity trackers**.

Components

Each type of system unit has the same basic components, including system board, microprocessor, and memory.

SYSTEM BOARD

BonD80/Shutterstock

The **system board** (**mainboard** or **motherboard**) controls all communication for the computer system. All external and internal devices and components connect to it.

- **Sockets** provide connection points for **chips (silicon chips, semiconductors, integrated circuits)**. Chips are mounted on **chip carriers**.
- **Slots** provide connection points for specialized cards or circuit boards.
- **Bus lines** provide pathways to support communication.

ktsdesign/Shutterstock

To efficiently and effectively use computers, you need to understand the functionality of the basic components in the system unit: system board, microprocessor, memory, expansion slots and cards, bus lines, ports, and cables. Additionally, you need to understand how data and programs are represented electronically.

MICROPROCESSOR

In most personal computers, the **central processing unit (CPU)**, or **processor**, is contained on a single chip called the **microprocessor**. It has two basic components: a **control unit** and an **ALU**.

Microprocessor Chips

A **word** is the number of bits that can be accessed by the microprocessor at one time. **Clock speed** represents the number of times the CPU can fetch and process data or instructions in a second.

Multicore processors can provide multiple independent CPUs. **Parallel processing** requires programs that allow multiple processors to work together to run large complex programs.

Specialty Processors

Specialty processors include **graphics coprocessors**, also known as **GPU** or **graphics processing unit** (processes graphic images), and processors in automobiles (monitor fuel efficiency, satellite entertainment, and tracking systems).

MEMORY

Memory holds data, instructions, and information. There are three types of memory chips.

RAM

RAM (random-access memory) chips are called temporary or volatile storage because their contents are lost if power is disrupted.

- **Cache memory** is a high-speed holding area for frequently used data and information.
- **DIMM (dual in-line memory module)** is used to expand memory.
- **Virtual memory** divides large programs into parts that are read into RAM as needed.

ROM

ROM (read-only memory) chips are nonvolatile storage and control essential system operations.

Flash Memory

Flash memory does not lose its contents when power is removed.

EXPANSION CARDS AND SLOTS

Andrey Solovev/Alamy Stock Photo

SD cards are fingernail-size expansion cards used primarily for digital cameras and mobile devices. They connect to expansion slots within these devices.

SIM (Subscriber Identity Module) Cards are card that allows a customer to access a cell phone network. SIM cards are replaced by **eSIMs,** which store the same information on a microchip in the phone.

Many computers allow users to expand their systems by providing **expansion slots** on their system boards to accept **expansion cards**.

Examples of expansion cards include **graphics cards**, **network interface cards (NIC; network adapter cards)**, and **wireless network cards**.

BUS LINES

Bus lines, also known as **buses**, provide data pathways that connect various system components. **Bus width** is the number of bits that can travel simultaneously.

System buses connect CPU and memory. **Expansion buses** connect CPU and slots.

Expansion Buses

Three principal expansion bus types:

- **USB (universal serial bus)** can connect from one USB device to another or to a common point (hub) and then onto the system board.
- **FireWire bus** is similar to USB bus but more specialized.
- **PCI Express (PCIe) bus** is widely used; provides a single dedicated path for each connected device.

PORTS

Siberian Art/Shutterstock

Ports are connecting sockets on the outside of the system unit.

Standard Ports

Four standard ports are as follows:

- **USB (universal serial bus)**—one USB port can connect several devices to system unit **USB-A** PCs; **USB-B** peripherals; **USB-C** gaining popularity.
- **HDMI (high-definition multimedia interface)**—provides high-definition video and audio.
- **Thunderbolt**—provides high-speed connections to up to seven Thunderbolt devices at once, such as external hard drives and monitors.
- **Ethernet**—high-speed networking port that has become a standard for many of today's computers.

Specialized Ports

Three specialty ports are **DisplayPort (DP)** for large monitors, and **DVI (Digital Video Interface)** for connecting to monitors, and **FireWire** for high-speed connection to devices such as camcorders and secondary storage.

Cables

Cables are used to connect external devices to the system unit via ports.

POWER SUPPLY

Power supply units convert AC to DC and power desktops. **AC adapters** power laptops and tablets and recharge batteries. Some cell phones use **wireless charging platforms**.

sashkin7/123RF

ELECTRONIC REPRESENTATION

Human voices create **analog** (continuous) signals; computers only recognize **digital** electronic signals.

Numeric Representation

Data and instructions can be represented electronically with a two-state or **binary system** of numbers (0 and 1). Each 0 or 1 is called a **bit**. A **byte** consists of 8 bits. A **hexadecimal system (hex)** uses 16 digits to represent binary numbers.

Character Encoding

Character encoding standards assign unique sequences of bits to each character. Three standards are as follows:

- **ASCII**—American Standard Code for Information Interchange. Historically used for personal computers.
- **EBCDIC**—Extended Binary Coded Decimal Interchange Code. Historically used for mainframe computers.
- **Unicode**—16-bit code, most widely used standard.

CAREERS in IT

Computer technicians repair and install computer components and systems. Certification in computer repair or an associate's degree from professional schools is required. Expected salary range is $27,000 to $58,000.

KEY TERMS

AC adapter
activity tracker
all-in-one desktop
analog
arithmetic-logic unit (ALU)
arithmetic operation
ASCII (American Standard Code for Information Interchange)
binary system
BIOS (basic input/output system)
bit
bus
bus line (115, 120)
bus width
byte
cable
cache memory
cell phone
central processing unit (CPU)
character encoding standards
chip
chip carrier
clock speed
computer technician
control unit
coprocessor
desktop
digital
DIMM (dual in-line memory module)
DisplayPort (DP)
DVI (Digital Video Interface) port
EBCDIC (Extended Binary Coded Decimal Interchange Code)
eSIM
Ethernet port
expansion bus
expansion card
expansion slot
FireWire bus
FireWire port
flash memory
gaming laptop
GPU (graphics processing unit)
graphics card
graphics coprocessor
hexadecimal system (hex)
High-Definition Multimedia Interface (HDMI) port
integrated circuit
laptop
logical operation
mainboard
memory
microprocessor
mini notebook
mini tablet
motherboard
multicore processor
network adapter card
network interface card (NIC)
parallel processing
PCI Express (PCIe)
peripheral
personal computer
port
power supply unit
processor
random-access memory (RAM)
read-only memory (ROM)
SD card
semiconductor
silicon chip
SIM Card
slot
smartphone
smartwatch
socket
system board
system bus
system chassis
system unit
tablet
tablet computer
Thunderbolt 3
tower computer
tower unit
two-in-one laptop
ultrabooks
ultraportables
Unicode
universal serial bus (USB)
universal serial bus (USB) port
Universal serial bus—A (USB-A)
Universal serial bus—B (USB-B)
Universal serial bus—C (USB-C)
virtual memory
wearable computer
wearable device
wireless charging platform
wireless network card
word

MULTIPLE CHOICE

Circle the correct answer.

1. This multiprocessor chip provides two or more separate and independent CPUs.
   - **a.** multicore
   - **b.** mobile
   - **c.** random access
   - **d.** graphics

2. This memory is volatile or loses its contents when power is turned off.
   - **a.** USB
   - **b.** random-access
   - **c.** flash
   - **d.** hex

3. This system board component provides a connection point for specialized cards or circuit boards.
   - **a.** slots
   - **b.** ports
   - **c.** sockets
   - **d.** buses

4. These provide connection points for chips.
   - **a.** microprocessors
   - **b.** sockets
   - **c.** Unicode
   - **d.** ROM

5. A type of memory that improves processing by acting as a temporary high-speed holding area between the memory and the CPU.
   - **a.** chassis
   - **b.** DVI
   - **c.** coprocessor
   - **d.** cache

6. A type of memory that provides a combination of features of RAM and ROM.
   - **a.** BIOS
   - **b.** PCIe
   - **c.** flash
   - **d.** system

7. Similar in appearance to some SD cards, these cards allow a customer's mobile device to access a cell phone network.
   - **a.** ASCII
   - **b.** GPU
   - **c.** NIC
   - **d.** SIM

8. This bus connects the CPU to memory on the system board.
   - **a.** tablet
   - **b.** peripheral
   - **c.** port
   - **d.** system

9. This port can be used to connect many USB devices to the system.
   - **a.** PCIe
   - **b.** USB
   - **c.** ALU
   - **d.** EBCDIC

10. A socket for external devices to connect to the system unit.
    - **a.** bus
    - **b.** chassis
    - **c.** port
    - **d.** tablet

MATCHING

Match each numbered item with the most closely related lettered item. Write your answers in the spaces provided.

a. system unit
b. cell phone
c. system board
d. word
e. chip
f. virtual
g. network
h. bus
i. eSim
j. digital

____ 1. The most popular personal computer.
____ 2. The mainboard or motherboard is also known as _____.
____ 3. This container houses most of the electrical components for a computer system.
____ 4. In a personal computer system, the central processing unit is typically contained on a single _____.
____ 5. This type of memory divides large programs into parts and stores the parts on a secondary storage device.
____ 6. Also known as NIC, this adapter card is used to connect a computer to a(n) _____.
____ 7. This provides a pathway to connect parts of the CPU to each other.
____ 8. The number of bits that can be accessed by the CPU at one time.
____ 9. This Stores a customer's Subscriber Identity Module on a specialized microchip on the cell phone's system board.
____ 10. Computers can only recognize this type of electronic signal.

OPEN-ENDED

On a separate sheet of paper, respond to each question or statement.

1. Describe the five most common types of personal computers.
2. Describe system boards, including sockets, chips, chip carriers, slots, and bus lines.
3. Discuss microprocessor components, chips, and specialty processors.
4. Define computer memory, including RAM, ROM, and flash memory.
5. Define SD cards, expansion slots, cards (including graphics cards), network interface cards, wireless network cards, and SD cards and SIM cards.
6. Describe bus lines, bus width, system bus, and expansion bus.
7. Define ports, including standard and specialized ports. Give examples of each.
8. Describe power supply, including power supply units and AC adapters.
9. Discuss electronic data and instructions.

DISCUSSION

Respond to each of the following questions.

 ### Making IT Work for You: GAMING

Review the Making IT Work for You: Gaming on pages 110 and 111 and then respond to the following: (a) What type of gaming system do you think you would most enjoy: mobile, console, or PC? Why? (b) Do you own any specialized gaming hardware? If you do, describe the equipment. If you do not, do you have any interest in game playing? Why or why not? (c) Look online to learn the differences between the Microsoft, Sony, and Nintendo consoles. Summarize the differences between the consoles. If you were to purchase a console system, which one would you buy? Defend your choice.

 ### Privacy: CRYPTOPROCESSORS

Did you know that some systems have specialty processors that automatically encrypt data before storing? Review the Privacy box on page 115, and respond to the following: (a) Who do you think would need a cryptoprocessor? Be specific. (b) Do you think these processors would be worthwhile for you to protect your privacy? Why or why not? (c) Are there any reasons why you might not want all your data encrypted? Explain. (d) Do you think that all computer systems should be required to have cryptoprocessors? Defend your position.

 ### Ethics: JOB LOSS AND WORKING CONDITIONS

Many computer-related products are produced in other countries, where pay and working conditions are reported to be well below acceptable standards. Review the Ethics box on page 108, and then respond to the following: (a) What do you think about products produced in other countries like China? What are the advantages and disadvantages to consumers? Be specific. (b) What are the ethical issues? Be specific, and defend your list of issues. (c) Would you be willing to pay more for a computer produced entirely in the United States? More specifically, would you be willing to pay three times as much for a cell phone? Why or why not? (d) Do you think consumers have an ethical responsibility to know how goods are produced? More specifically, would your purchase decisions be affected by knowledge about the working conditions of those who make the product? Why or why not?

 ### Community: RECYCLING COMPUTER HARDWARE

Have you ever wondered what you should do with your old computers, monitors, and mobile devices? Review the Community box on page 113, and then respond to the following: (a) What do you typically do with your used or broken computers and mobile devices? (b) What are three alternatives to throwing these devices in the trash? (c) Using a search engine, find one nonprofit organization near you that will accept used computers. List the name and URL. (d) Visit the waste management or recycling page of your local government's website. If it does not have a website, contact it. What is its recommended procedure for discarding your computers and other electronic devices?

Design Elements: Concept Check icon: Dizzle52/Getty Images

chapter 6
Input and Output

Elnur/Shutterstock

CHAPTER 6: Input and Output

Why should I read this chapter?

HQuality/Shutterstock

Input and output devices have seen staggering advances in recent years. A typical cell phone now has dozens of input sensors and display options. In the future, input devices such as tiny attachments to clothes or eyewear will likely be as common as cell phones. Augmented reality displays or wearable glasses will integrate what you see with extensive databases of relevant information.

This chapter covers the things you need to know to be prepared for this ever-changing digital world, including:

- Keyboards—discover how wireless and virtual keyboards might improve the speed and ease of typing.
- Intuitive input devices—learn about advances in touch screens and voice recognition systems.
- Display technology—discover how webcams and digital whiteboards can create videos and share presentations.
- Evolving output devices—learn about advances in UHDTVs, 3D printers, and e-books.

Learning Objectives

After you have read this chapter, you should be able to:

1. Define input.
2. Describe keyboard including types and features of keyboards.
3. Identify different pointing devices, including touch screens, game controllers, and styluses.
4. Describe scanning devices, including optical scanners, RFID readers, and recognition devices.
5. Recognize image-capturing and audio-input devices.
6. Define output.
7. Identify different monitor features and types, including different monitor types and e-books.
8. Define printing features and types, including inkjet, 3D, and cloud printers.
9. Recognize different audio-output devices, including headphones and headsets.
10. Define combination input and output devices, including multifunctional devices, VR head-mounted displays and controllers, drones, and robots.
11. Explain ergonomics and ways to minimize injuries.

Introduction

"Hi, I'm James, and I'm a technical writer. I'd like to talk with you about input and output devices . . . all those devices that help us to communicate with a computer. I'd also like to talk about emerging technologies such as virtual reality."

Dragon Images/Shutterstock

It is hard to imagine anything more human than creating and enjoying music. It is a little surprising that something as inhuman as computers has become so instrumental in creating and listening to music. How can a collection of wires and microchips capture and reproduce the human art of music?

The process of recording and playing music on a computer relies on microphones to input the sounds into the computer and speakers to output the sounds out of the computer. These are just some of the inputs and outputs that allow humans to interact with computers. You need to understand what input and output devices are and how they are used to work in an office, socialize online, or even buy groceries.

To efficiently and effectively use computers, you need to know about the most commonly used input devices, including keyboards, pointing devices, touch screens, game controllers, scanners, digital cameras, voice recognition, and audio-input devices. Additionally, you need to know about the most commonly used output devices, including monitors, printers, and audio and video output devices. You also need to be aware of combination input and output devices such as multifunctional devices, virtual reality devices, drones, and robots.

What Is Input?

Figure 6-1 Virtual keyboard
Tero Vesalainen/Shutterstock

Input is any data or instructions entered into a computer. It can come directly from you or from other sources. You provide input whenever you use system or application programs. For example, when sending a text message from a cell phone, you enter data in the form of numbers and letters and issue commands such as *send message* or *attach photo*. You also can enter data and issue commands by pointing to items or using your voice. Other sources of input include scanned or photographed images.

Input devices are hardware used to translate words, numbers, sounds, images, and gestures that people understand into a form that the system unit can process. For example, when writing an e-mail, you typically use a keyboard to enter text and a mouse to issue commands. In addition to keyboards and mice, there are a wide variety of other input devices. These include pointing, scanning, image-capturing, and audio-input devices.

Keyboard Entry

Figure 6-2 Laptop keyboard
Alexey Boldin/Shutterstock

When you type into your cell phone or laptop, you are using one of the most common input devices, a **keyboard**. As mentioned in **Chapter 5**, keyboards convert numbers, letters, and special characters that people understand into electrical signals. These signals are sent to, and processed by, the system unit. Most keyboards use an arrangement of keys given the name QWERTY. This name reflects the keyboard layout by taking the letters of the first six alphabetic characters found on the top row of keys displaying letters.

tips

Do you wish your virtual keyboard had the feel and sound of a physical keyboard? Look for these features in your cell phone or tablet's settings to give your virtual keyboard a more physical presence.

1. **The sound of a keyboard**—Cell phones can mimic the clicking noise of a physical keyboard by playing a clicking sound when you type on the virtual keyboard.
   - On an iOS device, turn on the system setting "Keyboard Clicks."
   - On an Android device, turn on the system setting "Touch sounds."

2. **The feel of a keyboard**—Cell phones can mimic the sense of pressing a physical key through a technology called haptics. Haptics is the use of sound and vibration to simulate a physical sensation.
   - On an iOS device, turn on the system setting "System Haptics."
   - On an Android device, turn on the system setting "Vibrate on touch."

Keyboards

There are a wide variety of keyboard designs, from small cell phone keyboards designed for short messages to desktop keyboards designed for longer computing sessions. There are three basic categories of keyboards: virtual, laptop, and traditional.

- **Virtual keyboards**—these keyboards are used primarily with cell phones and tablets. Unlike other keyboards, virtual keyboards do not have physical keys. Rather, the keys are displayed on a screen and selected by touching their image on the screen. (See **Figure 6-1**.)

- **Laptop keyboards**—these keyboards are attached to the laptop system unit and designed to fold up with the laptop monitor to easily fit in a backpack or briefcase. Many high-end tablets offer laptop-like features by including a laptop keyboard that can be removed from the tablet or double as a protective case for the tablet. Laptop keyboards come in a variety of configurations, depending on the manufacturer and the size of the laptop. These keyboards include all the keys found on a typical virtual keyboard, as well as extra keys, such as function and navigation keys. (See **Figure 6-2**.)

- **Traditional keyboards**—these full-size keyboards are widely used on desktops and larger computers. The standard U.S. traditional keyboard has 101 keys, including extra keys, such as function keys, navigation keys, and a numeric keypad. Some traditional keyboards include a few additional special keys. For example, the Windows keyboard includes a Windows key to directly access the Start menu. Some keys, such as the Caps Lock key, are **toggle keys**. These keys turn a feature on or off. Others, such as the Ctrl key, are **combination keys**, which perform an action when held down in combination with another key. (See **Figure 6-3**.)

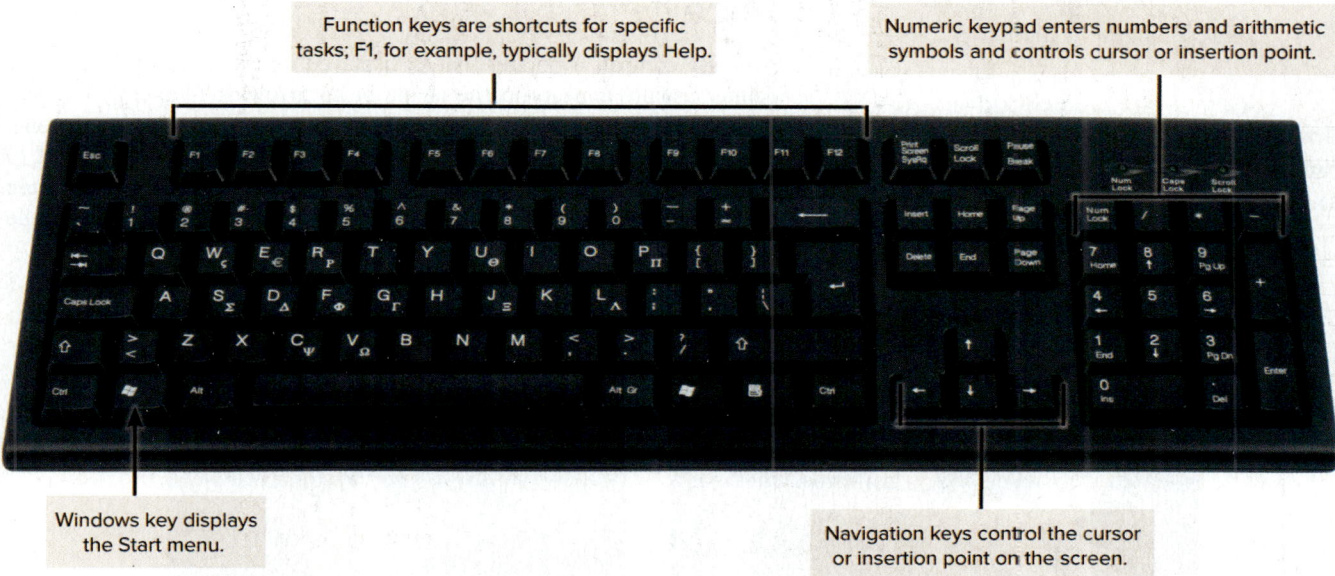

Figure 6-3   Traditional keyboard
Georgios Kollidas/Shutterstock

concept check

- What is input? What are input devices?
- List and compare the three categories of keyboards.
- What are toggle keys? What are combination keys?

Pointing Devices

Pointing is one of the most natural of all human gestures. **Pointing devices** provide an intuitive interface with the system unit by accepting physical movements or gestures such as a finger pointing or moving across a screen and converting these movements into machine-readable input. There are a wide variety of pointing devices, including the touch screen, mouse, and game controller.

Touch Screens

A **touch screen** allows users to select actions or commands by touching the screen with a finger or stylus. A **stylus** is a penlike device typically used with tablets and mobile devices. (See **Figure 6-4**.) Often, a stylus interacts with the computer through handwriting recognition software. **Handwriting recognition software** translates handwritten notes into a form that the system unit can process.

Figure 6-4   Stylus
ilkercelik/E+/Getty Images

Figure 6-5   Multitouch screen
George Dolgikh/Shutterstock

Multitouch screens can be touched with more than one finger, which allows for interactions such as rotating graphical objects on the screen with your hand or zooming in and out by pinching and stretching your fingers. Multitouch screens are commonly used with cell phones, tablets, and laptops, as well as some desktops. (See **Figure 6-5**.)

Mice

A **mouse** controls a pointer that is displayed on the monitor. The **mouse pointer** usually appears in the shape of an arrow. It frequently changes shape, however, depending on the application. A mouse can have one, two, or more buttons, which are used to select command options and to control the mouse pointer on the monitor. Some mice have a **wheel button** that can be rotated to scroll through information that is displayed on the monitor. (See **Figure 6-6**.)

Figure 6-6   Cordless mouse with wheel button
iHumnoi/Shutterstock

Figure 6-7   Touch pad
Naked King/iStock/Getty Images Plus

community

While the standard input devices work well for many of us, not everyone has the same physical or motor control abilities. Engineers and scientists have developed different pointing devices, keyboards, and even gamepads to make sure that everyone in our communities can write e-mails, surf the web, and play video games.

These devices vary. Some are designed to be used without using your hands; others are controlled by your mouth or eyes. Others make it easier to use your hands with limited mobility, such as gamepads and joysticks with large programmable buttons.

Have you used an input device designed for improved accessibility, or seen one used? How can accessible input devices improve someone's quality of life and connection with their community?

Some mice connect to the system board through a cord, while a **cordless** or **wireless mouse** uses radio waves or infrared light waves to communicate with the system unit. These devices require a battery that must be recharged or replaced. Alternatively, USB mice connect directly to the system unit through a USB cord and do not require a battery.

Like a mouse, a **touch pad** is used to control the mouse pointer and to make selections. Unlike a mouse, however, a touch pad operates by moving or tapping your finger on the surface of a pad. These devices are widely used instead of a mouse with laptops and some types of mobile devices. (See **Figure 6-7**.)

Game Controllers

Game controllers are devices that provide input to computer games. While keyboards and traditional mice can be used as game controllers, the four most popular and specialized game controllers are joysticks, gaming mice, gamepads, and motion-sensing devices. (See **Figure 6-8**.)

- **Joysticks** control game actions by users varying the pressure, speed, and direction of a control stick.
- **Gaming mice** are similar to traditional mice with higher precision, faster responsiveness, and programmable buttons.
- **Gamepads** are designed to be held by two hands and provide a wide array of inputs, including motion, turning, stopping, and firing.
- **Motion-sensing devices** control games by user movements. For example, Nintendo's Joy-Con controller accepts user movements to control games on the Nintendo Switch.

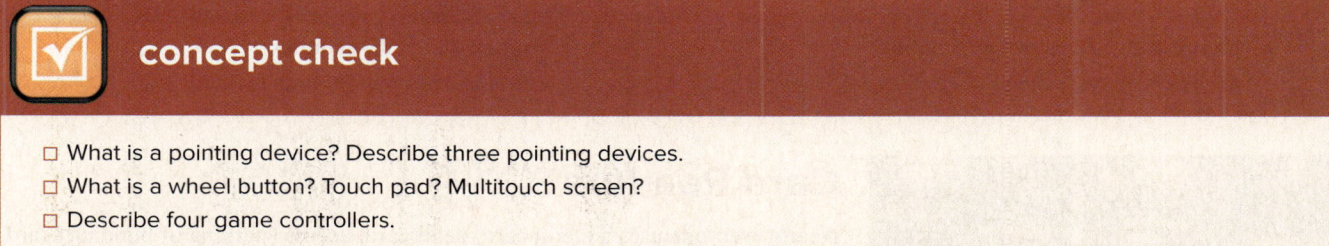

			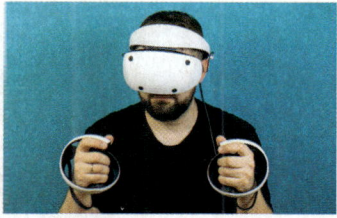
Joystick	Gaming mouse	Gamepad	Motion-sensing device

Figure 6-8   Game controllers
Boltenkoff/Shutterstock; Dario Lo Presti/Shutterstock; Aldeca Productions/Shutterstock; ESOlex/Shutterstock

concept check

- What is a pointing device? Describe three pointing devices.
- What is a wheel button? Touch pad? Multitouch screen?
- Describe four game controllers.

Scanning Devices

Scanning devices convert scanned text and images into a form that the system unit can process. There are five types of scanning devices: optical scanners, card readers, bar code readers, RFID readers, and character and mark recognition devices.

Optical Scanners

An **optical scanner**, also known simply as a **scanner**, accepts documents consisting of text and/or images and converts them to machine-readable form. These devices do not recognize individual letters or images. Rather, they recognize light, dark, and colored areas that make up individual letters or images. Typically, scanned documents are saved in files that can be further processed, displayed, printed, or stored for later use.

There are four basic types of optical scanners: flatbed, document, portable, and 3D.

- A **flatbed scanner** is much like a copy machine. The image to be scanned is placed on a glass surface, and the scanner records the image.
- A **document scanner** is similar to a flatbed scanner except that it can quickly scan multipage documents. It automatically feeds one page of a document at a time through a scanning surface. (See **Figure 6-9**.)
- A **portable scanner** is typically a handheld device that slides across the image, making direct contact.
- **3D scanners** use lasers, cameras, or robotic arms to record the shape of an object. Unlike 2D scanners, most 3D scanners cannot recognize light, dark, and colored areas. Instead, 3D scanners recognize the shape of the object they are scanning. (See **Figure 6-10**.)

privacy

Audio inputs have become incredibly common in our devices. With the increase in smart devices, it can be easy to overlook all the devices that listen to us for voice commands. While it is well recognized that cell phones have microphones, many would be surprised to learn that smart thermostats and smart refrigerators include microphones. One of the ways that our devices improve their voice recognition is to always be listening and recording our conversations and to send these conversations to researchers and scientists for analysis and study. Privacy advocates are concerned that people who own smart devices may not realize they are being recorded and studied. Companies and researchers point out that for a device to respond to our voice, it must be listening to us and that privacy settings allow users to customize how much of their information they want to share. How many microphones are in your home? Are you concerned about how your conversations might be used by the devices that listen to you?

Optical scanners are powerful tools for a wide variety of end users, including graphics and advertising professionals who scan images and combine them with text.

Figure 6-9   Document scanner
Thomas Northcut/Photodisc/Getty Images

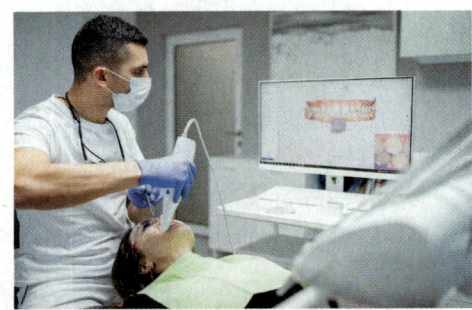

Figure 6-10   3D scanner
ArtistGNDphotography/Getty Images

Figure 6-11   Contactless payment
Igor Kardasov/Alamy Stock Photo

Card Readers

Nearly everyone uses a credit card, debit card, access (parking or building) card, and/or some type of identification card. These cards typically have the user's name, some type of identification number, and signature on the card. Additionally, encoded information is often stored on the card. **Card readers** interpret this encoded information.

Although there are several different types, by far the most common is the **magnetic card reader**. The encoded information is stored on a thin magnetic strip located on the back of the card. When the card is swiped through the magnetic card reader, the information is read. Many credit cards, known as **chip cards**, include additional security in the form of a microchip embedded in the credit card. This chip contains encrypted data that makes it nearly impossible for criminals to forge a duplicate card. Some chips require that you insert the card into a specialized reader, while others feature contactless payment, and only require that you hold the card near the reader. (See **Figure 6-11**.)

Figure 6-12   Cell phone reading a bar code
Aunging/Shutterstock

Bar Code Readers

You are probably familiar with **bar code readers** or **scanners** from grocery stores. These devices are either handheld **wand readers** or **platform scanners**. They contain photoelectric cells that scan or read **bar codes**, or the black and white patterns printed on product containers. Bar codes can be found in many places, from stickers on fruit to labels on prescription medicine bottles.

There are a variety of different codes, including UPC and QR codes.

- **UPCs (Universal Product Codes)** are a one-dimensional bar code made up of vertical zebra strip marks that uniquely identify a product. These identifying codes are used to automate the process to check out customers and maintain inventory records.

- **QR Codes** are a two-dimensional bar code typically made up of a grid of black and white squares. The 2D nature of QR codes means that they can be read from any angle and contain more information than a UPC. One of the most widely used types of QR Codes is the **MaxiCode**, a QR Code used by the United Parcel Service (UPS) and others to automate the process of routing and tracking packages.

Cell phones with the appropriate app can also scan codes. (See **Figure 6-12**.) For example, after scanning the bar code from a product you are thinking of buying, the app Price Check by Amazon will provide in-store and online price comparisons as well as other customer product reviews.

RFID Readers

RFID (radio-frequency identification) tags are tiny chips that can be embedded in almost everything. They can be found in consumer products, driver's licenses, passports, and any number of other items. (See **Figure 6-13**.) These chips contain electronically stored information that can be read using an **RFID reader** located several yards away and several tags can be read simultaneously. This makes RFID tags ideal for identifying pets with RFID tags embedded under the skin and for tracking multiple items quickly—such as a shopping bag full of clothes.

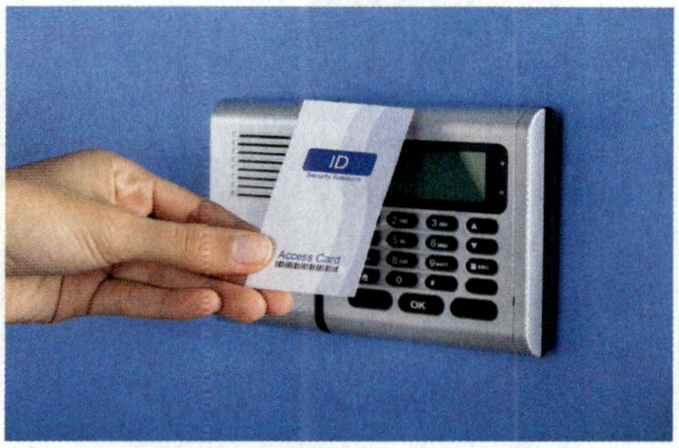

Character and Mark Recognition Devices

Character and mark recognition devices are scanners that are able to recognize special characters and marks. They are specialty devices that are essential tools for certain applications. Three types are

- **Magnetic-ink character recognition (MICR)**—used by banks to automatically read those unusual numbers on the bottom of checks and deposit slips. A special-purpose machine known as a reader/sorter reads these numbers and provides input that allows banks to efficiently maintain customer account balances.
- **Optical-character recognition (OCR)**—uses special preprinted characters that can be read by a light source and changed into machine-readable code. A common OCR device is the handheld wand reader. (See **Figure 6-14**.) These are used in department stores to read retail price tags by reflecting light on the printed characters.
- **Optical-mark recognition (OMR)**—senses the presence or absence of a mark, such as a pencil mark. OMR is often used to score standardized multiple-choice tests.

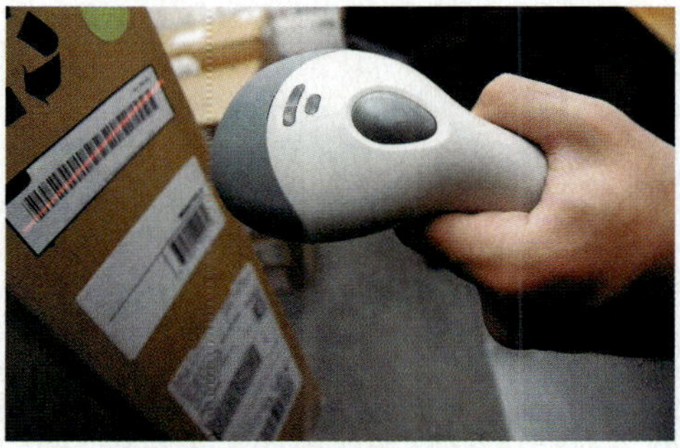

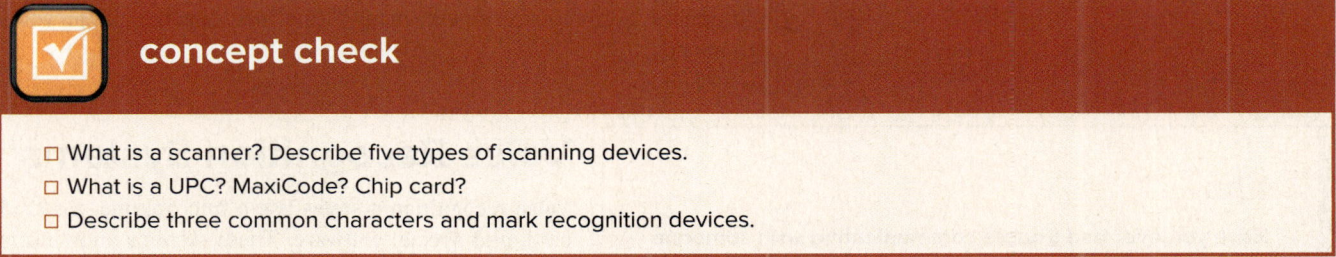

Figure 6-13 RFID reader
Andrey_Popov/Shutterstock

Figure 6-14 Wand reader
Anatoly Vartanov/123RF

concept check

☐ What is a scanner? Describe five types of scanning devices.
☐ What is a UPC? MaxiCode? Chip card?
☐ Describe three common characters and mark recognition devices.

Image-Capturing Devices

Optical scanners, like traditional copy machines, can make a copy from an original. For example, an optical scanner can make a digital copy of a photograph. *Image-capturing devices,* on the other hand, create or capture original images. These devices include digital cameras and webcams.

Digital Cameras

Digital cameras capture images digitally and store the images on a memory card or in the camera's memory. Most digital cameras are also able to record video as well. (See **Figure 6-15**.) Originally, digital cameras were only dedicated devices. Today, many digital cameras are embedded in other devices, such as cell phones and tablets. Digital cameras provide a fast and easy way to create photos to share over social media and e-mail.

Figure 6-15 Digital camera
caliNN and DiaNNA/Shutterstock

Figure 6-16 Attached webcam
Erlon Silva - TRI Digital/Moment/Getty Images

ethics

You may have heard of instances where webcams were used to broadcast the activities of individuals who did not know they were being recorded. For example, in a famous court case, a university student was prosecuted for using a webcam on an open laptop to secretly record his roommate's intimate activities. In other situations, public webcams have recorded embarrassing footage of people who were not aware of the camera. It has been argued that capturing and then broadcasting a person's image without his or her knowledge and consent is unethical. What do you think?

Webcams

Webcams are specialized digital video cameras that capture images and send them to a computer for broadcast over the Internet. Webcams are built into most cell phones and tablets. Desktop and laptop webcams are either built-in or attached to the computer's monitor. (See **Figure 6-16**.) Popular videoconferencing apps, such as Zoom, Microsoft Teams, and Apple's FaceTime, use webcams to allow you to communicate in real time with friends and family using live video.

Audio-Input Devices

Audio-input devices convert sounds into a form that can be processed by the system unit. By far the most widely used audio-input device is the microphone. Audio input can take many forms, including the human voice and music.

tips

Have you ever had trouble communicating with someone who does not speak English? If so, Google Translate may be just what you need.

1. Go to translate.google.com.
2. Using the buttons at the top, select the language you will be speaking, followed by the language you want your words translated to.
3. Click the microphone icon in the box on the left, and begin speaking clearly into your microphone. In a few seconds, you will see the translated text in the box on the right.
4. Click the speaker icon in the box on the right to hear the translation.

Note: While Google Translate is always improving, it is not perfect. For perfect translations, consult with a native speaker.

Voice Recognition Systems

Voice recognition systems use a microphone, a sound card, and special software. These systems allow users to operate computers and other devices, as well as to create documents, using voice commands. As discussed in **Chapter 4**, most cell phones include a virtual assistant that uses voice recognition to accept voice commands to control operations. Apple devices come with Siri, Windows devices come with Cortana, Amazon devices come with Alexa, and Android devices come with Google Assistant. There are even devices on the market that exclusively offer access to virtual assistants with voice recognition, such as Amazon's Echo, Apple's HomePod, and Google's Home. These voice recognition systems can perform any number of operations, including scheduling events on your calendar, composing simple text messages, and looking up facts on the web. Specialized portable voice recorders are widely used by doctors, lawyers, and others to record dictation. These devices are able to record for several hours before connecting to a computer running voice recognition software to edit, store, and print the dictated information. Some systems are even able to translate dictation from one language to another, such as from English to Japanese.

concept check

☐ How are image-capturing devices different from an optical scanner?
☐ Describe two image-capturing devices.
☐ What are voice recognition systems? Siri? Cortana? Google Assistant?

What Is Output?

Output is processed data or information, typically taking the form of text, graphics, photos, audio, and/or video. For example, when you create a presentation using a presentation graphics program, you typically input text and graphics. You also could include photographs, voice narration, and even video. The output would be the completed presentation.

Output devices are any hardware used to provide or to create output. They translate information that has been processed by the system unit into a form that humans can understand. There are a wide range of output devices. The most widely used are monitors, printers, and audio-output devices.

Monitors

The most frequently used output device is the **monitor**. Also known as **display screens**, monitors present visual images of text and graphics. Monitors vary in size, shape, and cost. Almost all, however, have some basic distinguishing features.

Features

The most important characteristic of a monitor is its clarity. **Clarity** refers to the quality and sharpness of the displayed images. It is a function of several monitor features, including resolution, dot pitch, contrast ratio, active display area, and aspect ratio.

- **Resolution** is one of the most important features. Images are formed on a monitor by a series of dots, or **pixels (picture elements)**. (See **Figure 6-17**.) Resolution is expressed as a grid of these dots or pixels. For example, many monitors today have a resolution of 1,920 pixel columns by 1,080 pixel rows for a total of 2,073,600 pixels. The higher a monitor's resolution (the more pixels), the clearer the image produced. See **Figure 6-18** for the most common monitor resolutions.
- **Refresh rate** is a measure of how quickly new images are displayed on the monitor. The higher the refresh rate, the more smooth the action on the monitor will appear. Refresh rate is measured in hertz (Hz), which is the number of times per second the monitor can display a new image. The most common refresh rate for monitors is 60 Hz, which means a new image can be displayed on the monitor 60 times in one second. This refresh rate is fine for most computer usage, such as browsing the web or working on documents. However, gaming computers often have refresh rates more than double that rate. A common gaming computer refresh rate is 144 Hz, which displays 144 images on the screen every second. This faster refresh rate will result in less motion blur when images move quickly across the screen.

Figure 6-17 Monitor resolution
tarczas/Shutterstock

Standard	Pixels
HD 720	1,280 × 720
HD 1080	1,920 × 1,080
WQXGA	2,560 × 1,600
UHD 4K	3,840 × 2,160
UHD 8K	8,192 × 4,608

Figure 6-18 Resolution standards

- **Dot (pixel) pitch** is the distance between each pixel. The ideal dot pitch depends on how you use your monitor. Most newer desktop monitors have a dot pitch below 0.30 mm (30/100th of a millimeter). Cell phones, which are designed to be viewed more closely than desktop monitors, can have dot pitches below 0.05 mm (5/1,000th of a millimeter). The lower the dot pitch (the shorter the distance between pixels), the clearer the images.
- **Contrast ratios** indicate a monitor's ability to display images. It compares the light intensity of the brightest white to the darkest black. The higher the ratio, the better the monitor. Good monitors typically have contrast ratios above 1,000:1.
- **Active display area,** or size, is measured by the diagonal length of a monitor's viewing area. Laptop monitors are commonly between 13 and 17 inches, and desktop monitors are commonly larger between 19 and 34 inches.
- **Aspect ratio** indicates the proportional relationship between a display's width and height. Typically, this relationship is expressed by two numbers separated by a colon (:). Many older, more square-shaped monitors have a 4:3 aspect ratio. Almost all newer monitors have a 16:9 aspect ratio designed to display widescreen content.

tips

Do your photos and videos look different depending on what screen you use? It may be due to different video settings. TV and computer monitors come from the factory with standard settings. To get the best performance from a monitor, you should calibrate the display.

1 Before you calibrate. Many factors can change a monitor's response to calibration. To ensure that you have a proper baseline before fine-tuning your device, consider these factors before calibration:

- Warm up the TV or monitor. Allow the display to warm up for half an hour before calibration.
- Set the display resolution to its default (or native) resolution.
- Calibrate in a room with ambient light. Direct light on the screen can make it difficult to accurately assess your display's picture quality.

2 Select your picture mode. Found within the display or monitor's settings, a picture mode is a collection of preset profiles that adjust the display to work well with different types of content. With categories like cinema, sports, games, and dynamic, these preset configurations change display settings, including color temperature, brightness, and contrast, to customize the display to the viewer's preference. For true color representation, look for THX or ISF modes; failing that, use cinema or theater mode.

3 Use software tools. To truly maximize your display for its surrounding environment, you will need to fine-tune its calibration. For computer monitors, both Windows and macOS come with free monitor calibration tools in the control panel or system preferences of your computer. Many TV manufacturers offer free setup tools as well—check the manual and software that came with your TV. Finally, you can purchase calibration tools, such as the THX Tune-up app from THX.

Another important monitor feature is the ability to accept touch or gesture input such as finger movements, including swiping, sliding, and pinching. Although most older monitors do not support touch input, it is becoming a standard feature of newer monitors. For a guide on purchasing a monitor, see **the Computer Buyer's Guide** in the Appendix.

Flat-Panel Monitors

Flat-panel monitors are the most widely used type of monitor today. You will find flat-panel monitors providing output on laptops, desktops, cellphones, and tablets. (See **Figure 6-19**.)

Almost all flat-panel displays are backlit, meaning that a common source of light is dispersed over all the pixels on the screen. There are three basic types of flat-panel monitors: LCD, LED, and OLED.

- **LCD (liquid crystal display)** is widely used for older monitors and is typically less expensive. (See **Figure 6-19**.)
- **LED (light-emitting diode)** monitors use similar technology with a more advanced backlighting technology. They produce better-quality images, are slimmer, and are more environmentally friendly as they require less power and use fewer toxic chemicals to manufacture. Most new monitors are LED.
- **OLED (organic light-emitting diode)** monitors replace the LED monitor's backlighting technology with a thin layer of organic compound that produces light. By eliminating the backlight, OLED monitors can be even thinner with better power efficiency and contrast ratios.

E-book Readers

E-books (electronic books) are traditional printed books in electronic format. These books are available from numerous sources, including many public and private libraries, bookstore websites, and the cloud. **E-book readers (e-readers)** are dedicated mobile devices for storing and displaying e-books and other electronic media, including electronic newspapers and magazines.

E-book readers have displays that are typically 6 inches and use a technology known as e-ink. **E-ink** produces images that reflect light like ordinary paper, making the display easy to read. Two well-known e-book readers are Amazon's Kindle and Walmart's Kobo eReaders (See **Figure 6-20**.)

Figure 6-19   **LCD monitor**
MishAl/Shutterstock

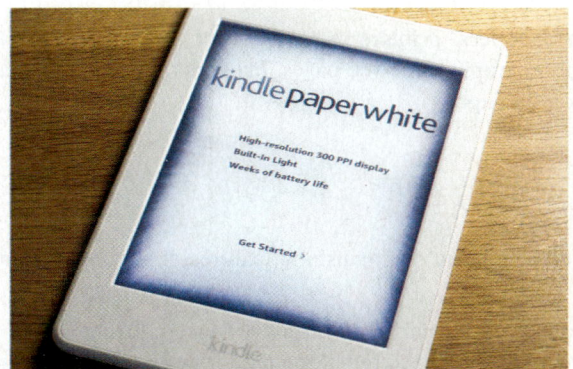

Figure 6-20   **E-book reader**
Stephen Frost/Alamy Stock Photo

Figure 6-21   **Digital whiteboard**
Gorodenkoff/Shutterstock

Other Monitors

There are several other types of monitors. Some are used for more specialized applications, such as making presentations and watching television.

- **Digital** or **interactive whiteboards** are specialized devices with a large display connected to a computer or projector. The computer's desktop is displayed on the digital whiteboard and controlled using a special pen, a finger, or some other device. Digital whiteboards are widely used in classrooms and corporate boardrooms. (See **Figure 6-21**.)
- **Flexible screens** allow digital devices to display images on nonflat surfaces. Early uses of flexible screens included cell phones with screens that wrap around the edges of the phone and curved monitors whose edges wrap toward the viewer. Recent innovations include foldable screens, such as the Samsung Galaxy Fold, a cell phone with a screen that unfolds to become a tablet. (See **Figure 6-22**.)
- **Digital projectors** project the images from a traditional monitor onto a screen or wall. This is ideal for presentations or meetings when several people need to see the screen at the same time. These projectors can be as small as a coffee cup and can project images larger than a standard computer monitor. Unfortunately, projected images can be difficult to see in bright rooms, so they are best used in dark rooms with curtains or no windows. (See **Figure 6-23**.)

Figure 6-22   **Flexible screen**
Angel Garcia/Bloomberg via Getty Images

Figure 6-23   **Digital projector**
SaveLight Studio/Shutterstock

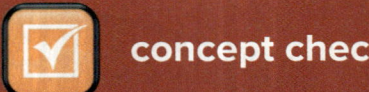

concept check

☐ Define these monitor features: clarity, resolution, dot pitch, contrast ratios, size, and aspect ratio.
☐ Describe LCD, LED, and OLED monitors.
☐ What are e-book readers, digital whiteboards, flexible screens, and digital projectors?

Printers

Even as many individuals, schools, and businesses are trying to go paperless, printers remain one of the most-used output devices. You probably use a printer to print homework assignments, photographs, and web pages. **Printers** translate information that has been processed by the system unit and present the information on paper.

Features

There are many different types of printers. Almost all, however, have some basic distinguishing features, including resolution, color capability, speed, memory, duplex printing, and connectivity.

- **Resolution** for a printer is similar to monitor resolution. It is a measure of the clarity of images produced. Printer resolution, however, is measured in **dpi (dots per inch)**. (See **Figure 6-24**.) Most printers designed for personal use average 1,200 by 4,800 dpi. The higher the dpi, the better the quality of images produced.

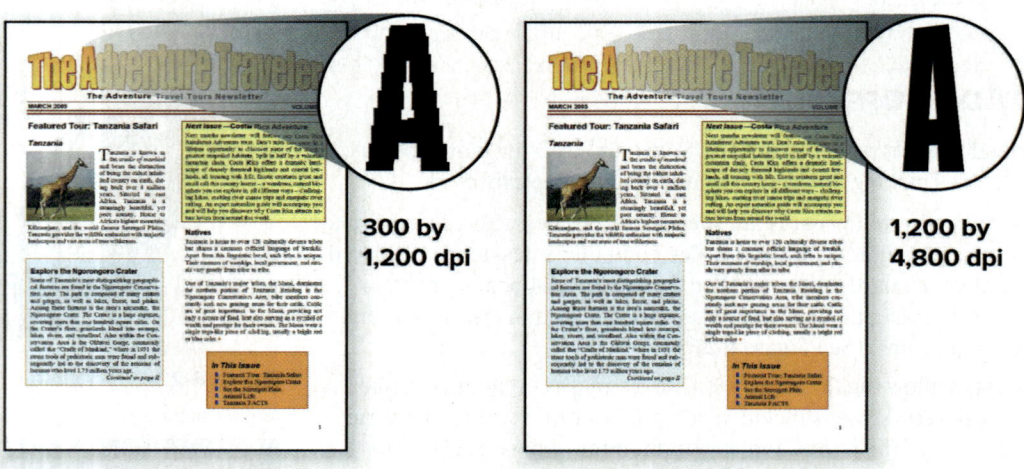

Figure 6-24   **DPI comparison**

- **Color capability** is provided by most printers today. Users typically have the option to print either with just black ink or with color. Because it is more expensive to print in color, most users select black ink for letters, drafts, and homework. The most common black ink selection is **grayscale**, in which images are displayed using many shades of gray. Color is used more selectively for final reports containing graphics and for photographs.
- **Speed** is measured in the number of pages printed per minute. Typically, printers for personal use average 15–19 pages per minute for single-color (black) output and 5–12 pages per minute for color output.
- **Memory** within a printer is used to store printing instructions and documents waiting to be printed. The more memory, the faster it will be able to print large documents.
- **Duplex printing** allows automatic printing on both sides of a sheet of paper. Although not currently a standard feature for all printers, it will likely become standard in the future as a way to reduce paper waste and to protect the environment.
- **Connectivity** is the ability of the printer to connect to a network, eliminating the need for a computer to be attached by a cable to the printer and making it easier for multiple computers to share one printer. Many printers include the ability to connect to a network over Wi-Fi or Ethernet.

Inkjet Printers

Inkjet printers spray ink at high speed onto the surface of paper. This process produces high-quality images in a variety of colors, making it ideal for printing photos. (See **Figure 6-25**.) Inkjet printers are relatively inexpensive and are the most widely used printers. In addition, they are reliable and quiet. The most costly aspect of inkjet printers is replacing the ink cartridges. For this reason, most users specify black ink for the majority of print jobs and use the more expensive color printing for select applications. Typical inkjet printers produce 15–19 pages per minute of black-only output and 5–12 pages of color output.

Laser Printers

The **laser printer** uses a technology similar to that used in a photocopying machine. Laser printers use a laser light beam to produce images with excellent letter and graphics quality. More expensive than inkjet printers, laser printers are faster, cost less per page, and have a very high dpi.

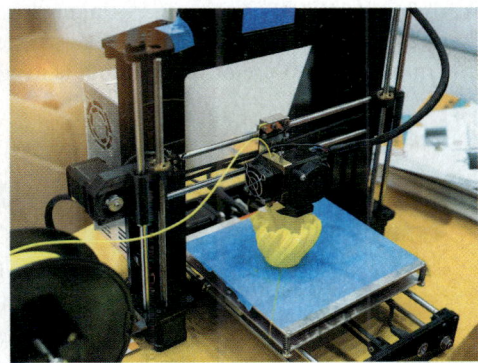

Figure 6-25 Inkjet printer
Hugh Threlfall/Alamy Stock Photo

There are two categories of laser printers. **Personal laser printers** are less expensive and are used by a single user. They typically can print 15–17 pages a minute. **Shared laser printers** typically support color, are more expensive, and are used (shared) by a group of users. Shared laser printers typically print over 50 pages a minute.

3D Printers

3D printers, also known as **additive manufacturing**, create three-dimensional shapes by adding very thin layer after layer of material until the final shape is fully formed. (See **Figure 6-26**.) A variety of different processes and materials can be used to create each layer. One of the most common extrudes is a liquid plastic or other substance through a nozzle similar to an inkjet printer.

3D printers are controlled by data describing the shape of the object to be created. This data typically comes from a file created by a 3D modeling program or from scanning a physical model using a 3D scanner. Specialized programs then take this data and further process it to create data describing hundreds or thousands of horizontal layers that, when placed one on top of another, form the shape of the intended object. The printer uses this data by extruding a foundation layer to very exact specifications. Successive layers are then created and attached to the layer below it until the product is finalized. The layers are so thin and so precise that they blend together, making a final product that shows no sign of the individual layers.

Figure 6-26 3D printer
asharkyu/Shutterstock

Commercial 3D printers have been used for decades. Their cost, however, limited them to specialized manufacturing and research applications. For example, auto manufacturer Volkswagen uses 3D printers in its production lines to produce gear knobs and custom tailgate lettering. Recently, the cost of 3D printers has dropped to as low as $200, making them available to individuals.

Other Printers

There are several other types of printers. These include thermal printers, and plotters:

- **Thermal printers** use heat elements to produce images on heat-sensitive paper. These printers are widely used with ATMs and gasoline pumps to print receipts.
- **Plotters** are special-purpose printers for producing a wide range of specialized output. Using output from graphics tablets and other graphical input devices, plotters create maps, images, and architectural and engineering drawings. Plotters are typically used by graphic artists, engineers, and architects to print out designs, sketches, and drawings.

concept check

☐ Discuss these printer features: resolution, color capability, speed, memory, connectivity, and duplex printing.
☐ Compare inkjet, laser, and 3D printers.
☐ Discuss cloud, thermal, and plotter printers.

Audio-Output Devices

Figure 6-27   Headphones
Alexander Demyanenko/Shutterstock

Audio-output devices translate audio information from the computer into sounds that people can understand. The most widely used audio-output devices are **speakers** and **headphones**. (See **Figure 6-27**.) These devices connect to a sound card within the system unit. This connection can be by cable to an audio jack on the system unit, or the connection can be wireless. Wireless connections typically use **Bluetooth** technology. This type of connection requires special Bluetooth-enabled speakers and/or headphones. Bluetooth will be discussed further in **Chapter 8**. The sound card is used to capture as well as play back recorded sounds. Audio-output devices are used to play music, vocalize translations from one language to another, and communicate information from the computer system to users. To learn more about headphones, see the Making IT Work for You: Headphones on page 148.

Creating voice output is not anywhere near as difficult as recognizing and interpreting voice input. In fact, voice output is quite common. Some of the most common applications and devices that offer voice output are cell phone virtual assistants, car GPS directions, and translation software dialogue. Perhaps the most powerful use of voice output is its ability to assist people with disabilities with reading and communication.

Combination Input and Output Devices

Figure 6-28   Headset
Third of november/Shutterstock

Many devices combine input and output capabilities. Sometimes this is done to save space. Other times it is done for very specialized applications. Common combination devices include headsets, multifunctional devices, virtual reality head-mounted displays and controllers, drones, and robots.

Headsets

Headsets combine the functionality of microphones and headphones. The microphone accepts audible input, and headphones provide audio output. Headsets are an integral part of multiplayer video game systems. (See **Figure 6-28**.)

Multifunctional Devices

Multifunctional devices (MFD) typically combine the capabilities of a scanner, printer, fax, and copy machine. These multifunctional devices offer a cost and space advantage. Compared to purchasing a scanner, printer, fax, and copy machine individually, the MFD is less expensive and takes up less space, but the MFD is also more complex and more likely to break down. Further, a device built for a single purpose (such as a scanner) is more likely to have professional-level features and perform faster than an MFD. The output quality for any one function is often not quite as good as that of the separate single-purpose devices. The reliability of multifunctional devices suffers because problems with one of the functional parts can make the entire device inoperable. Even so, multifunctional devices are widely used in home and small business offices.

Virtual Reality Head-Mounted Displays and Controllers

Virtual reality (VR) is an artificial, or simulated, reality created in 3D by computers. It strives to create a virtual or **immersive experience** by using specialized hardware that includes a head-mounted display and controller. (See **Figure 6-29**.)

- **VR head-mounted displays** have earphones for immersive sound, stereoscopic screens to present 3D images, and gyroscopic sensors to interpret head orientation.
- **VR controllers** have sensors that collect data about your hand movements. Coupled with software, this interactive sensory equipment lets you immerse yourself in a computer-generated world.

Figure 6-29 Virtual reality head-mounted display and controllers
Gorodenkoff/Shutterstock

There are any number of applications for virtual reality head-mounted displays and controllers. Automobile manufacturers use virtual reality to evaluate what it would be like to sit in the cars they have designed but not yet built. Recently released virtual reality hardware for gaming consoles and high-end PCs look to revolutionize the way we work and play. Virtual reality promises to become commonplace in the near future.

Drones

Drones, or **unmanned aerial vehicles (UAVs)**, were once too expensive for anything but military budgets. However, today's drones are inexpensive, faster, and smarter, making them a valuable tool and fun high-tech toy. (See **Figure 6-30**.) Most drones take input from a controller from either a radio joystick or a Wi-Fi-connected tablet or laptop. The drones act as an output device, sending back video and sound to the user. The resulting combination of video and aerial maneuverability has made drones a popular choice for a wide variety of activities ranging from amateur cinematography to civil engineering. Drone use is on the rise by all types of users. Drones are used in agriculture to apply pesticides and monitor livestock, in disaster relief to look for people in distress, and by the post office to deliver mail. Amazon Prime's drone delivery service offers delivery of packages less than five pounds in less than one hour. This service is being tested in limited areas but may be available in your town soon.

Figure 6-30 Drone
aerogondo/iStock/Getty Images Plus

Robots

Like drones, **robots** have become relatively inexpensive with expanding capabilities. Robots use cameras, microphones, and other sensors as inputs. Based on these inputs, robotic outputs can be as complex as exploring damaged nuclear reactors to as simple as taking a photo. Recent improvements in robotics software and decreases in hardware prices have made robots more prevalent in industrial and hobbyist workshops. Robots can be found almost everywhere, including vacuuming floors in homes, assembling cars in factories, and aiding surgeons in hospitals. (See **Figure 6-31**.)

Figure 6-31 Industrial robots assemble automobiles
Suwin/Shutterstock

Making IT work for you

HEADPHONES

For many computer users, headphones have become an indispensable accessory. From cell phone headsets with microphones for phone conversations to high-end ergonomic headphones for extended music listening, there are a lot of options when choosing the right headphones. Here are the main features you need to consider when deciding on what headphones are right for you: style, connection, and special features.

Over-ear headphones

Maksym Bondarchuk/Shutterstock

- **Style**

  There are three basic styles of headphones: over-ear, in-ear, and on-ear.

  Over-ear headphones are the largest headphones, with cups that surround the ear and soft linings to block out external sounds. The large size of the headphones allows for larger speakers with greater sound quality. Additionally, the soft linings are comfortable for long sessions listening to music. However, the larger size makes them difficult for travel or exercise, and their ability to block out external noise can cause you to miss important external sounds, such as an alarm or siren. Consequently, these headphones are popular with music enthusiasts and professionals and are primarily used indoors.

  In-ear headphones are the smallest of the headphones, with tiny speakers that insert into the ear canal or rest just outside of it. These headphones are ideal for travel and exercise as they are small, are easily packed, and stay in place when walking or running. However, the smaller size often comes at the cost of lower sound quality. Like the over-ear headphones, in-ear headphones block many external sounds—which makes them easier to hear but can be dangerous, especially for pedestrians and people who exercise outside or near traffic.

  On-ear headphones offer a form factor similar to over-ear headphones, except they do not have a cup that surrounds the ear, the speaker simply presses against the ear. This allows for more outside noise to be heard, which (depending on your use) can be valuable or dangerous. Compared to in-ear headphones, the larger speakers of on-ear headphones allow for greater volume and sound quality, at the cost of portability.

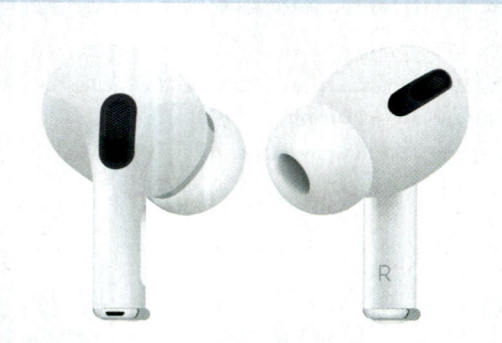

Wireless in-ear headphones

rvlsoft/Shutterstock

- **Connection**

  Traditionally, headphones have wires that transmit music from the digital device to the headphones, but recently wireless headphones have become very popular.

  Wireless headphones use digital compression techniques to send music from the digital device to the headphones. This can result in a loss of audio quality. Wireless headphones also need their own batteries. For small headphones that fit in the ear, the small size requires small batteries with shorter battery life.

  Although there are many wireless headphone connection technologies, the most common is Bluetooth, which is compatible with most laptops, tablets, and cell phones. Finally, wireless headphones offer the freedom of not being physically connected to your digital device; however, the wireless connection is not perfect. There is a limit to how far you can be from the digital device, and external factors such as walls and other wireless signals can cause wireless connections to be interrupted or dropped.

- **Special Features**

  Microphones. Although a strict definition of headphones would only include audio-output devices, many manufacturers sell headphones with microphones. Even though these are more accurately classified as a headset, you will often find them advertised as headphones with a microphone. A headset is particularly useful for making phone calls, contributing to videoconferences, and playing online video games.

  Water resistance. The small, intricate workings of headphones are particularly susceptible to water damage. If you expect to be using headphones in a damp environment or you expect to sweat while using your headphones, you will want headphones that offer water resistance. Keep in mind that water-resistant is not waterproof. Water-resistant headphones can handle being splashed with water but are not designed to be submerged.

  Active noise cancellation. Specialized headphones with unique hardware can listen to external sounds and produce sound waves that cancel out these external noises. This active noise cancellation can make a noisy environment sound quiet. This feature is often expensive and requires batteries. Nonetheless, active noise cancellation headphones are popular among travelers who want to drown out jet engines and road noise. Do not confuse this feature with noise isolation, which is simply blocking out sound with foam or padding.

  Transparency mode. Unlike active noise cancellation, which reduces external sounds, transparency mode increases external sound. Outside noise is mixed with the audio playing on the headphones to give the illusion that the headphones are not blocking out any external noise (the headphones sound as though they are *transparent*). This gives the user a greater awareness of their environment. There are many reasons why you might want to hear your environment clearly, including people who exercise near traffic and people who want to be able to hear others without removing their headphones.

Over-ear headphones with active noise cancellation

Hugh Threlfall/Alamy Stock Photo

 **concept check**

☐ What are the two most widely used audio-output devices? What is Bluetooth?
☐ What are headsets? Multifunctional devices?
☐ What is virtual reality? VR head-mounted display? VR controller? Robot? Drone?

Ergonomics

People use computers to enrich their personal and private lives. There are ways, however, that computers can make people less productive and even harm their health. Anyone who frequently uses a computer can be affected. As a result, there has been great interest in a field known as ergonomics.

Ergonomics (pronounced "er-guh-nom-ix") is defined as the study of human factors related to things people use. It is concerned with fitting the task to the user rather than forcing the user to contort to do the task. For computer users and manufacturers, this means designing input and output devices to increase ease of use and to avoid health risks.

Sitting in front of a screen in awkward positions for long periods may lead to physical problems such as eyestrain, headaches, and back pain. Computer users can alleviate these problems by taking frequent rest breaks and by using well-designed computer furniture. Some recommendations by ergonomics experts are illustrated in **Figure 6-32**.

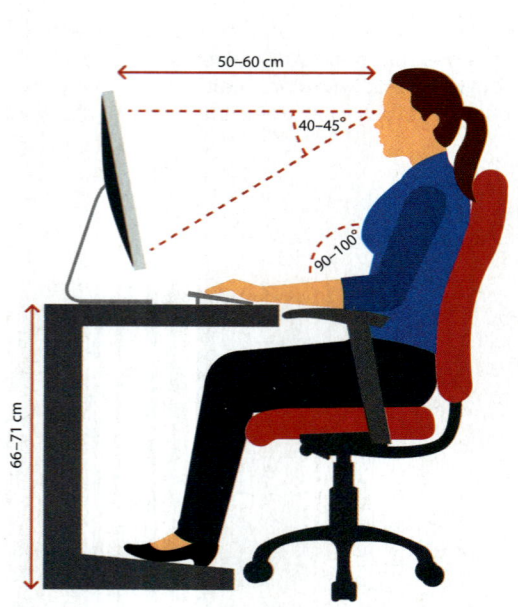
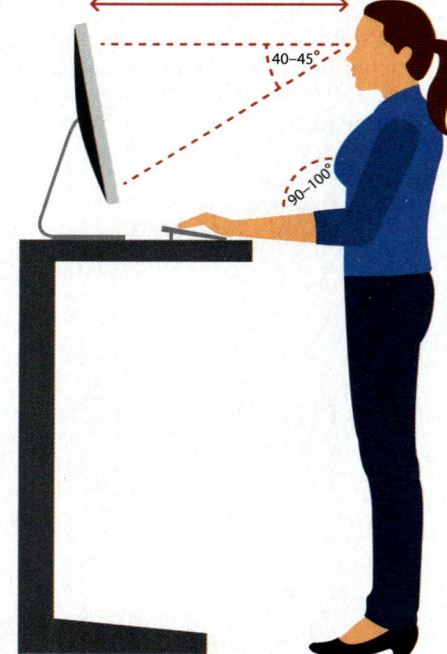

Figure 6-32 Ergonomic recommendations
MaanasShutterstock

Other recommendations to avoid physical discomfort are

- **Eyestrain and headache:** To make the computer easier on the eyes, take a 15-minute break every hour or two. Keep everything you're focusing on at about the same distance. For example, the computer screen, keyboard, and a document holder containing your work might be positioned about 20 inches away. Clean the screen of dust from time to time. One popular method for reducing eyestrain on digital devices is the use of blue-light-blocking eyeglasses and screen covers. However, the American Association of Ophthalmology has found no evidence that blue-light-blocking filters reduce eye strain and do not recommend any special eyewear for computer use.

- **Back and neck pain:** To help avoid back and neck problems, make sure your equipment is adjustable. You should be able to adjust your chair for height and angle, and the chair should have good back support. The monitor should be at eye level or slightly below eye level. Use a footrest, if necessary, to reduce leg fatigue.
- **Repetitive strain injury: Repetitive strain injury (RSI)** is any injury that is caused by fast, repetitive work that can generate neck, wrist, hand, and arm pain. RSI is by far the greatest cause of workplace illnesses, resulting in compensation claims totaling billions of dollars and lost productivity every year. One particular type of RSI, **carpal tunnel syndrome**, found among heavy computer users, consists of damage to nerves and tendons in the hands. Some victims report the pain is so intense that they cannot open doors or shake hands and that they require corrective surgery. Ergonomically correct keyboards have been developed to help prevent injury from heavy computer use. (See **Figure 6-33**.) In addition to using ergonomic keyboards, you should take frequent short rest breaks and gently massage your hands.

Figure 6-33 Ergonomic keyboard
Dmitriy Melnikov/123RF

Portable Computers

Although these recommendations apply to all personal computers, the design of portable computers, including cell phones, tablets, and laptops, presents some specific ergonomic challenges.

- **Cell phones:** Today, cell phones are more likely to be used for texting than talking. As a result, thumbs are often used to type on a tiny keyboard. The result can be a pain at the base or in the muscles of the thumb or wrist. This problem can be minimized by keeping wrists straight (not bent), head up, and shoulders straight and frequently resting thumbs by using other fingers.
- **Tablets:** Almost all tablets use a virtual keyboard and are designed to be held in your hands, flat on a table, or slightly angled. These design features cause the user to improperly align his or her head to the viewing surface, often causing neck and back pain. This problem, sometimes referred to as *tablet hunch,* can be minimized by taking frequent breaks, moving around while working, using a tablet cover or stand that allows the screen to be tipped at various angles, and using an external keyboard.
- **Laptops:** Almost all laptops have attached keyboards and screens. Unfortunately, it is impossible to optimally position either for safe ergonomic use. When the screen is positioned appropriately at eye level, the keyboard is too high. When the keyboard is appropriately positioned, the screen is too low. To minimize the negative impact, raise the level of the screen by using books or reams of paper under the laptop and attach an external keyboard to be used at waist level.

By using our devices with our ergonomic needs in mind, we can reduce stress and strain on the body and ensure our health and comfort into the future.

concept check

- What is ergonomics? How does it relate to input and output devices?
- What can be done to minimize eyestrain, headache, back pain, and neck pain?
- What is RSI? What is carpal tunnel syndrome?

Careers in IT

"Now that you've learned about input and output devices, I'd like to tell you about my career as a technical writer."

Dragon Images/Shutterstock

Technical writers prepare instruction manuals, technical reports, and other scientific or technical documents, including creating the manuals and instructions for most input and output devices. Most technical writers work for computer software firms, government agencies, or research institutions. They translate technical information into easily understandable instructions or summaries. As new technology continues to develop and expand, the need for technical writers who can communicate technical expertise to others is expected to increase.

Technical writing positions typically require an associate's or a bachelor's degree in communications, journalism, or English and a specialization in, or familiarity with, a technical field. However, individuals with strong writing skills sometimes transfer from jobs in the sciences to positions in technical writing.

Technical writers can expect to earn an annual salary in the range of $55,000 to $69,000. Advancement opportunities can be limited within a firm or company, but there are additional opportunities in consulting.

A LOOK TO THE FUTURE

The Internet of Things

Rawpixel.com/Shutterstock

As we have discussed, the Internet of Things (IoT) promises many exciting innovations. Chances are that you are already surrounded by the first wave of IoT technology. Your cell phone can monitor and share your location, your smartwatch counts and can share how many steps you took today, and your web-connected car can track and share your driving habits. Recent innovations are just the tip of the iceberg as we grapple with a new world where everything has a sensor and every action is recorded. Where will these monitoring tools show up next? What information will they record? How will that information be used?

In the future, most items will likely include a chip that uniquely identifies that item and shares that data to applications running on the Internet. Want to know if that Gucci bag is real or a knock-off? Using the embedded chip in the bag and connecting through Gucci's phone app, you will be able to verify that you are buying the genuine article. Were you just about to throw that delicate cashmere sweater into the washing machine on the hot cycle? Not to worry, your smart washer/dryer will be able to read the sweater's chip and inform you that you could accidentally shrink your favorite sweater.

Your clothes aren't the only items that will be marked with a chip. The food you purchase will also be uniquely identified. Your refrigerator will be able to read those chips to tell what food you are eating, how frequently, which foods are in danger of spoiling, and which foods need to be replenished. Can't find the mayonnaise? Your fridge will know where it is and can guide you to it. Running low on milk? Don't worry—your fridge will recognize when it is time to buy milk and will add it to your grocery list.

The IoT will extend beyond the home. As you drive to the grocery store to pick up the milk your refrigerator ordered, cameras on billboards along the road identify the make and model of your vehicle and tailor their advertising to your interests. Throughout your shopping trip, cameras are used to identify your race, gender, and age to highlight products and information that appeal to your demographic. Your smart grocery cart accesses your grocery list and guides you around the store, pointing out sale items and deals. As you pick items up and place them into the cart, the cart automatically updates your bill. No need to go to the cashier, just exit the building and your smart grocery cart will charge your credit card automatically.

Privacy advocates raise concerns related to the IoT. With sensors recording your location, shopping habits, diet, activity, and even what you are wearing, companies and governments will have an unprecedented detailed view of people's lives. Obviously, this can make shopping faster, laundry easier, and cooking less hassle, but it can also let companies know what you eat, what you wear, and how much you spend—information that you may not want to share. Today, there are concerns about cell phones and connected televisions having the capabilities to record your activities in your own home. In the future, as more devices and products become interconnected, some caution that you may need to worry about your dinner spying on you and your socks giving away your secrets!

What do you think? Is the world moments away from Internet-connected clothing, or is this more science fiction? Are the losses in privacy worth the gains in convenience?

VISUAL SUMMARY | Input and Output

KEYBOARDS

Georgios Kollidas/Shutterstock

Input is any data or instructions that are used by a computer. **Input devices** translate words, numbers, sounds, images, and gestures that people understand into a form that the system unit can process. These include keyboards and pointing, scanning, image-capturing, and audio-input devices.

Keyboards convert numbers, letters, and special characters that people understand into electrical signals. These signals are sent to, and processed by, the system unit.

Keyboards

There are three basic categories of keyboards: virtual, laptop, and traditional.

- **Virtual keyboards** are primarily used with cell phones and tablets. Does not have a physical keyboard. Keys displayed on screen and selected by touching a key's image.
- **Laptop keyboards** are used on laptop computers. Smaller than a traditional keyboard with fewer keys. Includes all the keys found on virtual keyboard plus extra keys, such as function and navigation keys.
- **Traditional keyboards** are used on desktops and larger computers. Standard keyboard has 101 keys. **Toggle keys** turn features on and off. **Combination keys** perform actions when combinations of keys are held down.

POINTING DEVICES

ilkercelik/E+/Getty Images

Pointing devices provide an intuitive interface with the system unit by accepting physical movements or gestures and converting them into machine-readable input.

Touch Screens

Touch screens allow users to select actions by touching the screen with a finger or penlike device. A **stylus** is a penlike device that uses pressure to draw images on a screen. **Handwriting recognition software** translates handwritten notes into a form that the system unit can process. **Multitouch screens** accept multiple-finger commands.

Mouse

A **mouse** controls a pointer that is displayed on the monitor. The **mouse pointer** usually appears in the shape of an arrow. Some mice have a **wheel button** that rotates to scroll through information on the monitor. A **cordless** or **wireless mouse** uses radio waves or infrared light waves. A **touch pad** operates by touching or tapping a surface. It is widely used instead of a mouse with laptops and some types of mobile devices.

Game Controllers

Game controllers provide input to computer games. Widely used controllers include **gaming mice, joysticks, gamepads,** and **motion-sensing devices.**

To efficiently and effectively use computers, you need to be aware of the most commonly used input and output devices. These devices are translators for information into and out of the system unit. Input devices translate words, sounds, and actions into symbols the system unit can process. Output devices translate symbols from the system unit into words, images, and sounds that people can understand.

SCANNING DEVICES

Aunging/Shutterstock

Scanning devices move across text and images to convert them into a form that the system unit can process.

Optical Scanners

An **optical scanner (scanner)** converts documents into machine-readable form. The four basic types are **flatbed**, **document**, **portable**, and **3D**.

Card Readers

Card readers interpret encoded information located on a variety of cards. The most common is the **magnetic card reader** that reads information from a thin magnetic strip on the back of a card. **Chip cards** contain microchips to encrypt data and improve security.

Bar Code Readers

Bar code readers or **scanners** (either handheld **wand readers** or **platform scanners**) read **bar codes** on products. There are a variety of different codes, including the **UPC** and **Maxi-Code**.

RFID Readers

RFID readers read **RFID (radio-frequency identification) tags**. These tags are widely used for tracking lost pets, production, and inventory and for recording prices and product descriptions.

Character and Mark Recognition Devices

Character and mark recognition devices are scanners that are able to recognize special characters and marks. Three types are **magnetic-ink character recognition (MICR)**, **optical-character recognition (OCR)**, and **optical-mark recognition (OMR)**.

IMAGE-CAPTURING DEVICES

caliNN and DiaNNA/Shutterstock

Image-capturing devices create or capture original images. These devices include digital cameras and webcams.

Digital Cameras

Digital cameras record images digitally and store them on a memory card or in the camera's memory. Most digital cameras record video too. Today, many digital cameras are embedded in other devices, such as cell phones and tablets.

Webcams

Webcams are specialized digital video cameras that capture images and send them to a computer for broadcast over the Internet. Webcams are built into many cell phones and tablets, while others are attached to the computer monitor.

AUDIO-INPUT DEVICES

Audio-input devices convert sounds into a form that can be processed by the system unit. By far the most widely used audio-input device is the microphone.

Voice Recognition Systems

Voice recognition systems use a microphone, a sound card, and special software. Siri, Cortana, and Google Assistant are digital assistants that use voice recognition. Specialized portable voice recorders are widely used by doctors, lawyers, and others to record dictation. Some systems are able to translate dictation from one language to another, such as from English to Japanese.

MONITORS

Angel Garcia/Bloomberg via Getty Images

Output is processed data or information. **Output devices** are hardware used to provide or create output. **Monitors (display screens)** are the most-used output device.

Features

Monitor **clarity** is a function of **resolution** (matrix of **pixels**, or **picture elements**), refresh rate, **dot pitch**, **contrast ratio**, **active display area**, and **aspect ratio**.

Monitor Types

Three basic types are **LCD (liquid crystal display)**, **LED (light-emitting diode)**, and **OLED (organic light-emitting diode)**.

E-book Readers

E-books (electronic books) are traditional printed books in electronic format. **E-book readers (e-readers)** store and display e-books and other electronic media. They use **e-ink** technology.

Other Monitors

Other types of monitors include:

- **Digital (interactive) whiteboards** are specialized devices with a large display connected to a computer or projector.
- **Flexible screens** allow digital devices to display images on nonflat surfaces.
- **Digital projectors** project the images from a traditional monitor onto a screen or wall.

PRINTERS

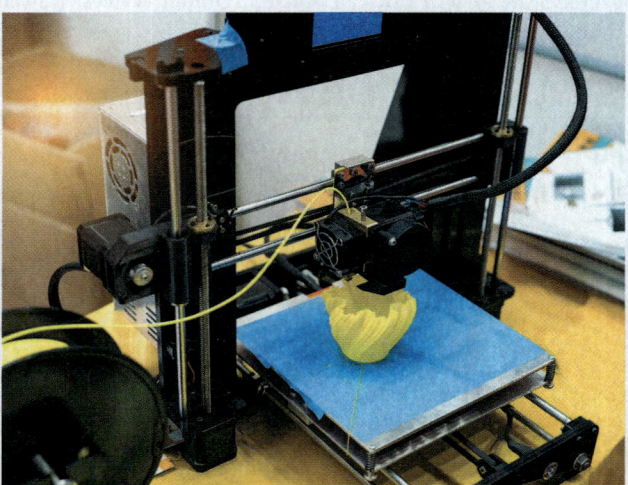

asharkyu/Shutterstock

Printers translate information processed by the system unit and present the information on paper.

Features

Basic features include **resolution** measured in **dpi (dots per inch)**, color capability (most common black ink selection is **grayscale**), speed, memory, **duplex** (both sides of paper) **printing, and connectivity**.

Inkjet Printers

Inkjet printers spray ink at high speed onto the surface of paper. They are the most widely used type of printer, reliable, quiet, and inexpensive. The most costly aspect of inkjet printers is replacing the ink cartridges.

Laser Printers

Laser printers use technology similar to a photocopying machine. There are two categories: **personal** and **shared**.

3D Printers

3D printers (additive manufacturing) create objects by adding layers of material onto one another. They have been available for decades; however, recent price reductions have increased their popularity.

Other Printers

There are several other types of printers, including **thermal printers** (use heat elements to produce images on heat-sensitive paper) and **plotters** (use data from graphics tablets and other graphical devices).

AUDIO-OUTPUT DEVICES

Alexander Demyanenko/Shutterstock

Audio-output devices translate audio information from the computer into sounds that people can understand. The most widely used are **speakers** and **headphones**. These devices connect either by cable to an audio jack on the system unit or by a wireless connection. **Bluetooth** technology is widely used to connect wireless devices.

COMBINATION INPUT AND OUTPUT DEVICES

Combination devices combine input and output capabilities. Devices within this category include:

- **Headsets** combine the functionality of microphones and headphones; integral part of serious video game systems.
- **Multifunctional devices (MFDs)** typically combine capabilities of scanner, printer, fax, and copy machine.
- **Virtual reality (VR)** creates 3D-simulated **immersive experiences.** VR hardware includes **head-mounted displays** and **controllers.**
- **Drones (unmanned aerial vehicles, UAVs)** take input from a controller and send output video and sound to the user.
- **Robots** use cameras, microphones, and other sensors as inputs; outputs can be as complex as exploring damaged nuclear reactors to as simple as taking a photo.

ERGONOMICS

Dmitriy Melnikov/123RF

Ergonomics is the study of human factors related to things people use.

Recommendations

Some recommendations to avoid physical discomfort associated with heavy computer use:

- To avoid eyestrain and headache, take a 15-minute break every hour, keep everything you're focusing on at the same distance, and clean the screen periodically.
- To avoid back and neck pain, use adjustable equipment; adjust chairs for height, angle, and back support; position monitors at eye level or slightly below. Use a footrest, if necessary, to reduce leg fatigue.
- To avoid **repetitive strain injury (RSI)** and **carpal tunnel syndrome,** use ergonomically correct keyboards; take frequent, short rest breaks; and gently massage hands.

Portable Computers

The design of portable computers presents ergonomic challenges.

- Laptops do not allow correct positioning of keyboard and screen; raise level of laptop and use external keyboard.
- Tablets with virtual keyboards cause improper alignment of user's head; take frequent breaks, move while working, use cover or stand, and use external keyboard.
- Cell phones require extensive use of thumbs; keep wrists straight, head up and shoulders straight, and use other fingers.

CAREERS in IT

Technical writers prepare instruction manuals, technical reports, and other documents. An associate's or a bachelor's degree in communication, journalism, or English and a specialization in, or familiarity with, a technical field are required. Expected salary range is $55,000 to $69,000.

KEY TERMS

3D printer
3D scanner
active display area
additive manufacturing
aspect ratio
bar code
bar code reader
bar code scanner
Bluetooth
card reader
carpal tunnel syndrome
chip card
clarity
combination key
connectivity
contrast ratio
cordless mouse
digital camera
digital projector
digital whiteboard
display screen
document scanner
dot pitch
dots per inch (dpi)
drone
duplex printing
e-book reader
e-book
e-ink
electronic book
e-reader
ergonomics
flatbed scanner
flat-panel monitor
flexible screen
game controller
gamepad
gaming mouse
Google Cloud Print
grayscale
handwriting recognition software
headphone
headset
immersive experience
inkjet printer
input
input device
interactive whiteboard
joystick
keyboard
laptop keyboard
laser printer
light-emitting diode (LED)
liquid crystal display (LCD)

magnetic card reader
magnetic-ink character recognition (MICR)
MaxiCode
monitor
motion-sensing device
mouse
mouse pointer
multifunctional device (MFD)
multitouch screen
optical scanner
optical-character recognition (OCR)
optical-mark recognition (OMR)
organic light-emitting diode (OLED)
output
output device
personal laser printer
picture element
pixel
pixel pitch
platform scanner
plotter
pointing device
portable scanner
printer
refresh rate
repetitive strain injury (RSI)
resolution (monitor)
resolution (printer)
RFID reader
RFID (radio-frequency identification) tag
robot
scanner
scanning device
shared laser printer
speaker
stylus
technical writer
thermal printer
toggle key
touch pad
touch screen
traditional keyboard
Universal Product Code (UPC)
unmanned aerial vehicle (UAV)
virtual keyboard
virtual reality (VR)
voice recognition system
VR controller
VR head-mounted display
wand reader
webcam
wheel button
wireless mouse

MULTIPLE CHOICE

Circle the correct answer.

1. A bar code system used by many electronic cash registers.
   - a. RSI
   - b. UPC
   - c. FDIC
   - d. RFID
2. This device records images digitally on a memory card or in its memory.
   - a. digital camera
   - b. MaxiCode
   - c. laser
   - d. ergonomics
3. Pressing this key turns a feature on or off.
   - a. QWERTY
   - b. liquid crystal
   - c. toggle key
   - d. MICR
4. These are special-purpose printers for creating maps, images, and architectural and engineering drawings.
   - a. plotters
   - b. headsets
   - c. e-book readers
   - d. whiteboards
5. Another name for an unmanned aerial vehicle (UAV).
   - a. OMR
   - b. drone
   - c. liquid crystal
   - d. ergonomics
6. A penlike device commonly used with tablets, PCs, and PDAs.
   - a. UPC
   - b. stylus
   - c. handheld wand
   - d. monitor
7. An input device that controls a pointer that is displayed on the monitor.
   - a. handheld wand
   - b. dot pitch
   - c. mouse
   - d. RFID
8. A monitor feature that is measured by the diagonal length of the viewing area.
   - a. MaxiCode
   - b. HDTV
   - c. resolution
   - d. active display area
9. Bar code readers use either handheld wand readers or platform _____.
   - a. scanners
   - b. resolution rate
   - c. OLED
   - d. multitouch
10. The distance between each pixel.
    - a. resolution
    - b. ergonomics
    - c. dot pitch
    - d. LCD

MATCHING

Match each numbered item with the most closely related lettered item. Write your answers in the spaces provided.

a. QWERTY
b. mouse
c. multitouch
d. scanners
e. MICR
f. microphone
g. contrast ratio
h. e-book readers
i. immersive
j. ergonomics

____ 1. Flatbed and document are types of _____.
____ 2. The most widely used audio-input device.
____ 3. The monitor feature that indicates the ability to display by comparing the light intensity of the brightest white to the darkest black.
____ 4. Most keyboards use this arrangement of keys.
____ 5. Virtual reality strives to create this type of experience.
____ 6. This is the study of human factors related to the things people use.
____ 7. The device that controls a pointer displayed on the monitor.
____ 8. The type of screen that can be touched with more than one finger and supports zooming in and out by pinching and stretching your fingers.
____ 9. Mobile devices that use e-ink displays to store and display electronic media.
____ 10. Device used by banks to automatically read those unusual numbers on the bottom of checks and deposit slips.

OPEN-ENDED

On a separate sheet of paper, respond to each question or statement.

1. Define input and input devices.
2. Describe the different types of keyboard, pointing, scanning, image-capturing, and audio-input devices.
3. Define output and output devices.
4. Describe the features and different types of monitors and printers.
5. Describe audio-output devices, including Bluetooth technology.
6. Discuss combination input and output devices, including multifunctional devices, headsets, drones, robots, and virtual reality head-mounted display and controllers.
7. Define ergonomics, describe ways to minimize physical discomfort, and discuss design issues with portable computers.

DISCUSSION

Respond to each of the following questions.

 ## Making IT Work for You: HEADPHONES

Review the Making IT Work for You: Headphones on pages 148 and 149, and then respond to the following: (a) What activities do you do while you have headphones on? What type of headphones would best suit those activities? (b) Go online and investigate two headphones that interest you. How are they different? How are they the same? (c) What is the most important feature of a pair of headphones to you, and why?

 ## Community: INPUT/OUTPUT ACCESSIBILITY

Review the Community box on page 136 and respond to the following questions: (a) Consider the disability of a friend, loved one, or acquaintance. How would such a disability make using common inputs and outputs difficult? (b) What do you do regularly on your devices that would become more difficult if you could not easily and conveniently use the input/outputs of that device? (c) Investigate the accessibility input/output options for your devices. What options are available to people with disabilities?

 ## Privacy: DEVICES WITH MICROPHONES AND CAMERAS

Review the Privacy box on page 137 and respond to the following questions: (a) How many digital devices are in your home with microphones and/or cameras? Consider that digital devices include laptops, tablets, desktops, TVs, video game consoles, cell phones, and wearable devices. (b) Consider where you might keep digital devices. If such a device had a camera, what could it see? If such a device had a microphone, what could it hear? Would any of those cameras or microphones be able to record private information? (c) If you had a digital device that was susceptible to being hacked by cybercriminals, what could you do to protect private information?

 ## Ethics: WEBCAMS

Every day, thousands of webcams continuously broadcast images to the Internet. Review the Ethics box on page 140, and respond to the following: (a) Do you think recording and broadcasting images without permission is an ethical or a privacy concern? Why or why not? (b) Do you object to being recorded in public, within a retail store, or in a private home? Explain. (c) Do you think police should have access to webcam videos? Should concerned parents? Should jealous spouses? Why or why not? Defend your responses.

Design Elements: Concept Check icon: Dizzle52/Getty Images

chapter 7
Secondary Storage

wutzkohphoto/Shutterstock

CHAPTER 7: Secondary Storage

Why should I read this chapter?

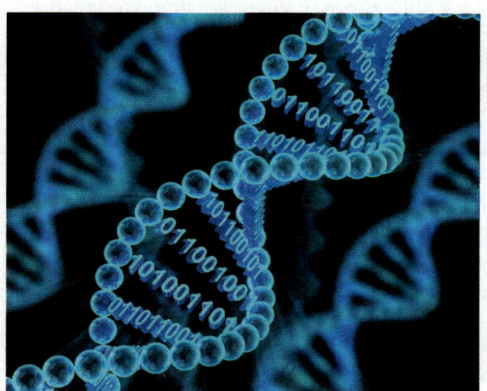
ymgerman/Shutterstock

Secondary storage is vital to keeping our data safe, secure, and available. If you run out of secondary storage, your cell phone won't take videos, your laptop won't install apps, and your tablet will slow to a crawl. The future of secondary storage looks to hold even more information in smaller packages using holograms or even organic molecules to store data.

This chapter covers the things you need to know to be prepared for this ever-changing digital world, including:

- Hard drives—get the right hard drive on your computer to meet all your needs.
- Optical discs—share digital information on Blu-ray, CD, or DVD.
- Solid-state storage—make your portable electronics faster and use less power.
- Cloud storage—store your information safely and securely on the Internet.

Learning Objectives

After you have read this chapter, you should be able to:

1. Distinguish between primary and secondary storage.
2. Identify the important characteristics of secondary storage, including media, capacity, storage devices, and access speed.
3. Define solid-state storage, including solid-state drives, flash memory cards, and USB drives.
4. Describe hard-disk platters, tracks, sectors, and cylinders.
5. Compare performance enhancements, including disk caching, RAID, hybrid drives, file compression, and file decompression.
6. Define optical storage, including compact discs, digital versatile discs, and Blu-ray discs.
7. Compare internal, external, and network drives.
8. Define cloud storage and cloud storage services.
9. Describe mass storage, mass storage devices, enterprise storage systems, and storage area networks.

Introduction

"Hi, I'm Nicole, and I'm a disaster recovery specialist. I'd like to talk with you about secondary storage, one of the most critical parts of any computer system. I'd also like to talk about various cloud storage services."

jeffbergen/Getty Images

Secondary storage is how we store, protect, and transport our most important data. Families save precious photos to secondary storage. Students back up important homework assignments to secondary storage. And when your cell phone runs out of secondary storage, you can no longer download new apps or take photos or videos.

Secondary storage devices have always been an indispensable element in any computer system. They have similarities to output and input devices. Like output devices, secondary storage devices receive information from the system unit in the form of the machine language of 0s and 1s. Rather than translating the information, however, secondary storage devices save the information in machine language for later use. Like input devices, secondary storage devices send information to the system unit for processing. However, the information, because it is already in machine form, does not need to be translated. It is sent directly to memory (RAM), where it can be accessed and processed by the CPU.

Understanding the uses and different types of secondary storage will help you make smart decisions when purchasing a new cell phone or laptop, help you to protect your data from disaster or theft, and improve the usefulness of the devices you already own. You need to know the capabilities, limitations, and uses of hard disks, solid-state drives, optical discs, cloud storage, and other types of secondary storage. Additionally, you need to be aware of specialty storage devices for portable computers and to be knowledgeable about how large organizations manage their extensive data resources.

Storage

An essential feature of every computer is the ability to save, or store, information. As discussed in **Chapter 5**, random-access memory (RAM) holds or stores data and programs that the CPU is presently processing. Before data can be processed or a program can be run, it must be in RAM. For this reason, RAM is sometimes referred to as **primary storage**.

Unfortunately, most RAM provides only temporary or volatile storage. That is, if the computer is turned off or loses power due to a power failure, all the information stored on RAM is erased. This volatility results in a need for more permanent or nonvolatile storage for data and programs. RAM also has the disadvantage of being relatively small and unable to hold all the videos, photos, and applications that we regularly use.

Figure 7-1 A solid-state drive
Maxx-Studio/Shutterstock

Secondary storage provides permanent or nonvolatile storage with far larger capacity than RAM. Using **secondary storage devices** such as a solid-state drive, data and programs can be retained after the computer has been shut off. This is accomplished by *writing* files to and *reading* files from secondary storage devices. Writing is the process of saving information *to* the secondary storage device. Reading is the process of accessing information *from* secondary storage. The most important characteristics of secondary storage are:

- **Storage Media** are the physical material that holds the data and programs.
- **Capacity** measures how much a particular storage medium can hold.
- **Storage devices** are hardware that reads data and programs from storage media. These devices are typically referred to as **drives**. Most also write to storage media.
- **Access speed** measures the amount of time required by the storage device to retrieve data and programs.

Most cell phones, tablets, and laptops have solid-state storage drives. Desktop computers often have internal hard-disk drives. (See **Figure 7-1**.)

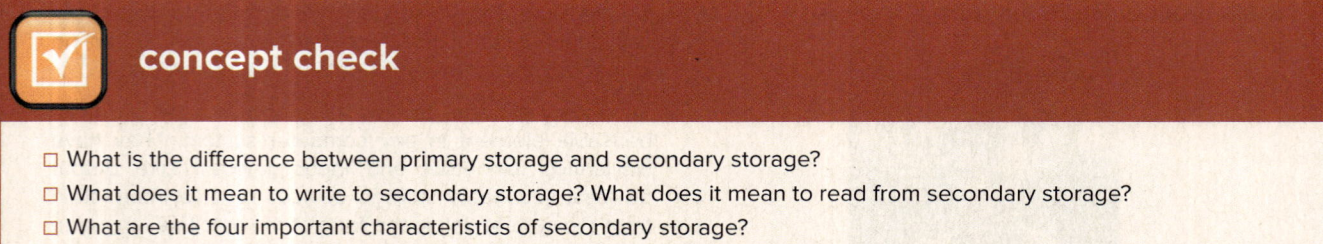

concept check

☐ What is the difference between primary storage and secondary storage?
☐ What does it mean to write to secondary storage? What does it mean to read from secondary storage?
☐ What are the four important characteristics of secondary storage?

Solid-State Storage

Solid-state storage devices provide access to **flash memory**, commonly known as **solid-state storage**. As we discussed in **Chapter 5**, flash memory offers a combination of features of RAM and ROM. Like RAM, it can be updated, and like ROM it does not lose information when a computer is turned off. Flash memory is a little slower than traditional memory but much faster than other secondary storage drives.

Solid-State Drives

Solid-state drives (SSDs) store and retrieve data and information much in the same manner as RAM. As we discussed in **Chapter 5**, characters are represented by positive (+) and negative (−) charges using the ASCII, EBCDIC, or Unicode binary codes. For example, the letter A would require a series of eight charges. (See **Figure 7-2**.)

> **community**
>
> Criminals often target communities with malware and viruses by using infected USB flash drives. These infected drives are handed out for free at conventions, "accidentally" left behind at the library, or simply dropped on the ground. When the victim finds the drive and plugs it into their computer, the drive automatically installs malware and viruses. What data, photos, videos, and passwords do you store on your computer that could be at risk from an infected USB drive? Is the risk of an infected computer worth the benefit of plugging in an unknown USB drive?

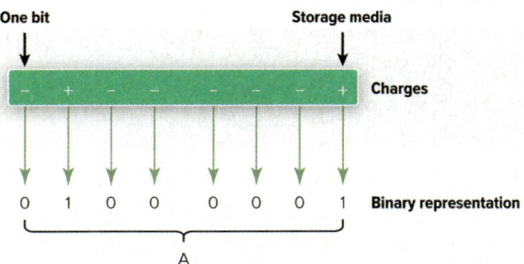

Figure 7-2 Charges representing the letter A

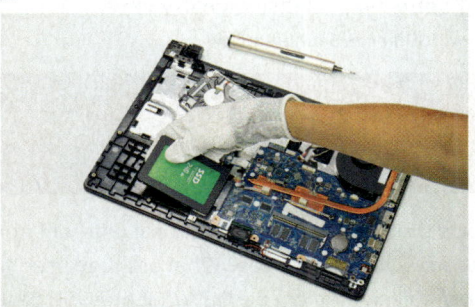

Figure 7-3 Internal solid-state drive
NMStudio789/Shutterstock

SSDs are fast and durable and require little power, making them popular for cell phones, tablets, and laptops. SSDs are the most common internal storage in desktop computers. (See **Figure 7-3**.) The speed and durability of SSDs make these drives especially popular as internal storage in gaming computers.

Flash Memory Cards

Flash memory cards are small solid-state storage devices widely used in portable devices. Some of the cards are used within devices such as cell phones, laptops, and GPS navigation systems. Other cards provide external or removable storage. Flash memory cards are commonly used with digital cameras and cell phones to expand their internal storage. (See **Figure 7-4**.)

USB Flash Drives

USB flash drives, or simply **flash drives**, are so compact that they can be transported on a key ring. These small, solid-state drives conveniently connect directly to a computer's USB port to transfer files and can have capacities ranging from 1 GB to 2 TB and higher, with a broad price range to match. (See **Figure 7-5**.) Due to

> **tips**
>
> Have you ever accidentally deleted or lost important files from your flash drive? Do you have a USB flash drive that is no longer being recognized by your computer? Here are a few suggestions that might help:
>
> **1 Recovery/undelete software.** If you accidentally deleted files from a flash drive, it is unlikely that you can recover them using your operating system or searching through your recycle bin. Fortunately, there are several recovery (or undelete) programs that might help, and some are even free. For example, two free programs are *Disk Drill* and *Recuva*. These programs will scan your flash drives for deleted files and offer you a chance to recover the ones you want back.
>
> **2 Testing USB ports.** If your computer does not recognize your USB flash drive, there could be a problem with your USB port. Try plugging another device into that same port to see if it works. If this device does not work, then your computer's USB is most likely defective and needs to be replaced. If the device works, then most likely your USB flash drive is damaged and you should try the professional recovery services discussed in the next step.
>
> **3 Professional recovery services.** For damaged flash drives, there is a possibility that your data could be recovered by several companies that are dedicated to data recovery. Although the fees are high, they can rescue data from the actual memory chip, even if the drive or supporting circuits are damaged.

their convenient size and large capacities, flash drives have replaced older external storage for transporting data and information among computers, specialty devices, and the Internet. As wireless and cloud technologies improve, physical media devices such as flash drives are becoming less common for transferring data.

Figure 7-4　Flash memory card
wk1003mike/Shutterstock

Figure 7-5　USB flash drive
kyoshino/E+/Getty Images

concept check

- What is solid-state storage? How is it different from RAM? ROM?
- What are solid-state drives? What are they used for?
- What are flash memory cards? What are USB flash drives? What are they used for?

Hard Disks

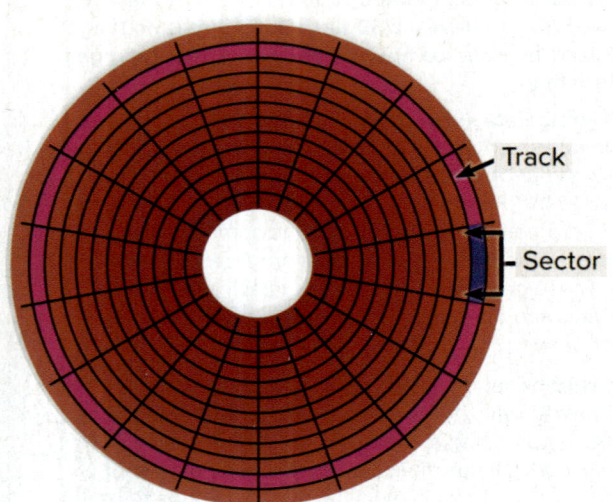

Figure 7-6　Tracks and sectors

Unlike solid-state drives, which have no moving parts, **hard-disk drives** rotate and have read/write heads that move in and out. These moving parts allow hard-disk drives to use inexpensive technology to provide large-capacity storage; however, they also make hard disks less durable and slower than solid-state drives.

Hard disks save files by altering the magnetic charges of the disk's surface to represent 1s and 0s. Hard disks retrieve data and programs by reading these charges from the magnetic disk. **Density** refers to how tightly these charges can be packed next to one another on the disk.

Hard disks use rigid metallic **platters** that are stacked one on top of another. Hard disks store and organize files using tracks, sectors, and cylinders. **Tracks** are rings of concentric circles on the platter. Each track is divided into invisible wedge-shaped sections called **sectors**. (See **Figure 7-6**.) A **cylinder** runs through each track of a stack of platters. Cylinders are necessary to differentiate files stored on the same track and sector of different platters. When a hard disk is formatted, tracks, sectors, and cylinders are assigned. Solid state drives use a similar organization method, but instead of a circle of tracks and sectors, they are organized in a grid of pages and blocks (**see Figure 7-7**).

There are three basic types of storage drives: internal, external, and network.

Internal Storage

An **internal storage** is located inside the system unit. These hard disks are able to store and retrieve large quantities of information quickly. They are used to store programs and data files. For example, nearly every personal computer uses its internal hard disk to store its operating system and major applications such as Word and Excel.

To ensure adequate performance of your internal storage and the safety of your data, you should perform routine maintenance and periodically make backup copies of all important files.

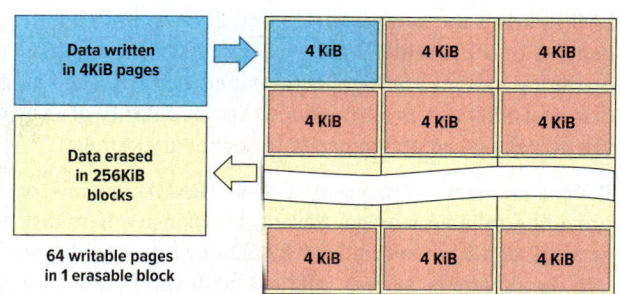

Figure 7-7 SSD drive using 4KiB pages and 256KiB blocks.

External Hard Disks

Although internal hard disks provide fast access, they have a fixed amount of storage and cannot be easily removed from the system unit. External hard disks provide slower access and are typically connected to a USB port on the system unit and are easily removed. Like internal storage, external hard disks have a fixed amount of storage. However, because each removable storage can be easily replaced by another removable hard disk, a single port on the system unit can provide access to an unlimited amount of storage. (See **Figure 7-8**.)

External storage uses the same basic technology as internal storage and is used primarily to complement an internal storage. Because they are easily removed, they are particularly useful to protect or secure sensitive information. Other uses for external storage include backing up the contents of the internal storage and providing additional storage capacity.

Figure 7-8 External hard drive
Stephen Frost/Alamy Stock Photo

Network Drives

While internal and external storage drives exist local to the system unit, either within the system unit or nearby, **network drives** place the storage on a network and can be located across the world from the system unit. This is a popular solution for individuals and businesses sharing files where access speed is not essential, but capacity and durability are crucial.

Performance Enhancements

Four ways to improve the performance of storage are disk caching, hybrid drives, redundant arrays of inexpensive disks, and file compression/decompression.

Disk caching improves hard-disk performance by anticipating data needs. It performs a function similar to cache memory discussed in **Chapter 5**. While cache memory improves processing by acting as a temporary high-speed holding area between memory and the CPU, disk caching improves processing by acting as a temporary high-speed holding area between a secondary storage device and memory. Disk caching requires a combination of hardware and software. During idle processing time, frequently used data is automatically identified and read from the hard disk into the disk cache. When needed, the data is then accessed directly from memory. The transfer rate from memory is much faster than from the hard disk. As a result, overall system performance is often increased by as much as 30 percent.

privacy

Diminishing secondary storage prices have an unexpected impact on privacy. The availability of cheap digital storage has resulted in a permanent digital record of our lives available for all to see on the Internet. Once an image, video, or message is released on the Internet, it is very difficult to remove. Some argue that we all have a "right to be forgotten" and that major Internet companies like Instagram and Google should help people permanently remove records of embarrassing or unpleasant moments. Others say that the Internet is a record of our past and we can't choose to only hold on to the positive things. What do you think?

Hybrid drives are storage drives that contain both solid-state storage and hard disks in an attempt to gain the speed and power benefits of SSDs while still having the low cost and large capacity of hard drives. Typically, these systems use SSD to store the operating system and applications and hard disks to store videos, music, and documents. The wide availability and lowering prices of SSD drives have reduced the popularity of hybrid drives, but they remain popular with users who need an inexpensive way to gain speed and capacity in secondary storage.

Redundant arrays of inexpensive disks (RAID) improve performance by expanding external storage, improving access speed, and providing reliable storage. Several inexpensive hard-disk drives are connected to one another. These connections can be by a network or within specialized RAID devices. (See **Figure 7-9**.) The connected hard-disk drives are related or grouped together, and the computer system interacts with the RAID system as though it were a single large-capacity hard-disk drive. The result is expanded storage capability, fast access speed, and high reliability. For these reasons, RAID is often used by Internet servers and large organizations.

File compression and **file decompression** increase storage capacity by reducing the amount of space required to store data and programs. File compression is not limited to hard-disk systems. It is frequently used to compress files on DVDs, CDs, and flash drives as well. File compression also helps to speed up transmission of files from one computer system to another. Sending and receiving compressed files across the Internet is a common activity.

File compression programs scan files for ways to reduce the amount of required storage. One way is to search for repeating patterns. The repeating patterns are replaced with a token, leaving enough tokens so that the original can be rebuilt or decompressed. These programs often shrink files to a quarter of their original size.

Windows and Mac operating systems provide compression and decompression utilities. Windows machines can compress the data on an entire drive, improving the storage efficiency of the drive, at the cost of speed and energy necessary to decompress every file on the drive before using it. For more advanced compression schemes, you can use specialized utilities such as WinZip. For a summary of performance enhancement techniques, see **Figure 7-10**.

Figure 7-9 RAID storage device
Santi Nanta/Shutterstock

Technique	Description
Disk caching	Uses cache and anticipates data needs
Hybrid drive	Uses both SSD and hard disks
RAID	Linked, inexpensive hard-disk drives
File compression	Reduces file size
File decompression	Expands compressed files

Figure 7-10 Performance enhancement techniques

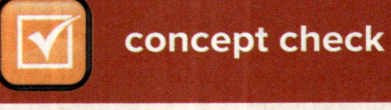 **concept check**

- What are the three types of drives? Briefly describe each.
- What is density? What are tracks, sectors, and cylinders?
- List and describe four ways to improve the performance of hard disks.

Optical Discs

Optical discs can hold over 128 GB (gigabytes) of data. (See **Figure 7-11**.) That is the equivalent of millions of typewritten pages or a medium-sized library, all on a single disc.

In optical disc technology, a laser beam alters the surface of a plastic or metallic disc to represent data. Unlike hard disks, which use magnetic charges to represent 1s and 0s, optical discs use reflected light. The 1s and 0s are represented by flat areas called **lands** and bumpy areas called **pits** on the disc surface. The disc is read by an **optical disc drive** using a laser that projects a tiny beam of light on these areas. The amount of reflected light determines whether the area represents a 1 or a 0.

Like hard disks, optical discs use tracks and sectors to organize and store files. Unlike the concentric tracks and wedge-shaped sectors used for hard disks, however, optical discs typically use a single track that spirals outward from the center of the disc. This single track is divided into equal-sized sectors.

Figure 7-11 Optical disc
limpido/Shutterstock

The most widely used optical discs are CD, DVD, and Blu-ray discs.

- **Compact discs (CDs)** were the first widely available optical format for PC users, but have largely been replaced by DVDs and Blu-rays. Typically, CD drives store 700 MB (megabytes) of storage. Optical discs that store music are often CDs.

- **Digital versatile discs (DVDs)** are the standard optical discs in PCs. DVDs are very similar to CDs except that typical DVD discs can store 4.7 GB (gigabytes)—seven times the capacity of CDs. Optical discs that store movies or software are often DVDs. DVD drives and CD drives look very similar.

- **Blu-ray discs (BDs)** are the newest form of optical storage designed to store **hi-def (high-definition)** video. The name Blu-ray comes from a special blue-colored laser used to read the discs that gives them a typical capacity of 50 GB—10 times the capacity of DVDs. Optical discs that store hi-def video and the newest video games are often Blu-ray discs. The newest Blu-ray discs, **Ultra HD Blu-rays (UHD BD)**, are able to play back 4K video content and store up to 100 GB of data.

Each of these optical discs has three basic formats: read only, write once, and rewritable.

- **Read-only (ROM for read-only memory) discs** are discs that cannot be written on or erased by the user. Optical discs that you buy in a store, such as music CDs, DVD movies, and Blu-ray video games, are often read only.

- **Write-once (R for recordable) discs** can be written on once but read many times. This type of permanent writing to storage is sometimes referred to as WORM (write once, read many).

- **Rewritable (RW for rewritable or RAM for random-access memory) discs** are able to be written to and read from multiple times.

tips

Today's cloud services allow your cell phone to access important files from anywhere; however, it can use a lot of data. Going over your cell phone plan's data limits can be expensive and frustrating. Even if you have an unlimited data plan, if you are a heavy Internet user, your service provider may slow your Internet connection. If you'd like to reduce your data usage, consider the following suggestions.

1. **Monitor your data usage.** Find out what apps are using the most data by checking your operating system's data usage statistics.
   - For Android cell phones: Click on the *Settings* icon from the home screen, then choose *Wireless & Networks* and then *Data Usage*. You may need to scroll down to see how much data each app is using.
   - For iOS cell phones: Click on the *Settings* icon from the Home screen and tap on *Cellular*. You may need to scroll down to see how much data each app is using.

2. **Limit background data use.** Many apps will use your data plan even when the app is not open. Your mobile OS can limit which apps can use your data in the background.
   - For Android cell phones: Click on the *Settings* icon from the home screen, then choose *Data Usage*, and then tap *Data Saver*. This will automatically block all apps from downloading data while active. Finally, select *Unrestricted data access* and select any apps you want to allow to use background data.
   - For iOS cell phones: Click on the *Settings* icon from the Home screen and choose *General* and then *Background App Refresh*. This lists your apps, and you can select which can use background data and which cannot.

Format	Typical Capacity	Description
CD	700 MB	Once the standard optical disc
DVD	4.7 GB	Current standard
BD	25 GB	Hi-def format, large capacity
UHD BD	100 GB	4K video

Figure 7-12 Types of optical discs

There are limitations to the optical storage. Rewritable discs will stop being readable after the disc has been rewritten over 1,000 times. Scratches and damage to the disc will eventually make the disc unreadable. Furthermore, there are some less common optical storage techniques that can increase the storage capacity of the disc, such as writing on both sides of the disc or storing information on several layers sandwiched together on one side of the disc.

For a summary of the different types of optical discs, see **Figure 7-12**.

concept check

- How is data represented on optical discs?
- Compare CD, DVD, and BD formats.
- Compare ROM, R, and RW discs.

Cloud Storage

Recently, many applications that would have required installation on your computer to run have moved to the web. As we discussed in **Chapter 2**, this is known as **cloud computing**, where the Internet acts as a "cloud" of servers that supply applications to clients as a *service* rather than a *product*. Additionally, these servers provide **cloud storage**, also known as **online storage**.

The popular cloud storage service Google Drive comes with the online office suite Google Docs. If you have used Google Docs to create a word-processing document or a spreadsheet, you have already used cloud computing. (See **Figure 7-13**.) The service provider's server runs the applications, and your computer displays results. The applications and data can be accessed from any Internet-ready device. This means that even devices with little storage, memory, or processing power, such as a cell phone, can run the same powerful applications as a desktop computer.

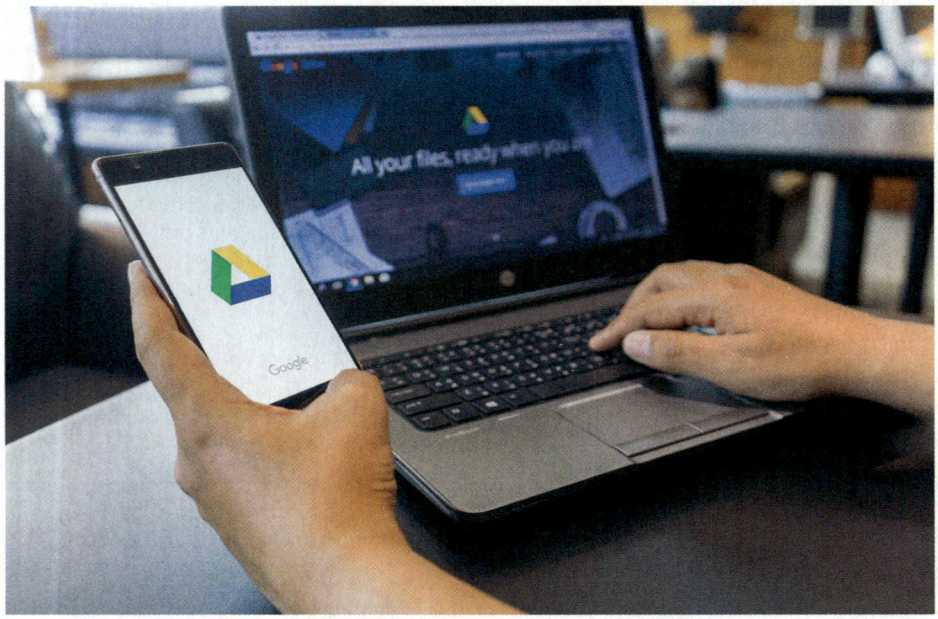

Figure 7-13 Google Drive Docs
Nopparat Khokthong/Shutterstock

The benefits to this arrangement are numerous:

- Maintenance—The cloud service will take care of backups, encryption, and security.
- Hardware upgrades—The cloud service will never run out of disk space and can replace failed hard disks without interruption to the user.
- File syncing and collaboration—Users can share documents, spreadsheets, and files with others from anywhere with an Internet connection. One user can sync files across multiple devices, and all users will have the most recent version of their documents.

Of course, there are some disadvantages of cloud storage:

- Access speed—The data transfer rate is dependent upon the speed of your Internet connection, which most likely is not as fast as a user's internal network.
- File security—Users are dependent upon the cloud service's security procedures, which may not be as effective as your own.

ethics

Cloud storage has created some interesting legal and ethical questions regarding the storage of sensitive and privileged information. Who is responsible for maintaining security and privacy of sensitive and confidential information? For example, an attorney or doctor is legally and ethically required not to share your personal data. But often such data is stored insecurely in the cloud. There are many instances of lawyers' emails and doctors' files being hacked and released on the Internet, but who is responsible? Is it your attorney, your doctor, the company providing the cloud service, or do you somehow have the responsibility to protect personal sensitive information?

Numerous websites provide cloud storage services. (See **Figure 7-14**.) To learn more about how you could use cloud storage, see Making IT Work for You: Cloud Storage on pages 172 and 173.

Company	Location
Dropbox	dropbox.com
Google	drive.google.com
Microsoft	onedrive.live.com
Amazon	amazon.com/gp/drive
Apple	icloud.com

Figure 7-14 Cloud storage services

concept check

- What is cloud computing?
- What is cloud storage?
- What are some of the advantages and disadvantages of cloud storage?

Mass Storage Devices

It is natural to think of secondary storage media and devices as they relate to us as individuals. It may not be as obvious how important these matters are to organizations. **Mass storage** refers to the tremendous amount of secondary storage required by large organizations.

Mass storage devices are specialized high-capacity secondary storage devices designed to meet organizational demands for data storage. These mass storage solutions allow large corporations and institutions to centralize their maintenance and security of data, thereby reducing costs and personnel.

Making IT work for you

CLOUD STORAGE

Do you find that you take a lot of photos and videos on your phone, and your storage space is running low? Are you working on a group project and finding it difficult to keep everyone updated with the most recent version of documents and files? Are you looking for a safe, secure location to store backups and important files? If so, cloud storage may be the solution you are looking for. Here are some things to consider when choosing a cloud storage option.

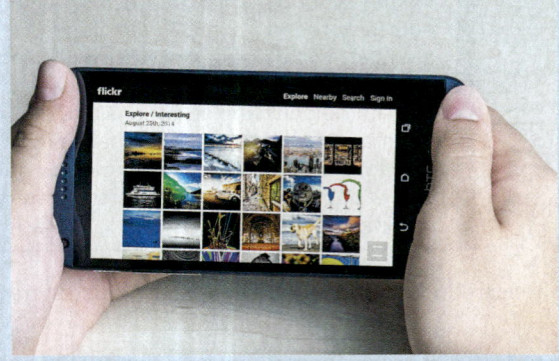

Cloud storage service Flickr is designed to store and display digital photos.

OlegDoroshin/Shutterstock

- **What will you store?**

  The types of files that you store can have a big impact on determining the best cloud storage service for you. The following suggests the best file service for you based on the types of files you typically store.

  - If you primarily store photos, then consider the cloud services of Flickr and Adobe Creative Cloud. They feature online tools to edit, share, and search photos.
  - If you primarily store music, then consider the cloud services of Google Play Music and iTunes Match. They feature online tools to listen to music and create customizable playlists.
  - If you primarily store documents, then consider the cloud services Adobe Document Cloud and Microsoft's One Drive. They feature online tools to view and edit documents.
  - If you primarily need storage to back up your system programs, consider the cloud services of Backblaze and iDrive. They have apps that back up your devices' data, making backups seamless and easy.

  Also, what you store will impact how much storage you need. If you are only looking to store documents and text files, you will not need much storage space; however, videos and photo albums can take up a lot more storage. Different services offer different pricing plans and have special offers depending on what types of files you store—the best cloud storage plan for you will tailor itself to your storage needs.

Apple's iCloud works with the iWorks office suite.

nikkimeel/123RF

- **What tools will you use?**

  If your storage needs are mostly sharing and working on documents, your best cloud storage choice may be determined by the software you use to create those documents. Most office software suites are designed to work seamlessly with specific cloud storage services. Examples of popular office suites and their corresponding cloud services include:

  - If you primarily use Microsoft Office applications, then consider Microsoft OneDrive cloud service.
  - If you primarily use Google Docs, then consider Google Drive cloud service.
  - If you primarily use Apple iWork applications, then consider Apple iCloud cloud service.

- **What hardware do you have?**

  Your hardware choices can influence your best cloud storage option. Apple's iCloud works best on Apple devices. Chromebooks and Android phones are designed around using Google Drive, and Windows PCs have OneDrive designed into the operating system. Also, not just the manufacturer of your device, but the type of device can impact your choice as well. For example, if you do most of your computing on a mobile device, Google Drive has apps and tools that are designed to work with touch screens and mobile Internet connections. However, if you spend most of your time offline, you may find that Google Drive is not as robust as other cloud storage options.

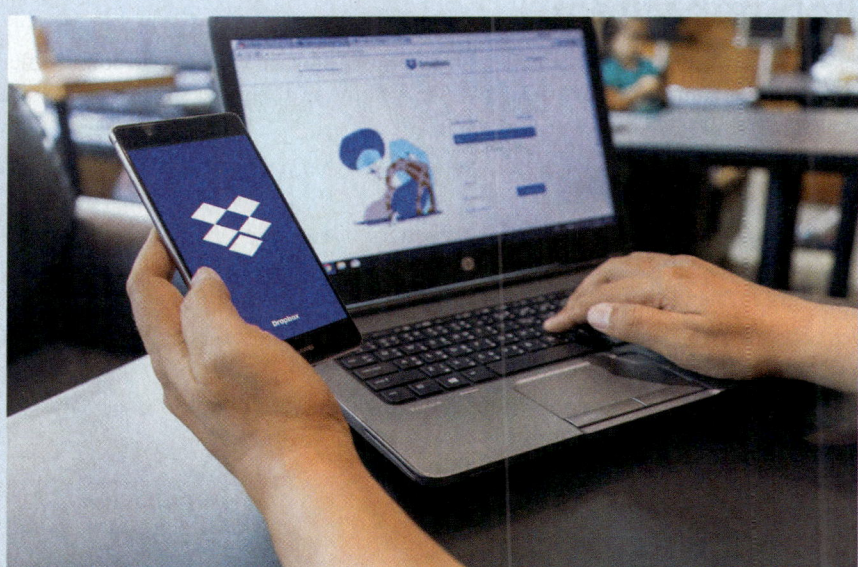

Microsoft OneDrive works well on Windows OS and with Microsoft Office products.

Nopparat Khokthong/Shutterstock

- **How do you want to share?**
One of the best features of cloud storage is the ease of sharing files. Consider who you share with and what you want to share before choosing a cloud storage provider. For example, if you are working on a group project at school, and the school uses Microsoft OneDrive, Microsoft OneDrive might be the easiest solution. If you share files at work, but only want the recipient to be able to read but not write to a file, you may need the more robust security features found in a service like box.com. Sharing photos and videos on Google and Dropbox is easy if the recipient has a Google or Dropbox account; otherwise, it might take a little more effort. Also, consider the technical knowledge of the people you share with. When you share a document, others can read it, but you may decide if they can also make changes to the document (referred to as edit privileges). Denying others edit privileges can insure that your document is not accidentally altered or deleted—but many users will be confused as to why they cannot make edits. Some cloud services are designed to be used by businesses with complex security and backup features. Such features would be valuable to a company of trained professionals sharing sensitive documents, but for many less technical people, it is an unnecessary complication that makes sharing photos and files more difficult. Apple's iCloud makes organizing and sharing photos simple for a typical iPhone user, but that simplicity might be too limiting for a professional photographer. Facebook may store your low-resolution photos for free, but if you want to share a photo, the recipient must sign up for Facebook.

Many standard apps are included free with a mobile OS, including:

- email apps—to view and reply to work and personal emails.
- to-do list and calendar apps to organize and plan work and personal activities.
- alarm and timer apps—to help you wake up or remember to do things.
- messaging and phone call apps to communicate with coworkers and friends.

There are many options in the cloud storage market—finding the right one for you will require understanding how you want to use cloud storage. Recognizing the features and limitations of each service is a start to finding the right place for your family photos or last year's tax records. Finally, most services offer a free trial before purchasing or committing fully—try out a service with different files, on different devices, and sharing with different people to find out which is best for you.

Enterprise Storage System

Most large organizations have established a strategy called an **enterprise storage system** to promote efficient and safe use of data across the networks within their organizations. (See **Figure 7-15**.) Some of the mass storage devices that support this strategy are as follows:

- **File servers**—dedicated computers with very large storage capacities that provide users access to fast storage and retrieval of data.
- **Network attached storage (NAS)**—a type of file server designed for homes and small businesses. NAS is less expensive, easier to set up, and easier to manage than most file servers. However, it does not include powerful management tools and features found in many large-scale file servers.
- **RAID systems**—larger versions of the specialized devices discussed earlier in this chapter that protect data by constantly making backup copies of files moving across the organization's networks.
- **Organizational cloud storage**—high-speed Internet connection to a dedicated remote storage facility. These facilities contain banks of file servers to offer enormous amounts of storage.

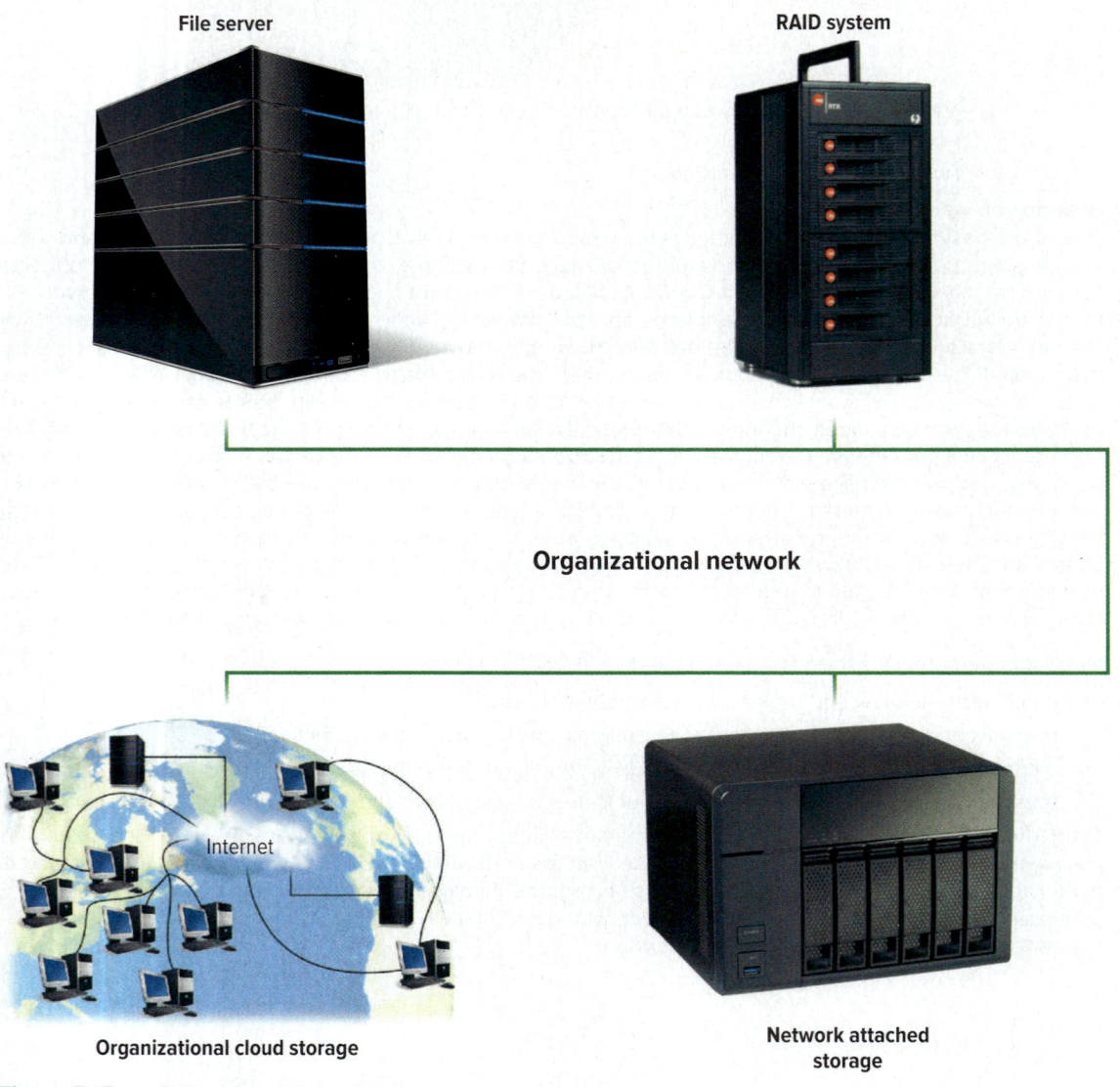

Figure 7-15 Enterprise storage system

(top-left): Gravvi/Shutterstock; (top-right): Copyright 2015, CRU Acquisition Group, LLC. All Rights Reserved.; (bottom-right): 300dpi/Shutterstock.

Storage Area Network

An innovative and growing mass storage technology is **storage area network (SAN)** systems. SAN is an architecture to link remote computer storage devices, such as enterprise storage systems, to computers such that the devices are as available as locally attached drives. In a SAN system, the user's computer provides the file system for storing data, but the SAN provides the disk space for data.

The key to a SAN is a high-speed network, connecting individual computers to mass storage devices. Special file systems prevent simultaneous users from interfering with each other. SANs provide the ability to house data in remote locations and still allow efficient and secure access.

concept check

- Define mass storage and mass storage devices.
- What is an enterprise storage system?
- What is a storage area network system?

Careers in IT

"Now that you've learned about secondary storage, let me tell you a little bit about my career as a disaster recovery specialist."

jeffbergen/Getty Images

Disaster recovery specialists are responsible for recovering systems and data after a disaster strikes an organization. In addition, they often create plans to prevent and prepare for such disasters. A crucial part of that plan is to use storage devices and media in order to ensure that all company data is backed up and, in some cases, stored off-site.

Employers typically look for candidates with a bachelor's or associate's degree in information systems or computer science. Experience in this field is usually required, and additional skills in the areas of networking, security, and database administration are desirable. Disaster recovery specialists should possess good communication skills and be able to handle high-stress situations.

Disaster recovery specialists can expect to earn an annual salary of $63,000 to $112,000. Opportunities for advancement typically include upper-management positions. With so many types of threats facing organizations, demand for these types of specialists is expected to grow.

A LOOK TO THE FUTURE

Next-Generation Storage

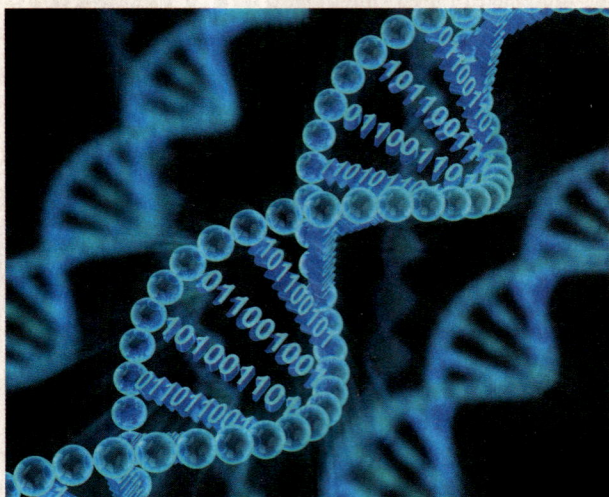

ymgerman/Shutterstock

Every year, cell phones improve. One of their greatest features is the incredible cameras we can now carry in our pockets. As we record longer videos at greater quality, we need more secondary storage to hold them. Scientists are looking to a future where you may carry years worth of videos or walk around with an entire copy of the Internet in your pocket and where every moment of your day will be recorded in vastly larger secondary storage devices than the ones we have today. Such future storage will require new ways of thinking and astounding technological advancements that are currently being studied. In the future, your cell phone may record data using heat rays, lasers, or even DNA.

Our cell phones, tablets, and laptops have secondary storage that uses magnets to store information. A magnetic field writes data on the surface of a magnetized material by flipping a magnetic charge on the surface of the material. The retained magnetic charge on the surface represents one bit of information. The magnetic charge on the surface must be kept far apart from the other bits of information stored elsewhere on the surface. If the two bits are stored too closely together, the magnetic field of one bit could overwrite the data in the other bit. This is one of the major limitations to how much space is required to store information digitally.

To overcome this limitation, scientists are exploring the use of Samsung's Heat-Assisted Magnetic Recording (HAMR) technologies. A HAMR writes data on a special heat-sensitive magnetic surface. The surface can be written on with a magnetic field—but only when the surface is heated. When not heated, the surface will not change. Thus, closely packed bits of information will not overwrite each other as long as the material is not heated. Writing to the surface is accomplished by preheating an area using a precision laser. If this technology can be brought to market, a hard disk could store 50 times more information.

While HAMR technology may allow us to store more information in every square inch of a magnetic surface, Intel's Optane flash memory is exploring the possibility of storing data not on the surface but in 3D multilevel cells (MLCs). Optane uses flash memory technologies to create nonvolatile memory stacks that store information both horizontally and vertically. These dense stacks of bits can be accessed much faster than traditional flash memory—at speeds rivaling RAM access speeds.

The greatest innovations in secondary storage will likely require a major change in the way we store data. Many of the most promising and futuristic secondary storage devices do not use magnets to store information. In nature, living creatures store genetic information on strands of sequenced cells, called DNA. Columbia University used encoding techniques developed at Netflix to store 215 petabytes in one gram of DNA. In 2019, researchers at the University of Washington automated the read and write processes used to record data on DNA—bringing the possibility of DNA secondary storage even closer.

Obviously, the technology to read and write DNA data is neither cheap nor common, but when it comes to storing data, nature has been doing it longer and better than we have. Perhaps in the future, your family photos won't be stored on DVDs, but instead on genetic material. Do you think that the expenses associated with DNA processing will ever reduce enough to make DNA data storage useful, or will we continue with hard drives and flash memory?

VISUAL SUMMARY | Secondary Storage

STORAGE

Maxx-Studio/Shutterstock

RAM is **primary storage**. Most RAM is volatile, meaning that it loses its contents whenever power is disrupted. **Secondary storage** provides nonvolatile storage. Secondary storage retains data and information after the computer system is turned off.

Writing is the process of saving information to **secondary storage devices (drives)**. Reading is the process of accessing information.

Important characteristics include the following:

- **Media**—actual physical material that retains data and programs.
- **Capacity**—how much a particular storage medium can hold.
- **Storage devices**—hardware that reads and writes to storage media.
- **Access speed**—time required to retrieve data from a secondary storage device.

SOLID-STATE STORAGE

kyoshino/E+/Getty Images

Solid-state storage devices provide access to **flash memory**, also known as **solid-state storage**. Flash memory is slower than traditional memory but faster than other secondary storage drives. **Solid-state storage** devices have no moving parts and provide access to **flash memory (solid-state storage)**.

Solid-State Drives

Solid-state drives (SSDs) work much like RAM. They are fast and durable, require little power, are relatively expensive, and are lower in capacity than Hard Disc Drives (HDDs). SSDs can provide internal storage or can be located outside the system unit.

Flash Memory Cards

Flash memory cards are small solid-state storage devices that are widely used with portable devices including digital cameras, cell phones, and navigation systems.

USB Flash Drives

USB flash drives (flash drives) are so small that they fit onto a key ring. These drives connect to a computer's USB port and are widely used to transfer data and information between computers, specialty devices, and the Internet.

To efficiently and effectively use computers, you need to be aware of the different types of secondary storage. You need to know their capabilities, limitations, and uses. There are four widely used storage media: hard disk, solid state, optical disc, and cloud storage.

HARD DISKS

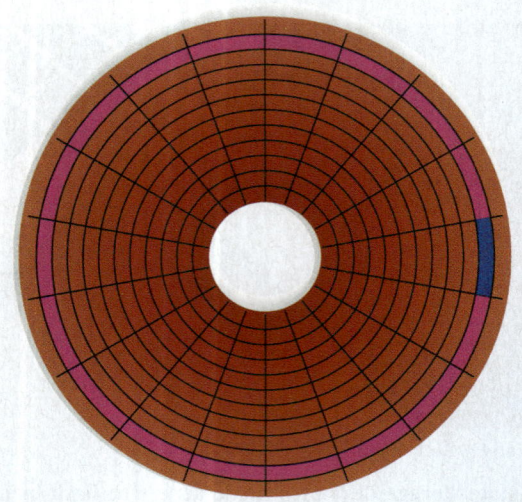

Hard disks use rigid metallic **platters** that provide a large amount of capacity. They store data and programs by altering the electromagnetic charges on the platter's surface. Files are organized according to

- **Tracks**—concentric rings on the platter.
- **Sectors**—wedge-shaped sections.
- **Cylinders**—run through each track of a stack of platters.

Density refers to how tightly electromagnetic charges can be packed next to one another on the disk.

Internal Storage

Internal storage is located within the system unit. They are used to store programs and data files.

External Storage

External storage connects to a port on the system unit, is not as fast as internal drives, and is removable, providing unlimited storage using a single port.

HARD DISKS

Santi Nanta/Shutterstock

Network Drives

Network drives are located on a network anywhere in the world.

Performance Enhancements

Four ways to improve hard-disk performance are disk caching, hybrid drives, RAID, and file compression and decompression.

- **Disk caching**—provides a temporary high-speed holding area between a secondary storage device and the CPU; improves performance by anticipating data needs and reducing time to access data from secondary storage.
- **Hybrid drives**—contain both solid-state storage and hard disks in an attempt to gain the speed and power benefits of SSDs while still having the low cost and large capacity of hard drives.
- **RAID (redundant array of inexpensive disks)**—several inexpensive hard-disk drives are connected together; improves performance by providing expanded storage, fast access, and high reliability.
- **File compression and decompression**—files are compressed before storing and then decompressed before being used again; improves performance through efficient storage.

OPTICAL DISCS

limpido/Shutterstock

Optical discs use laser technology. 1s and 0s are represented by **pits** and **lands**. **Optical disc drives** project light and measure the reflected light.

The most widely used optical discs are the following:

- **Compact discs (CDs)** were the first; typical storage 700 MB; often used to store music.
- **Digital versatile discs (DVDs)** are standard optical discs; typical storage 4.7 GB; often used to store movies and software.
- **Blu-ray discs (BDs)** are designed to store **hi-def (high-definition)** video; typical storage 50 GB; often used to store hi-def video and video games; **Ultra HD Blu-rays (UHD BD)** are the newest and able to use 4K video content and store up to 100 GB of data.

Each of these optical discs has three basic formats:

- **Read-only** (**ROM** for **read-only memory**) discs cannot be written on or erased by the user.
- **Write-once** (**R** for **recordable**) discs can be written on once and read many times (**WORM, write once read many**).
- **Rewritable** (**RW** for **rewritable** or **RAM** for **random-access memory**) discs are similar to write-once discs except that the disc surface is not permanently altered when data is recorded.

CLOUD STORAGE

Nopparat Khokthong/Shutterstock

With **cloud computing,** the Internet acts as a "cloud" of servers. **Cloud storage (online storage)** is supplied by these servers.

- Cloud servers provide storage, processing, and memory.
- Advantages for users include less maintenance, fewer hardware upgrades, and easy file sharing and collaboration.
- Disadvantages for users include slower access speed and less control over file security.

MASS STORAGE DEVICES

Mass storage refers to the tremendous amount of secondary storage required by large organizations. **Mass storage devices** are specialized high-capacity secondary storage devices.

Most large organizations have established a strategy called an **enterprise storage system** to promote efficient and safe use of data.

Mass storage devices that support this strategy are **file servers, network attached storage (NAS), RAID systems,** and **organizational cloud storage**. A **storage area network (SAN)** is a method of using enterprise-level remote storage systems as if they were local to your computer.

CAREERS in IT

Disaster recovery specialists are responsible for preparing for and recovering from data and system disasters in an organization. Bachelor's or associate's degree in information systems or computer science, experience, and additional skills in the areas of networking, security, and database administration are desirable. Expected salary range is $63,000 to $112,000.

KEY TERMS

- access speed
- Blu-ray disc (BD)
- capacity
- cloud computing
- cloud storage
- compact disc (CD)
- cylinder
- density
- digital versatile disc or digital video disc (DVD)
- disaster recovery specialist
- disk caching
- drives
- enterprise storage system
- external storage
- file compression
- file decompression
- file server
- flash drive
- flash memory
- flash memory card
- hard-disk drive
- hi-def (high-definition)
- hybrid drive
- internal storage
- land
- mass storage
- mass storage devices
- network attached storage (NAS)
- network drive
- online storage
- optical disc
- optical disc drive
- organizational cloud storage
- pit
- platter
- primary storage
- RAID system
- random-access memory (RAM) disc
- read-only memory (ROM) disc
- recordable (R) disc
- redundant array of inexpensive disks (RAID)
- rewritable (RW) disc
- secondary storage
- secondary storage device
- sector
- solid-state drive (SSD)
- solid-state storage
- storage area network (SAN)
- storage device
- storage media
- track
- Ultra HD Blu-ray (UHD BD)
- USB flash drive
- write-once disc
- write once read many (WORM)

MULTIPLE CHOICE

Circle the correct answer.

1. Provides permanent or nonvolatile storage.
    - **a.** secondary storage
    - **b.** RAM
    - **c.** primary storage
    - **d.** NAS
2. Hardware that reads data and programs from storage media.
    - **a.** storage devices
    - **b.** density
    - **c.** sectors
    - **d.** disk caching
3. Concentric rings on a hard-disk platter.
    - **a.** sectors
    - **b.** tracks
    - **c.** vectors
    - **d.** platters
4. Each track is divided into invisible wedge-shaped sections called _____.
    - **a.** tracks
    - **b.** sectors
    - **c.** vectors
    - **d.** platters
5. Increases storage capacity by reducing the amount of space required to store data and programs.
    - **a.** file server
    - **b.** file compression
    - **c.** storage area network
    - **d.** hybrid drive
6. Optical disc most common on today's personal computers.
    - **a.** DVD
    - **b.** flash memory card
    - **c.** NAS
    - **d.** sector
7. Which of the following is NOT an important characteristic of secondary storage?
    - **a.** storage media
    - **b.** capacity
    - **c.** access speed
    - **d.** file compression
8. Similar to internal hard-disk drives except they use solid-state memory.
    - **a.** DVDs
    - **b.** cloud storage
    - **c.** solid-state drives
    - **d.** enterprise storage
9. Mass storage device widely used for home and small business storage.
    - **a.** cloud storage
    - **b.** flash drive
    - **c.** Blu-ray
    - **d.** network attached storage
10. Architecture to link remote storage devices to computers such that the devices are as available as locally attached drives.
    - **a.** flash drive
    - **b.** RAID
    - **c.** platter
    - **d.** storage area network

MATCHING

Match each numbered item with the most closely related lettered item. Write your answers in the spaces provided.

a. cloud
b. density
c. disk caching
d. enterprise storage system
e. flash memory cards
f. mass storage devices
g. storage media
h. primary storage
i. solid state
j. tracks

_____ 1. Another name for online storage is _____ storage.
_____ 2. Commonly used on cameras and cell phones to expand their internal storage.
_____ 3. This hard-disk performance enhancement anticipates data needs.
_____ 4. The name for the concentric circles on a hard disk.
_____ 5. This type of storage uses pits and lands to represent 1s and 0s.
_____ 6. RAM is sometimes referred to as _____.
_____ 7. The actual physical material that holds the data and programs.
_____ 8. This measures how tightly the magnetic charges can be packed next to one another on the disk.
_____ 9. An organizational strategy to promote efficient and safe use of data across the networks.
_____ 10. Specialized high-capacity secondary storage devices designed to meet organizational demands.

OPEN-ENDED

On a separate sheet of paper, respond to each question or statement.

1. Compare primary storage and secondary storage, and discuss the most important characteristics of secondary storage.
2. Discuss solid-state storage, including solid-state drives, flash memory, and USB drives.
3. Discuss hard disks and storage, including density, platters, tracks, sectors, cylinders, internal, external, and performance enhancements.
4. Discuss optical discs, including pits, lands, CDs, DVDs, Blu-ray, and hi-def.
5. Discuss cloud computing and cloud storage.
6. Describe mass storage devices, including enterprise storage systems, file servers, network attached storage, RAID systems, organizational cloud storage, and storage area network systems.

DISCUSSION

Respond to each of the following questions.

 ## Making IT Work for You: CLOUD STORAGE

Have you ever found yourself e-mailing files back and forth between two of your computers or with others as a way to transport them? Review the Making IT Work for You: Cloud Storage on pages 172 and 173. Then respond to the following: (a) Have you ever used Dropbox or a similar service? If so, what service have you used, and what do you typically use it for? If you have not used Dropbox or a similar service, describe how and why you might use one. (b) If you do not have a Dropbox account, set up a free one and create a Dropbox folder. Use Dropbox to either (1) access a file from another computer or (2) share a file with one of your classmates. Describe your experience. (c) Try a few of Dropbox's features, and describe your experience with these features. (d) Do you see yourself using Dropbox on an everyday basis? Why or why not?

 ## Privacy: RIGHT TO BE FORGOTTEN

As a generation grows up with social media, a surplus of youthful indiscretions is now stored on the Internet for all to see. Review the Privacy box on page 167 and respond to the following: (a) Is there a photo or video of you on the Internet that you would prefer not to be publicly available? Have you said or done things that, if recorded and posted on social media, could have a negative impact on a job interview? (b) Do you have the right to decide what photos of you are posted on the Internet by others? Why or why not? (c) Does someone else have the right to tell you what to do with the photos you take, even if they are in the photo? Why or why not? (d) Should social media companies remove photos, videos, or messages if someone is embarrassed by them? Should social media companies have the right to remove your photos, videos, or messages if someone is embarrassed by the content? Justify your answer.

 ## Ethics: CLOUD STORAGE AND CONFIDENTIALITY

When individuals and businesses store files using cloud services, they expect the cloud company to behave ethically by providing adequate security to protect confidential files. What if this expectation is not met? Review the Ethics box on page 171, and then respond to the following: (a) Would you be comfortable if your attorney stored digital copies of your legal documents in the cloud? What about your doctor or psychologist? Why or why not? (b) Who should be responsible if files stored in the cloud are stolen or viewed by hackers or unethical employees? Who would suffer the consequences? Defend your position. (c) Should laws be created that require cloud storage companies to operate ethically and to assume responsibility for security and confidentiality of stored data? Why or why not? (d) Cloud computers are not necessarily located within the borders of the United States and therefore may not be subject to the same regulations as U.S.-based computers. Do you think that all U.S. companies should be required to keep their cloud servers in this country? Defend your response. (e) How do you feel about storing personal and confidential information in the cloud? Do you currently do it? Why or why not?

 ## Community: DANGEROUS USB FLASH DRIVES

Review the Community box on page 165 and then respond to the following: (a) Have you (or someone you know) ever attended an event where USB flash drives were handed out for free? Have you (or someone you know) ever found a USB flash drive on the ground? (b) Often, USB flash drives are handed out at booths at convention centers and other community gatherings. How closely monitored are such booths? Would it be possible for someone to get a free drive, infect it, and then return it to the booth among the other drives being handed out? (c) Are there any circumstances where you would trust an unknown USB flash drive? What are the potential costs to plugging in an unknown drive? What are the potential benefits?

Design Elements: Concept Check icon: Dizzle52/Getty Images

chapter **8**

Communications and Networks

Blue Planet Studio/Shutterstock

CHAPTER 8: Communications and Networks

Why should I read this chapter?

Elnur/Shutterstock

Communication networks are the backbone of nearly every aspect of modern digital life. In the future, telepresence (the ability to fully experience the reality of a different place without actually being there) will be commonplace. For example, doctors will routinely perform surgery on patients located halfway around the world!

This chapter covers the things you need to know to be prepared for this ever-changing digital world, including:

- Mobile computing—become a digital road warrior using cellular data networks and GPS.
- Wireless networks—use your digital devices in smarter and safer ways by understanding Wi-Fi, satellites, and Bluetooth.
- Wired networks—learn about coaxial and fiber-optic cables so you can make smart decisions about home Internet connections.

Learning Objectives

After you have read this chapter, you should be able to:

1. Explain connectivity, the wireless revolution, and communication systems.
2. Describe wireless and physical communication channels.
3. Differentiate between connection devices and services, including cellular, dial-up, DSL, cable, and satellite.
4. Describe data transmission factors, including bandwidth and protocols.
5. Define networks and key network terminology, including network interface cards and network operating systems.
6. Describe different types of networks, including local, home, wireless, personal, metropolitan, and wide area networks.
7. Describe network architectures, including topologies and strategies.
8. Explain the organizational issues related to Internet technologies and network security.

Introduction

"Hi, I'm Michael, and I'm a network administrator. I'd like to talk with you about computer communications and networks. I'd also like to talk about technologies that support mobile computing including global positioning systems, Wi-Fi, and 4G and 5G networks."

Dmitry Kalinovsky/Shutterstock

At one time, network-connected computers were only found at the largest businesses and universities. Today, with cell phones, we carry network-connected computers in our pockets. Smart appliances in your kitchen mean your refrigerator may be connected to the Internet. The Internet of Things means your watch or your thermostat may be using networks to communicate. The communications and information options we have at our fingertips have changed how we react and relate to the world around us.

As the power and flexibility of our communication systems have expanded, the sophistication of the networks that support these systems has become increasingly critical and complex. The network technologies that handle our cellular, business, and Internet communications come in many different forms. Satellites, broadcast towers, telephone lines, even buried cables and fiber optics carry our telephone messages, e-mail, and text messages. These different networks must be able to efficiently and effectively integrate with one another.

To efficiently and effectively use computers, you need to understand the concept of connectivity, wireless networking, and the elements that make up network and communications systems. Additionally, you need to understand the basics of communications channels, connection devices, data transmission, network types, network architectures, and organizational networks.

Communications

Computer Communication is the process of sharing data, programs, and information between two or more computers. We have discussed numerous applications that depend on communication systems to share data among devices, including

- **Texting**—provides very efficient direct text communication between individuals using short electronic messages.
- **E-mail**—provides a fast, efficient alternative to traditional mail by sending and receiving electronic documents.
- **Videoconferencing**—provides an alternative to meeting in person using audio communications (such as a phone call) often with video to see the people you talk to.
- **Electronic commerce**—buying and selling goods electronically.

In this chapter, we will focus on the communication systems that support these and many other applications. Connectivity, the wireless revolution, and communication systems are key concepts and technologies for the 21st century.

Connectivity

Connectivity is a concept related to using computer networks to link people and resources. For example, connectivity means that you can connect your cell phone to other devices and information sources from almost anywhere. With this connection, you are linked to the world of larger computers and the Internet. This includes hundreds of thousands of web servers and their extensive information resources. Thus, being able to efficiently and effectively use computers becomes a matter of knowing not only about connectivity through networks to cell phones but also about larger computer systems and their information resources.

The Wireless Revolution

Figure 8-1 Wireless revolution
pics five/Shutterstock

The single most dramatic change in connectivity and communications since the development of the Internet has been the widespread use of mobile devices like cell phones and tablets with fast wireless Internet connectivity. Students, parents, teachers, businesspeople, and others routinely talk and communicate with these devices. It is estimated that over 4.6 billion cell phones are in use worldwide. This wireless technology allows individuals to stay connected with one another from almost anywhere at any time.

So what's the revolution? Although wireless technology was originally used primarily for voice communications, today's mobile computers support e-mail, web access, social networking, and a variety of Internet applications. In addition, wireless technology allows a wide variety of nearby devices to communicate with one another without any physical connection. Wireless communications allow you to share a high-speed printer, share data files, and collaborate on working documents with a nearby co-worker without having your computers connected by cables or telephone. High-speed Internet wireless technology allows individuals to connect to the Internet and share information from almost anywhere in the world. (See **Figure 8-1**.) The revolution has just begun, and where it goes from here will be decided by you and others who understand the power of communication systems.

Communication Systems

Communication systems are electronic systems that transmit data from one location to another. Whether wired or wireless, every communication system has four basic elements. (See **Figure 8-2**.)

- **Sending and receiving devices.** These are often a computer or specialized communication device. They originate (send) as well as accept (receive) messages in the form of data, information, and/or instructions.
- **Connection devices.** These devices act as an interface between the sending and receiving devices and the communication channel. They convert outgoing messages into packets that can travel across the communication channel. They also reverse the process for incoming messages.

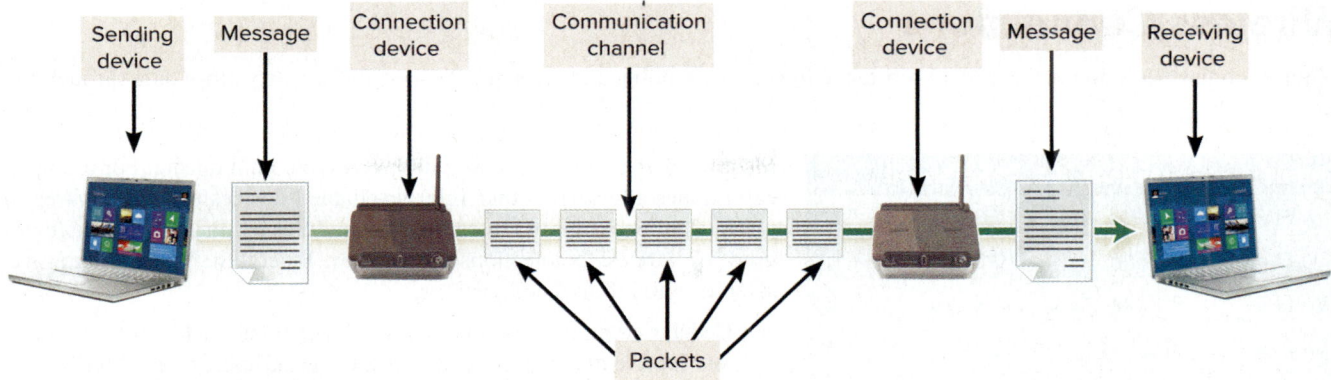

Figure 8-2 Basic elements of a communication system
Sergey Peterman/Shutterstock

- **Data transmission specifications.** These are rules and procedures that coordinate the sending and receiving devices by precisely defining how the message will be sent across the communication channel.
- **Communication channel.** This is the actual connecting or transmission medium that carries the message. This medium can be a physical wire or cable, or it can be wireless.

For example, if you wanted to send an e-mail to a friend, you could create and send the message using your cell phone, the *sending device*. Your modem is on a microchip located inside your cell phone. It functions as the *connection device* that would modify and format the message so that it could travel efficiently across *communication channels,* such as cell phone towers. The specifics describing how the message is modified, reformatted, and sent would be described in the *data transmission specifications.* After your message traveled across the channel, a connection device, such as a modem connected to a desktop computer, would reformat it so that it could be displayed on your friend's computer, the *receiving device.* (Note: This example presents the basic communication system elements involved in sending e-mail. It does not and is not intended to demonstrate all the specific steps and equipment involved in an e-mail delivery system.)

> **ethics**
>
> As eavesdropping tools become more sophisticated, there is concern that law enforcement and government agencies will monitor everyone's Internet and cell phone activity. In the private sector, companies are increasingly using network tools and software to monitor the activity of their employees. Many websites also track your activity, and government officials have often requested these records during the course of an investigation. Some believe that it is unethical for government and businesses to engage in such monitoring and tracking. Do you agree?

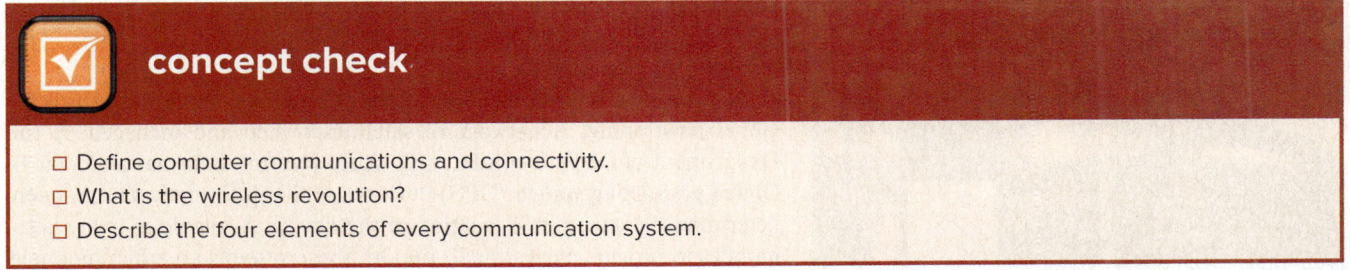

concept check

☐ Define computer communications and connectivity.
☐ What is the wireless revolution?
☐ Describe the four elements of every communication system.

Communication Channels

Communication channels are an essential element of every communication system. These channels actually carry the data from one computer to another. There are two categories of communication channels, wireless and wired. Wireless communications send information through the air, such as when you use a cell phone. Wired communications require a physical connection and are more common with desktop computers.

Wireless Connections

Wireless connections do not use a solid substance to connect sending and receiving devices. Rather, they move data through the air.

Name	Standard	Max Speed
Wi-Fi 3	802.11g	0.054 Gbps
Wi-Fi 4	802.11n	0.6 Gbps
Wi-Fi 5	802.11ac	3.5 Gbps
Wi-Fi 6	802.11ax	9.6 Gbps
Wi-Fi 7	802.11be	46 Gbps

Figure 8-3   Wi-Fi standards

Most wireless connections use radio waves to communicate. For example, cell phones and many other Internet-enabled devices use radio waves to place telephone calls and to connect to the Internet. Primary technologies used for wireless connections are cellular, Bluetooth, Wi-Fi, microwave, satellite, and infrared connections.

- **Cellular** communication uses multiple antennae (**cell towers**) to send and receive data within relatively small geographic regions (**cells**). Most cell phones and other mobile devices use cellular networks.
- **Bluetooth** is a short-range radio communication standard that transmits data over short distances of up to approximately 33 feet. Bluetooth is widely used for wireless headsets, printer connections, and handheld devices.
- **Wi-Fi (wireless fidelity)** uses high-frequency radio signals to transmit data. A number of standards for Wi-Fi exist, and each can send and receive data at a different speed. (See **Figure 8-3**.) Most home and business wireless networks use Wi-Fi.
- **Microwave** communication uses high-frequency radio waves. It is sometimes referred to as line-of-sight communication because microwaves can only travel in a straight line. Because the waves cannot bend with the curvature of the earth, they can be transmitted only over relatively short distances. Thus, microwave is a good medium for sending data between buildings in a city or on a large college campus. For longer distances, the waves must be relayed by means of microwave stations with microwave dishes or antennas. (See **Figure 8-4**.)

Figure 8-4   Microwave dish
Mariusz Burcz/Alamy Stock Photo

- **Satellite** communication uses satellites orbiting about 22,000 miles above the earth as microwave relay stations. Many of these are offered by Intelsat, the International Telecommunications Satellite Consortium, which is owned by 114 governments and forms a worldwide communication system. Satellites orbit at a precise point and speed above the earth. They can amplify and relay microwave signals from one transmitter on the ground to another. Satellites can be used to send and receive large volumes of data. **Uplink** is a term relating to sending data to a satellite. **Downlink** refers to receiving data from a satellite. The major drawback to satellite communication is that the data connection can be interrupted by bad weather or when inside a building.

Figure 8-5   GPS navigation
Aleksey Boldin/123RF

One of the most interesting applications of satellite communications is for global positioning. A network of satellites owned and managed by the Department of Defense continuously sends location information to earth. **Global positioning system (GPS)** devices use that information to uniquely determine the geographic location of the device. Available in many automobiles to provide navigational support, these systems are often mounted into the dash with a monitor to display maps and speakers to provide spoken directions. Most of today's cell phones and tablets use GPS technology for handheld navigation. (See **Figure 8-5**.)

- **infrared** connections use infrared light waves to communicate over short distances. Like microwave transmissions, infrared is a line-of-sight communication. Because light waves can only travel in a straight line, sending and receiving devices must be in clear view of one another without any obstructions blocking that view. One of the most common infrared devices is the TV remote control.

Physical Connections

Physical connections use a solid medium to connect sending and receiving devices. These connections include twisted-pair, coaxial, and fiber-optic cables.

- **Twisted-pair cable** consists of pairs of copper wire that are twisted together. Both landline **telephone lines** and **Ethernet cables** use twisted pair. (See **Figure 8-6**.) Ethernet cables are often used in networks to connect a variety of components to the system unit.
- **Coaxial cable**, a high-frequency transmission cable, replaces the multiple wires of telephone lines with a single solid-copper core. (See **Figure 8-7**.) In terms of the number of telephone connections, a coaxial cable has over 80 times the transmission capacity of twisted pair. Coaxial cable is used to deliver television signals as well as to connect computers in a network.
- **Fiber-optic cable** transmits data as pulses of light through tiny tubes of glass. (See **Figure 8-8**.) The data transmission speeds of fiber-optic cables are incredible; recently, speeds of 1 petabit per second were measured (a petabit is 1 million gigabits). Compared to coaxial cable, it is lighter, faster, and more reliable at transmitting data. Fiber-optic cable is rapidly replacing twisted-pair cable telephone lines.

community

Much of the developing world does not have access to the Internet. SpaceX's Starlink looks to change that by launching a network of over 4,000 low-orbit satellites to provide Internet to the unserved and underserved communities of the world. Proponents of the network suggest that the citizens of developing nations will be greatly benefited by the essential services that the Internet can provide, such as education and health care. Opponents point out that the majority of developing nation citizens cannot afford the devices necessary to connect to the Internet. What do you think? Will bringing the Internet to developing countries help the average citizen, or will this only benefit the wealthiest in these countries?

Figure 8-6 Ethernet cable
Kamolrat/Shutterstock

Figure 8-7 Coaxial cable
DestinaDesign/Shutterstock

Figure 8-8 Fiber-optic cable
zentilia/Shutterstock

concept check

- What are communication channels? List three physical connections.
- What is cellular communication? Wi-Fi? Microwave communication?
- What is satellite communication? GPS? Infrared?

Connection Devices

At one time nearly all computer communication used telephone lines. However, because the telephone was originally designed for voice transmission, telephone lines were designed to carry **analog signals**, which are continuous electronic waves. Computers, in contrast, send and receive **digital signals**. (See **Figure 8-9**.) These represent the presence or absence of an electronic pulse—the binary signals of 1s and 0s we mentioned in **Chapter 5**. To convert the digital signals to analog signals and vice versa, you need a modem.

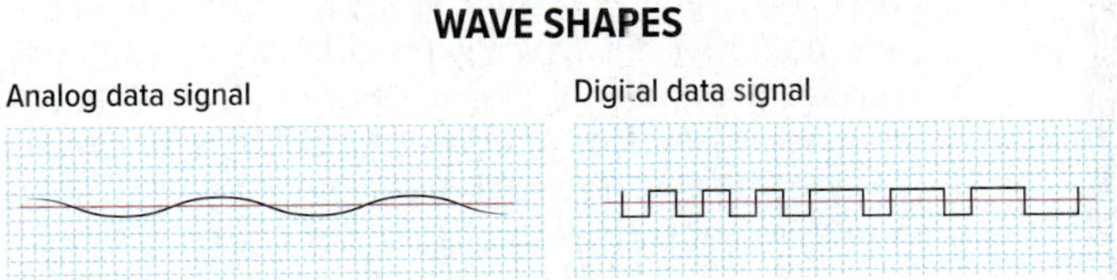

WAVE SHAPES

Analog data signal   Digital data signal

Figure 8-9 Analog and digital signals
Designua/Shutterstock

Modems

The word **modem** is short for *modulator-demodulator*. **Modulation** is the name of the process of converting from digital to analog. **Demodulation** is the process of converting from analog to digital. The modem enables digital personal computers to communicate across different media, including telephone wires, cable lines, and radio waves.

Unit	Speed
Mbps	Million bits per second
Gbps	Billion bits per second
Tbps	Trillion bits per second

Figure 8-10 Typical transfer rates

The speed with which modems transmit data varies. This speed, called **transfer rate**, is typically measured in millions of bits **(megabits) per second (Mbps)**, and some are fast enough to measure in billions of bits (gigabits) per second (Gbps). (See **Figure 8-10**.) The higher the speed, the faster you can send and receive information. For example, a typical 4K 2-hour movie requires 28 GB of data. This would take over 3 hours at 20 Mbps; however, it would take less than 4 minutes at 1 Gbps.

There are three commonly used types of modems: DSL, cable, and wireless. (See **Figure 8-11**.)

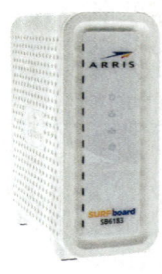 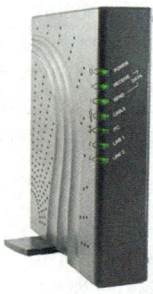 

Figure 8-11 Basic types of modems
(left): Keith Homan/Alamy Stock Photo; (middle): blue_iq/E+/Getty Images; (right): Ascannio/Shutterstock

tips

When it comes to the Internet, faster is better—but also more expensive. How fast do you need your Internet to be? There are a few factors to consider.

1. **What video devices will use your Internet?** Video is one of the heaviest demands on Internet bandwidth. If you have a 4K TV, you are going to want to watch 4K movies. A 4K movie typically requires an Internet connection of 25 Mbps or faster. Lower resolution devices, such as most tablets and cell phones require less bandwidth to display video.

2. **How many people use the Internet at the same time?** If multiple family members watch multiple 4K movies at the same time—you will need 25 Mbps *for each movie being watched*. Also consider other bandwidth-heavy activities such as listening to music, posting or viewing social media, and videoconferencing.

3. **What are your upload and download needs?** Most of our use of the Internet is downloading music, movies, and TV. We do not tend to upload nearly as often. Consequently, Internet providers typically provide much higher download speeds than upload speeds. One application that does require high upload speeds is videoconferencing. Each videoconferencing session requires download speeds of 10–25 Mbps and upload speeds of at least 3 Mbps.

- A **DSL (digital subscriber line)** modem uses standard phone lines to create a high-speed connection directly to your phone company's offices. These devices are usually external and connect to the system unit using either a USB or an Ethernet port.
- A **cable modem** uses the same coaxial cable as your television. Like a DSL modem, a cable modem creates high-speed connections using the system unit's USB or Ethernet port.
- A **wireless modem** is also known as a **WWAN (wireless wide area network) modem**. Almost all computers today have built-in wireless modems. For those that do not, wireless adapter cards are available that plug into USB or special card ports.

Connection Service

For years, large corporations have been leasing special high-speed lines from telephone companies. Originally, these were copper lines, known as **T1** lines, that could be combined to form higher-capacity options known as **T3** or **DS3** lines. These lines have largely been replaced by faster **optical carrier (OC)** lines.

Although the special high-speed lines are too costly for most individuals, Internet service providers (as discussed in Chapter 2) do provide affordable connections. For years, individuals relied on **dial-up services** using existing telephones and telephone modems to connect to the Internet. This type of service has been replaced by higher-speed connection services, including cellular, DSL, cable, and satellite services.

- **Cellular service providers**, including Verizon, AT&T, Sprint, and T-Mobile, support voice and data transmission to wireless devices using cellular networks. The progress of mobile telecommunication is measured in generations. The first generation (1G) started in the 1980s and has progressed to today's fifth generation (5G):
    - **First generation (1G)** could only transmit analog signals used for voice communications.
    - **Second generation (2G)** used radio signals for cellular calls and introduced texting.
    - **Third generation (3G)** improved data speeds and introduced widespread use of the Internet on cell phones.
    - **Fourth generation (4G)** uses **Long Term Evolution (LTE)** connections for increased Internet speeds, making streaming videos and music popular on cell phones.
    - **Fifth generation (5G)** is the newest and fastest network, with speeds rivaling home Internet connections. The 5G network is a new technology and may not be available yet in your area as service providers update old antennae and add new antennae to the network.
- **Digital subscriber line (DSL) service** is provided by telephone companies using existing telephone lines to provide high-speed connections. **ADSL (asymmetric digital subscriber line)** is one of the most widely used types of DSL. DSL is much faster than dial-up.
- **Cable service** is provided by cable television companies using their existing television cables. These connections are usually faster than DSL.
- **Fiber-optic service (FiOS)** often offers faster speeds, than DSL and cable. However, FiOS requires upgraded technologies and infrastructure—which means it may not be available in all areas. Current providers of FiOS include Google and Verizon.
- **Satellite connection services** use satellites to provide wireless connections. Can offer speeds competitive with DSL and cable, but satellite connections can be impacted by weather and are often more expensive than other Internet options. However, satellite connections can be established in remote areas where the infrastructure for other types of Internet may not exist.

To learn more about how you can use mobile communications, see Making IT Work for You: The Mobile Office on pages 192 and 193.

concept check

- What is the function of a modem? Compare the three types of modems.
- What is a connection service? Compare the five high-speed connection services.
- Describe the five generations of mobile telecommunications.

Data Transmission

Several factors affect how data is transmitted. These factors include bandwidth and protocols.

Making IT work for you

THE MOBILE OFFICE

With the proper devices and software, you can take your work away from the office and move your studies outside the classroom. Mobile tools can allow you to set up an effective workspace at a coffee shop, in an airport, or at the kitchen table. A mobile office gives you the flexibility to get work done whenever you have a spare moment, wherever you may be.

Two of the most important features of the mobile office are your Internet connection when on the go and the tools to replicate the resources of working in person. For accessing the Internet, there are three popular options: mobile hotspot devices, personal hotspots, and public Wi-Fi.

Mobile hotspot devices, such as the Nighthawk LTE Mobile Hotspot Router, create an Internet-connected Wi-Fi network on the go.

Virrage Images/Shutterstock

- **Mobile Hotspot Device**
  Many cell phone service providers offer standalone devices that connect to a 4G or 5G network and allow multiple devices near it to access the Internet via Wi-Fi connection. Most home office setups use a lot of data on your data connection. When choosing a hotspot device from a cell phone service provider, consider the provider's data options and pricing.

- **Personal Hotspot**
  Many cell phones can act as mobile hotspot devices, allowing nearby tablets and laptops to connect to the Internet. Sometimes called tethering, devices can connect to the cell phone via Bluetooth, Wi-Fi, or USB cable and, through that connection, access the Internet. Personal hotspots offer a simple mobile Internet solution without the additional cost of purchasing a mobile hotspot device. Unfortunately, using a personal hotspot will drain your cell phone's battery more quickly, will increase your phone's data plan usage, and may require some extra effort to set up. Contact your cell phone provider to learn about the costs and processes associated with setting up a personal hotspot.

A personal hotspot allows devices to connect to the Internet over your phone's data plan.

Aleksey H/Shutterstock

- **Public Wi-Fi**
  Even if you can't use the Internet on the go—you can probably get online when you arrive. Many public locations, especially airports, hotels, and coffee shops, offer free or low-cost Internet connections. However, you should always practice good security habits when using public Wi-Fi. Do not assume that the connection is secure and avoid entering sensitive data when on public Wi-Fi. One way to increase your security on public Wi-Fi is the use of a Virtual Private Network (VPN) service, discussed later in this chapter.

An Internet connection is valueless if you don't have the right software to take advantage of it. The two most common mobile office tools are cloud storage to have access to your files from anywhere, and videoconferencing software to communicate and collaborate with your teammates.

- **Cloud Storage**
  When on the go, you want your documents to be available, up-to-date, and secure. With an Internet connection, you can connect to cloud storage services that will hold your important documents. Through cloud storage, you can collaborate with co-workers and always be confident you have the most up-to-date version of the document. Further, if you lose or damage your laptop, your documents are stored in the cloud and will not be lost. Finally, cloud storage syncs across multiple devices, so you can review a spreadsheet on your cell phone, make changes on your laptop, and then share it with co-workers on your tablet. Popular cloud services are Google Drive, Microsoft OneDrive, and Apple's iCloud.

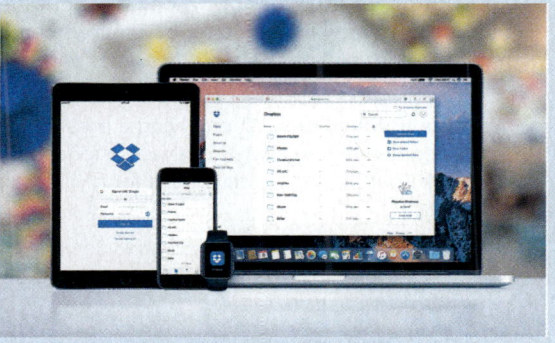

Dropbox is a cloud storage service that works easily with multiple operating systems.

Alexey Boldin/Shutterstock

The Microsoft videoconferencing tool, Skype, is a popular way to communicate with friends, family, and co-workers.

Andriy Popov/123RF

- **Videoconferencing**
  Videoconferencing can allow you to have a face-to-face meeting without being in the same room—or even the same country. A camera, microphone, and speakers will be needed to videoconference—most cell phones, tablets, and laptops come with these accessories.

  Here are the most popular videoconferencing apps:

  - Microsoft Teams is a videoconferencing tool that works well with other Microsoft products such as OneDrive and Microsoft Office 365.
  - FaceTime is a videoconferencing application from Apple that has a simple interface and is integrated into all Apple devices.
  - Zoom is a popular videoconferencing app that offers many features valuable to large meetings and online classrooms.
  - Meet is Google's solution to videoconferencing. It includes many features to organize and effectively run a meeting with several participants, including extensive integration with other Google apps, such as Gmail, Google Calendar, Google Drive, and Google Workspace.

tips

One of the best things about a cell phone is its network connection. However, always being connected can be stressful. To customize your cell phone's notifications and take a break from your connectivity, consider the following:

 Mute notifications. When giving a presentation or trying to spend some time away from your phone, it can be nice to mute your phone's notifications for a while. Here's how to do it.

- For Android cell phones: Swipe down from the top of the screen to display the notification shade and turn on the *Do Not Disturb* button.
- For iOS cell phones: Swipe down from the top right corner of the screen to display the control center and tap the crescent moon button.

 Disable an app's notifications. If you have an app that is annoying you with notifications, you can limit or disable its ability to send notifications.

- For Android cell phones: Click on the *Settings* icon from the home screen, then choose *Apps & Notifications*. Tap on *Notifications and App Notifications*. From here, you can click on an app to customize its notification settings.
- For iOS cell phones: Tap on the *Settings* icon from the home screen, then select *Notifications*. A list of your apps, with a toggle switch next to each app, will be displayed. Switch the toggle switch to *off* to disable that app's notifications.

Bandwidth

Bandwidth is a measurement of the width or capacity of the communication channel. Effectively, it means how much information can move across the communication channel in a given amount of time. For example, to transmit text documents, a slow bandwidth would be acceptable. However, to effectively transmit video and audio, a wider bandwidth is required. There are four categories of bandwidth:

- **Voiceband**, also known as **low bandwidth**, is used for landline telephone communication. At one time, personal computers with telephone modems and dial-up service used this bandwidth. Although effective for transmitting text documents, it is too slow for many types of transmission, including high-quality audio and video.
- **Medium band** is used in special leased lines to connect midrange computers and mainframes, as well as to transmit data over long distances. This bandwidth is capable of very high-speed data transfer.
- **Broadband** is widely used for DSL, cable, and satellite connections to the Internet. Several users can simultaneously use a single broadband connection for high-speed data transfer.
- **Baseband** is widely used to connect individual computers that are located close to one another, such as the computers in an office building. Like broadband, it is able to support high-speed transmission. Unlike broadband, however, baseband can only carry a single signal at a time.

Protocols

For data transmission to be successful, sending and receiving devices must follow a set of communication rules for the exchange of information. These rules for exchanging data between computers are known as **protocols**.

As discussed in **Chapter 2**, **https**, or **hypertext transfer protocol secure**, is widely used to protect the transfer of sensitive information. Another widely used Internet protocol is **TCP/IP (transmission control protocol/Internet protocol)**. The essential features of this protocol involve (1) identifying sending and receiving devices and (2) breaking information into small parts, or packets, for transmission across the Internet.

- **Identification:** Every computer on the Internet has a unique numeric address called an **IP address (Internet protocol address)**. Similar to the way a postal service uses addresses to deliver mail, the Internet uses IP addresses to deliver e-mail and to locate websites. Because these numeric addresses are difficult for people to remember and use, a system was developed to automatically convert text-based addresses to numeric IP addresses. This system uses a **domain name server (DNS)** that converts text-based addresses to IP addresses. For example, the URL www.mhhe.com is easy for a person to remember and enter, but before a connection can be made, a DNS converts this to an IP address, such as 65.39.69.50, that computers can use to locate that website on the Internet. (See **Figure 8-12**.)

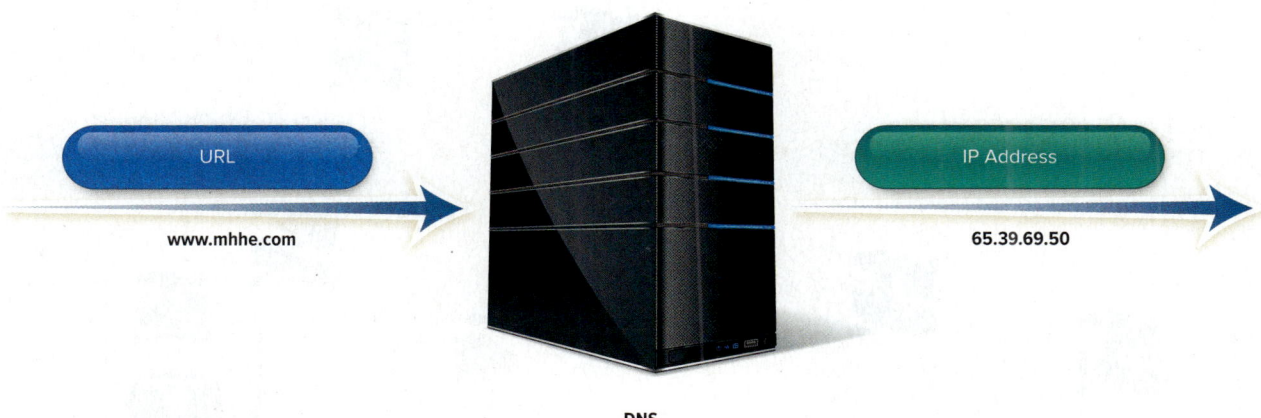

Figure 8-12 DNS converts text-based addresses to numeric IP addresses
Gravvi/Shutterstock

- **Packetization:** Information sent or transmitted across the Internet usually travels through numerous interconnected networks. Before the message is sent, it is reformatted or broken down into small parts called **packets**. Each packet is then sent separately over the Internet, possibly traveling different routes to one common destination. At the receiving end, the packets are reassembled into the correct order.

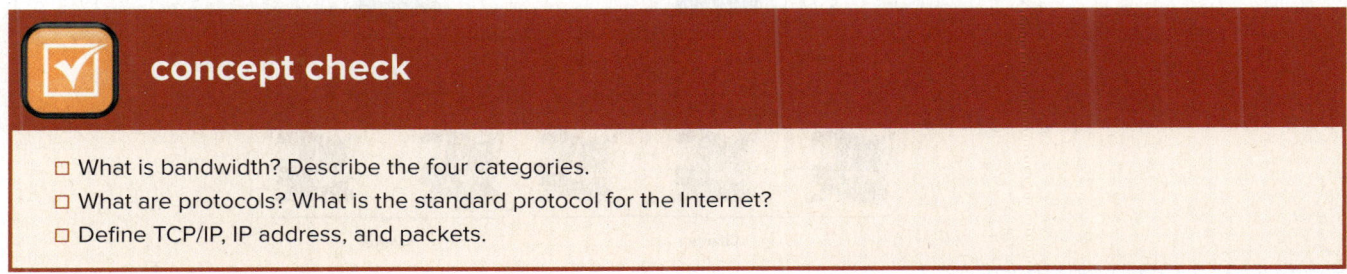

☐ What is bandwidth? Describe the four categories.
☐ What are protocols? What is the standard protocol for the Internet?
☐ Define TCP/IP, IP address, and packets.

Networks

A **computer network** is a communication system that connects two or more computing devices so that they can exchange information and share resources. Networks can be set up in different arrangements to suit users' needs. (See **Figure 8-13**.)

Terms

There are a number of specialized terms that describe computer networks. These terms include

- **Node**—any device that is connected to a network. It could be a computer, printer, or data storage device.
- **Client**—a node that requests and uses resources available from other nodes. Typically, a client is a user's personal computing device.
- **Server**—a node that shares resources with other nodes. Dedicated servers specialize in performing specific tasks. Depending on the specific task, they may be called an application server, communication server, database server, file server, printer server, or web server.
- **Directory server**—a specialized server that manages resources, such as user accounts, for an entire network.
- **Host**—any computer system connected to a network that provides access to its resources.
- **Router**—a node that forwards or routes data packets from one network to their destination in another network.
- **Switch**—central node that coordinates the flow of data by sending messages directly between sender and receiver nodes. A **hub** previously filled this purpose by sending a received message to all connected nodes, rather than just the intended node.

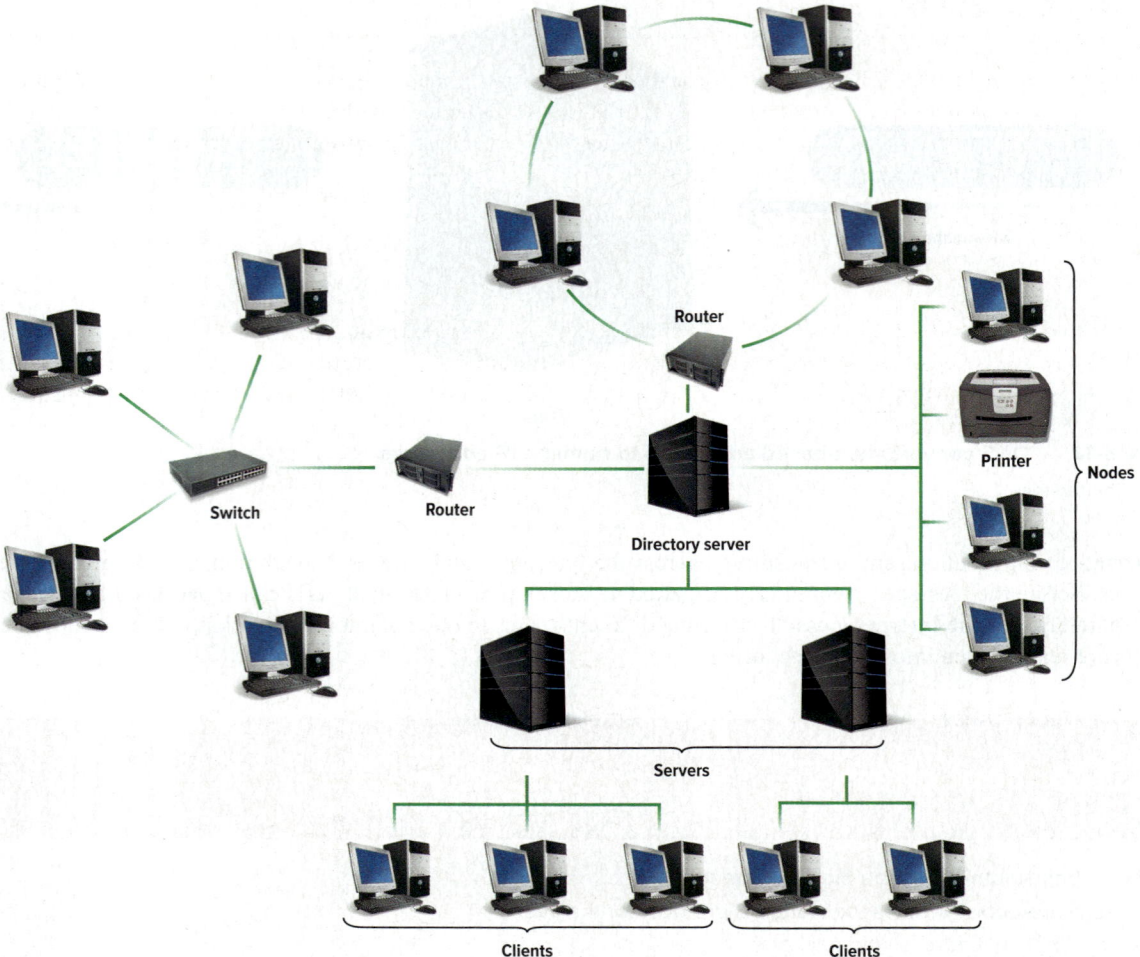

Figure 8-13 Computer network
Gravvi/Shutterstock

- **Network interface cards (NICs)**—as discussed in **Chapter 5**, these are expansion cards located within the system unit that connect the computer to a network. Sometimes referred to as a LAN adapter.
- **Network operating systems (NOSs)**—control and coordinate the activities of all computers and other devices on a network. These activities include electronic communication and the sharing of information and resources. Popular NOSs include Cisco NX-OS and Juniper Network's Junos OS.
- **Network administrator**—a computer specialist responsible for efficient network operations and implementation of new networks.

A network may consist only of personal computers, or it may integrate personal computers or other devices with larger computers. Networks can be controlled by all nodes working together equally or by specialized nodes coordinating and supplying all resources. Networks may be simple or complex, self-contained, or dispersed over a large geographic area.

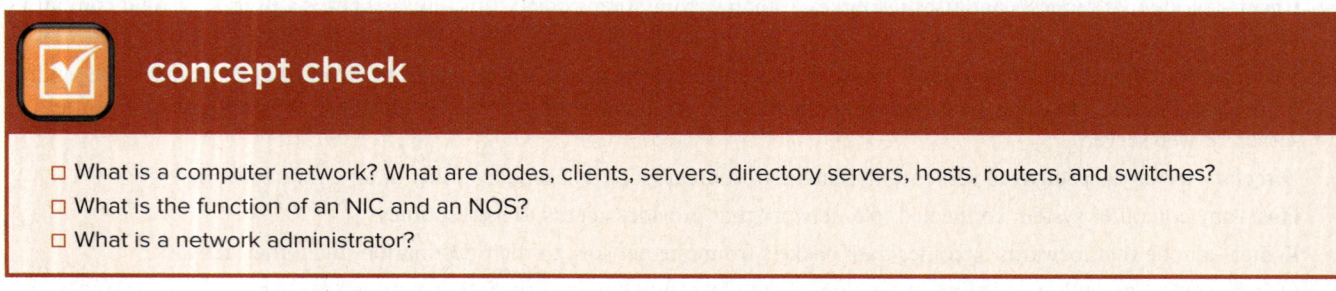

concept check

- What is a computer network? What are nodes, clients, servers, directory servers, hosts, routers, and switches?
- What is the function of an NIC and an NOS?
- What is a network administrator?

Network Types

Clearly, different types of channels—wired or wireless—allow different kinds of networks to be formed. Telephone lines, for instance, may connect communications equipment within the same building or within a home. Networks also may be citywide and even international, using both cable and wireless connections. Local area, metropolitan area, and wide area networks are distinguished by the size of the geographic area they serve.

Local Area Networks

Networks with nodes that are in close physical proximity—within the same building, for instance—are called **local area networks (LANs)**. Typically, LANs span distances less than a mile and are owned and operated by individual organizations. LANs are widely used by colleges, universities, and other types of organizations to link personal computers and to share printers and other resources. For a simple LAN, see **Figure 8-14**.

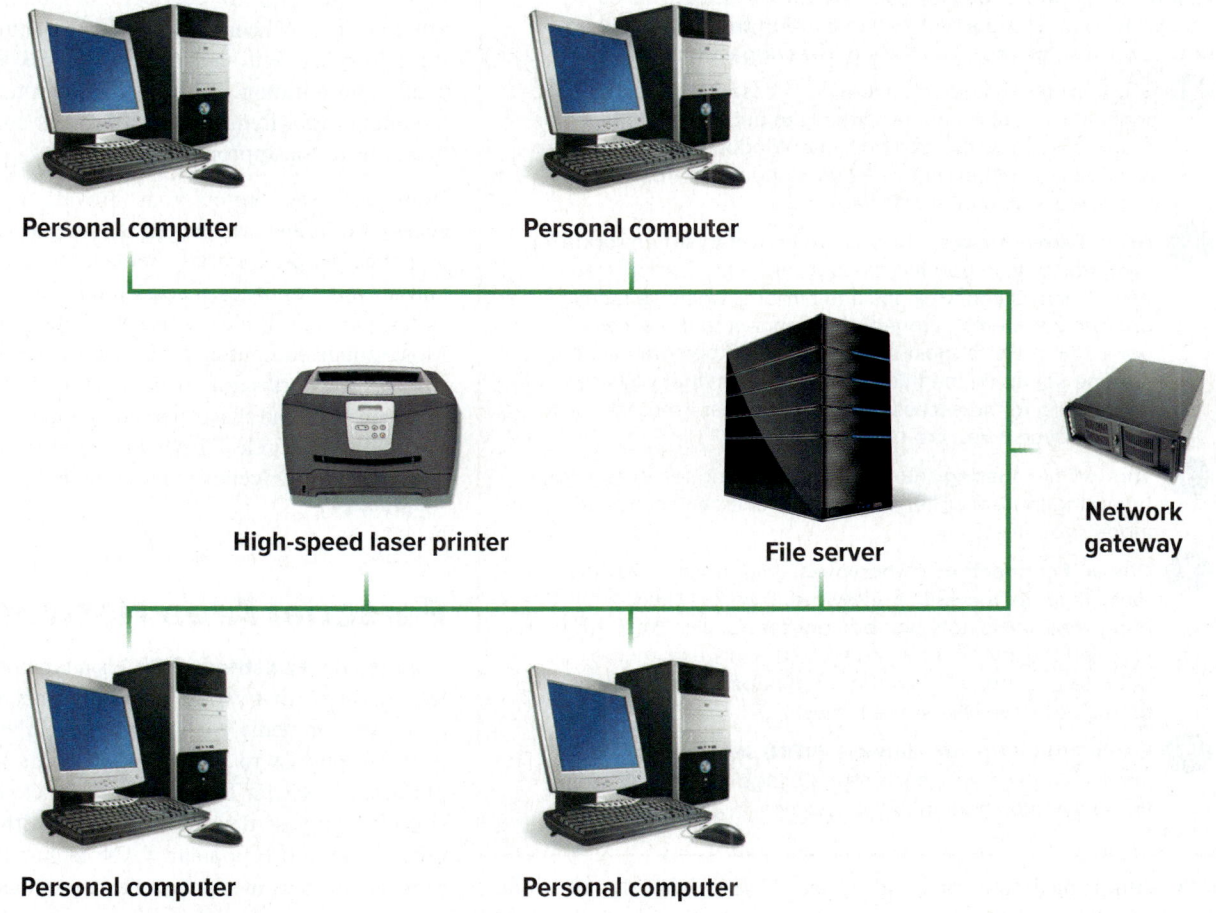

Figure 8-14 Local area network

Gravvi/Shutterstock

The LAN represented in **Figure 8-14** is a typical arrangement and provides two benefits: economy and flexibility. People can share costly equipment. For instance, the four personal computers share the high-speed laser printer and the file server, which are expensive pieces of hardware. Other equipment or nodes also may be added to the LAN—for instance, more personal computers, a mainframe computer, or optical disc storage devices. Additionally, the **network gateway** is a device that allows one LAN to be linked to other LANs or to larger networks. For example, the LAN of one office group may be connected to the LAN of another office group.

There are a variety of different standards or ways in which nodes can be connected to one another and ways in which their communications are controlled in a LAN. The most common standard is known as **Ethernet**. LANs using this standard are sometimes referred to as Ethernet LANs.

Home Networks

Figure 8-15 Wireless adapter

magraphics/123RF

Although LANs have been widely used within organizations for years, they are now being commonly used by individuals in their homes and apartments. These LANs, called **home networks**, allow different computers to share resources, including a common Internet connection. Computers can be connected in a variety of ways, including electrical wiring, telephone wiring, and special cables. One of the simplest ways, however, is without cables, or wireless.

tips

Do you use your laptop to connect to wireless networks at school or in public places such as coffee shops, airports, or hotels? If so, it is important to use caution to protect your computer and your privacy. Here are a few suggestions:

1. **Use a firewall.** A personal firewall is essential when connecting your computer directly to public networks. Some firewalls, such as the one built into Windows 8 and Windows 10, will ask whether a new network should be treated as home, work, or a public network.

2. **Avoid fake hotspots.** Thieves are known to set up rogue (or fake) hotspots in popular areas where users expect free Wi-Fi, such as coffee shops and airports. Because many operating systems automatically connect to the access point with the strongest signal, you could be connecting to the one set up by the thief. Always confirm that you are connecting to the access point of that establishment. Ask an employee if you are unsure.

3. **Turn off file sharing.** Turning off file-sharing features in your operating system will ensure that no one can access or modify your files.

4. **Check if connection is encrypted.** If the hotspot you are using is protected with a password, then it is likely encrypted. If it is not, then be very careful with websites you visit and the information you provide. Only use secure websites, as indicated by the "s" of https or a small padlock to the left of the URL in the browser.

5. **Use a Virtual Private Network (VPN) service.** A VPN service can encrypt your Internet usage to prevent others from spying on your Internet activities.

Wireless LAN

A wireless local area network is typically referred to as a **wireless LAN (WLAN)**. It uses radio frequencies to connect computers and other devices. All communications pass through the network's centrally located **wireless access point** or **base station**. This access point interprets incoming radio frequencies and routes communications to the appropriate devices.

Wireless access points that provide Internet access are widely available in public places such as coffee shops, libraries, bookstores, colleges, and universities. These access points are known as **hotspots** and typically use Wi-Fi technology. Most mobile computing devices have an internal wireless network card to connect to hotspots. You can also purchase personal hotspots that create a local wireless LAN and connect to the Internet through cellular data connections (see **Figure 8-15**).

Personal Area Networks

A **personal area network (PAN)** is a type of wireless network that works within a very small area—your immediate surroundings. PANs connect cell phones to headsets, keyboards to cell phones, and so on. These networks make it possible for wireless devices to interact with each other. The most popular PAN technology is Bluetooth, with a maximum range of around 33 feet. Virtually all wireless peripheral devices available today use Bluetooth, including the controllers on popular game systems like the PlayStation and Wii.

Metropolitan Area Networks

Metropolitan area networks (MANs) span distances up to 100 miles. These networks are frequently used as links between office buildings that are located throughout a city.

Unlike a LAN, a MAN is typically not owned by a single organization. Rather, it is owned either by a group of organizations or by a single network service provider that provides network services for a fee.

Wide Area Networks

Wide area networks (WANs) are countrywide and worldwide networks. These networks provide access to regional service (MAN) providers and typically span distances greater than 100 miles. They use microwave relays and satellites to reach users over long distances—for example, from Los Angeles to Paris. Of course, the widest of all WANs is the Internet, which spans the entire globe.

The primary difference between a PAN, LAN, MAN, and WAN is the geographic range. Each may have various combinations of hardware, such as personal computers, midrange computers, mainframes, and various peripheral devices.

For a summary of network types, see **Figure 8-16**.

Type	Description
PAN	Personal area network; connects digital devices, such as Bluetooth headphones
Home	Local area network for home and apartment use; typically wireless
WLAN	Wireless local area network; all communication passes through access point
LAN	Local area network; located within close proximity
MAN	Metropolitan area network; typically spans cities with coverage up to 100 miles
WAN	Wide area network for countrywide or worldwide coverage

Figure 8-16 Types of networks

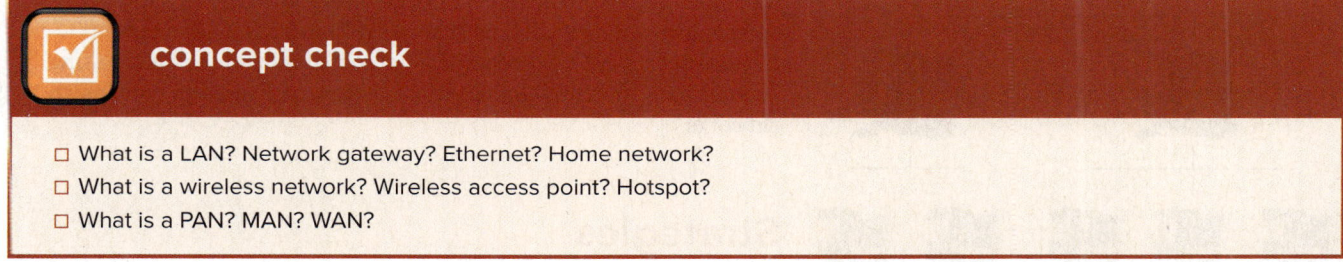

concept check

- What is a LAN? Network gateway? Ethernet? Home network?
- What is a wireless network? Wireless access point? Hotspot?
- What is a PAN? MAN? WAN?

Network Architecture

Network architecture describes how a network is arranged and how resources are coordinated and shared. It encompasses a variety of different network specifics, including network topologies and strategies. Network topology describes the physical arrangement of the network. Network strategies define how information and resources are shared.

Topologies

A network can be arranged or configured in several different ways. This arrangement is called the network's **topology**. The most common topologies are

- **Bus network**—each device is connected to a common cable called a **bus** or **backbone**, and all communications travel along this bus. (See **Figure 8-14**)
- **Ring network**—each device is connected to two other devices, forming a ring. (See **Figure 8-17**.) When a message is sent, it is passed around the ring until it reaches the intended destination.
- **Star network**—each device is connected directly to a central network switch. (See **Figure 8-18**.) Whenever a node sends a message, it is routed to the switch, which then passes the message along to the intended recipient. The star network is the most widely used network topology today. It is applied to a broad range of applications from small networks in the home to very large networks in major corporations.
- **Tree network**—each device is connected to a central node, either directly or through one or more other devices. The central node is connected to two or more subordinate nodes that in turn are connected to other subordinate nodes, and so forth, forming a treelike structure. (See **Figure 8-19**.) This network, also known as a **hierarchical network**, is often used to share corporate-wide data.

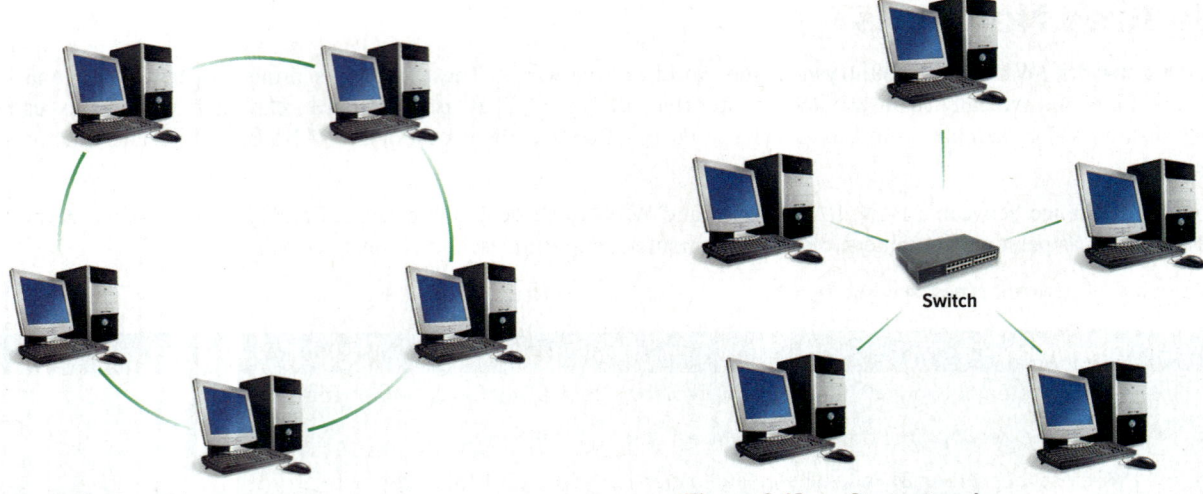

Figure 8-17 Ring network

Figure 8-18 Star network

Figure 8-19 Tree network

Figure 8-20 Mesh network

- **Mesh network**—this topology is the newest type and does not use a specific physical layout (such as a star or a tree). Rather, the mesh network requires that each node have more than one connection to the other nodes. (See **Figure 8-20**.) The resulting pattern forms the appearance of a mesh. If a path between two nodes is somehow disrupted, data can be automatically rerouted around the failure using another path. Wireless technologies are frequently used to build mesh networks.

Strategies

Every network has a **strategy**, or way of coordinating the sharing of information and resources. Two of the most common network strategies are client/server and peer-to-peer.

Client/server networks use central servers to coordinate and supply services to other nodes on the network. The server provides access to resources such as web pages, databases, application software, and hardware. (See **Figure 8-21**.) This strategy is based on specialization. Server nodes coordinate and supply specialized services, and client nodes request the services. Commonly used server operating systems are Windows Server, macOS X Server, Linux, and Solaris.

Client/server networks are widely used on the Internet. For example, each time you open a web browser, your computer (the client) sends out a request for a specific web page. This request is routed over the Internet to a server. This server locates and sends the requested material back to your computer.

One advantage of the client/server network strategy is the ability to handle very large networks efficiently. Another advantage is the availability of powerful network management software to monitor and control network activities. The major disadvantages are the cost of installation and maintenance.

In a **peer-to-peer (P2P) network**, nodes have equal authority and can act as both clients and servers. The most common way to share games, movies, and music over the Internet is to use a P2P network. For example, special file-sharing software such as BitTorrent can be used to obtain files located on another personal computer and can provide files to other personal computers.

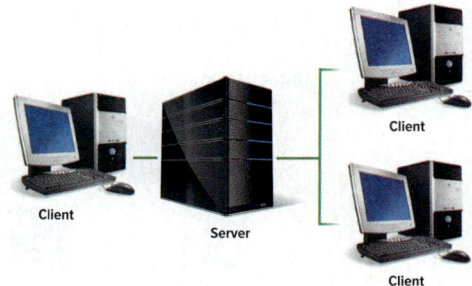

Figure 8-21 Client/server network
Gravvi/Shutterstock

P2P networks are rapidly growing in popularity as people continue to share information with others around the world. The primary advantage is that they are easy and inexpensive (often free) to set up and use. One disadvantage of P2P networks is the lack of security controls or other common management functions. For this reason, few businesses use this type of network to communicate sensitive information.

concept check

- What is a network topology?
- Compare bus, ring, star, tree, and mesh topologies.
- What is a network strategy?
- Compare client/server and peer-to-peer strategies.

Organizational Networks

Computer networks in organizations have evolved over time. Most large organizations have a complex and wide range of different network configurations, operating systems, and strategies. These organizations face the challenge of making these networks work together effectively and securely.

Internet Technologies

Many organizations today employ Internet technologies to support effective communication within and between organizations using intranets and extranets.

- An **intranet** is a *private* network within an organization that resembles the Internet. Like the *public* Internet, intranets use browsers, websites, and web pages. Typical applications include electronic telephone directories, e-mail addresses, employee benefit information, internal job openings, and much more. Employees find surfing their organizational intranets to be as easy and as intuitive as surfing the Internet.
- An **extranet** is a *private* network that connects *more than one* organization. Many organizations use Internet technologies to allow suppliers and others limited access to their networks. The purpose is to increase efficiency and reduce costs. For example, an automobile manufacturer has hundreds of suppliers for the parts that go into making a car. By having access to the car production schedules, suppliers can schedule and deliver parts as they are needed at the assembly plants. In this way, operational efficiency is maintained by both the manufacturer and the suppliers.

Network Security

Large organizations face the challenge of ensuring that only authorized users have access to network resources, sometimes from multiple geographic locations or across the Internet. Securing large computer networks requires specialized technology. Three technologies commonly used to ensure network security are firewalls, intrusion detection systems, and virtual private networks.

- A **firewall** consists of hardware and software that control access to a company's intranet and other internal networks. Most use software or a special computer called a **proxy server**. All communications between the company's internal networks and the outside world pass through this server. By evaluating the source and the content of each communication, the proxy server decides whether it is safe to let a particular message or file pass into or out of the organization's network. (See **Figure 8-22**.)

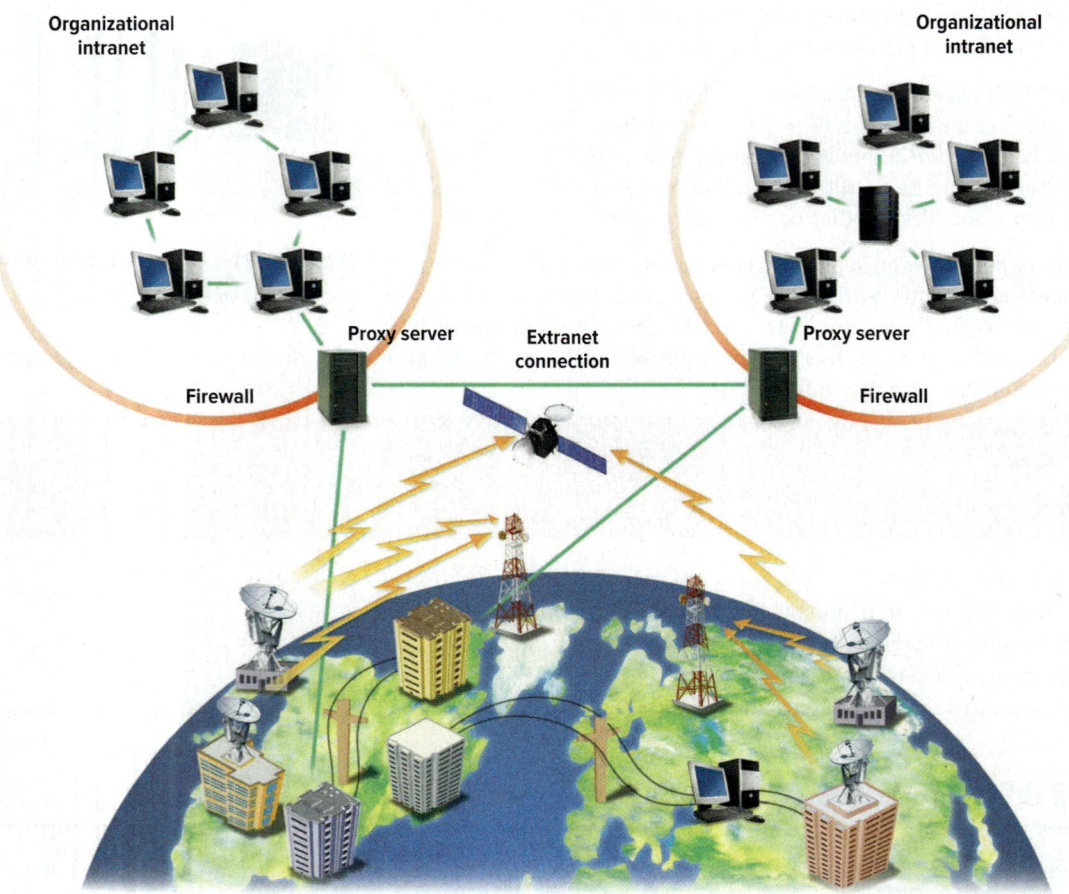

Figure 8-22   Intranets, extranets, firewalls, and proxy servers

privacy

In this age of social media and cell phones, it can be difficult to maintain your privacy on the Internet—but have you considered how to maintain the privacy of your Internet connection? When traveling it is common to use unfamiliar Wi-Fi connections—at a hotel, airport, or coffee shop—but what confidence do you have that your Internet activities won't be monitored? Proponents of monitoring point out that public spaces offer no guarantee of privacy and a business has the right to monitor Internet traffic to deny inappropriate web pages. Privacy advocates counter that many of the most common Internet activities include private data—such as sharing photos and receiving e-mails—and that in offering public Wi-Fi to perform common Internet tasks, it is reasonable for a person to have an expectation of privacy. What do you think?

- **Intrusion detection systems (IDSs)** work with firewalls to protect an organization's network. These systems use sophisticated software to analyze all incoming and outgoing network traffic. Microsoft's Azure Security Center is an IDS that uses artificial intelligence to detect and respond to attacks. These systems can learn and adapt to new attacks, recognizing an intruder and disabling their access much faster than previous techniques.

- **Virtual private networks (VPNs)** create a secure private connection between a remote user and an organization's internal network. Special VPN protocols create the equivalent of a dedicated line between a user's home or laptop computer and a company server. The connection is heavily encrypted, and, from the perspective of the user, it appears that the workstation is actually located on the corporate network. Another common use for VPNs is for individuals to use VPNs to protect their Internet data when on public hotspots when traveling.

Like organizations, end users have security challenges and concerns. We need to be concerned about the privacy of our personal information. In the next chapter, we will discuss personal firewalls and other ways to protect personal privacy and security.

concept check

☐ What are Internet technologies? Compare intranets and extranets.
☐ What is a firewall? What is a proxy server?
☐ What are intrusion detection systems?
☐ What are virtual private networks?

Careers in IT

"Now that you have learned about computer communications and networks, let me tell you about my career as a network administrator."

Dmitry Kalinovsky/Shutterstock

Network administrators manage a company's LAN and WAN networks. They may be responsible for design, implementation, and maintenance of networks. Duties usually include maintenance of both hardware and software related to a company's intranet and Internet networks. Network administrators are typically responsible for diagnosing and repairing problems with these networks. Some network administrators' duties include planning and implementation of network security as well.

Employers typically look for candidates with a bachelor's or an associate's degree in computer science, computer technology, or information systems, as well as practical networking experience. Experience with network security and maintenance is preferred. Also, technical certification may be helpful or required in obtaining this position. Because network administrators are involved directly with people in many departments, good communication skills are essential.

Network administrators can expect to earn an annual salary of $61,000 to $76,000. Opportunities for advancement typically include upper-management positions. This position is expected to be among the fastest-growing jobs in the near future.

A LOOK TO THE FUTURE

Telepresence Lets You Be There without Actually Being There

Khakimullin Aleksandr/Shutterstock

In 2019, a global pandemic fundamentally changed how we work, educate, and socialize. Videoconferencing, technologies bloomed into massive industries as people learned to work and play through socially distanced online interactions. While the pandemic may be over, these new technologies remain and are growing in use. How will our communications improve as technology and networks continue to grow in power and availability? The future of communication is a world with seamless telepresence.

Telepresence is the use of technology to create the illusion that you are actually somewhere else. Unlike a videoconference, where a small screen conveys the images and sounds of another location, telepresence attempts to replicate the foreign location entirely. These technologies include improved video and audio with immersive screens and multiple directional speakers. But it goes further than that: Telepresence will allow you to touch things in this other location—to interact with the world there through robotics and even feel this other location. In the future, telepresence will likely include robots that will allow you to walk the halls of your office or other remote locations from the comfort of your home. Technology will continue to evolve to improve our lives as we look to the future.

Today's early implementations, such as Cisco TelePresence, use very high-definition video, acoustically tuned audio systems, and high-speed networks to create a very convincing videoconferencing experience. Google's Project Starline uses ultra-high resolution monitors and cameras to generate 3D models of telepresence participants. It then combines the 3D model and video to generate a 3D experience for the viewer.

However, telepresence could someday go beyond the simple voice and videoconferencing available today, with robotics allowing the telecommuter to change location and pick up objects. The ability to manipulate things through robotic arms is only part of the illusion of being in another place—you must also feel and touch the distant objects. Called haptic technology, specialized gloves and surfaces can give the user touch sensations that mimic holding or touching a distant object.

One of the areas where telepresence robotics are already being used is in hospitals. Ava Robotics has developed the RP-VITA, a telepresence robot that has multiple cameras, high-resolution screens, and complex robotics to allow a doctor to care for and consult with patients and doctors from around the world. Scientists at the National Institutes of Health are investigating the use of augmented reality videoconferencing to assist in the triage and care of patients entering a hospital. Future technologies include specialized cameras that monitor temperature, heart rate, and respiration and haptic robotics that can assist in diagnostic testing.

Telepresence robots will also play a role in the home. Futurists imagine a day when a telepresence robot is as common as the vacuum cleaner. The home telepresence robot might be taken over by skilled technicians, such as a plumber or electrician. The telepresence robot would maneuver around the home, reviewing your pipes or wiring, and perform simple tests, make recommendations, and offer quotes on work to be done. A mechanic might use a telepresence robot to perform standard vehicle tune-ups or repairs. What do you think—could telepresence technology improve enough to convince you that you are somewhere else, or would the gloves and headset be too distracting?

VISUAL SUMMARY | Communications and Networks

COMMUNICATIONS

pics five/Shutterstock

Communications are the process of sharing data, programs, and information between two or more computers. Applications include texting, e-mail, videoconferencing, and electronic commerce.

Connectivity

Connectivity is a concept related to using computer networks to link people and resources. You can link or connect to large computers and the Internet, providing access to extensive information resources.

The Wireless Revolution

Mobile devices like cell phones and tablets have brought dramatic changes in connectivity and communications. These wireless devices are becoming widely used for computer communication.

Communication Systems

Communication systems transmit data from one location to another. There are four basic elements:

- Sending and receiving devices originate or accept messages.
- Connection devices act as an interface between sending and receiving devices and the communication channel.
- Data transmission specifications are rules and procedures for sending and receiving data.
- Communication channel is the actual connecting or transmission medium for messages.

COMMUNICATION CHANNELS

Mariusz Burcz/Alamy Stock Photo

Communication channels carry data from one computer to another.

Wireless Connections

Wireless connections do not use a solid substance to connect devices. Most use radio waves.

- **Bluetooth**—transmits data over short distances; widely used for a variety of wireless devices.
- **Wi-Fi (wireless fidelity)**—uses high-frequency radio signals; most home and business wireless networks use Wi-Fi.
- **Microwave**—line-of-sight communication; used to send data between buildings; longer distances require microwave stations.
- **Cellular**—uses **cell towers** to send and receive data within relatively small geographic regions or **cells.**
- **Satellite**—uses microwave relay stations; **GPS (global positioning system)** tracks geographic locations.
- **Infrared**—uses light waves over a short distance; line-of-sight communication.

Physical Connections

Physical connections use a solid medium to connect sending and receiving devices. Connections include **twisted-pair cable (telephone lines** and **Ethernet cables), coaxial cable,** and **fiber-optic cable.**

To efficiently and effectively use computers, you need to understand the concepts of connectivity, the wireless revolution, and communication systems. Additionally, you need to know the essential parts of communication technology, including channels, connection devices, data transmission, networks, network architectures, and network types.

CONNECTION DEVICES

(left): Keith Homan/Alamy Stock Photo; (middle): blue_iq/E+/Getty Images; (right): Ascannio/Shutterstock

Many communication systems use standard telephone lines and **analog signals.** Computers use **digital signals.**

Modems

Modems modulate and **demodulate. Transfer rate** is measured in **megabits per second** or **gigabits per second.** Three types are **wireless (wireless wide area network, WWAN), DSL,** and **cable.**

Connection Services

T1, T3 (**DS3**), and **OC** (**optical carrier**) lines provide support for very-high-speed, all-digital transmission for large corporations. **Cellular service providers** support voice and data transmission using cellular networks. These mobile telecommunications networks have gone through different generations—**1G,** using analog radio signals; **2G,** using digital radio signals; **3G,** beginning of cell phones; **4G,** using **LTE (Long Term Evolution)**; and **5G,** the newest and fastest network.

Other more affordable technologies include **dial-up, ADSL** (the most widely used type of **DSL (digital subscriber line)**), **cable, fiber-optic service (FiOS),** and **satellite.**

DATA TRANSMISSION

Bandwidth measures a communication channel's width or capacity. Four bandwidths are **voiceband, medium band, broadband,** and **baseband. Protocols** are rules for exchanging data. Internet protocols include **https** and **TCP/IP. IP addresses (Internet protocol addresses)** are unique numeric Internet addresses. **DNS (domain name server)** converts text-based addresses to and from numeric IP addresses. **Packets** are small parts of messages.

NETWORKS

Computer networks connect two or more computers. Some specialized network terms include

- **Node**—any device connected to a network.
- **Client**—node requesting resources.
- **Server**—node providing resources.
- **Directory server**—specialized node that manages resources.
- **Host**—any computer system that provides access to its resources over a network.
- **Router**—a node that forwards data packets from one network to another network.
- **Switch**—node that coordinates direct flow of data between other nodes. **Hub** is an older device that directed flow to all nodes.
- **NIC (network interface card)**—LAN adapter card for connecting to a network.
- **NOS (network operating system)**—controls and coordinates network operations.
- **Network administrator**—network specialist responsible for network operations.

NETWORK TYPES

Networks can be citywide or even international, using both wired and wireless connections.

- **Local area networks (LANs)** connect nearby devices. **Network gateways** connect networks to one another. **Ethernet** is a LAN standard. These LANs are called Ethernet LANs.
- **Home networks** are LANs used in homes.
- **Wireless LANs (WLANs)** use a **wireless access point (base station)** as a hub. Hotspots provide Internet access in public places.
- **Personal area networks (PANs)** are wireless networks for Bluetooth headphones, cell phones, and other wireless devices.
- **Metropolitan area networks (MANs)** link office buildings within a city, spanning up to 100 miles.
- **Wide area networks (WANs)** are the largest type. They span states and countries or form worldwide networks. The Internet is the largest wide area network in the world.

NETWORK ARCHITECTURE

Network architecture describes how networks are arranged and resources are shared.

Topologies

A network's **topology** describes the physical arrangement of a network.

- **Bus network**—each device is connected to a common cable called a **bus** or **backbone.**
- **Ring network**—each device is connected to two other devices, forming a ring.
- **Star network**—each device is connected directly to a central network switch; most common type today.
- **Tree (hierarchical) network**—a central node is connected to subordinate nodes forming a treelike structure.
- **Mesh network**—newest; each node has two or more connecting nodes.

Strategies

Every network has a **strategy**, or way of sharing information and resources. Common network strategies include client/server and peer-to-peer.

- **Client/server (hierarchical) network**—central computers coordinate and supply services to other nodes; based on specialization of nodes; widely used on the Internet; able to handle very large networks efficiently; powerful network management software available.
- **Peer-to-peer network**—nodes have equal authority and act as both clients and servers; widely used to share games, movies, and music over the Internet; easy to set up and use; lacks security controls.

ORGANIZATIONAL NETWORKS

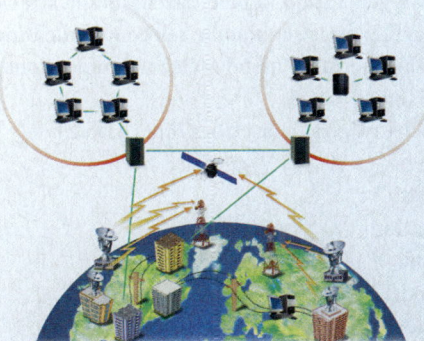

Internet Technologies

Internet technologies support effective communication using intranets and extranets.

- **Intranet**—private network within an organization; uses browsers, websites, and web pages. Typical applications include electronic telephone directories, e-mail addresses, employee benefit information, internal job openings, and much more.
- **Extranet**—like intranet except connects *more than one* organization; typically allows suppliers and others limited access to their networks.

Network Security

Three technologies commonly used to ensure network security are firewalls, intrusion detection systems, and virtual private networks.

- **Firewall**—controls access; all communications pass through a **proxy server.**
- **Intrusion detection systems (IDSs)**—work with firewalls; use sophisticated statistical techniques to recognize and disable network attacks.
- **Virtual private network (VPN)**—creates secure private connection between remote user and organization's internal network.

CAREERS in IT

Network administrators manage a company's LAN and WAN networks. Bachelor's or associate's degree in computer science, computer technology, or information systems and practical networking experience required. Expected salary range is $61,000–$76,000.

KEY TERMS

1G (first-generation mobile telecommunications)
2G (second-generation mobile telecommunications)
3G (third-generation mobile telecommunications)
4G (fourth-generation mobile telecommunications)
5G (fifth-generation mobile telecommunications)
analog signal
asymmetric digital subscriber line (ADSL)
backbone
bandwidth
base station
baseband
Bluetooth
broadband
bus
bus network
cable modem
cable service
cell
cell tower
cellular
cellular service provider
client
client/server network
coaxial cable
communication channel
communication system
computer network
connectivity
demodulation
dial-up service
digital signal
digital subscriber line (DSL)
digital subscriber line (DSL) service
directory server
domain name server (DNS)
downlink
DS3
Ethernet
Ethernet cable
extranet
fiber-optic cable
fiber-optic service (FiOS)
firewall
gigabits per second (Gbps)
global positioning system (GPS)
hierarchical network
home network
host
hotspot
https (hypertext transfer protocol secure)
hub
infrared
intranet
intrusion detection system (IDS)
IP address (Internet protocol address)
local area network (LAN)
low bandwidth
LTE (Long Term Evolution)
medium band
megabits per second (Mbps)
mesh network
metropolitan area network (MAN)
microwave
modem
modulation
network administrator
network architecture
network gateway
network interface card (NIC)
network operating system (NOS)
node
optical carrier (OC)
packet
peer-to-peer (P2P) network
personal area network (PAN)
protocol
proxy server
ring network
router
satellite
satellite connection service
server
star network
strategy
switch
T1
T3
telephone line
topology
transfer rate
transmission control protocol/Internet protocol (TCP/IP)
tree network
twisted-pair cable
uplink
virtual private network (VPN)
voiceband
wide area network (WAN)
Wi-Fi (wireless fidelity)
wireless access point
wireless LAN (WLAN)
wireless modem
wireless wide area network (WWAN) modem

MULTIPLE CHOICE

Circle the correct answer.

1. Type of network topology in which each device is connected to a common cable called a backbone.
   - a. star
   - b. bus
   - c. peer-to-peer
   - d. ring

2. A widely used Internet protocol.
   - a. TCP/IP
   - b. WAN
   - c. LAN
   - d. Bluetooth

3. Uses high-frequency radio waves.
   - a. NIC
   - b. mesh
   - c. VPN
   - d. Wi-Fi

4. Signals that are continuous electronic waves.
   - a. analog
   - b. digital
   - c. backbone
   - d. mesh

5. Rules for exchanging data between computers.
   - a. modem
   - b. https
   - c. extranet
   - d. protocols

6. Any device that is connected to a network.
   - a. ADSL
   - b. node
   - c. optical carrier
   - d. hotspot

7. A computer specialist responsible for efficient network operations and implementation of new networks.
   - a. network administrator
   - b. wireless access point
   - c. cellular service provider
   - d. hub

8. This network, also known as a hierarchical network, is often used to share corporate-wide data.
   - a. bus
   - b. tree
   - c. mesh
   - d. ring

9. In this network, nodes have equal authority and can act as both clients and servers.
   - a. bus
   - b. tree
   - c. peer-to-peer
   - d. mesh

10. This works with firewalls to protect an organization's network.
    - a. uplink
    - b. FiOS
    - c. intrusion detection system
    - d. DSL

MATCHING

Match each numbered item with the most closely related lettered item. Write your answers in the spaces provided.

a. connectivity
b. IDSs
c. Bluetooth
d. transfer rate
e. broadband
f. IP address
g. NIC
h. network gateway
i. hotspots
j. topologies

____ 1. The bandwidth typically used for DSL, cable, and satellite connections to the Internet.
____ 2. Sometimes referred to as a LAN adapter, these expansion cards connect a computer to a network.
____ 3. The speed with which a modem transmits data is called its _____.
____ 4. A device that allows one LAN to be linked to other LANs or to larger networks.
____ 5. The concept related to using computer networks to link people and resources.
____ 6. These systems use sophisticated software and artificial intelligence to protect an organization's network.
____ 7. Every computer on the Internet has a unique numeric address called a(n) _____.
____ 8. Typically using Wi-Fi technology, these wireless access points are available from public places such as coffee shops, libraries, bookstores, colleges, and universities.
____ 9. A short-range radio communication standard that transmits data over short distances of up to approximately 33 feet.
____ 10. Bus, ring, star, tree, and mesh are five types of network _____.

OPEN-ENDED

On a separate sheet of paper, respond to each question or statement.

1. Define communications, including connectivity, the wireless revolution, and communication systems.
2. Discuss communication channels, including wireless connections and physical connections.
3. Discuss connection devices, including modems (DSL, cable, and wireless modems) and connection services (cellular, DSL, ADSL, cable, and satellite).
4. Discuss data transmission, including bandwidths (voice band, medium band, broadband, and baseband) as well as protocols (IP addresses, domain name servers, and packetization).
5. Discuss networks by identifying and defining specialized terms that describe computer networks.
6. Discuss network types, including local area, home, wireless, personal, metropolitan, and wide area networks.
7. Define network architecture, including topologies (bus, ring, star, tree, and mesh) and strategies (client/server and peer-to-peer).
8. Discuss organization networks, including Internet technologies (intranets and extranets) and network security (firewalls, proxy servers, intrusion detection systems, and virtual private networks).

DISCUSSION

Respond to each of the following questions.

Making IT Work for You: THE MOBILE OFFICE

Review the Making IT Work for You: The Mobile Office on pages 192 and 193 and then respond to the following: (a) What types of professions would get the most out of a mobile office? Why? What type of office work is best done at work? What office work can be handled while driving? Or on an airplane? (b) Do you have a mobile Internet device, such as a cell phone or tablet? (If you do not own any, research a device on the Internet that you feel would be most beneficial to you.) Can your mobile Internet device be used as a hotspot? Does your service provider allow you to use that feature? (c) Have you ever connected to a public Wi-Fi? Did you share any private information, including videos, e-mails, credit card numbers, or passwords, over a public Wi-Fi? How would you protect your data in the future when on a public Wi-Fi?

Privacy: UNAUTHORIZED NETWORK INTRUSION

Public Wi-Fi is extremely convenient, but also offers several risks. Review the Privacy box on page 202 and then respond to the following: (a) Does a company offering public Wi-Fi have the right to monitor the traffic on its own network? Does it have the right to read the e-mails that are downloaded on that network? Where would you draw the line, and why? (b) What data should the government collect about its citizens' Internet use, and how long should the government hold on to that data? Would your opinion change if you knew terrorists were using the Internet to plan crimes? Would your opinion change if you knew politicians were using collected data to influence election outcomes? How will you improve your cybersecurity when using the Internet in public?

Ethics: ELECTRONIC MONITORING

Review the Ethics box on page 187 and respond to the following: (a) Is it unethical for an organization or corporation to use programs to monitor communications on its network? Why or why not? (b) Is it unethical for a government agency (such as the FBI) to monitor communications on the Internet or gather your records from the websites you visit? Why or why not? (c) Do you feel that new laws are needed to handle these issues? How would you balance the needs of companies and the government with the needs of individuals? Explain your answers.

Community: INTERNET FOR THE DEVELOPING WORLD

Review the Community box on page 189 and then respond to the following: (a) What daily challenges face the citizens of developing countries? How could an Internet connection assist in overcoming these challenges? (b) Many individuals in developing countries cannot afford the devices that connect to the Internet. How can a community work together to make such devices available? (c) Within your own community, are there any community programs that provide Internet-accessible devices to those who cannot afford them? How could such a program be applied to the communities of the developing world?

Design Elements: Concept Check icon: Dizzle52/Getty Images

chapter 9
Privacy, Security, and Ethics

tsingha25/Shutterstock

CHAPTER 9: Privacy, Security, and Ethics

Why should I read this chapter?

ImageFlow/Shutterstock

Our every click, every post, and every like online is recorded, analyzed, and turned to a profit. The details of your social media usage can be packaged and sold to bankers, employers, and ad agencies. While your social media account may be free, the data they collect on you is valuable. It will affect which advertisements you see, what interest rates you pay on credit, and what price you see when you shop for goods online.

What rights do you have to your information? What laws protect you? This chapter covers the things you need to know to be prepared for this ever-changing world, including the following:

- Cybercrime—protect yourself from viruses, Internet scams, and identity theft.
- Privacy rights—learn what companies can legally record about your Internet usage and how they use that information.
- Safe computing—avoid embarrassment and worse by knowing the way Facebook and social networking sites share your information.
- Protect yourself by being aware of what information you are sharing, how that information might be used, and what rights you have to be private and secure on the Internet.

Learning Objectives

After you have read this chapter, you should be able to:

1. Describe the impact of large databases, private networks, the Internet, and the web on privacy.
2. Discuss online identity and the major laws on privacy.
3. Discuss cybercrimes, including identity theft, Internet scams, data manipulation, ransomware, and denial of service.
4. Describe social engineering and malicious software, including crackers, malware, viruses, worms, and Trojan horses.
5. Discuss malicious hardware, including zombies, botnets, rogue Wi-Fi networks, and infected USB flash drives.
6. Detail ways to protect computer security, including restricting access, encrypting data, anticipating disasters, and preventing data loss.
7. Discuss computer ethics, including copyright law, software piracy, digital rights management, the Digital Millennium Copyright Act, as well as cyberbullying, plagiarism, and ways to identify plagiarism.

Introduction

"Hi, I'm Ann, and I'm an IT security analyst. I'd like to talk with you about privacy, security, and ethics, three critical topics for anyone who uses computers today. I would also like to talk about how you can protect your privacy, ensure your security, and act ethically."

Roman Samborskyi/
Shutterstock

There are more than 1 billion personal computers in use today. What are the consequences of the widespread presence of this technology? Does technology make it easy for others to invade our personal privacy? When we apply for a loan or for a driver's license, or when we check out at the supermarket, is that information about us being distributed and used without our permission? When we use the web, is information about us being collected and shared with others? How can criminals use this information for ransom, blackmail, or vandalism?

This technology prompts lots of questions—very important questions. Perhaps these are some of the most important questions for the 21st century. To efficiently and effectively use computers, you need to be aware of the potential impact of technology on people and how to protect yourself on the web. You need to be sensitive to and knowledgeable about personal privacy and organizational security.

People

As we have discussed, information systems consist of people, procedures, software, hardware, data, and the Internet. This chapter focuses on people. Although most everyone agrees that technology has had a very positive impact on people, it is important to recognize the negative, or potentially negative, impacts as well.

Effective implementation of computer technology involves maximizing its positive effects while minimizing its negative effects. The most significant concerns are

- **Privacy:** What are the threats to personal privacy, and how can we protect ourselves?
- **Security:** How can access to sensitive information be controlled, and how can we secure hardware and software?
- **Ethics:** How do the actions of individual users and companies affect society?

Let us begin by examining privacy.

Privacy

To appreciate the impact that the digital revolution is having on our world, you need to look no further than the cell phone in your pocket. A cell phone is a marvel that provides a camera, microphone, and minuscule sensors ready to share your life with the world. With these new tools come new challenges to old ways of thinking. What does it mean to preserve privacy in a social media world where every experience is recorded and preserved? Who owns the memories we create with these devices? Who has the right to share them? To delete them? To profit from them?

Privacy concerns the collection and use of data about individuals. There are three primary privacy issues:

- **Accuracy** relates to the responsibility of those who collect data to ensure that the data is correct.
- **Property** relates to who owns data.
- **Access** relates to the responsibility of those who have data to control who is able to use that data.

Big Data

Today, almost all human events are recorded digitally. For example, making a phone call, posting a video to social media, or wearing a smartwatch creates a digital record. A digital record of a phone call can include who you called, when you called, where you were when you made the call, and even the contents of the conversation itself. This results in an unprecedented amount of digital information being stored. This stored information is often referred to as **big data**.

Large organizations are constantly compiling information about us. The federal government alone has over 2,000 databases. For example, credit card companies maintain user databases that track cardholder purchases, payments, and credit records. Supermarket scanners in grocery checkout counters record what we buy, when we buy it, how much we buy, and the price. Financial institutions, including banks and credit unions, record how much money we have, what we use it for, and how much we owe. Search engines record the search histories of their users, including search topics and sites visited. Social networking sites collect every entry.

This collection of data can be searched to find all the action of one person—creating a **digital footprint** that reveals a highly detailed account of your life. A vast industry of data gatherers known as **information resellers** or **information brokers** now exists that collects, analyzes, and sells such personal data. (See **Figure 9-1**.)

Your digital footprint can reveal more than you might wish to make public and have an impact beyond what you might imagine. This raises many important issues, including

- **Collecting public, but personally identifying, information:** What if people anywhere in the world could view detailed images of you, your home, or your vehicle? Using detailed images captured with a specially equipped van, Google's Street View project allows just that. Street View makes it possible to take a virtual tour of many cities and neighborhoods from any computer with a connection to the Internet. (See **Figure 9-2**.) Although the images available on Street View are all taken in public locations, some have objected to the project as being an intrusion on their privacy.

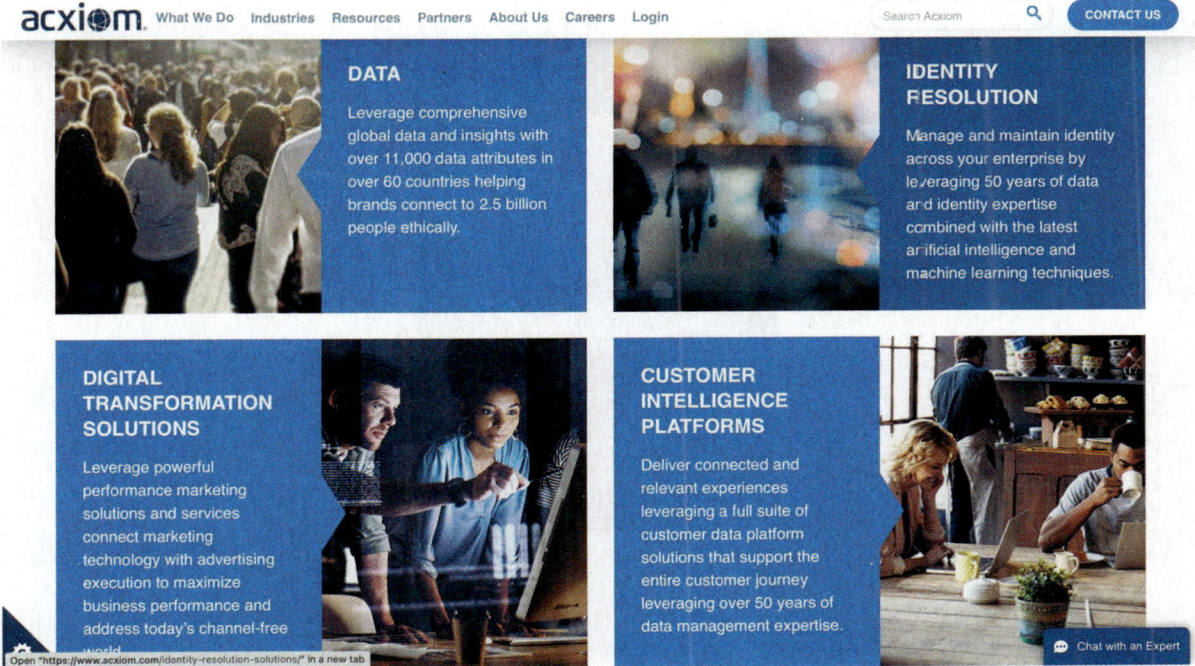

Figure 9-1 Information reseller's website

Acxiom

As digital cameras and webcams become cheaper and software becomes more sophisticated, it is likely that many more issues involving personal privacy in public spaces will need to be addressed. Such a combination of computing technologies could, for example, make real-time tracking of individuals in public places possible.

- **Spreading information without personal consent:** How would you feel if an employer were using your Facebook, Twitter, or other social networking profiles to make decisions about hiring, placement, promotion, and firing? It is a common practice today for many organizations.

  As we have discussed in **Chapter 2**, social networking is designed for open sharing of information among individuals who share a common interest. Unfortunately, this openness can put individuals using social networking sites at risk. In fact, some have lost their jobs after posting unflattering remarks about their supervisor or after discussing their dislike of their current job. Deeper analysis of your social networking profile may reveal even more about yourself than you intend.

  The potential to inadvertently share personal information extends past what you might post to your social networking site. As an example, without your knowledge or permission, a social networking friend might tag or identify you in a photo on his or her site. Once tagged, that photo can become part of your digital footprint and available to others without your consent.

> **community**
>
> Sharing personal information on social media is a necessary part of creating and participating in an online community. However, many individuals do not fully understand the complex sharing and privacy policies of these networks. This often causes unintentional sharing with people outside their intended social circle. The social networks themselves have come under fire from privacy groups, who say that these companies use complex settings and policies to get users to share more information than intended. This information is in turn shared with advertisers. Do you think social networking companies are doing enough to ensure the privacy and security of the communities they serve?

- **Spreading inaccurate information:** How would you like to be turned down for a home loan because of an error in your credit history? This is much more common than you might expect. What if you could not find a job or were fired from a job because of an error giving you a serious criminal history? This can and has happened due to simple clerical errors. In one case, an arresting officer while completing an arrest warrant incorrectly recorded the Social Security number of a criminal. From that time onward, this arrest and the subsequent conviction became part of another person's digital footprint. This is an example of **mistaken identity** in which the digital footprint of one person is switched with another.

  It's important to know that you have some recourse. The law allows you to gain access to those records about you that are held by credit bureaus. Under the **Freedom of Information Act**, you are also entitled to look at your records held by government agencies. (Portions may be deleted for national security reasons.)

Figure 9-2   Google Street View
incamerastock/Alamy

Private Networks

Suppose you use your company's electronic mail system to send a co-worker an unflattering message about your supervisor or to send a highly personal message to a friend. Later you find the boss has been reading these e-mails. In fact, many businesses search employees' electronic mail and computer files using **employee-monitoring software**. These programs record virtually everything you do on your computer. Many people incorrectly assume that their employers won't or can't monitor employee activity on work computers and networks. You should never do anything on a work computer or a work network connection that you wouldn't want shared with your employer. If you are employed and would like to know your company's current policy on monitoring electronic communication, contact your human relations department.

 concept check

- Describe how big data can affect our privacy.
- What is big data? Information resellers? Digital footprint?
- List three important issues related to digital footprints. What is mistaken identity? What is the Freedom of Information Act?
- What are private networks? What is employee-monitoring software? Is it legal?

The Internet and the Web

When you send a message or browse the web, do you have any concerns about privacy? Most people do not. They think that as long as they are using their own computer and are selective about disclosing their names or other personal information, then little can be done to invade their personal privacy. Experts call this the **illusion of anonymity** that the Internet brings.

As we discussed in **Chapter 8**, every computer on the Internet is identified by a unique number known as an IP address. IP addresses can be used to trace Internet activities to their origin, allowing computer security experts and law enforcement officers to investigate computer crimes such as unauthorized access to networks or sharing copyright files without permission.

Some websites are designed to be hidden from standard search engines. These websites make up the **deep web** and allow communication in a secure and anonymous manner. One part of the deep web is hidden websites that make up the **dark web**. These websites use special software that hides a user's IP address and makes it nearly impossible to identify who is using the site. The ability to communicate anonymously attracts criminals who want to sell drugs, share child pornography, or profit from the poaching of endangered animals. This same anonymity allows people in countries where political dissent is dangerous and free speech is censored to communicate, plan, and organize toward a more free and open society without fear of jail or execution.

When you browse the web, your browser stores critical information, typically without you being aware of it. This information, which contains records about your Internet activities, includes history and temporary Internet files.

- **History files** include the locations, or addresses, of sites that you have recently visited. This history file can be displayed by your browser in various locations, including the address bar (as you type) and the *History* page. To view your browsing history on your Android or iOS cell phone, see **Figure 9-3**.

- **Temporary Internet files**, also known as the **browser cache**, contain web page content and instructions for displaying this content. Whenever you visit a website, these files are saved by your browser. If you leave a site and then return later, these files are used to quickly redisplay web content.

Another way your web activity can be monitored is with **cookies**. Cookies are small data files that are deposited on your hard disk from websites you have visited. Based on your browser's settings, these cookies can be accepted or blocked. Websites that use cookies will typically ask you for permission to store your information when you first open the website. Although cookies are harmless in and of themselves, what makes them a potential privacy risk is that they can store information about you, your preferences, and your browsing habits. The information stored generally depends on whether the cookie is a first-party or a third-party cookie.

- A **first-party cookie** is one that is generated (and then read) only by the website you are currently visiting. Many websites use first-party cookies to store information about the current session, your general preferences, and your activity on the site. The intention of these cookies is to provide a personalized experience on a particular site. For example, when you are shopping online, you might fill your cart with items and then leave the site without making a purchase. Cookies allow you to return to that site and have the site remember what was in your cart.

- A **third-party cookie** is usually generated by an advertising company that is affiliated with the website you are currently visiting. These cookies are used by the advertising company to keep track of your web activity as you move from one site to the next. For this reason, they are often referred to as **tracking cookies**. Critics of this practice claim that your privacy is being violated because your activity is being recorded across multiple websites. Defenders of this practice argue that these

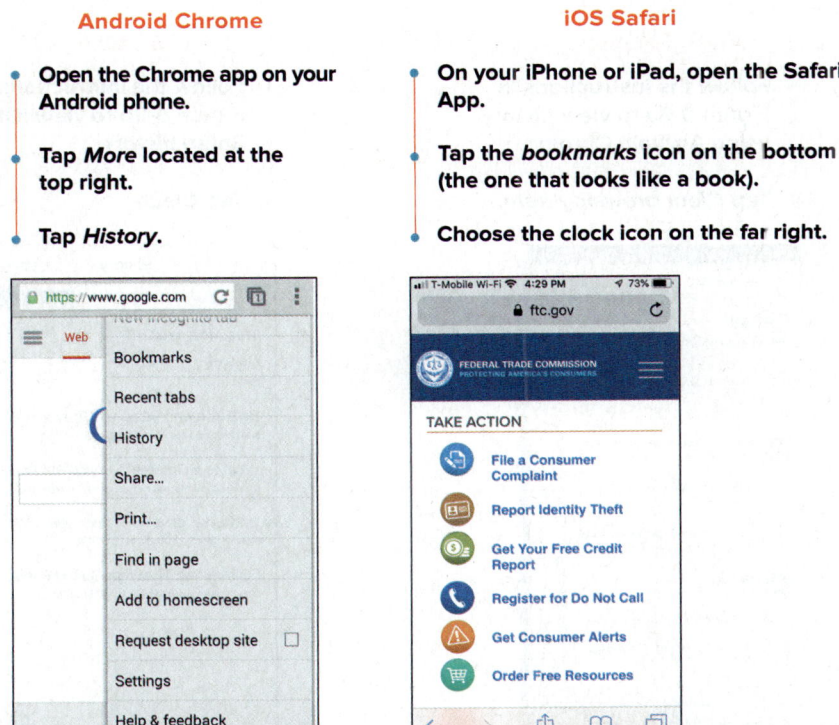

Figure 9-3 Viewing history files using Android Chrome and iOS Safari

(a) Google INC. (b) Apple Inc.

cookies are beneficial because they help websites deliver ads that interest you. For example, suppose you visit four different websites that employ the same advertising agency. The first three sites are about cars, but the fourth is a search engine. When you visit the fourth site, you will likely see a car advertisement because your cookie showed that you had been visiting car-related websites.

> ## privacy
>
> Google's Chrome web browser erases any record of your web activities from your computer while using Incognito Mode. Many believe this means that Google will not record your activities on the web. However, it does not erase your activities recorded on the computers that run the websites that you visit. For example, Google's websites will record information about your activities even when you are in Incognito Mode. What do you think? Does Google Chrome's promise of private browsing include the information that Google collects on its websites?

Some users are not comfortable with the idea of web browsers storing so much information in the form of temporary Internet files, cookies, and history. For this reason, browsers now offer users an easy way to delete their browsing history. To see how to delete browsing histories on your Android or iOS cell phone, see **Figure 9-4**. In addition, most browsers also offer a **privacy mode**, which ensures that your browsing activity is not recorded on your computer. For example, Google Chrome provides **Incognito Mode** accessible from the Chrome menu, and Safari provides **Private Browsing** accessible from the *Safari* option on the main menu.

Although these web browser files can concern many individuals, several other threats could potentially violate your privacy. **Web bugs**, which are invisible images or HTML code hidden within a web page or e-mail message, can be used to transmit information without your knowledge. When a user opens an e-mail containing a web bug, information is sent back to the source of the bug. The receiving server will now know that this e-mail address is active. One of the most common web bugs is used by companies that sell active mailing lists to spammers. Because of this deception, many e-mail programs now block images and HTML code from unknown senders. It is up to the user to decide whether or not to allow such content to be displayed for current and future messages.

The most dangerous type of privacy threat comes in the form of spyware. The term **spyware** is used to describe a wide range of programs that are designed to secretly record and report an individual's activities on the Internet. Some of these programs can even make changes to your browser in order to deceive you and manipulate what you see online. **Computer monitoring software** is perhaps the most invasive and dangerous type of spyware. One type of computer monitoring software, known as a **keylogger**, records every activity and keystroke made on your computer system, including credit card numbers, passwords, and e-mail messages. Computer monitoring software can be deposited onto your hard drive without your knowledge by a malicious website or by someone installing the program directly onto your computer. Although such software is deadly in the hands of criminals, it can be legally used by companies monitoring employees or law enforcement officials who are collecting evidence.

Android Chrome

- Follow the instructions in Figure 9-3a to view history using Android Chrome.
- Tap *Clear browsing data.*

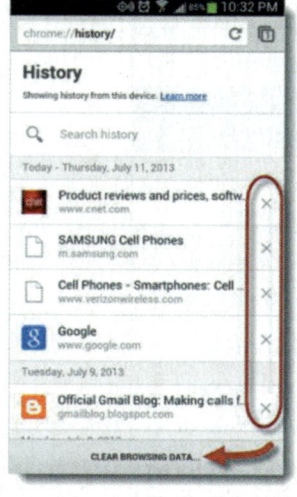

(a)

iOS Safari

- Follow the instructions in Figure 9-3b to view iOS Safari history.
- Tap *Clear.*

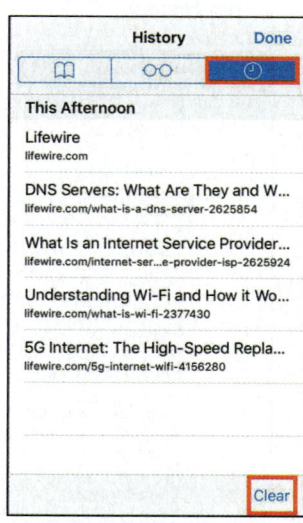

(b)

Figure 9-4 Deleting history files using Android Chrome and iOS Safari
(a) Google INC. (b) Apple Inc.

Unfortunately, many spyware programs go undetected, largely because users have no idea they are infected. Spyware will run in the background, invisible to the average user. Other times, it disguises itself as useful software, such as a security program. Spyware is alarmingly common on laptops and desktops, but recent spyware is designed to work on other types of computers. For example, Pegasus is a spyware designed to work on iOS and Android cell phones. The financial impact to individuals, companies, and financial institutions is estimated at billions of dollars.

One of the best defenses against spyware is to exercise caution when visiting new websites and downloading software from an unknown source. Another defense involves using a category of software known as **antispyware** or **spy removal programs**, which are designed to detect and remove various types of privacy threats. (See **Figure 9-5**.) For a list of some of these programs, see **Figure 9-6**.

ethics

Data encryption protects your bank accounts and private e-mails, but it also allows criminals to hide evidence from police. Some government agencies want to limit the power of data encryption to make it easier to find and jail criminals. Privacy advocates argue that limiting encryption gives criminals and dictators tools that will be used to steal data and suppress freedoms. Security advocates respond that to protect citizens from criminals, the police must be able to search encrypted devices when warranted. Should governments be given a skeleton key to bypass encryption? Should unbreakable encryption be illegal?

Figure 9-5   Antispyware
Piotr Swat/Alamy Stock Photo

Online Identity

Another aspect of Internet privacy comes from **online identity**, the information that people voluntarily post about themselves online. With the popularity of social networking, blogging, and photo- and video-sharing sites, many people post intimate details of their lives without considering the consequences. Although it is easy to think of online identity as something shared between friends, the archiving and search features of the web make it available indefinitely to anyone who cares to look.

Program	Website
Ad-Aware	adaware.com
Norton Security	norton.com
Windows Defender	microsoft.com
AVG Antitrack	avg.com

Figure 9-6   Antispyware programs

There are many number of cases of people who have lost their jobs on the basis of posts on social networking sites. These job losses range from a teacher (using off-color language and photos showing drinking) to a chief financial officer of a major corporation (discussing corporate dealings and financial data). The cases include college graduates being refused a job because of Facebook posts. How would you feel if information you posted about yourself on the web kept you from getting a job?

Major Laws on Privacy

Internationally, many countries are working to codify digital privacy into law. For example, the European Union's **General Data Protection Regulation (GDPR)** regulates the processing of personal data for all organizations that collect or process data about EU citizens, regardless of the nation of origin of the organization. Some federal laws governing privacy matters have been created. For example, the **Gramm-Leach-Bliley Act** protects personal financial information, the **Health Insurance Portability and Accountability Act (HIPAA)** protects medical records, and the **Family Educational Rights and Privacy Act (FERPA)** restricts disclosure of educational records. At the more local level, individual states also regulate and protect privacy. For example, the **California Consumer Privacy Act (CCPA)** allows consumers to see all data a company has collected on them and any third parties that data has been shared with.

Most of the information collected by private organizations is not covered by existing laws. However, as more and more individuals become concerned about controlling who has the right to personal information and how that information is used, companies and lawmakers will respond.

concept check

- What is the illusion of anonymity? Define and compare history files and temporary Internet files. What is Privacy mode?
- What is a cookie? A first-party cookie? A third-party cookie?
- What is a web bug? Spyware? Keylogger? Antispyware? Online identity?
- Describe three federal laws to protect privacy.

Security

tips

Identity theft is a growing problem, and can be financially devastating if you are a victim. Thieves are after anything that can help them steal your identity, from your Social Security number and date of birth to account information and passwords. Here are some steps to help protect your identity:

1. Be careful what you post on the Internet. Never post personal information on forums or social networking areas that are public or in response to an e-mail from someone you do not know or trust.
2. Only do business on the Internet with companies you know to be legitimate.
3. When selling a computer, be sure to completely remove all personal information from the hard drive. To ensure that your personal information is erased, consider using a free erasure software tool, such as Dban (dban.org).
4. Monitor your credit. Each year, you are entitled to a free personal credit report from each of the three major credit reporting agencies. Monitor your credit by requesting a report every four months from a different reporting agency. The official site for this service is www.annualcreditreport.com.

Personal security protects us from crime and danger in the physical world. For example, bank security protects our savings, home security protects our families, and airport security protects our safety. The digital revolution brings with it new types of crimes and dangers. To protect ourselves from these new threats, computer **security** needs to protect our information, hardware, and software from unauthorized use, as well as preventing or limiting the damage from intrusions, sabotage, and natural disasters.

Cybercrime

Cybercrime or **computer crime** is any criminal offense that involves a computer and a network. Cybercrimes are often in the news, especially **cyberterrorism**, which is a politically motivated cybercrime. It was recently estimated that cybercrime affects over 400 million people and costs over $400 billion each year. Cybercrimes can take various forms, including identity theft, Internet scams, data manipulation, ransomware, and denial of service attacks.

- **Identity theft** is the illegal assumption of someone's identity for the purposes of economic gain. Stolen identities are used to steal credit cards and mail and to commit other crimes. It is estimated that identity thieves stole $43 billion and victimized 25 million individuals in 2022.

- **Internet scams** are scams using the Internet. Internet scams have created financial and legal problems for many thousands of people. Almost all the scams are initiated by a mass mailing to unsuspecting individuals. Recently, Interpol busted a criminal network of 40 people across Nigeria, Malaysia, and South Africa accused of using Internet scams to steal more than $60 million. See **Figure 9-7** for a list of common types of Internet scams.
- **Data manipulation** is the unauthorized access of a computer network and copying files to or from the server. This can be as simple as making a post on Facebook when logged in as someone else or as complex as feeding a company false reports to change its business practices. Unlike other cybercrimes, data manipulation can occur for months, even years, without the victims being aware of the security breach, making it hard to detect.
- **Ransomware** is malicious software that encrypts your computer's data and ransoms the password to the user. Ransomware criminals have targeted food processing plants, hospitals, and police stations, endangering the health and safety of millions. Most recently, a ransomware attack on the Colonial Pipeline caused fuel shortages that grounded airplanes and closed gas stations across the southeastern United States in the summer of 2021. It is widely believed that there may be many more ransomware attacks than reported, as companies that pay the ransom may not be transparent about the security breach.
- **Denial of service (DoS) attacks** attempt to slow down or stop a computer system or network by flooding a computer or network with requests for information and data. These requests can come from a single computer issuing repeated requests. Widely used today is a variation known as **distributed denial of service (DDoS)**, which coordinates several computers making repeated requests for service. The targets of these attacks are usually Internet service providers (ISPs) and specific websites. Once under attack, the servers at the ISP or the website become overwhelmed with these requests for service and are unable to respond to legitimate users. As a result, the ISP or website is effectively shut down.

With cybercrime on the rise and the high-profile victims of cybercrime in the news, it is important to understand the tools the cybercriminal uses to harm his victims. Knowledge of these tools will help you make smarter choices and protect yourself from becoming a victim. The tools of the cybercriminal include social engineering, malicious software, and malicious hardware. For a summary of cybercrimes, see **Figure 9-8**.

Type	Description
Phishing	Communications in which a criminal pretends to be from an official organization and tricks you into giving them sensitive data, such as passwords and bank account numbers. Often these communications include a link to a website that looks like an official log-in screen but in fact is a fake website designed to trick people into giving up their username and password.
Advanced-fee scam	A classic e-mail scam. The recipient receives an e-mail from a wealthy foreigner in distress who needs your bank account information to safely store his or her wealth, and for your troubles you will receive a large amount of money. Of course, once the scammer has your bank account information, your accounts will be drained and he or she will disappear.
Greeting card scam	An e-mail or social media communication informs you that a friend has sent you a greeting card and you need to download software to view it. In fact, the software is malware that can steal your data and infect your computer.
Bank loan/credit card scam	Criminals acting as bank or credit card officials offer you unusually good deals on bank loans or credit cards—but these are just attempts to get you to pay huge "processing fees" and to get your personal information.
Lottery scam	An e-mail informs you that you have won the lottery and to claim your prize, you need to pay processing fees. Criminals will take the processing fees, but you will not receive any lottery winnings.

Figure 9-7   **Common Internet scams**

Type	Description
Identity theft	Illegal assumption of someone's identity for economic gain
Internet scam	Scams over the Internet usually initiated by mass e-mail
Data manipulation	Unauthorized access to a computer network and copying of files
Ransomware	Encrypts data on a user's computer and then ransoms password to access encrypted data
Denial of service (DoS)	Slows down or stops a computer system or network by flooding it with repeated requests for information and/or data

Figure 9-8   **Common cybercrimes**

concept check

- What is cybercrime?
- What are identity theft and Internet scams?
- What are data manipulation, ransomware, and denial of service attacks?

privacy

Did you know that the World Wide Web contains websites purposely hidden from you? Some websites are designed to be hidden from standard search engines and allow people to communicate in a secure and anonymous manner. These hidden websites make up the dark web and require special software that makes it nearly impossible to identify who is using it. The ability to communicate anonymously attracts criminals who want to sell drugs, share child pornography, or profit from the poaching of endangered animals. This same anonymity allows people in countries where political dissent is dangerous and free speech is censored to communicate, plan, and organize toward a more free and open society without fear of jail or execution. Prior to reading this Privacy section, were you aware of the dark web? Do you think it should be stopped? Do you think it should be regulated?

Social Engineering

Often the least secure parts of a network or computer are the humans who work with it. **Social engineering** is the practice of manipulating people to divulge private data. For example, a criminal may call you at work, pretending to be an IT worker who needs your password and user name, or they may "friend" you on social media to get access to private information. Social engineering has played a key role in identity theft, Internet scams, and data manipulation. One of the most common social engineering techniques is **phishing** (pronounced "fishing"). Phishing attempts to trick Internet users into thinking a fake but official-looking website or e-mail is legitimate. Phishing has grown in sophistication, replicating entire websites, like PayPal, to try to lure users into divulging their financial information.

Malicious Software

A **cracker** is a computer criminal who creates and distributes malicious programs or **malware**, which is short for **malicious software**. Malware is specifically designed to damage or disrupt a computer system. The three most common types of malware are viruses, worms, and Trojan horses.

- **Viruses** are programs that migrate through networks and operating systems, and mostly attach themselves to other programs and databases. Although some viruses are relatively harmless, many can be quite destructive. Once activated, these destructive viruses can alter and/or delete files.

- **Worms** are programs that simply replicate themselves over and over again. Once active in a network, the self-replicating activity clogs computers and networks until their operations are slowed or stopped. Unlike a virus, a worm typically does not attach itself to a program or alter and/or delete files. Worms, however, can carry a virus. Once a virus has been deposited by a worm onto an unsuspecting computer system, the virus will either activate immediately or lie dormant until some future time. In 2004, the worst computer worm in history, MyDoom, is estimated to have resulted in $38,000,000,000 in damages.

- **Trojan horses** are programs that appear to be harmless; however, they contain malicious programs. Trojan horses are not viruses. Like worms, however, they can be carriers of viruses. The most common types of Trojan horses appear as free computer games and free antivirus software that can be downloaded from the Internet. When a user installs one of these programs, the Trojan horse also secretly installs a virus on the computer system. The virus then begins its mischief. Some of the most common Trojan horses claim to provide free antivirus programs. When a user downloads one of these programs, the Trojan horse first installs a virus that locates and disables any existing virus protection programs before depositing other viruses.

Malicious Hardware

Criminals use computer hardware to steal information, infect computers with malicious software, and disrupt computer systems. The most common malicious hardware includes zombie botnets, rogue Wi-Fi hotspots, and infected USB flash drives.

- **Zombies** are computers infected by a virus, worm, or Trojan horse that allows them to be remotely controlled for malicious purposes. A collection of zombie computers is known as a **botnet**, or **robot network**. Botnets harness the combined power of many zombies for malicious activities like password cracking, denial of service attacks, or sending junk e-mail. Because they are formed by many computers distributed across the Internet, botnets are hard to shut down even after they are detected. Unfortunately for individual computer owners, it also can be difficult to detect when a personal computer has been compromised.

- **Rogue Wi-Fi hotspots** imitate free Wi-Fi networks. These rogue networks operate close to the legitimate free hotspots and typically provide stronger signals that many users unsuspectingly connect to. Once connected, the rogue networks capture any and all information sent by the users to legitimate sites, including user names and passwords.
- **Infected USB flash drives** contain viruses and other malicious software. Crackers typically leave these drives in public spaces in the hope that others will find them, plug them into their computer, and become infected. Infected drives have also been found distributed for free at conferences and slipped into people's mailboxes.

concept check

☐ What is social engineering? What is phishing?

☐ What is malicious software? A cracker? Viruses? Worms? Trojan horses?

☐ What is malicious hardware? Zombies? Botnets? Rogue Wi-Fi hotspots? Infected USB flash drives?

Measures to Protect Computer Security

There are numerous ways in which computer systems and data can be compromised and many ways to ensure computer security. The **Computer Fraud and Abuse Act** makes it a crime for unauthorized persons even to view—let alone copy or damage—data using any computer across state lines. It also prohibits unauthorized use of any government computer or a computer used by any federally insured financial institution. Offenders can be sentenced up to 20 years in prison and fined up to $100,000. However, the best protection for computer security is to be prepared. Some of the principal measures to ensure computer security are restricting access, encrypting data, anticipating disasters, and preventing data loss.

Restricting Access

Security experts are constantly devising ways to protect computer systems from access by unauthorized persons. Sometimes security is a matter of putting guards on company computer rooms and checking the identification of everyone admitted. However, the most common way to restrict access is the use of a password. **Passwords** are secret words or phrases (including numbers, letters, and special characters) that must be keyed into a computer system to gain access. For many applications on the web, users assign their own passwords.

The strength of a password depends on how easily it can be guessed. A **dictionary attack** uses software to try thousands of common words sequentially to gain unauthorized access to a user's account. For this reason, words, names, and simple numeric patterns make weak or poor passwords. Strong passwords have at least 12 characters and use a combination of letters, numbers, and symbols.

Newer technology allows access to be restricted by passwords that do not require secret words or phrases. Windows 11 includes an application, **Picture Password**, that accepts a series of gestures over a picture of the user's choice to gain access. Other times, **biometric scanning** devices such as fingerprint and iris (eye) scanners are used to access restricted data. (See **Figure 9-9**.) Numerous applications use face recognition to allow access to a computer system. For example,

tips

Security professionals warn that most people use passwords that are too easily guessed. They categorize passwords as weak (easily guessed) or strong (difficult to guess). A weak password can be guessed in a matter of seconds; a strong one would take years to crack. Make sure you have a strong password by employing the following tips when creating a password.

1. Use a password with at least eight characters. The fewer the characters, the easier it is to guess.
2. Do not use your username, real name, or company name in your password. The first thing a cracker will try is variations of your personal information.
3. Do not use a complete word. Using a computer, a cracker can easily try every word in the dictionary as a password guess.
4. Do not reuse passwords. If a user's password for one account is compromised, crackers will attempt to use that password on the user's other accounts.
5. Create a password that contains at least one of each of the following characters: uppercase letter, lowercase letter, number, and symbol.

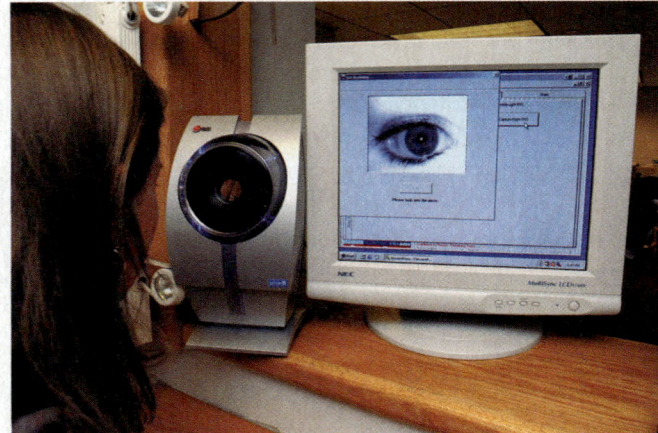

Figure 9-9   Biometric scanning devices
(left): Khongtham/Shutterstock (right): Michael Dwyer/Alamy Stock Photo

Microsoft's Windows 11 and Apple's iOS 15 include **facial recognition** software, which uses specialized cameras to identify users and automatically log them in. There are also several face recognition apps for mobile devices, including True Key by McAfee and Intel Security Group.

As mentioned in previous chapters, individuals and organizations use a variety of ways to perform and automate important security tasks:

ethics

When you lock your cell phone, it encrypts your data so that no one else can see your data without unlocking your phone. This also allows criminals to hide information on their cell phones from the police. Some software companies have discovered flaws in cell phone security and sell software that exploit these flaws and break open locked cell phones. U.S. law enforcement agencies have used this software to break open criminals' cell phones.

Unfortunately this software is also purchased and used by dictators and tyrants to spy on their populace and imprison political dissenters. Does a company that discovers a security flaw have an ethical responsibility to not exploit the flaw, but instead work with cell phone companies to eliminate the flaw? Should there be limitations on the tools U.S. agencies can purchase, if that purchase helps a company that helps dictators?

- **Security suites** provide a collection of utility programs designed to protect your privacy and security while you are on the web. These programs alert users when certain kinds of viruses and worms enter their system. Two of the most widely used are Avast! Free Antivirus and Microsoft Windows Defender. Unfortunately, new viruses are being developed all the time, and not all viruses can be detected. The best way to stay current is through services that keep track of viruses on a daily basis. For example, Symantec, McAfee, and Microsoft all track the most serious virus threats. (See **Figure 9-10**.)

- **Firewalls** act as a security buffer between a corporation's private network and all external networks, including the Internet. All electronic communications coming into and leaving the corporation must pass through the company's firewall, where they are evaluated. Security is maintained by denying access to unauthorized communications.

- **Password managers** help you create strong passwords. Additionally, they will store all your passwords in one location and automatically provide the appropriate password when requested from one of your favorite sites. This avoids many of the mistakes people make in generating and remembering passwords. However, this master list of passwords is protected by one "master" password. If you forget or reveal this master password, you open yourself up to considerable risk.

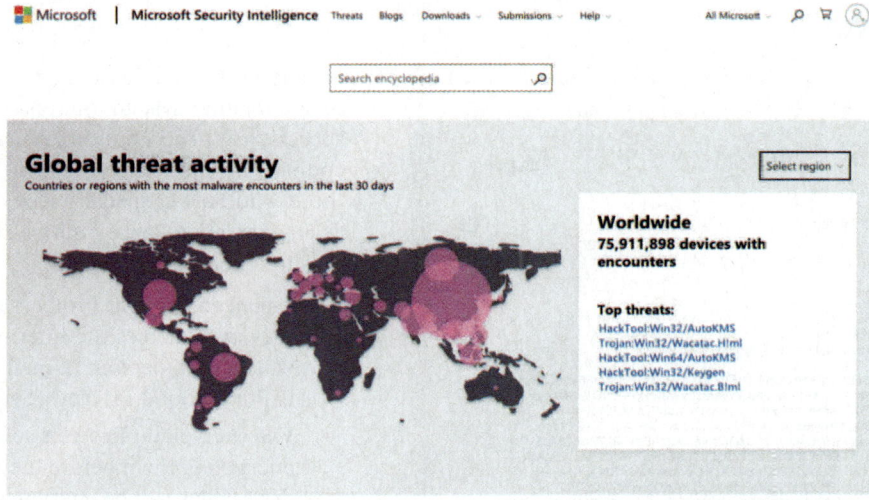

Figure 9-10 Tracking viruses
Microsoft Corporation

- **Authentication** is the process of ensuring the identity of a user. When you enter a password to gain access to a site, that is a single authentication of your identity. For added security, some security systems require multiple authentications.
Multi-factor authentication uses multiple types (or factors) of data to verify your identity. These types of data include knowledge (such as a password or the answer to a security question), possession (such as your cell phone or credit card), and biometric data (such as a fingerprint or voice print).

The most common type of multi-factor authentication is two-factor authentication—where two types (or factors) of data are used to verify your identity. A common form of two-factor authentication is when a person logs into an online account, they first enter their username and password. This first factor (the password) is followed up with a text message to the user's cellphone with a short code (the possession of the cell phone is the second factor) that must be entered before they are allowed access to the account. Furthermore, the short code (sent to the user's cell phone) is called a one-time password (OTP) or one-time PIN. An OTP can only be used once and often expire if not used quickly.

Two-step authentication uses one type of authentication twice (such as asking for two knowledge-based authentications such as a password and the answer to a security question). While both types are more secure than a single authentication, multi-factor authentication is considered more secure than two-step authentication—and both are more secure than a single authentication.

> **tips**
>
> Has your cell phone ever been lost or stolen? If so, then you not only are without the device, but you also have lost any data (photos, contact lists, etc.). To avoid this, you need to take some action *before* your cell phone goes missing. Consider the following suggestions.
>
> **Find your phone.** Set up your phone so that you can see a map of where your phone is and remotely turn on an alarm, secure the device, and even erase the contents of your cell phone.
> - For Android cell phones: Click on the *Settings* icon from the home screen and then choose *Security & Lock Screen* and *Device administrators*. Finally, tap on *Find My Device*. If your phone goes missing, go to android.com/find to find or remotely secure your device.
> - For iOS cell phones: Click on the *Settings* icon from the home screen and then select your Apple ID at the top of the screen. Select *iCloud* and turn on the button for *Find My iPhone*. If your phone goes missing, sign in to icloud.com/find to find or remotely secure your device.
>
> **Back up your phone.** Automatically back up your cell phone data to the cloud, so even if you cannot recover your phone, your data will not be lost. To enable these features, you will need an iCloud or Google Drive account.
> - For Android cell phones: Click on the *Settings* icon from the home screen and then choose *Backup*. Turn on the button for *Backup to Google Drive*.
> - For iOS cell phones: Click on the *Settings* icon from the home screen and then select your *Apple ID* at the top of the screen. Select *iCloud* and turn on the button for *iCloud Backup*.

Encrypting Data

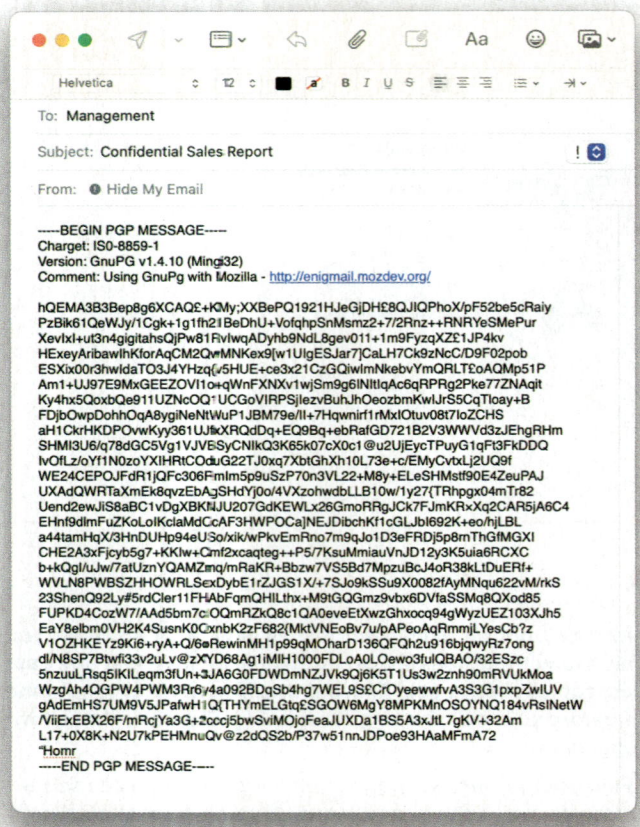

Figure 9-11 Encrypted e-mail
Enigmail

Whenever information is sent over a network or stored on a computer system, the possibility of unauthorized access exists. The solution is **encryption**, the process of coding information to make it unreadable except to those who have a special piece of information known as an **encryption key**, or, simply, a **key**. Some common uses for encryption include

- **E-mail encryption:** Protects e-mail messages as they move across the Internet. One of the most widely used personal e-mail encryption programs is Pretty Good Privacy. (See **Figure 9-11**.)

- **File encryption:** Protects sensitive files by encrypting them before they are stored on a hard drive. Files can be encrypted individually, or specialized software can be used to encrypt all files automatically each time they are saved to a certain hard drive location. (See **Figure 9-12**.)

- **Website encryption:** Secures web transactions, especially financial transactions. Web pages that accept passwords or confidential information like a credit card number are often encrypted. As we discussed in **Chapters 2** and **8**, **https (hypertext transfer protocol secure)** is the most widely used Internet protocol. This protocol requires that the browser and the connecting site encrypt all messages, providing a safer and more secure transmission.

- **Virtual private networks: Virtual private networks (VPNs)** encrypt connections between company networks and remote users such as workers connecting from home. This connection creates a secure virtual connection to a company LAN across the Internet.

- **Wireless network encryption:** Restricts access to authorized users on wireless networks. **WPA3 (Wi-Fi Protected Access)** is the newest **wireless network encryption** for home wireless networks. WPA3 is typically established for a wireless network through the network's wireless router. Although the specifics vary between routers, WPA3 is usually set through the router's settings options.

Anticipating Disasters

Companies (and even individuals) should prepare themselves for disasters. **Physical security** is concerned with protecting hardware from possible human and natural disasters. **Data security** is concerned with protecting software and data from unauthorized tampering or damage. Most large organizations have **disaster recovery plans** describing ways to continue operating until normal computer operations can be restored.

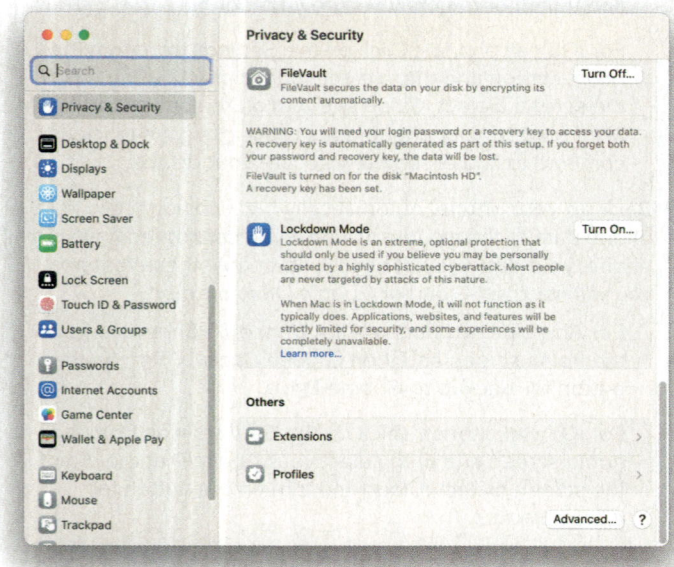

Figure 9-12 File encryption
Apple, Inc.

Preventing Data Loss

Equipment can always be replaced. A company's *data,* however, may be irreplaceable. Most companies have ways of trying to keep software and data from being tampered with in the first place. They include careful screening of job applicants, guarding of passwords, and auditing of data and programs from time to time. Some systems use redundant storage to prevent loss of data even when a hard drive fails. We discussed RAID in **Chapter 7**, which is a commonly used type of redundant storage. Backup batteries protect against data loss due to file corruption during unexpected power outages.

Making frequent backups of data is essential to prevent data loss. Backups are often stored at an off-site location to protect data in case of theft, fire, flood, or other disasters. Students and others often use flash drives and cloud storage, as discussed in **Chapter 7**, to back up homework and important papers. Incremental backups store multiple versions of data at different points in time to prevent data loss due to unwanted changes or accidental deletion.

To see what you should do to protect yourself, see Making IT Work for You: Security and Technology on page 228.

See **Figure 9-13** for a summary of the different measures to protect computer security.

Measure	Description
Restricting access	Limit access to authorized persons using such measures as passwords, picture passwords, and biometric scanning.
Encrypting data	Code all messages sent over a network.
Anticipating disasters	Prepare for disasters by ensuring physical security and data security through a disaster recovery plan.
Preventing data loss	Routinely copy data and store it at a remote location.

Figure 9-13 Measures to protect computer security

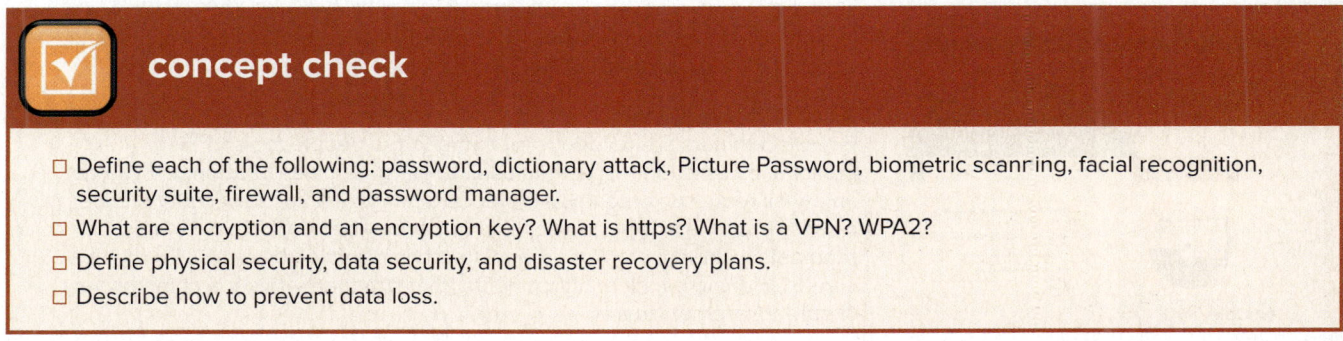

concept check

- Define each of the following: password, dictionary attack, Picture Password, biometric scanning, facial recognition, security suite, firewall, and password manager.
- What are encryption and an encryption key? What is https? What is a VPN? WPA2?
- Define physical security, data security, and disaster recovery plans.
- Describe how to prevent data loss.

Ethics

New technologies create new opportunities and interactions for everyone. For a criminal, these new opportunities may include new computer crimes to commit. For a police officer, this means new cybercrime laws and new ways to serve and protect. As we explore the impact of digital identities and digital lives, we must reexamine our standards of moral conduct, or **ethics**, in the context of these new opportunities and interactions. **Computer ethics** are guidelines for the morally acceptable use of computers in our society. Ethical treatment is critically important to us all, and we are all entitled to ethical treatment. This includes the right to keep personal information, such as credit ratings and medical histories, from getting into unauthorized hands. These issues, largely under the control of corporations and government agencies, were covered earlier in this chapter, and many more have been addressed in the Ethics boxes throughout this book. Now we'll examine three important issues in computer ethics where average users have a role to play.

Making IT work for you

SECURITY AND TECHNOLOGY

The news is filled with instances of cybercrime. From hacked e-mail to stolen identities, it's a dangerous digital world. Thankfully, there are a lot of precautions you as an individual can and should take to make sure that you aren't the victim of high-tech criminals.

Here are some guidelines for keeping yourself safe from computer crime.

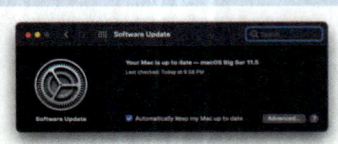
Apple Inc.

- **Update software**
  Your operating system, web browser, and applications are constantly improving to reduce the risk of malware infection. Be sure that your devices are up to date. Look into the setting of your PC, tablet, and phone—most devices have the option to check for software updates and to opt in to automatically updating software.

Backblaze

- **Regularly back up your data**
  Criminals may infect your computer with ransomware—software that encrypts your data and demands a payment for the password to get the data back. Without regular backups, you may lose data due to an accident or ransomware criminal. Much data can't be replaced—precious photos, important documents, and months of work on projects. However, with a backup, you can restore your data within minutes. Most operating systems offer automated cloud backups that are easy to set up and effortless once in place.

- **Be careful when browsing**
  Always think before you click. Don't download software from sites you are not familiar with or don't trust. Don't assume that a pop-up alert or warning is legitimate—scams and malware are often disguised as legitimate plug-ins or updates. When updating software, use the operating system's app store.

Google, INC.

- **Be alert to e-mail scams**
  E-mail scammers are notorious for trying to get people to send them personal information that is then used to steal identities (phishing). If you get an e-mail requesting money, a credit card number, a Social Security number, a password, or any other personal information—be very cautious—most reputable institutions will not request such information by e-mail. Further, be careful opening e-mail or attachments from strangers—this is a common way to spread Internet viruses.

- **Use antivirus software**
  There are many free or inexpensive antivirus software packages available. Be sure to update the antivirus software regularly (or automate the update process) to make sure the software is up to date.

- **Set up strong passwords**
  Common passwords, such as "password" or "love," are a criminal's best friend. To make your passwords as secure as possible, do not use common words or phrases and use a combination of lowercase and uppercase letters, symbols, and numbers. You can also have secure passwords suggested to you by a password management system—macOS comes with Keychain and Internet browsers Google Chrome and Microsoft Edge include password managers.

- **Protect against data breaches**
  A data breach is when criminals break into a company's database and sell or leak the account information found there. To reduce the risk of having your accounts infiltrated by hackers, do not reuse the same password on multiple accounts.

- **Enable two-factor authentication**
  Most security-minded applications or websites have the option for two-factor authentication, especially online banking, financial services, webmail, social media sites, and cloud services. However, you must often go the extra step of enabling this process. Two-factor authentication is one of the best ways to ensure that your privacy and identity remain secure.

Cyberbullying

A fairly recent and all-too-common phenomenon, **cyberbullying** is the use of the Internet to send or post content intended to hurt or embarrass another person. Although not always a crime, it can lead to criminal prosecution. Cyberbullying includes

- harassment—such as sending threatening or unwanted messages to an individual after having been asked to stop.
- doxing—maliciously disclosing personal data about a person that could lead to harm to that person.
- false statements or images—posting lies or false images to slander a person. Improvements in artificial intelligence allow deep fakes to be more convincing, common, and easily created.
- outing—disclosing personal information about a person that could result in them being harmed.

Never participate in cyberbullying, and discourage others from participating in this dangerous and hateful activity. If you or someone you know is the victim of cyberbullying, go to stopbullying.gov for help and advice.

Copyright and Digital Rights Management

A **copyright** is a legal concept that gives content creators the right to control the use and distribution of their work. Materials that can be copyrighted include paintings, books, music, films, and even video games. Some users choose to make unauthorized copies of digital media, which violates copyright. For example, making an unauthorized copy of a digital music file for a friend might be a copyright violation.

Software piracy is the unauthorized copying and/or distribution of software. According to a recent study, software piracy costs the software industry over $60 billion annually. To prevent copyright violations, corporations often use **digital rights management (DRM)**. DRM encompasses various technologies that control access to electronic media and files. Typically, DRM is used to (1) control the number of devices that can access a given file and (2) limit the kinds of devices that can access a file. Although some companies see DRM as a necessity to protect their rights, some users feel they should have the right to use the media they buy—including movies, music, software, and video games—as they choose.

The **Digital Millennium Copyright Act** makes it illegal to deactivate or otherwise disable any anti-piracy technologies, including DRM technologies. The act also establishes that copies of commercial programs may not be legally resold or given away. It further makes it a crime to sell or to use programs or devices that are used to illegally copy software. This may come as a surprise to those who copy software, movies, or music from a friend or from the Internet. The law is clear: It is illegal to copy or download copyright-protected music and videos from the Internet without appropriate authorization.

Today, there are many legal sources for digital media. Television programs can be watched online, often for free, on television-network-sponsored sites. Sites like Pandora allow listeners to enjoy music at no cost. There are several online stores for purchasing music and video content. A pioneer in this area is Apple's iTunes Music Store. (See **Figure 9-14**.)

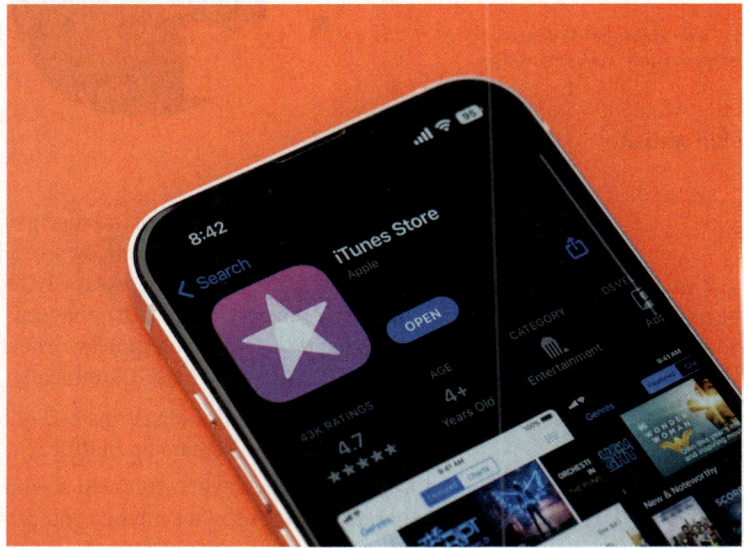

Figure 9-14   **iTunes Music Store**
Seemanta Dutta/Alamy Stock Photo

ethics

Net neutrality is the policy that ISPs should deliver all data to customers at the same speed, regardless of the content. Opponents of net neutrality would allow ISPs to charge companies to deliver some websites faster, arguing that ISPs should be allowed to charge however they like. Proponents of net neutrality argue that the Internet has always had equal and unbiased access to information and that allowing ISPs to vary access to websites based on content is a new form of censorship. What do you think? Is net neutrality an important defense of Internet freedom of speech, or is it government overreach looking to hinder free markets?

Plagiarism

Another ethical issue is **plagiarism**, which means representing some other person's work and ideas as your own without giving credit to the original source. Although plagiarism was a problem long before the invention of computers, computer technology has made plagiarism easier. For example, simply cutting and pasting content from a web page into a report or paper may seem tempting to an overworked student or employee.

Correspondingly, computer technology has made it easier than ever to recognize and catch **plagiarists**. For example, services such as Turnitin are dedicated to preventing Internet plagiarism. This service will examine the content of a paper and compare it to a wide range of known public electronic documents, including web page content. In this way, Turnitin can identify an undocumented paper or even parts of an undocumented paper. (See **Figure 9-15**.)

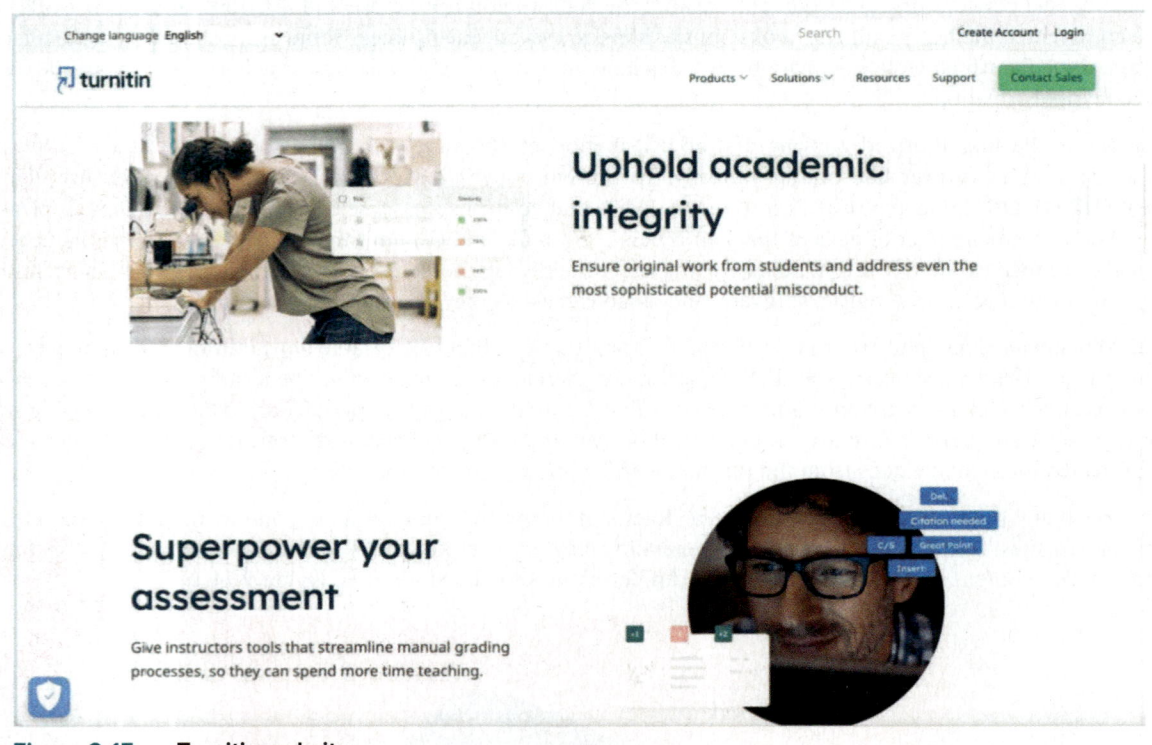

Figure 9-15 Turnitin website
Turnitin

ethics

You can use ChatGPT, or any AI, to create an essay. You must ask the program to create the essay, describing the features you want to see in the result. ChatGPT then interprets your question and picks and chooses data from the Internet that its programming thinks will best answer your question.

When ChatGPT selects data from sources on the Internet—it does not copy the words it finds, but it can output very similar text as its sources, with variations in word choice. Is ChatGPT plagiarizing Internet sources?

Recent developments in artificial intelligence have raised concern that plagiarists may use AI output as their own original work. One AI program, ChatGPT, can quickly generate essays based on the topics and writing styles the user requests. Despite the concerns of plagiarism, programs like ChatGPT can be a powerful tool for effective writing. Here are some guidelines for how to use ChatGPT ethically:

- Generate and organize ideas—instead of asking ChatGPT to write your essay, ask it to create an outline.
- Rework sentences—when stuck on a sentence, ask ChatGPT to generate variations of a sentence you provide.
- Include your own words and experiences—ChatGPT can help spark ideas and refine text, but it is your words and your experiences that should be reflected in your work.

There are also some things you should avoid when using ChatGPT:

- Do not directly copy and paste a ChatGPT response into an assignment and present it as your own.
- Do not assume the facts ChatGPT provide are accurate—ChatGPT learns based on text from the Internet. The Internet can contain rumors, conspiracy theories, and lies that ChatGPT will report as fact.

You can try out ChatGPT at chat.openai.com.

concept check

☐ What is the distinction between ethics and computer ethics?
☐ Define copyright, software privacy, digital rights management, and the Digital Millennium Copyright Act.
☐ What is cyberbullying? What is plagiarism? What is Turnitin and what does it do?

Careers in IT

"**Now that you have learned about privacy, security, and ethics, let me tell you about my career as an IT security analyst.**"

Roman Samborskyi/Shutterstock

IT security analysts are responsible for maintaining the security of a company's networks, systems, and data. Their goal is to ensure the confidentiality, integrity, and availability of information. These analysts must safeguard information systems against a variety of external threats, such as crackers and viruses, as well as be vigilant of threats that may come from within the company.

Employers typically look for candidates with a bachelor's or associate's degree in information systems or computer science. Experience in this field or in network administration is usually required. IT security analysts should possess good communication and research skills and be able to handle high-stress situations.

IT security analysts can expect to earn an annual salary of $64,000 to $77,000. Opportunities for advancement typically depend on experience. Demand for this position is expected to grow as malware, crackers, and other types of threats become more complex and prevalent.

A LOOK TO THE FUTURE

End of Anonymity

ImageFlow/Shutterstock

Do you use the Internet anonymously? Do you post photos or tweet without revealing your name? Some people anonymously post horrible, dangerous, and even criminal things on the Internet. Should anonymity protect us and others from public scrutiny and possible criminal prosecution? Should we sacrifice some or all of our anonymity for a safer, more civil discourse on the web?

Have you ever thought about how your digital actions might be linked to your real-world life? Imagine a future where your digital and real-world lives blend, and neither has space for anonymous interaction. The Internet has always had a challenging relationship with anonymity. Free speech advocates praise anonymous speech, allowing citizens in oppressive regimes to meet and share ideas without fear of government reprisal. In contrast, security advocates argue that the same tools that allow for free speech in dictatorships also allow terrorists to plan attacks against democracies. In the future, we are likely to see an ever-decreasing level of anonymity. Companies and Internet service providers already buy and sell your web browsing habits and online purchase history to better advertise and sell things to you. In the future, your browsing habits will likely follow you into the real world, and your real-world actions will filter into your digital persona.

In the future, when you visit a mall, security cameras will recognize you, track your path using facial recognition, and update your online identity. This personal digital footprint will be continuously updated based on social media postings, Internet browsing history, and credit card purchases. Sensors in your clothes will tell stores what brands you like and how much you spend.

Imagine, as computers analyze your path through the mall, a link is made between your recent social media posts about your new exercise routine. A sporting goods store just three shops ahead of you takes notice. Instantly, your cell phone buzzes with a coupon for the store with 25 percent off for first-time customers. It may seem like coincidence, but in a world without anonymity, there is no coincidence.

As you return home with your new treadmill, the sporting goods store posts to your Facebook wall, congratulating you on your first steps to a healthier you. It offers each of your Facebook friends a 20 percent off coupon to join you in making healthy life choices. Your Internet-connected treadmill posts your fitness progress on Twitter, describing the duration and intensity of your workouts. Your web-connected scale shows the results of your efforts. Your health insurance company, monitoring your exercise, gives you a discount on your insurance rates recent efforts. Your boss sees your treadmill posts and praises you at the next meeting, encouraging other employees to learn from your example and improve their lives through discipline and effort. However, this raises the question: What will your boss think if you stop running on the treadmill? Can you afford to take a day off if your insurance rates will skyrocket? What do you think: Is this world without anonymity better?

VISUAL SUMMARY | Privacy, Security, and Ethics

PRIVACY

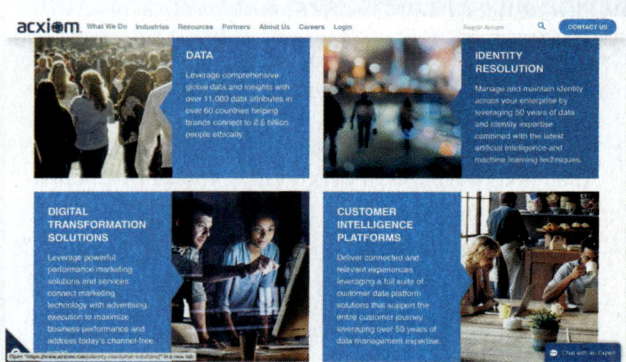

The three primary **privacy** issues: **accuracy**, **property**, and **access**.

Big Data

The ever-growing volume of data collected about us is often referred to as **big data**. **Information resellers (information brokers)** collect and sell personal data.

Mistaken identity occurs when the **digital footprint** of one person is switched with another. The **Freedom of Information Act** entitles individuals access to governmental records relating to them.

Private Networks

Many organizations monitor employee e-mail and computer files using special software called **employee-monitoring software**.

The Internet and the Web

Many people believe that, while using the web, little can be done to invade their privacy. This is called the **illusion of anonymity**. The **deep web** is comprised of websites that are hidden from standard search engines and provide secure and anonymous communication. The **dark web** is a part of the deep web comprised of sites that use special software to hide their IP addresses, making it nearly impossible to identify who is using it.

PRIVACY

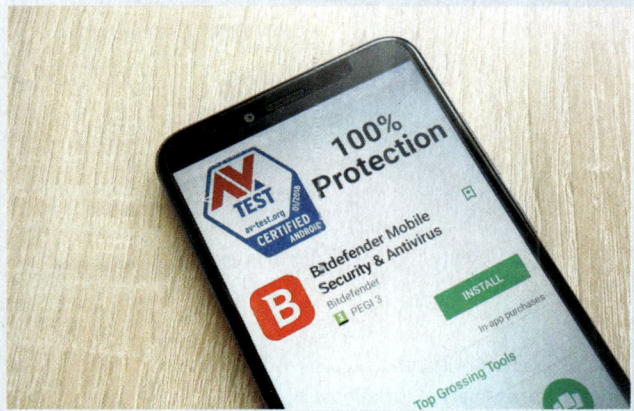

Piotr Swat/Alamy Stock Photo

Information stored by browsers includes **history files** (record of sites visited) and **temporary Internet files** or **browser cache** (contain website content and display instructions). **Cookies** store and track information. **Privacy mode (Private browsing)** ensures that your browsing activity is not recorded.

Spyware secretly records and reports Internet activities. **Keyloggers** (one type of **computer monitoring software**) records every activity and keystroke. **Antispyware (spy removal programs)** detects and removes various privacy threats.

Online Identity

Many people post personal information and sometimes intimate details of their lives without considering the consequences. This creates an **online identity**. With the archiving and search features of the web, this identity is indefinitely available to anyone who cares to look for it.

Major Laws on Privacy

The **Gramm-Leach-Bliley Act** protects personal financial information, the **Health Insurance Portability and Accountability Act (HIPAA)** protects medical records, and the **Family Educational Rights and Privacy Act (FERPA)** restricts disclosure of educational records.

To efficiently and effectively use computers, you need to be aware of the potential impact of technology on people. You need to be sensitive to and knowledgeable about personal privacy, organizational security, and ethics.

SECURITY

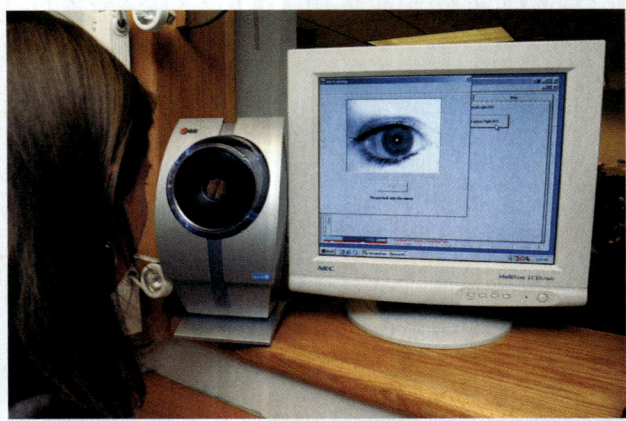

Michael Dwyer/Alamy Stock Photo

Computer **security** focuses on protecting information, hardware, and software from unauthorized use as well as preventing damage from intrusions, sabotage, and natural disasters.

Cybercrime

Cybercrime (computer crime) is a criminal offense involving a computer or network. (**Cyberterrorism** is a politically motivated cybercrime.) Some types of cybercrime are **identity theft**, **Internet scams**, **data manipulation**, **ransomware**, **denial of service (DoS) attack**, and **distributed denial of service (DDoS) attack**.

Social Engineering

Social engineering is the practice of manipulating people to divulge private data. **Phishing** attempts to trick people into believing a fake website is real.

Malicious Software

Crackers create and distribute **malware (malicious software)**. The three most common types are

- **Viruses** migrate through networks and operating systems; can alter and/or delete files.
- **Worms** repeatedly replicate themselves, clogging computers and **networks**.
- **Trojan horses** appear harmless but contain malicious programs.

SECURITY

Malicious Hardware

Cybercriminals can use computer hardware to steal information. Three types of malicious hardware:

- **Zombies**—remotely controlled computers used for malicious purposes; a collection is known as a **botnet (robot network)**.
- **Rogue Wi-Fi hotspots**—network that appears to be a legitimate free Wi-Fi hotspot. Once connected, any input by users is captured.
- **Infected USB flash drives**—USB drives that are often free that contain a virus.

Measures to Protect Computer Security

The **Computer Fraud and Abuse Act** makes it a crime for unauthorized persons to view data across state lines using computers and prohibits unauthorized use of computers owned by government or federally insured financial institution. Ways to protect computer security include limiting access, encrypting, and anticipating disasters.

- Access can be restricted through **biometric scanning** devices and **passwords** (**dictionary attacks** use thousands of words to attempt to gain access; **Picture Password** uses series of gestures; **facial recognition** limits access); **security suites, firewalls, password managers**, and **authentication** (**two-factor authentication** uses two types of data to verify your identity; **two-step authentication** uses one type of authentication twice).
- **Encrypting** is coding information to make it unreadable except to those who have the **encryption key**. **Hypertext transfer protocol secure (https)** requires browsers and websites to encrypt all messages. **Virtual private networks (VPNs)** encrypt connections between company networks and remote users. **WPA3 (Wi-Fi Protected Access)** is the most widely used wireless network encryption for home wireless networks.
- Anticipating disasters involves physical security, data security, and disaster recovery plans.
- Preventing data loss involves protecting data by screening job applicants, guarding passwords, and auditing and backing up data.

ETHICS

What do you suppose controls how computers can be used? You probably think first of laws. Of course, that is right, but technology is moving so fast that it is very difficult for our legal system to keep up. The essential element that controls how computers are used today is **ethics**.

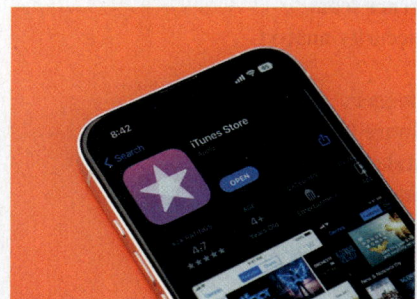

Seemanta Dutta/Alamy Stock Photo

Computer ethics are guidelines for the morally acceptable use of computers in our society. We are all entitled to ethical treatment. This includes the right to keep personal information, such as credit ratings and medical histories, from getting into unauthorized hands.

Cyberbullying

Cyberbullying is the use of the Internet to send or post content intended to hurt or embarrass another person. It includes sending repeated unwanted e-mails to an individual, ganging up on victims in electronic forums, posting false statements, maliciously disclosing personal data, and sending any type of threatening or harassing communication. It can lead to criminal prosecution.

Copyright and Digital Rights Management

A **copyright** is a legal concept that gives content creators the right to control use and distribution of their work. Materials that can be copyrighted include paintings, books, music, films, and even video games.

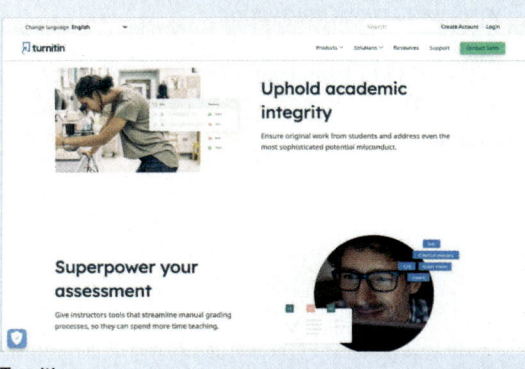

Turnitin

ETHICS

Software piracy is the unauthorized copying and distribution of software. The software industry loses over $60 billion annually to software piracy. Two related topics are the Digital Millennium Copyright Act and digital rights management.

- The **Digital Millennium Copyright Act** makes it illegal to deactivate or disable antipiracy technologies; to copy, resell, or give away commercial programs; or to sell or use programs or devices that are illegally copying software.
- **Digital rights management (DRM)** is a collection of technologies designed to prevent copyright violations. Typically, DRM is used to (1) control the number of devices that can access a given file and (2) limit the kinds of devices that can access a file.

Today, many legal sources for digital media exist, including

- Television programs that can be watched online, often for free, on television-network-sponsored sites.
- Sites like Pandora that allow listeners to enjoy music at no cost.
- Online stores that legally sell music and video content. A pioneer in this area is Apple's iTunes Music Store.

Plagiarism

Plagiarism is the illegal and unethical representation of some other person's work and ideas as your own without giving credit to the original source. Examples include cutting and pasting web content into a report or paper.

Recognizing and catching **plagiarists** is relatively easy. For example, services such as Turnitin are dedicated to preventing Internet plagiarism. This service examines a paper's content and compares it to a wide range of known public electronic documents including web page content. Exact duplication or paraphrasing is readily identified.

CAREERS in IT

IT security analysts are responsible for maintaining the security of a company's network, systems, and data. Employers look for candidates with a bachelor's or associate's degree in information systems or computer science and network experience. Expected salary range is $64,000 to $77,000.

KEY TERMS

access
accuracy
antispyware
authentication
big data
biometric scanning
botnet
browser cache
California Consumer Privacy Act (CCPA)
computer crime
computer ethics
Computer Fraud and Abuse Act
computer monitoring software
cookie
copyright
cracker
cyberbullying
cybercrime
cyberterrorism
dark web
data manipulation
data security
deep web
denial of service (DoS) attack
dictionary attack
digital footprint
Digital Millennium Copyright Act
digital rights management (DRM)
disaster recovery plan
distributed denial of service (DDoS)
employee-monitoring software
encryption
encryption key
ethics
facial recognition
Family Educational Rights and Privacy Act (FERPA)
firewall
first-party cookie
Freedom of Information Act
General Data Protection Regulation (GDPR)
Gramm-Leach-Bliley Act
Health Insurance Portability and Accountability Act (HIPAA)
history file
https (hypertext transfer protocol secure)
identity theft
illusion of anonymity
Incognito Mode
infected USB flash drive
information broker
information reseller
Internet scam
IT security analyst
key
keylogger
malware (malicious software)
mistaken identity
online identity
password
password manager
phishing
physical security
Picture Password
plagiarism
plagiarist
privacy
privacy mode
Private Browsing
property
ransomware
robot network
rogue Wi-Fi hotspot
security
security suite
social engineering
software piracy
spy removal program
spyware
temporary Internet file
third-party cookie
tracking cookie
Trojan horse
two-factor authentication
two-step authentication
virtual private network (VPN)
virus
web bug
wireless network encryption
worm
WPA3 (Wi-Fi Protected Access 3)
zombie

MULTIPLE CHOICE

Circle the correct answer.

1. Privacy concern that relates to the responsibility to ensure correct data collection.
   - a. accuracy
   - b. access
   - c. history
   - d. identity

2. Individuals who collect and sell personal data.
   - a. information brokers
   - b. Trojan horses
   - c. phishers
   - d. DRM

3. Small data files deposited on your hard disk from websites you have visited.
   - a. digital footprints
   - b. online identities
   - c. cookies
   - d. crackers

4. Wide range of programs that secretly record and report an individual's activities on the Internet.
   - a. spyware
   - b. viruses
   - c. zombies
   - d. web bugs

5. Malicious programs that damage or disrupt a computer system.
   - a. zombies
   - b. malware
   - c. cookies
   - d. firewalls

6. Infected computers that can be remotely controlled.
   - a. Trojan horses
   - b. zombies
   - c. crackers
   - d. WPA3

7. Used by scammers to trick Internet users with official-looking websites.
   - a. online identity
   - b. history file
   - c. phishing
   - d. Trojan horse

8. A type of scanning device such as fingerprint and iris (eye) scanner.
   - a. phishing
   - b. biometric
   - c. firewall
   - d. web bug

9. Process of coding information to make it unreadable except to those who have a key.
   - a. phishing
   - b. spyware
   - c. dark web
   - d. encryption

10. An ethical issue relating to using another person's work and ideas as your own without giving credit to the original source.
    - a. phishing
    - b. identity theft
    - c. key logging
    - d. plagiarism

MATCHING

Match each numbered item with the most closely related lettered item. Write your answers in the spaces provided.

a. access
b. digital footprints
c. history file
d. privacy
e. online identity
f. crackers
g. Trojan horses
h. cyberbullying
i. firewall
j. DRM

____ 1. Special hardware and software used to control access to a corporation's private network is known as a(n) _____.
____ 2. The information that people voluntarily post in social networking sites, blogs, and photo- and video-sharing sites is used to create their _____.
____ 3. Browsers store the locations of sites visited in a _____.
____ 4. The browser mode that ensures your browsing activity is not recorded.
____ 5. Highly detailed and personalized descriptions of individuals are electronic _____.
____ 6. To prevent copyright violations, corporations often use _____.
____ 7. The use of the Internet to send or post content intended to hurt or embarrass another person is known as _____.
____ 8. Computer criminals who create and distribute malicious programs.
____ 9. Programs that come into a computer system disguised as something else are called _____.
____ 10. The three primary privacy issues are accuracy, property, and _____.

OPEN-ENDED

On a separate sheet of paper, respond to each question or statement.

1. Define privacy, and discuss the impact of large databases, private networks, the Internet, and the web.
2. Define and discuss online identity and the major privacy laws.
3. Define security. Define cybercrime, social engineering, malicious software, and malicious hardware including identity theft, Internet scams, data manipulation, ransomware, DoS attacks, viruses, worms, Trojan horses, zombies, rogue Wi-Fi hotspots, and infected USB flash drives.
4. Discuss ways to protect computer security including restricting access, encrypting data, anticipating disasters, and preventing data loss.
5. Define computer ethics, copyright law, cyberbullying, and plagiarism.

DISCUSSION

Respond to each of the following questions.

 Making IT Work for You: SECURITY AND TECHNOLOGY

Review the Making IT Work for You: Security and Technology on page 228, and then respond to the following: (a) Check your phone, tablet, laptop, or desktop to find out what version OS it is using. (b) Go online and search to find out if that is the most recent OS. Is your OS up to date? (c) Does your device have the option to automatically download updates to the OS? (d) Does your device come with a password manager? Have you used it? Why or why not?

 Community: SOCIAL NETWORKING

Review the Community box on page 215, and then answer the following: (a) Do you use any social networking sites? If so, which ones and do you know their policy of sharing your posts and photos to others without your express consent? (b) Do you believe that social networks violate your privacy when they share information with advertisers? Why or why not? (c) Suppose that a high school teacher posts photos showing her drinking at a party. When the school board becomes aware of these photos, the teacher is suspended. Is this a violation of her rights? Why or why not?

 Privacy: PRIVATE BROWSING

Review the Privacy box on page 218 and then respond to the following: (a) Have you ever used the privacy mode on your web browser? What is the name of the company that makes your browser? What is the name your browser uses for its privacy mode? (b) Each browser defines private browsing a little differently. Search online to see how your browser explains how privacy mode works. How does your browser protect your privacy in privacy mode? (c) Privacy mode changes what is recorded on your device; however, computers other than your own are involved when you browse the web and may still record your activities. What companies or people would still know your activities if you search on Google? Check your work webmail? Purchase an item on Amazon?

 Ethics: ENCRYPTION

Review the Ethics box on page 219, and then respond to the following: (a) What is your opinion on unbreakable encryption? Should it be illegal? Defend your position. (b) If a phone was believed to hold information about a planned terrorist attack, would you support building a backdoor to encryption that would allow the government to search the phone? What if you knew that the backdoor technique would eventually be leaked and available to everyone? What if the government involved considered terrorism to include the possession of documents criticizing the government?

 Ethics: NET NEUTRALITY

Review the Ethics box on page 230, and then respond to the following: (a) Do you think that government agencies should be allowed to regulate ISP pricing policies? Why or why not? (b) Do you think the speed or ease of access of web pages can be a form of censorship? Why or why not? (c) If a government made anti-government websites run slower, would that be censorship? Why or why not? (d) If a company made its competitions' websites run slower, would that be censorship? Why or why not? (e) Should an ISP be allowed to watch you browse the Internet to decide which web pages should download quickly and which should download slowly? Defend your position.

 Ethics: PLAGIARISM

Review the Ethics box on page 230, and then respond to the following: (a) Do you think that the rewording of someone else's work is plagiarism? Why or why not? (b) If ChatGPT is just a tool that creates essays based on a user's questions—isn't the user actually creating the essay? Why or Why not? (c) When a teacher assigns an essay—what learning might they hope to inspire in their students? Be specific. (d) How can using ChatGPT help students achieve those learning goals? How can it harm those goals?

Tips: IDENTITY THEFT

Review the Tips box on page 220 and consider the following. Identity theft occurs when someone acquires your personal information and uses it to hijack your finances. A common scenario is a thief using your Social Security number to open a credit card account in your name. When the thief does not pay, it is your credit history that is blemished. Consider this scam thoroughly, and then respond to the following: (a) List three steps an individual should take to avoid identity theft. (b) List three steps a corporation that maintains your personal data in its information system should take to safeguard your data. (c) How can Internet activities contribute to the likelihood of identity theft? How can this be prevented?

Design Elements: Concept Check icon: Dizzle52/Getty Images

chapter 10
Information Systems

NicoElNino/Shutterstock

Why should I read this chapter?

jittawit21/Shutterstock

A simple online purchase requires a tremendous collection of carefully planned and executed jobs. If you buy some sports equipment online, your purchase becomes part of a complex stream of information that allows a company to succeed, employees to flourish, and you to get your sports equipment. The ability to gather, analyze, and respond to data, like the data generated when you make an online purchase, can mean the difference between a company giving employees bonuses or layoff notices. The best information systems automatically adjust to their environments and help organizations react to current and future changes.

This chapter covers the things you need to know to be prepared for this ever-changing digital world, including:

- Organizational flow—identify how information flows within an organization.
- Computer-based information systems—recognize the levels of information systems and how they help businesses make decisions.
- Other information systems—understand expert systems and how you can use them to make faster, smarter decisions.

Learning Objectives

After you have read this chapter, you should be able to:

1. Explain the functional view of an organization and describe each function.
2. Describe the management levels and the informational needs for each level in an organization.
3. Describe how information flows within an organization.
4. Describe computer-based information systems.
5. Distinguish among a transaction processing system, a management information system, a decision support system, and an executive support system.
6. Distinguish between office automation systems and knowledge work systems.
7. Explain the difference between data workers and knowledge workers.
8. Define expert systems and knowledge bases.

Introduction

"Hi, I'm Sue, and I'm an information systems manager. I'd like to talk with you about how organizations use computer information systems. I'd also like to talk about specialized knowledge work systems that assist managers, engineers, and scientists make decisions."

voronaman/Shutterstock

An **information system** is a collection of people, procedures, software, hardware, data, and connectivity (as we discussed in **Chapter 1**). They all work together to provide information essential to running an organization. This information is critical to successfully produce a product or service and, for profit-oriented enterprises, to derive a profit.

Why are computers used in organizations? At a basic level, computers are used to record events efficiently. A computer will record the details of a sale made in a store or update a database to recognize that an item sold has reduced inventory. However, at a deeper level, computers are used to make decisions. The data from sales in stores will decide employee bonuses. The data about inventory items will identify market trends and affect what is manufactured next season. In this chapter, you will learn how businesses collect and use information to make decisions.

To efficiently and effectively use computers within an organization, you need to understand how the information flows as it moves through an organization's different functional areas and management levels. You need to be aware of the different types of computer-based information systems, including transaction processing systems, management information systems, decision support systems, and executive support systems. You also need to understand the role and importance of databases to support each level or type of information system.

Organizational Information Flow

Computerized information systems do not just keep track of transactions and day-to-day business operations. They also support the vertical and horizontal flow of information within the organization. To understand this, we need to understand how an organization is structured. One way to examine an organization's structure is to view it from a functional perspective. That is, you can study the different basic functional areas in organizations and the different types of people within these functional areas.

As we describe these, consider how they apply to a hypothetical manufacturer of sporting goods, the HealthWise Group. This company manufactures equipment for sports and physical activities. Its products range from soccer balls to yoga mats. (See **Figure 10-1**.)

Like many organizations, HealthWise Group can be viewed from a functional perspective with various management levels. Effective operations require an efficient and coordinated flow of information throughout the organization.

Figure 10-1 Yoga mats
FatCamera/E+/Getty Images

Functions

Depending on the services or products they provide, most organizations have departments that specialize in one of five basic functions. These are accounting, marketing, human resources, production, and research. (See **Figure 10-2**.)

- **Accounting** records all financial activity from billing customers to paying employees. The HealthWise accounting department tracks all sales, payments, and transfers of funds. It also produces reports detailing the financial condition of the company.

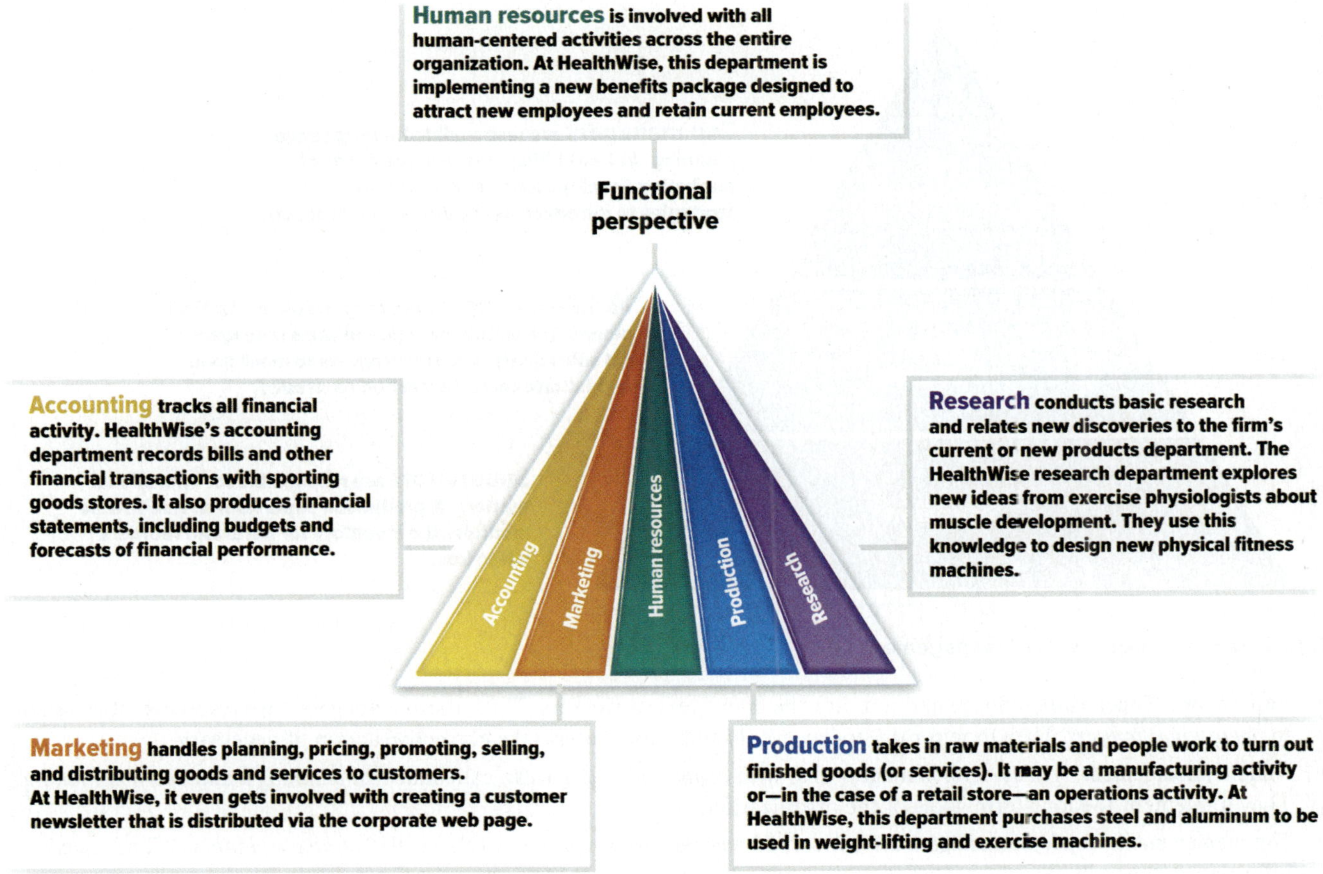

Human resources is involved with all human-centered activities across the entire organization. At HealthWise, this department is implementing a new benefits package designed to attract new employees and retain current employees.

Functional perspective

Accounting tracks all financial activity. HealthWise's accounting department records bills and other financial transactions with sporting goods stores. It also produces financial statements, including budgets and forecasts of financial performance.

Research conducts basic research and relates new discoveries to the firm's current or new products department. The HealthWise research department explores new ideas from exercise physiologists about muscle development. They use this knowledge to design new physical fitness machines.

Marketing handles planning, pricing, promoting, selling, and distributing goods and services to customers. At HealthWise, it even gets involved with creating a customer newsletter that is distributed via the corporate web page.

Production takes in raw materials and people work to turn out finished goods (or services). It may be a manufacturing activity or—in the case of a retail store—an operations activity. At HealthWise, this department purchases steel and aluminum to be used in weight-lifting and exercise machines.

Figure 10-2   The five functions of an organization

- **Marketing** plans, prices, promotes, sells, and distributes the organization's goods and services. HealthWise's goods include a wide range of products related to sports and other types of physical activity.
- **Human resources** focuses on people—hiring, training, promoting, employee relations, and any number of other human-centered activities within the organization. At HealthWise, human resources is responsible for implementing a new benefits package, for hiring new skilled workers, and much more.
- **Production** actually creates finished goods and services using raw materials and personnel. HealthWise manufacturers a variety of sports equipment, including yoga mats.
- **Research** identifies, investigates, and develops new products and services. For example, at HealthWise, scientists are investigating a light, inexpensive alloy for a new line of weight-training equipment.

Although the titles may vary, nearly every large and small organization has departments that perform these basic functions. Whatever your job in an organization, it is likely to be in one of these functional areas.

Management Levels

The foundation of any organization is not managers, but the employees who produce goods and services. These employees are the base of the organizational pyramid and consist of assemblers, painters, welders, drivers, and others. Above them are various levels of managers—people with titles such as supervisor, director, regional manager, and vice president. These are the people who do the planning, leading, organizing, and controlling necessary to see that the work gets done.

For example, the northwest district sales manager for HealthWise directs and coordinates all the salespeople in her area. Other job titles might be vice president of marketing, director of human resources, or production manager. In smaller organizations, these titles are often combined.

Management in many organizations is divided into three levels. These levels are supervisors, middle management, and top management. (See **Figure 10-3**.)

Figure 10-3 Three levels of management

- **Supervisors: Supervisors** manage and monitor the employees or workers. Thus, these managers have responsibilities relating to *operational matters*. They monitor day-to-day events and immediately take corrective action, if necessary.
- **Middle management: Middle-level managers** deal with *control, planning* (also called *tactical planning*), and *decision making*. They implement the long-term goals of the organization.
- **Top management: Top-level managers** are concerned with *long-range planning* (also called *strategic planning*). They need information that will help them plan the future growth and direction of the organization.

Information Flow

Each level of management has different information needs. Top-level managers need summary information describing the overall operations of the business. They also need information from outside the organization because top-level managers need to forecast and plan for long-range events. Middle-level managers need summarized information—weekly or monthly reports. They need to develop budget projections and to evaluate the performance of supervisors. Supervisors need detailed, very current, day-to-day information on their units so that they can keep operations running smoothly. (See **Figure 10-4**.)

Figure 10-4 Supervisors monitor day-to-day events
Hispanolistic/E+/Getty Images

To support these different needs, information *flows* in different directions. (See **Figure 10-5**.) For top-level managers, the flow of information from within the organization is both vertical and horizontal. The top-level managers, such as the chief executive officer (CEO), need information from below and from all departments. They also need information from outside the organization. For example, HealthWise is deciding whether to introduce a line of fitness trackers in the southwestern United States. The vice president of marketing must look at relevant data. Such data might include the number of people currently using the HealthWise fitness app and census data about the number of young people. It also might include sales histories on related fitness monitoring equipment.

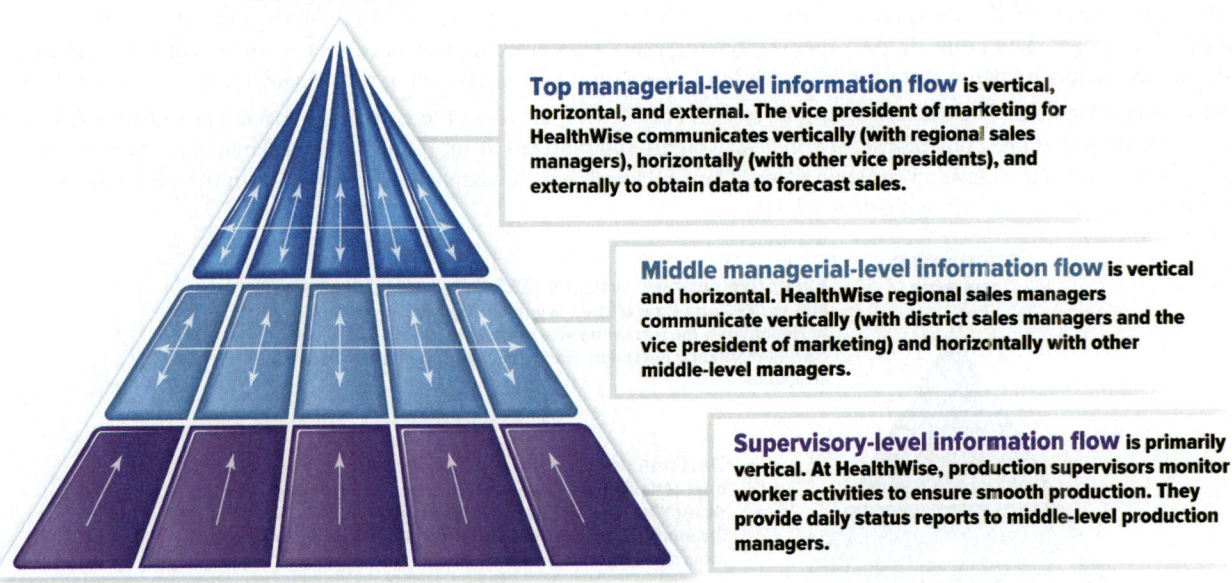

Figure 10-5 Information flow within an organization

For middle-level managers, the HealthWise information flow is both vertical and horizontal across functional lines. For example, the regional sales managers set their sales goals by coordinating with middle managers in the production department. They are able to tell sales managers what products will be produced, how many, and when. The regional sales managers also must coordinate with the strategic goals set by the top managers. They must set and monitor the sales goals for the supervisors beneath them.

For supervisory managers, information flow is primarily vertical. That is, supervisors communicate mainly with their middle managers and with the workers beneath them. For instance, at HealthWise, production supervisors rarely communicate with people in the accounting department. However, they are constantly communicating with production-line workers and with their own managers.

Now we know how many organizations are structured and how information flows within the organization. But how is a computer-based information system likely to be set up to support its needs? And what do you need to know to use it?

concept check

☐ What are the five basic functions within an organization?
☐ What are the three levels of management? Discuss each level.
☐ Describe the flow of information within an organization.

Computer-Based Information Systems

Almost all organizations have computer-based information systems. Large organizations typically have formal names for the systems designed to collect and use the data. Although different organizations may use different names, the most common names are transaction processing, management information, decision support, and executive support systems. (See **Figure 10-6**.)

- **Transaction processing system:** The transaction processing system (TPS) records day-to-day transactions, such as customer orders, bills, inventory levels, and production output. The TPS helps supervisors by generating databases that act as the foundation for the other information systems.
- **Management information system:** The management information system (MIS) summarizes the detailed data of the transaction processing system in standard reports for middle-level managers. Such reports might include weekly sales and production schedules.
- **Decision support system:** The decision support system (DSS) provides a flexible tool for analysis. The DSS helps middle-level managers and others in the organization analyze a wide range of problems, such as the effect of events and trends outside the organization. Like the MIS, the DSS draws on the detailed data of the transaction processing system.
- **Executive support system:** The executive support system (ESS), also known as the **executive information system (EIS)**, is an easy-to-use system that presents information in a very highly summarized form. It helps top-level managers oversee the company's operations and develop strategic plans. The ESS combines the databases generated from the TPS and the reports generated from the MIS with external data.

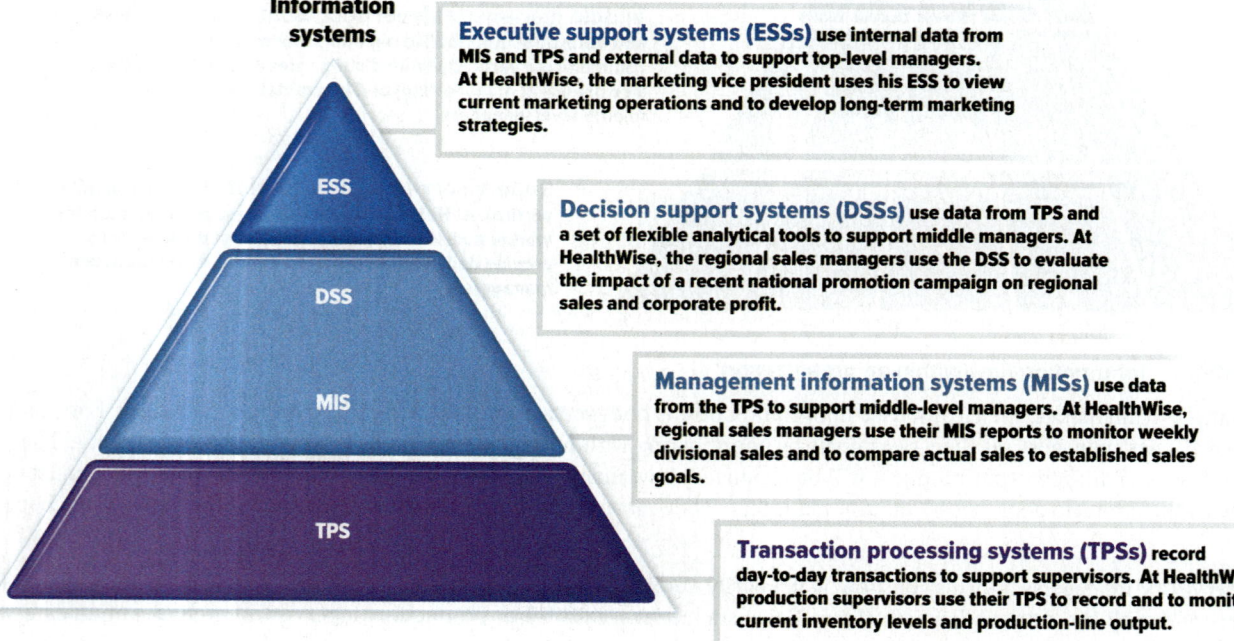

Figure 10-6   Types of computer-based information systems

concept check

☐ What is a transaction processing system? How does it help supervisors?
☐ What is a management information system? Decision support system? How are they different?
☐ What is an executive support system? Who uses it? What is it used for?

Transaction Processing Systems

A **transaction processing system (TPS)** helps an organization keep track of routine operations and records these events in a database. For this reason, some firms call this the **data processing system (DPS)**. The data from operations—for example, customer orders for HealthWise's products—makes up a database that records the transactions of the company. This database of transactions is used to support the MIS, DSS, and ESS.

One of the most essential transaction processing systems for any organization is in the accounting area. (See **Figure 10-7**.) Every accounting department handles six basic activities. Five of these are sales order processing, accounts receivable, inventory and purchasing, accounts payable, and payroll. All of these are recorded in the general ledger, the sixth activity.

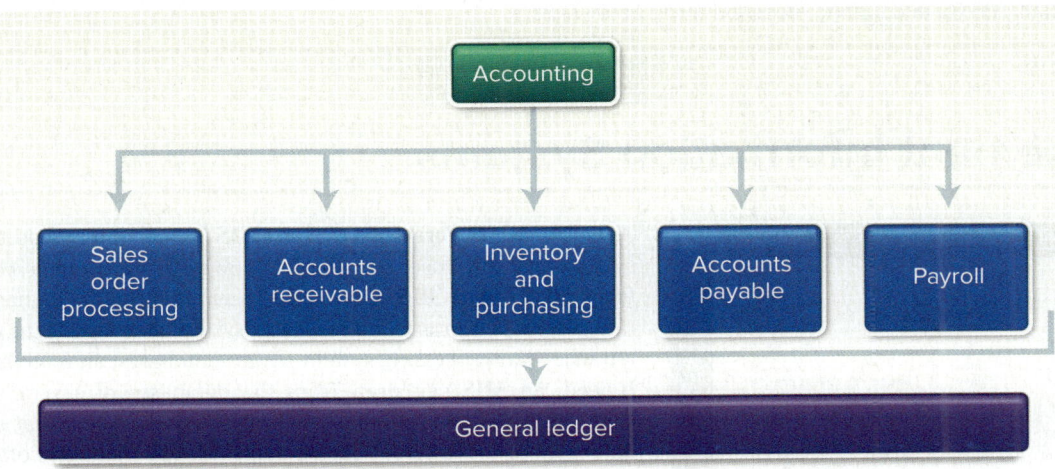

Figure 10-7   Transaction processing system for accounting

Let us take a look at these six activities. They will make up the basis of the accounting system for almost any organization you might work in.

- The **sales order processing** activity records the customer requests for the company's products or services. At HealthWise when an order comes in—a request for a set of barbells, for example—the warehouse is alerted to ship a product.
- The **accounts receivable** activity records money received from or owed by customers. HealthWise keeps track of bills paid by sporting goods stores and by gyms and health clubs to which it sells directly.
- The parts and finished goods that the company has in stock are called **inventory**. At HealthWise this would include all exercise machines, footballs, soccer balls, and yoga mats ready for sale in the warehouse. (See **Figure 10-8**.) An **inventory control system** keeps records of the number of each kind of part or finished good in the warehouse.

Purchasing is the buying of materials and services. Often a **purchase order** is used. This is a form that shows the name of the company supplying the material or service and what is being purchased.

Figure 10-8   Inventory control systems manage the merchandise in the warehouse
Andersen Ross/Brand X Pictures/Getty Images

- **Accounts payable** refers to money the company owes its suppliers for materials and services it has received. At HealthWise this would include materials such as steel and aluminum used to manufacture its exercise equipment.
- The **payroll** activity is concerned with calculating employee paychecks. Amounts are generally determined by the pay rate, hours worked, and deductions (such as taxes, Social Security, and medical insurance). Paychecks may be calculated from employee time cards or, in some cases, supervisors' time sheets.
- The **general ledger** keeps track of all summaries of all the foregoing transactions. A typical general ledger system can produce income statements and balance sheets. **Income statements** show a company's financial performance—income, expenses, and the difference between them for a specific time period. **Balance sheets** list the overall financial condition of an organization. They include assets (for example, buildings and property owned), liabilities (debts), and how much of the organization (the equity) is owned by the owners.

There are many other transaction processing systems that you come into contact with every day. These include automatic teller machines, which record cash withdrawals; online registration systems, which track student enrollments; and supermarket discount cards, which track customer purchases.

concept check

☐ What is the purpose of a transaction processing system?
☐ Describe the six activities of a TPS for accounting.
☐ Other than TPS for accounting, describe three other TPSs.

Management Information Systems

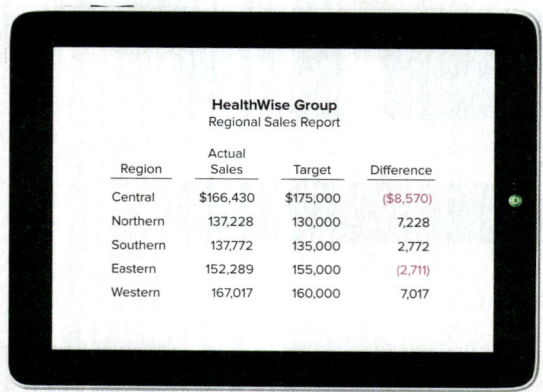

Figure 10-9 Management information system report

A **management information system (MIS)** is a computer-based information system that produces standardized reports in summarized structured form. (See **Figure 10-9**.) It is used to support middle managers. An MIS differs from a transaction processing system in a significant way. Whereas a transaction processing system *creates* databases, an MIS *uses* databases. Indeed, an MIS can draw from the databases of several departments. Thus, an MIS requires a *database management system* that integrates the databases of the different departments. Middle managers often need summary data drawn from across different functional areas.

An MIS produces reports that are *predetermined*. That is, they follow a predetermined format and always show the same kinds of content. Although reports may differ from one industry to another, there are three common categories of reports: periodic, exception, and demand.

- **Periodic reports** are produced at regular intervals—weekly, monthly, or quarterly, for instance. Examples are HealthWise's monthly sales and production reports. The sales reports from district sales managers are combined into a monthly report for the regional sales managers. For comparison purposes, a regional manager is also able to see the sales reports of other regional managers.
- **Exception reports** call attention to unusual events. An example is a sales report that shows that certain items are selling significantly above or below marketing department forecasts. For instance, if fewer exercise bicycles are selling than were predicted for the northwest sales region, the regional manager will receive an exception report. That report may be used to alert the district managers and salespeople to give this product more attention.
- The opposite of a periodic report, a **demand report** is produced on request. An example is a report on the numbers and types of jobs held by women and minorities. Such a report is not needed periodically, but it may be required when requested by the U.S. government. At HealthWise, many government contracts require this information. It is used to certify that HealthWise is within certain government equal-opportunity guidelines.

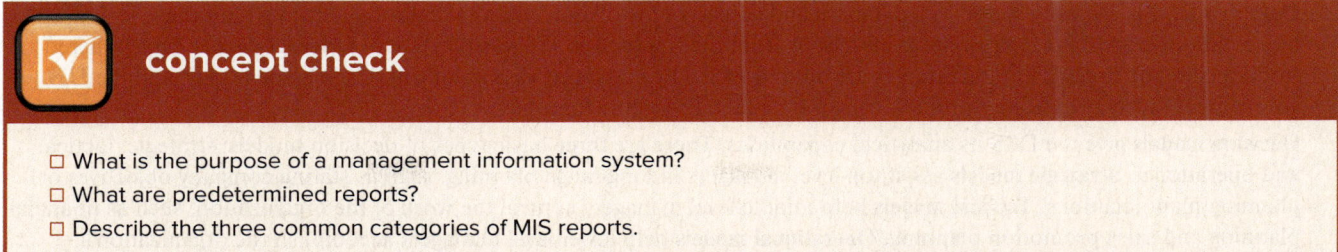

concept check

☐ What is the purpose of a management information system?
☐ What are predetermined reports?
☐ Describe the three common categories of MIS reports.

Decision Support Systems

Managers often must deal with unanticipated questions. For example, the HealthWise manager in charge of manufacturing might ask how an anticipated labor strike would affect production schedules. A **decision support system (DSS)** enables managers to get answers to such unexpected and generally nonrecurring kinds of problems.

A DSS is quite different from a transaction processing system, which simply records data. It is also different from a management information system, which summarizes data in predetermined reports. A DSS is used to analyze data. Moreover, it produces reports that do not have a fixed format. This makes the DSS a flexible tool for analysis.

A DSS must be easy to use—or most likely it will not be used at all. A HealthWise marketing manager might want to know which territories are not meeting their monthly sales quotas. To find out, the executive could ask or query the sales database for all "SALES < QUOTA." (See **Figure 10-10**.)

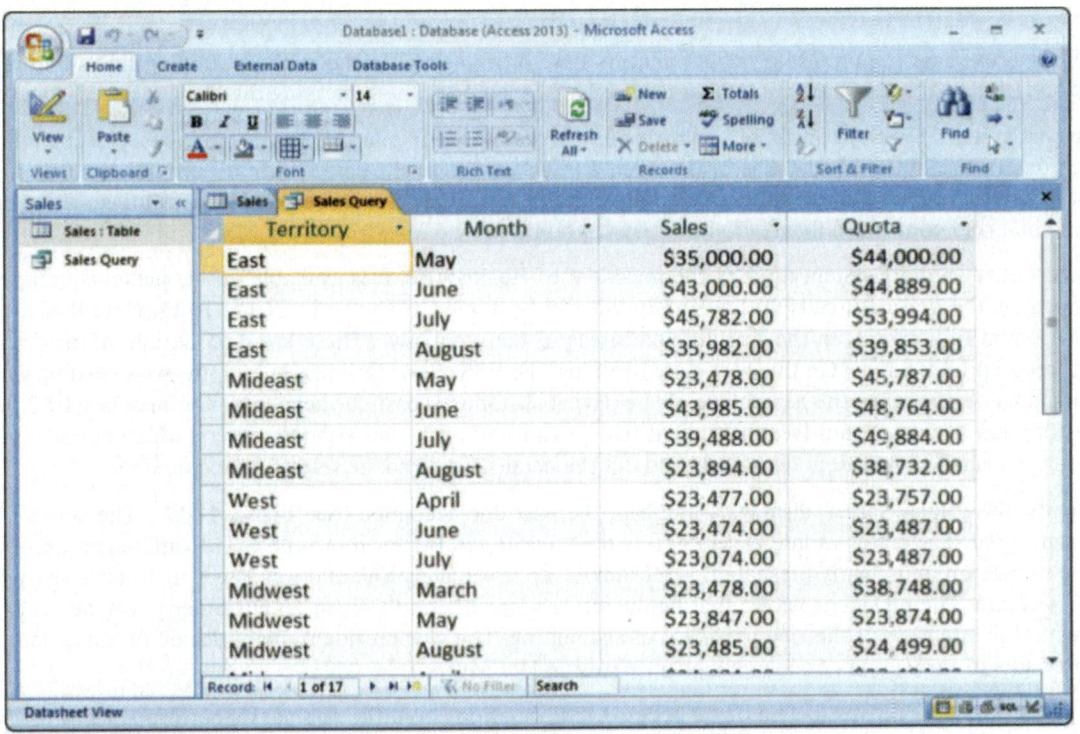

Figure 10-10 Decision support system query results for SALES < QUOTA
Microsoft Corporation

How does a decision support system work? Essentially, it consists of four parts: the user, system software, data, and decision models.

- The **user** could be you. In general, the user is someone who has to make decisions—a manager, often a middle-level manager.
- **System software** is essentially the operating system—programs designed to work behind the scenes to handle detailed operating procedures. In order to give the user a good, intuitive interface, the software typically is menu or icon driven. That is, the screen presents easily understood lists of commands or icons, giving the user several options.

- **Data** in a DSS is typically stored in a database and consists of two kinds. **Internal data**—data from within the organization—consists principally of transactions from the transaction processing system. **External data** is data gathered from outside the organization. Examples are data provided by marketing research firms, trade associations, and the U.S. government (such as customer profiles, census data, and economic forecasts).
- **Decision models** give the DSS its analytical capabilities. There are three basic types of decision models: strategic, tactical, and operational. **Strategic models** assist top-level managers in long-range planning, such as stating company objectives or planning plant locations. **Tactical models** help middle-level managers control the work of the organization, such as financial planning and sales promotion planning. **Operational models** help lower-level managers accomplish the organization's day-to-day activities, such as evaluating and maintaining quality control.

Some DSSs are specifically designed to support more than one or a team of decision makers. These systems, known as **group decision support systems (GDSS)**, include tools to support group meetings and collective work.

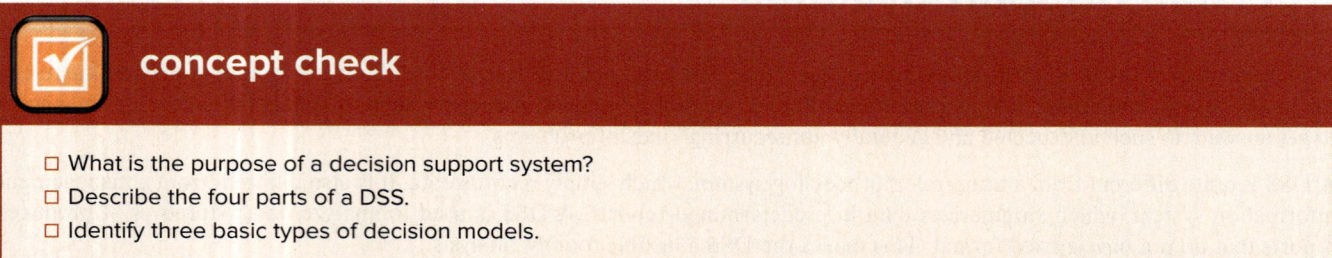

- What is the purpose of a decision support system?
- Describe the four parts of a DSS.
- Identify three basic types of decision models.

Executive Support Systems

Using a DSS requires some training. Many top managers have other people in their offices running DSSs and reporting their findings. Top-level executives also want something more concise than an MIS—something that produces very focused reports.

Executive support systems (ESSs) consist of sophisticated software that, like an MIS or a DSS, can present, summarize, and analyze data from an organization's databases. However, an ESS is specifically designed to be easy to use. This is so that a top executive with little spare time, for example, can obtain essential information without extensive training. Thus, information is often displayed in a very condensed form with informative graphics.

Consider an executive support system used by the president of HealthWise. It is available on his personal computer. The first thing each morning, the president calls up the ESS on his display screen as shown in **Figure 10-11**. Note that the screen gives a condensed account of activities in the five different areas of the company. (These are Accounting, Marketing, Production, Human Resources, and Research.) On this particular morning, the ESS shows business in four areas proceeding smoothly. However, in the first area, Accounting, the percentage of late-paying customers—past due accounts—has increased by 3 percent. Three percent may not seem like much, but HealthWise has had a history of problems with late payers, which has left the company at times strapped for cash. The president decides to find out the details. To do so, he selects 1. Accounting.

Within moments, the display screen displays a graph of the past due accounts. (See **Figure 10-12**.) The status of today's late payers is shown in red. The status of late payers at this time a year ago is shown in yellow. The differences between today and a year ago are significant and clearly presented. For example, approximately $60,000 was late 1 to 10 days last year. This year, over $80,000 was late. The president knows that he must take some action to speed up customer payments. (For example, he might call this to the attention of the vice president of accounting. The vice president might decide to implement a new policy that offers discounts to early payers or charge higher interest to late payers.)

ESSs permit a firm's top executives to gain direct access to information about the company's performance. Most provide direct electronic communication links to other executives. In addition, some ESSs have the ability to retrieve information from databases outside the company, such as business-news services. This enables a firm to watch for stories on competitors and stay current on relevant news events that could affect its business. For example, news of increased sports injuries caused by running and aerobic dancing, and the consequent decrease in people's interest in these activities, might cause HealthWise to alter its sales and production goals for its line of fitness-related shoes.

For a summary of the different types of information systems, see **Figure 10-13**.

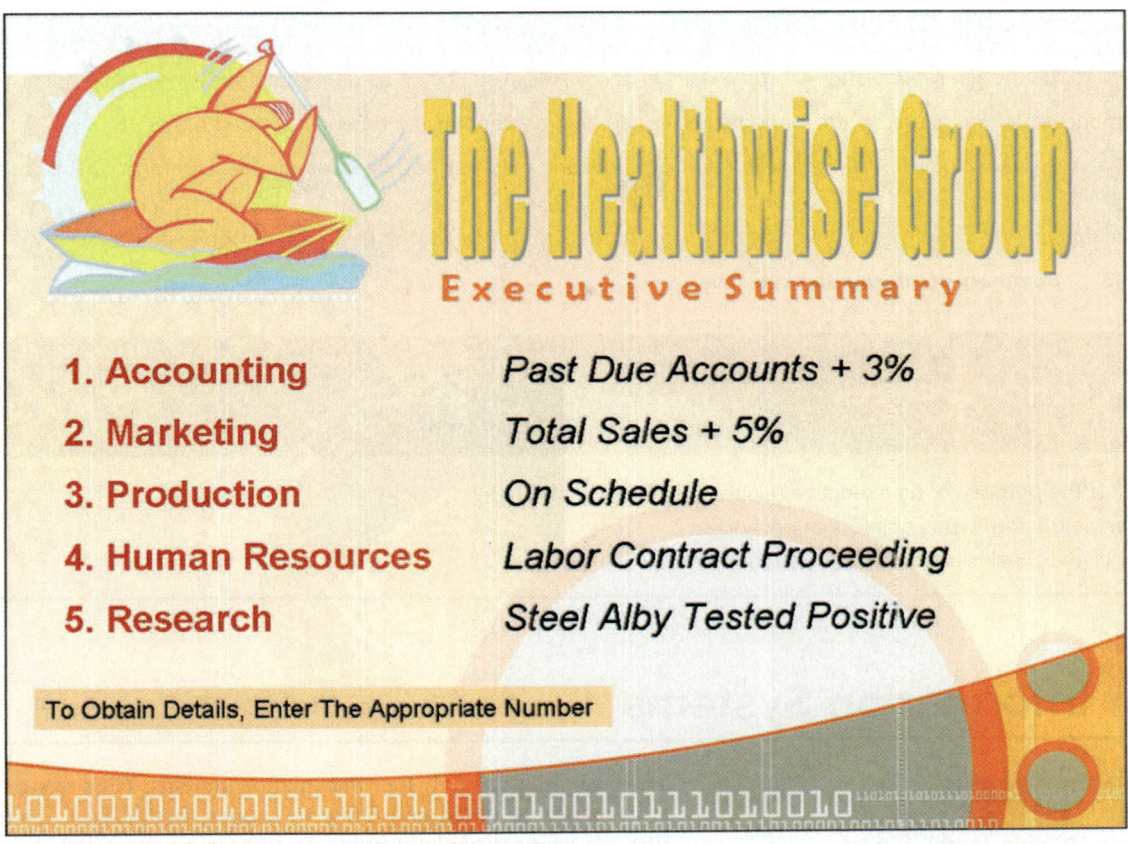

Figure 10-11 Opening screen for an executive support system
Microsoft Corporation

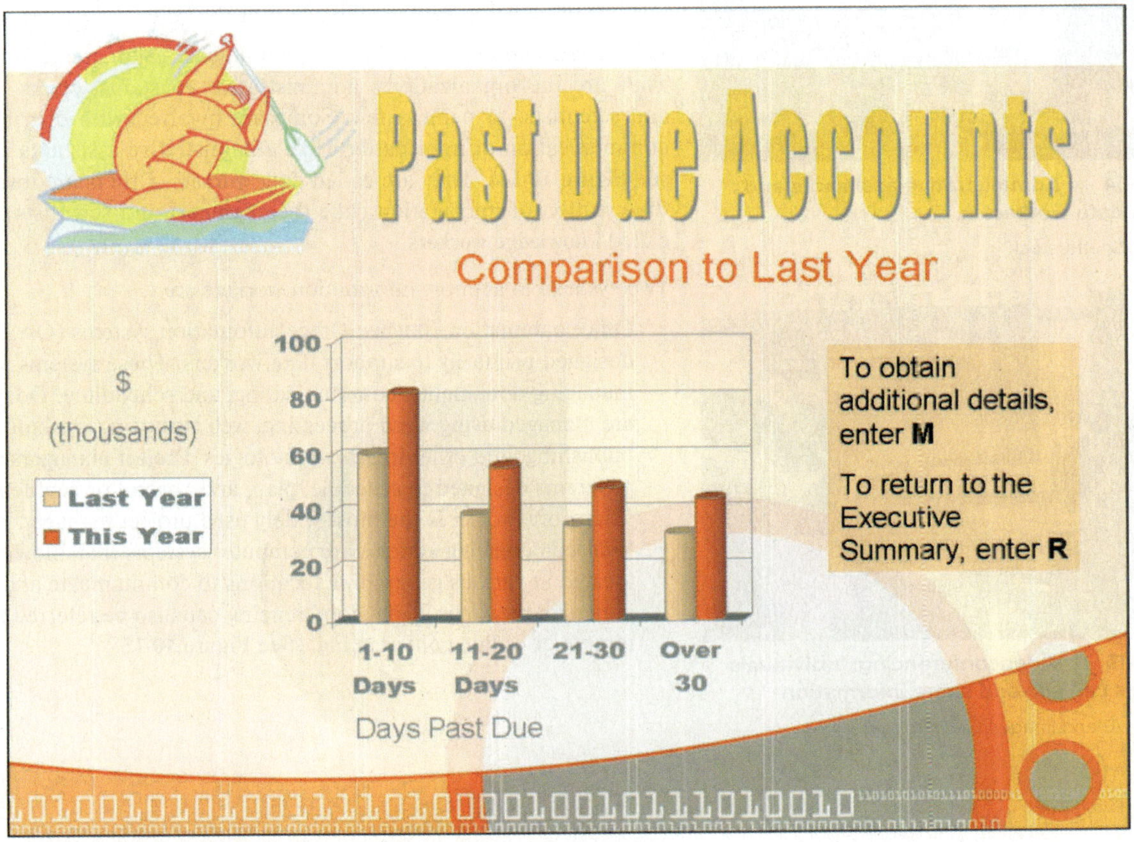

Figure 10-12 Graphic representation of past due accounts
Microsoft Corporation

Type	Description
TPS	Tracks routine operations and records events in databases; also known as data processing systems
MIS	Produces standardized reports (periodic, exception, and demand) using databases created by TPS
DSS	Analyzes unanticipated situations using data (internal and external) and decision models (strategic, tactical, and operational)
ESS	Presents summary information in a flexible, easy-to-use, graphical format designed for top executives

Figure 10-13 Summary of information systems

concept check

- What is the purpose of an executive support system?
- Describe the four types of information systems.
- How is an ESS similar to and different from an MIS or a DSS?

Other Information Systems

Figure 10-14 Administrative assistants and clerks are data workers
Stock-Asso/Shutterstock

We have discussed only four information systems: TPSs to support lower-level managers, MISs and DSSs to support middle-level managers, and ESSs to support top-level managers. There are many other information systems to support different individuals and functions. These systems compile data (facts) and organize that data into the context of the business's needs (creating information). The fastest growing are information systems designed to support information workers.

Information workers distribute, communicate, and create information. They are the organization's administrative assistants, clerks, engineers, and scientists, to name a few. Some are involved with distribution and communication of information (like administrative assistants and clerks; see **Figure 10-14**). They are called **data workers**. Others are involved with the creation of information (like the engineers and scientists). They are called **knowledge workers**.

Two systems to support information workers are

- **Office automation systems:** Office automation systems (OASs) are designed primarily to support data workers. These systems focus on managing documents, communicating, and scheduling. Documents are managed using word processing, web authoring, desktop publishing, and other image technologies. **Project managers** are programs designed to schedule, plan, and control project resources. Microsoft Project is the most widely used project manager. **Videoconferencing systems** are computer systems that allow people located at various geographic locations to communicate and have in-person meetings. Videoconferencing can also be referred to as remote or online conferencing. (See **Figure 10-15**.)

Figure 10-15 Videoconferencing: Individuals and groups can see and share information
Ariel Skelley/Blend Images/Getty Images

- **Knowledge work systems:** Knowledge workers use OAS systems. Additionally, they use specialized information systems called knowledge work systems (KWSs) to create information in their areas of expertise. For example, engineers involved in product design and manufacturing use **computer-aided design/computer-aided manufacturing (CAD/CAM) systems**. (See **Figure 10-16**.) These KWSs consist of powerful personal computers running special programs that integrate the design and manufacturing activities. CAD/CAM is widely used in the manufacture of automobiles and other products.

Figure 10-16   CAD/CAM: Knowledge work systems used by design and manufacturing engineers

Gorodenkoff/Shutterstock

Expert systems are another widely used knowledge work system.

Expert Systems

People who are expert in a particular area—certain kinds of medicine, accounting, engineering, and so on—are generally well paid for their specialized knowledge. Unfortunately for their clients and customers, these experts are expensive and not always available.

What if you were to somehow capture the knowledge of a human expert and make it accessible to everyone through a computer program? This is exactly what is being done with expert systems. **Expert systems**, also known as **knowledge-based systems**, are a type of artificial intelligence that uses a database to provide assistance to users. This database, known as a **knowledge base**, contains facts and rules to relate these facts distilled from a human expert. Users interact with an expert system by describing a particular situation or problem. The expert system takes the inputs and searches the knowledge base until a solution or recommendation is formulated. Expert systems are highly specialized software built by experienced professionals in their field. Industries where expert systems are popular include medicine, geology, architecture, and nature

 concept check

- What is an information worker?
- Who are data workers? What type of information system is designed to support them?
- Who are knowledge workers? What type of information system is designed to support them?
- What are expert systems? What is a knowledge base?

Careers in IT

"Now that you have learned about information systems, let me tell you about my career as an information systems manager."

voronaman/Shutterstock

Information systems managers oversee the work of programmers, computer specialists, systems analysts, and other computer professionals. They create and implement corporate computer policy and systems.

Most companies look for individuals with strong technical backgrounds, sometimes as consultants, with a master's degree in business. Employers seek individuals with strong leadership and excellent communication skills. Information systems managers must be able to communicate with people in technical and nontechnical terms. Information systems management positions are often filled by individuals who have been consultants or managers in previous positions. Those with experience in computer and network security will be in demand as businesses and society continue to struggle with important security issues.

Information systems managers can expect an annual salary of $126,000 to $154,000. Advancement opportunities typically include leadership in the field.

A LOOK TO THE FUTURE

ChatGPT: Changing the Workplace

jittawit21/Shutterstock

AI joins the workforce

The day of Artificial Intelligence (AI) is near. AI is used to drive vehicles, write news articles, and create art. Three industries are poised to see the greatest change in the next decade due to AI: health care, software development, and the arts.

The health care industry has often been at the forefront of technological innovation, and AI is no exception. IBM's WATSON, an AI program that competed on the TV show Jeopardy, has been used to reduce the administrative load at hospitals and doctors' offices. However, recent AI tools are far more advanced. In the future, AI's unique ability to identify patterns and draw relationships between multiple data sets will assist in diagnosing diseases. Medical professionals predict that AI will be used to tailor drugs to an individual's needs—considering factors such as size, family history, allergies, and complicating factors. Of course, not everyone would want a computer to diagnose their illness and prescribe them drugs. The integration of AI tools into the healthcare industry will be slow and careful. While many hospitals rely on AI tools to maintain records and reduce paperwork; so far, no AI programs are diagnosing patients.

Software programmers have been among the first to embrace the AI tools. AI tools, such as chatGPT, can offer quick, standardized solutions to basic programming tasks. In this way, AI can quickly generate the framework of programmer's code based on the descriptions the programmer inputs. Despite AI code being very impressive, AI is not ready to begin programming all on its own. AI tools develop programs based on the patterns and methods of similar programs it has reviewed. The result can look good at first glance, but upon closer inspection, it is clear that the AI has no real understanding of the task at hand. In this way, AI offers a jumping-off point, but human programmers are still needed to complete the task. In the future, it is expected that programmers will be valued for their ability to clearly and precisely define programming problems to AI, and AI will handle the development and execution of the code itself.

Artificial Intelligence is also creating an upheaval in the arts. Recently, the AI program Midjourney was used to win a state fair art contest. News Corp Australia has used AI to generate 3,000 news articles per week. For many, it would seem that AI is taking over the role of writers and artists. Others see the AI generated works as theft, simply the modification and combination of the works of human artists. The art world has long debated the line between innovation and plagiarism—and AI simply adds a new chapter to this debate. However, the tools are here to stay and not limited to words and paintings. Actors and musicians face similar questions as AI generates digital actors and AI generated harmonies. In the future, AI may become a collaborative tool—much in the way programmers use AI now—as a starting point for human-made art.

What do you think is the role of AI in the workplace? How could AI impact your career? How could you use AI to produce more, higher quality work? Is there anything a human can produce that an AI will never be able to accomplish?

VISUAL SUMMARY | Information Systems

ORGANIZATIONAL INFORMATION FLOW

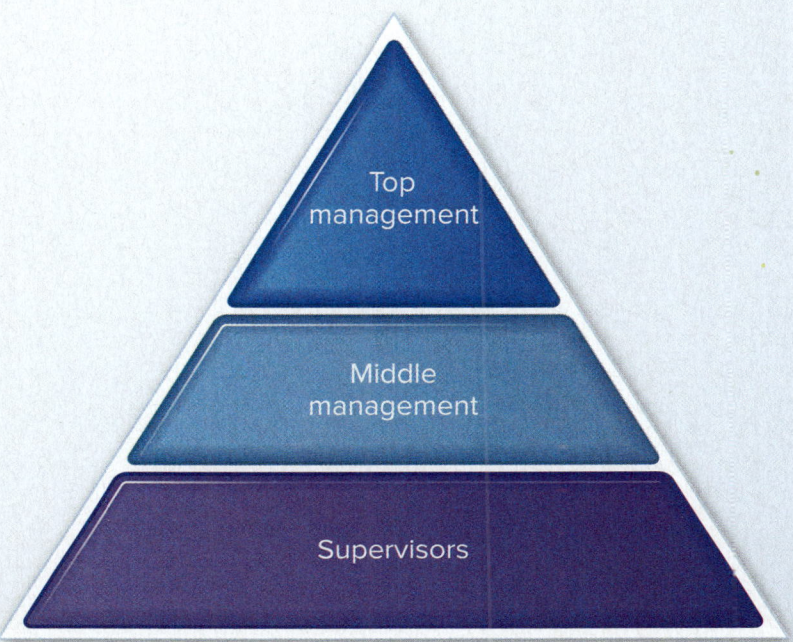

Information flows in an organization through functional areas and between management levels.

Functions

Most organizations have separate departments to perform five functions:

- **Accounting**—tracks all financial activities and generates periodic financial statements.
- **Marketing**—advertises, promotes, and sells the product (or service).
- **Production**—makes the product (or service) using raw materials and people to turn out finished goods.
- **Human resources**—finds and hires people; handles such matters as sick leave, retirement benefits, evaluation, compensation, and professional development.
- **Research**—conducts product research and development; monitors and troubleshoots new products.

Management Levels

The three basic management levels are

- **Top level**—concerned with long-range planning and forecasting.
- **Middle level**—deals with control, planning, decision making, and implementing long-term goals.
- **Supervisors**—control operational matters, monitor day-to-day events, and supervise workers.

Information Flow

Information flows within an organization in different directions.

- For **top-level managers,** the information flow is primarily upward from within the organization and into the organization from the outside.
- For **middle-level managers,** the information flow is horizontal and vertical within departments.
- For **supervisors,** the information flow is primarily vertical.

To efficiently and effectively use computers within an organization, you need to understand how information flows through functional areas and management levels. You need to be aware of the different types of computer-based information systems, including transaction processing systems, management information systems, decision support systems, and executive support systems.

INFORMATION SYSTEMS

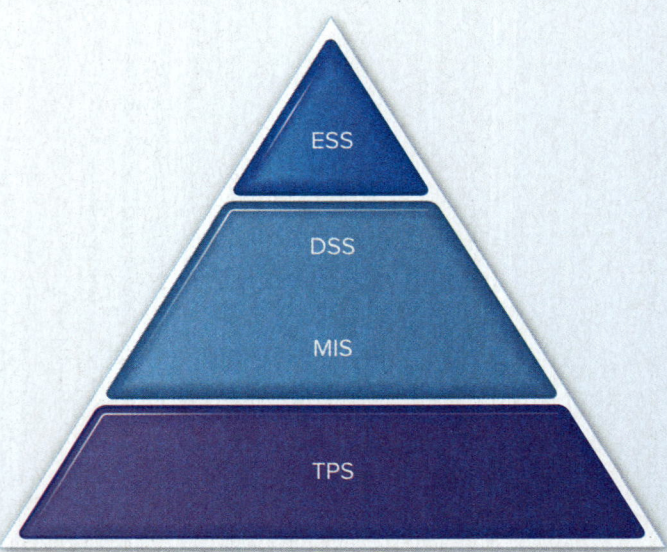

Transaction Processing Systems

Transaction processing systems (TPSs), sometimes called **data processing systems (DPSs)**, record day-to-day transactions. Accounting activities include **sales order processing**, **accounts receivable**, **inventory** and **purchasing**, **accounts payable**, **payroll**, and **general ledger**. A general ledger is used to produce **income statements** and **balance sheets**.

Management Information Systems

Management information systems (MISs) produce predetermined reports **(periodic, exception,** and **demand)**.

Decision Support Systems

Decision support systems (DSSs) focus on unanticipated questions; consist of the **user, system software, data (internal** and **external),** and **decision models**. Three types are **strategic, tactical,** and **operational**. Group decision support systems **(GDSS)** support a team of decision makers.

Executive Support Systems

Executive support systems (ESSs) are similar to MIS or DSS but easier to use. ESSs are designed specifically for top-level decision makers.

Other Information Systems

Many other systems are designed to support **information workers** who create, distribute, and communicate information. Three such systems are

- **Office automation systems (OASs),** which support **data workers** who are involved with distribution and communication of information. **Project managers** and **videoconferencing systems** are **OASs.**
- **Knowledge work systems (KWSs),** which support **knowledge workers,** who create information. Many engineers use **computer-aided design/computer-aided manufacturing (CAD/CAM) systems.**
- **Expert (knowledge-based) systems,** which are a type of knowledge work system. They use **knowledge bases** to apply expert knowledge to specific user problems.

CAREERS in IT

Information systems managers oversee a variety of other computer professionals. Strong leadership and communication skills are required; experience as a consultant and/ or manager is desired. Expected salary range is $126,000 to $154,000.

KEY TERMS

accounting
accounts payable
accounts receivable
balance sheet
computer-aided design/computer-aided manufacturing (CAD/CAM) system
data
data processing system (DPS)
data worker
decision model
decision support system (DSS)
demand report
exception report
executive information system (EIS)
executive support system (ESS)
expert system
external data
general ledger
group decision support system (GDSS)
human resources
income statement
information system
information systems manager
information worker
internal data
inventory
inventory control system
knowledge base
knowledge-based system
knowledge work system (KWS)
knowledge worker
management information system (MIS)
marketing
middle-level manager
office automation system (OAS)
operational model
payroll
periodic report
production
project manager
purchase order
purchasing
research
sales order processing
strategic model
supervisor
system software
tactical model
top-level manager
transaction processing system (TPS)
user
videoconferencing system

MULTIPLE CHOICE

Circle the correct answer.

1. The accounting activity that records the customer requests for the company's products or services is sales order _____.
   - a. research
   - b. accounting
   - c. processing
   - d. tracking

2. The accounting activity concerned with calculating employee paychecks is _____.
   - a. purchasing
   - b. payroll
   - c. TPS
   - d. processing

3. MIS produces this type of report.
   - a. support summary
   - b. boundary
   - c. inventory
   - d. standardized

4. This function plans, prices, promotes, sells, and distributes the organization's goods and services.
   - a. accounting
   - b. production
   - c. marketing
   - d. human resources

5. Select a managerial level where information flow is vertical and horizontal.
   - a. top level
   - b. middle level
   - c. supervisors
   - d. omega level

6. Type of worker who is involved with the distribution and communication of information.
   - a. data
   - b. knowledge
   - c. expert
   - d. accountant

7. Computer system that allows people located at various geographic locations to communicate and have in-person meetings.
   - a. MIS
   - b. videoconferencing
   - c. managerial
   - d. TPS

8. A computer-based information system that uses data from TPS to support middle-level managers.
   - a. ESS
   - b. DSS
   - c. MIS
   - d. GPS

9. A type of report that calls attention to unusual events.
   - a. periodic
   - b. emergency
   - c. inventory
   - d. exception

10. Type of software that works behind the scenes to handle detailed operating procedures.
    - a. system
    - b. ESS
    - c. accounting
    - d. supervisor

MATCHING

Match each numbered item with the most closely related lettered item. Write your answers in the spaces provided.

a. accounting
b. balance sheet
c. data
d. general ledger
e. information
f. MIS
g. periodic
h. project managers
i. received
j. top

____ 1. Accounts payable refers to money the company owes its suppliers for materials and services it has _____.
____ 2. This accounting activity keeps track of all summaries of all transactions.
____ 3. This managerial level has information flow that is vertical, horizontal, and external.
____ 4. This computer-based information system uses data from TPS and analytical tools to support middle managers.
____ 5. The type of worker involved with the distribution, communication, and creation of information.
____ 6. The type of program designed to schedule, plan, and control project resources.
____ 7. This accounting statement lists the overall financial condition of an organization.
____ 8. A type of report produced at regular intervals.
____ 9. A DSS consists of four parts: user, system software, decision models, and _____.
____ 10. This basic organizational function records all financial activity from billing customers to paying employees.

OPEN-ENDED

On a separate sheet of paper, respond to each question or statement.
1. Name and discuss the five common functions of most organizations.
2. Discuss the roles of the three kinds of management in a corporation.
3. What are the four most common computer-based information systems?
4. Describe the different reports and their roles in managerial decision making.
5. What is the difference between an office automation system and a knowledge work system?

DISCUSSION

Respond to each of the following questions.

 Expanding Your Knowledge: EXECUTIVE SUPPORT SYSTEMS

Research at least three different executive support systems using a web search. Review each, and then answer the following questions: (a) Which ESSs did you review? (b) What are the common features of ESSs? (c) What type of company is likely to use each? Provide some examples.

 Expanding Your Knowledge: COMPUTER-BASED INFORMATION SYSTEMS

In this chapter, you were introduced to the fictional company HealthWise, a sports and exercise equipment manufacturer. HealthWise used computer-based information systems to generate reports specific to its industry. Now consider a different company that makes something other than sports and exercise equipment. This company could be real or imagined. Answer the following questions about this company: (a) What does the company sell? (b) What decisions would top-level management need to make in this company? What information would be found on an ESS report that would help them make these decisions? (c) What decisions would middle-level management need to make in this company? What information would be found on a DSS report that would help them make these decisions? (d) What information would be gathered by TPS systems at this company? How would this information affect MIS reports? What decisions important to your company would MIS reports affect?

Design Elements: Concept Check icon: Dizzle52/Getty Images

chapter 11
Databases

NicoElNino/Shutterstock

Why should I read this chapter?

Blue Planet Studio/Shutterstock

Giant databases are recording your every digital action, from credit card purchases to Facebook check-ins. These databases can be used to tailor advertisements to your interests, predict your credit score, and even assess your safety at home. In the future, databases with DNA records of all U.S. citizens and every digital activity of a city may be able to predict crime and even identify likely criminals.

This chapter covers the things you need to know to be prepared for this ever-changing digital world, including:

- Different data-organization methods—understand the significance of relational, multidimensional, and hierarchical databases.
- Types of databases—identify the right database for an individual, company, distributed, or commercial situation.

Learning Objectives

After you have read this chapter, you should be able to:

1 Distinguish between the physical and logical views of data.
2 Describe how data is organized: characters, fields, records, tables, and databases.
3 Define key fields and how they are used to integrate data in a database.
4 Define and compare batch processing and real-time processing.
5 Describe databases, including the need for databases and database management systems (DBMSs).
6 Describe the five common database models: hierarchical, network, relational, multidimensional, and object-oriented.
7 Distinguish among individual, company, distributed, and commercial databases.
8 Describe strategic database uses and security concerns.

Introduction

"Hi, I'm Anthony. I'm a database administrator, and I'd like to talk with you about databases. I'd also like to talk about how organizations are using data warehouses and data mining to perform complex analyses and discover new information."

SeventyFour/Shutterstock

Like a library, secondary storage is designed to store information. How is this stored information organized? What are databases, and why do you need to know anything about them?

Only a few decades ago, a computer was considered to be an island with only limited access to information beyond its own hard disk. Now, through communication networks and the Internet, individual computers have direct electronic access to almost unlimited sources of information.

In today's world, almost all information is stored in databases. They are an important part of nearly every organization, including schools, hospitals, and banks. To effectively compete in today's world, you need to know how to find information and understand how it is stored.

To efficiently and effectively use computers, you need to understand data fields, records, tables, and databases. You need to be aware of the different ways in which a database can be structured and the different types of databases. Also, you need to know the most important database uses and issues.

Data

As we have discussed throughout this book, information systems consist of people, procedures, software, hardware, data, and the Internet. This chapter focuses on **data**, which can be defined as facts or observations about people, places, things, and events. More specifically, this chapter focuses on how databases store, organize, and use data.

Not long ago, data was limited to numbers, letters, and symbols recorded by keyboards. Now, data is much richer and includes

- Audio captured, interpreted, and saved using microphones and voice recognition systems.
- Music downloaded from the Internet and saved on cell phones, tablets, and other devices.
- Photographs captured by digital cameras, edited by image editing software, and shared with others over the Internet.
- Video captured by digital video cameras, TV tuner cards, and webcams.

There are two ways, or perspectives, to view data. These perspectives are the *physical view* and the *logical view*. The **physical view** focuses on the actual format and location of the data. As discussed in **Chapter 5**, data is recorded as digital bits that are typically grouped together into bytes that represent characters using a coding scheme such as Unicode. Typically, only very specialized computer professionals are concerned with the physical view. The other perspective, the **logical view**, focuses on the meaning, content, and context of the data. End users and most computer professionals are concerned with this view. They are involved with actually using the data with application programs. This chapter presents the logical view of data and how data is stored in databases.

concept check

☐ Describe some of the different types of data.
☐ What is the physical view of data?
☐ What is the logical view of data?

Data Organization

The first step in understanding databases is to learn how data is organized. In the logical view, data is organized into groups or categories. Each group is more complex than the one before. (See **Figure 11-1**.)

- **Character:** A **character** is the most basic logical data element. It is a single letter, number, or special character, such as a punctuation mark, or a symbol, such as $.
- **Field:** The next higher level is a **field**, or group of related characters. In our example, Brown is in the data field for the Last Name of an employee. It consists of the individual letters (characters) that make up the last name. A data field represents an **attribute** (description or characteristic) of some **entity** (person, place, thing, or object). For example, an employee is an entity with many attributes, including his or her last name.
- **Record:** A **record** is a collection of related fields. A record represents a collection of attributes that describe an entity. In our example, the payroll record for an employee consists of the data fields describing the attributes for one employee. These attributes are First Name, Last Name, Employee ID, and Salary.
- **Table:** A **table** is a collection of related records. For example, the Payroll Table would include payroll information (records) for the employees (entities).
- **Database:** A **database** is an integrated collection of logically related tables. For example, the Personnel Database would include all related employee tables, including the Payroll Table and the Benefits Table.

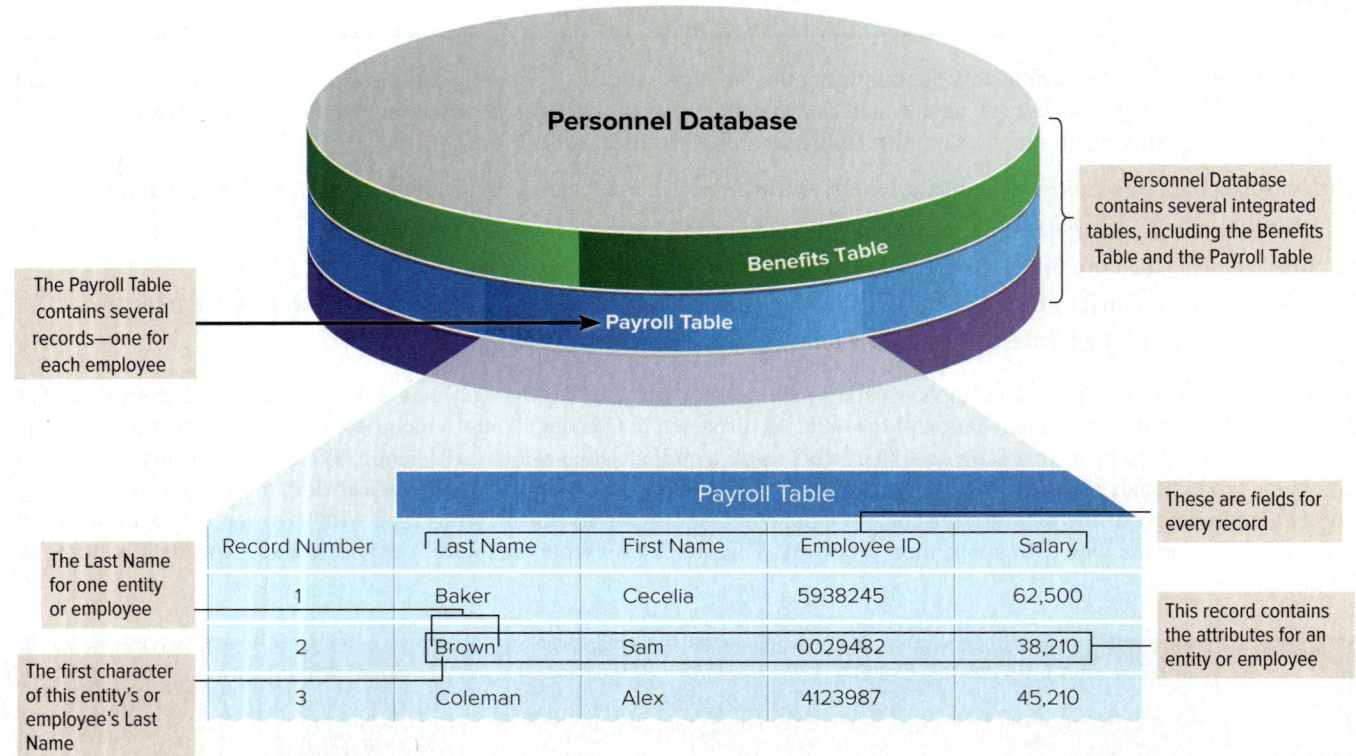

Figure 11-1 Logical data organization

Key Field

Each record in a table has at least one distinctive field, called the **key field**. Also known as the **primary key**, this field uniquely identifies the record. Tables can be related or connected to other tables by common key fields.

For most employee databases, a key field is an employee identification number. Key fields in different tables can be used to integrate the data in a database. For example, in the Personnel Database, both the Payroll and the Benefits tables include the field Employee ID. Data from the two tables could be related by combining all records with the same key field (Employee ID).

Batch versus Real-Time Processing

Traditionally, data is processed in one of two ways. These are batch processing, or what we might call "later," and real-time processing, or what we might call "now." These two methods have been used to handle common record-keeping activities such as payroll and sales orders.

- **Batch processing:** In **batch processing**, data is collected over several hours, days, or even weeks. It is then processed all at once as a "batch." If you have a credit card, your bill probably reflects batch processing. That is, during the month, you buy things and charge them to your credit card. Each time you charge something, an electronic copy of the transaction is sent to the credit card company. At some point in the month, the company's data processing department puts all those transactions (and those of many other customers) together and processes them at one time. The company then sends you a single bill totaling the amount you owe. (See **Figure 11-2**.)

- **Real-time processing: Real-time processing**, also known as **online processing**, occurs when data is processed at the same time the transaction occurs. For example, whenever you request funds at an ATM, real-time processing occurs. After you have provided account information and requested a specific withdrawal, the bank's computer verifies that you have sufficient funds in your account. If you do, then the funds are dispensed to you, and the bank immediately updates the balance of your account. (See **Figure 11-3**.)

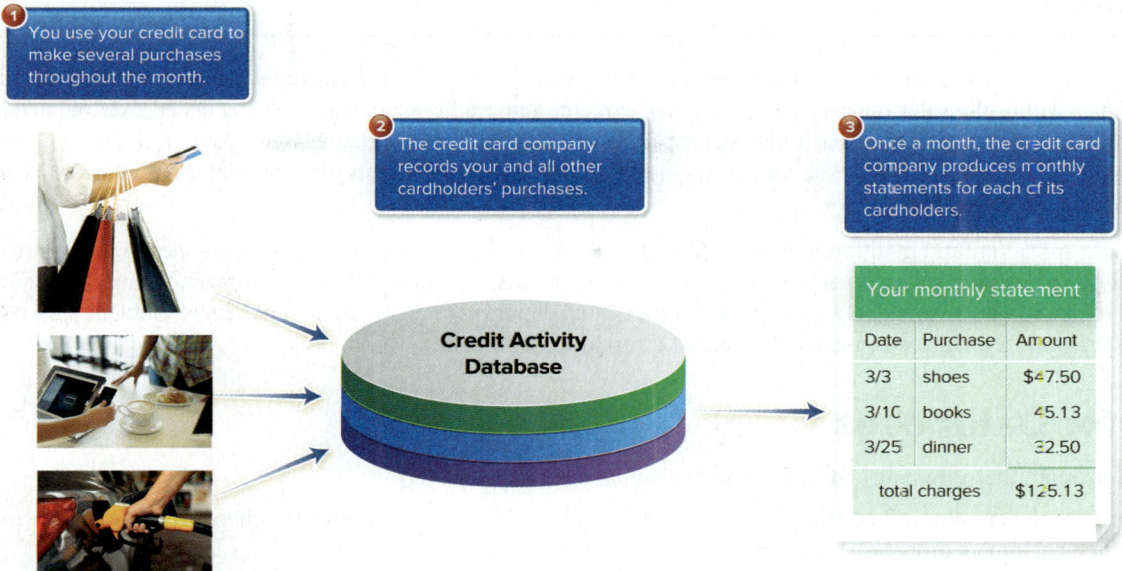

Figure 11-2 Batch processing: Monthly credit card statements
top: urfin/Shutterstock; **middle:** Roberto Westbrook/Image Source; **bottom:** RecCameraStock/Shutterstock

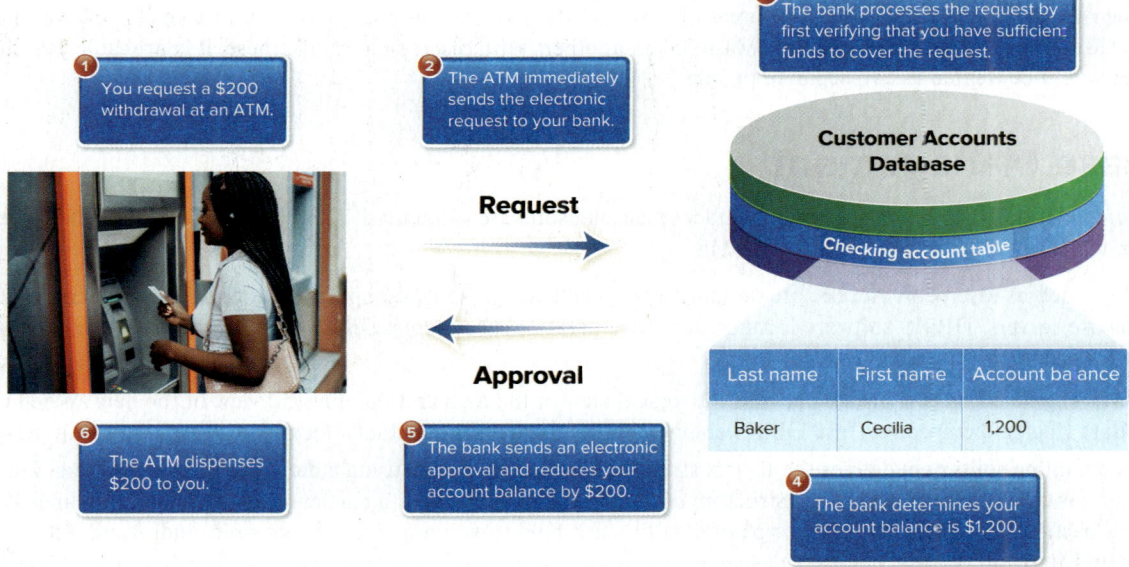

Figure 11-3 Real-time processing: ATM withdrawal
Zamrznuti tonovi/Shutterstock

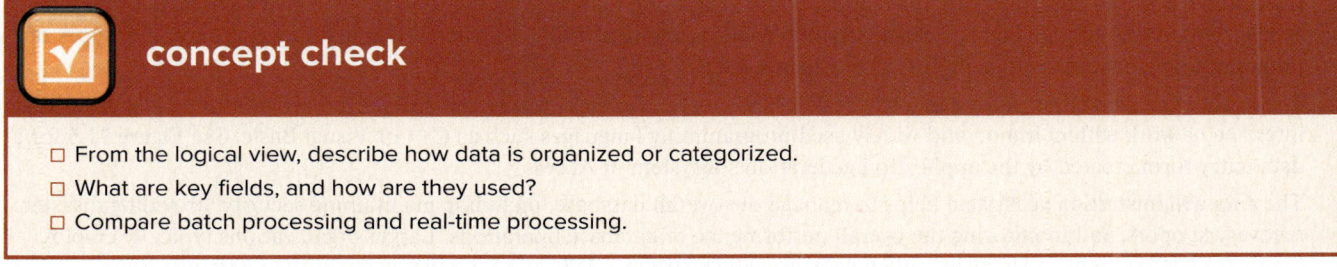

concept check

- From the logical view, describe how data is organized or categorized.
- What are key fields, and how are they used?
- Compare batch processing and real-time processing.

Databases

Many organizations have multiple files on the same subject or person. For example, a customer's name and address could appear in different files within the sales department, billing department, and credit department. This is called **data redundancy**. If the customer moves, then the address in each file must be updated. If one or more files are overlooked, problems will likely result. For example, a product ordered might be sent to the new address, but the bill might be sent to the old address. This situation results from a lack of **data integrity**.

Moreover, data spread around in different files is not as useful. The marketing department, for instance, might want to offer special promotions to customers who order large quantities of merchandise. To identify these customers, the marketing department would need to obtain permission and access to files in the billing department. It would be much more efficient if all data were in a common database. A database can make the needed information available.

Need for Databases

For an organization, there are many advantages to having databases:

- **Sharing:** In organizations, information from one department can be readily shared with others. Billing could let marketing know which customers ordered large quantities of merchandise.
- **Security:** Users are given passwords or access only to the kind of information they need. Thus, the payroll department may have access to employees' pay rates, but other departments would not.
- **Less data redundancy:** Without a common database, individual departments have to create and maintain their own data, and data redundancy results. For example, an employee's home address would likely appear in several files. Redundancy wastes storage space, increases data flaws, and decreases data compatibility across departments and applications.
- **Data integrity:** When there are multiple sources of data, each source may have variations. A customer's address may be listed as "Main Street" in one system and "Main St." in another. With discrepancies like these, it is probable that the customer would be treated as two separate people.

Database Management

In order to create, modify, and gain access to a database, special software is required. This software is called a **database management system**, which is commonly abbreviated **DBMS**.

Some DBMSs, such as Microsoft Access, are designed specifically for personal computers. Other DBMSs are designed for specialized database servers. DBMS software is made up of five parts or subsystems: *DBMS engine, data definition, data manipulation, application generation,* and *data administration.*

- The **DBMS engine** provides a bridge between the logical view of the data and the physical view of the data. When users request data (logical perspective), the DBMS engine handles the details of actually locating the data (physical perspective).
- The **data definition subsystem** defines the logical structure of the database by using a **data dictionary** or **schema**. This dictionary contains a description of the structure of data in the database. For a particular item of data, it defines the names used for a particular field. It defines the type of data for each field (text, numeric, time, graphic, audio, and video). For example, in Microsoft Access, database designers can use the design view to access the data dictionary form. This includes the field name, data type, and a short description of the contents of the field. (See **Figure 11-4**.)
- The **data manipulation subsystem** provides tools for maintaining and analyzing data. Maintaining data is known as **data maintenance**. It involves adding new data, deleting old data, and editing existing data. Analysis tools support viewing all or selected parts of the data, querying the database, and generating reports. Specific tools include **query-by-example** and a specialized programming language called **structured query language (SQL)**. (Structured query language and other types of programming languages will be discussed in **Chapter 13**.)
- The **application generation subsystem** provides tools to create data entry forms and specialized programming languages that interface or work with common and widely used programming languages such as C++ or Visual Basic. See **Figure 11-5** for a data entry form created by the application generation subsystem in Access.
- The **data administration subsystem** helps to manage the overall database, including maintaining security, providing disaster recovery support, and monitoring the overall performance of database operations. Larger organizations typically employ highly trained computer specialists, called **database administrators (DBAs)**, to interact with the data administration subsystem. Additional duties of database administrators include determining **processing rights** or determining which people have access to what kinds of data in the database.

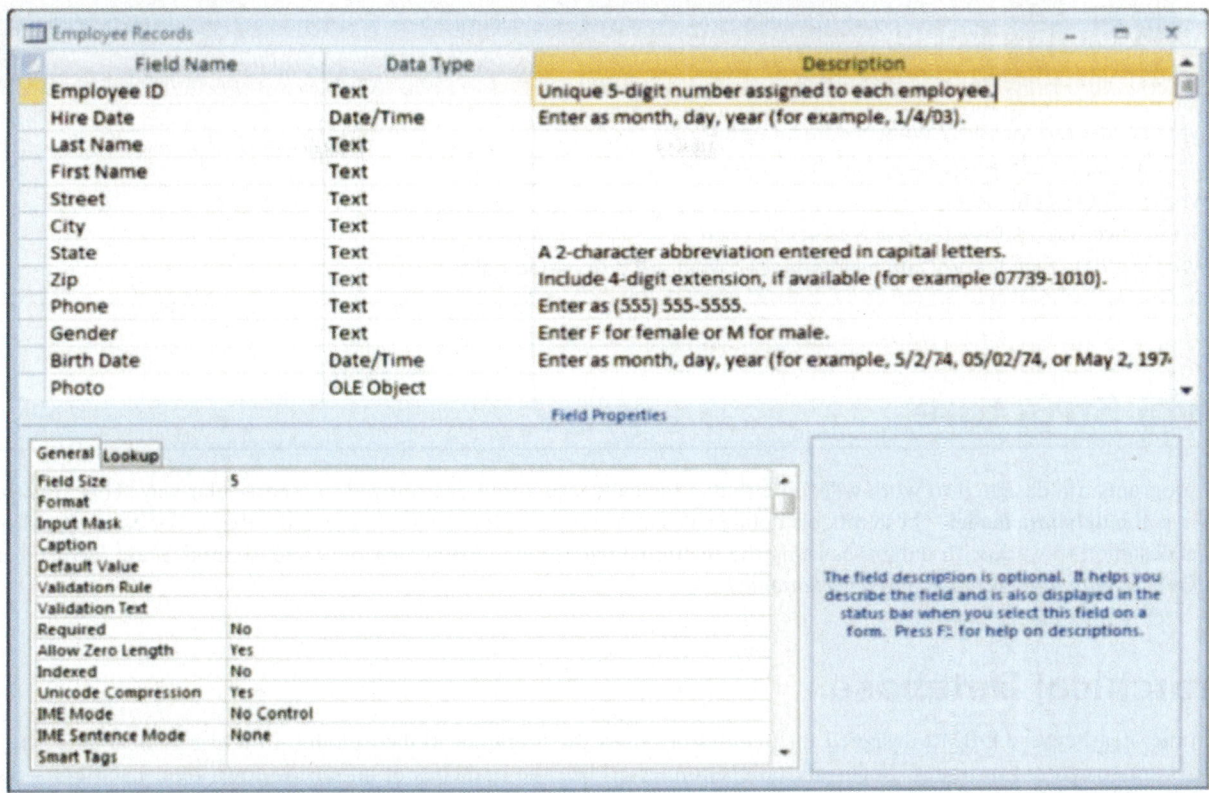

Figure 11-4 Access data dictionary form
Microsoft Corporation

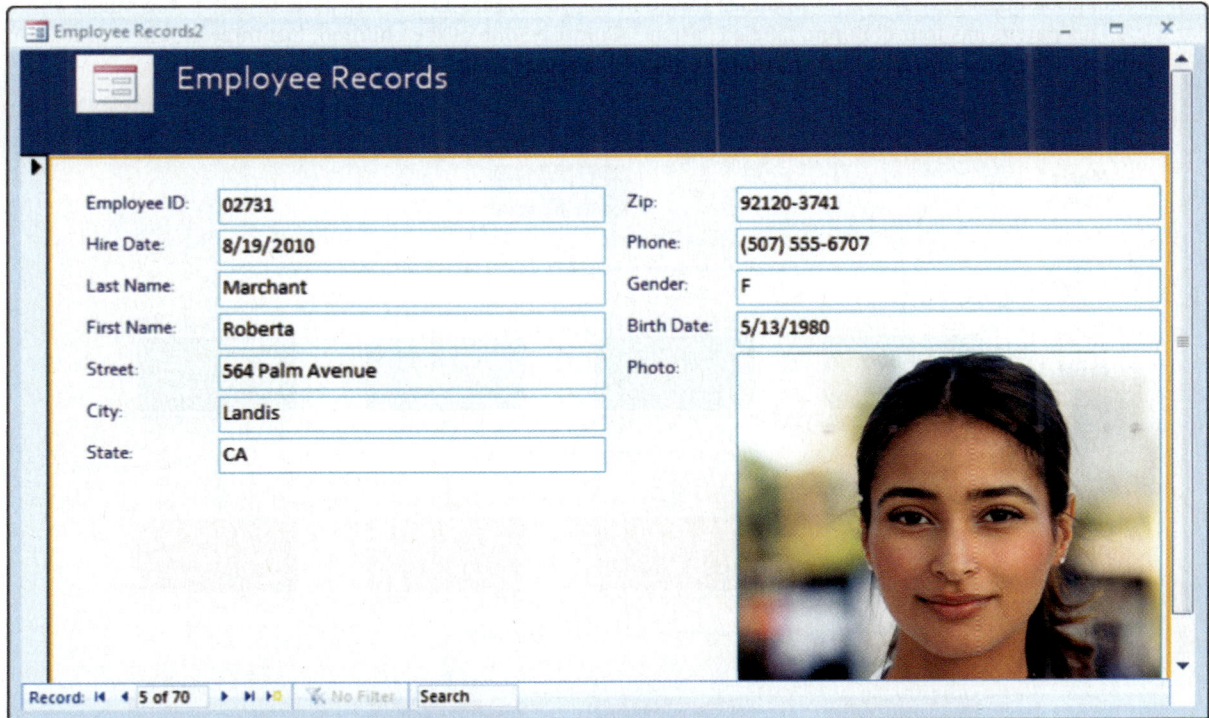

Figure 11-5 Access data entry form
Microsoft Corporation; El Nariz/Shutterstock

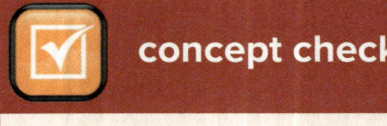

concept check

☐ What is data redundancy? What is data integrity?
☐ What are some of the advantages to having databases?
☐ What is DBMS software?
☐ List the five basic subsystems and describe each.
☐ What is a data dictionary? Data maintenance? What are processing rights?

DBMS Structure

DBMS programs are designed to work with data that is logically structured or arranged in a particular way. This arrangement is known as the **database model**. These models define rules and standards for all the data in a database. For example, Microsoft Access is designed to work with databases using the relational data model. Five common database models are *hierarchical, network, relational, multidimensional,* and *object-oriented.*

Hierarchical Database

At one time, nearly every DBMS designed for mainframes used the hierarchical data model. In a **hierarchical database**, fields or records are structured in nodes. **Nodes** are points connected like the branches of an upside-down tree. Each entry has one **parent node**, although a parent may have several **child nodes**. This is sometimes described as a **one-to-many relationship**. To find a particular field, you have to start at the top with a parent and trace down the tree to a child.

The nodes farther down the system are subordinate to the ones above, like the hierarchy of managers in a corporation. An example of a hierarchical database is a system to organize music files. (See **Figure 11-6**.) The parent node is the music library for a particular user. This parent has four children, labeled "artist." Coldplay, one of the children, has three children of its own. They are labeled "album." The *Greatest Hits* album has three children, labeled "song."

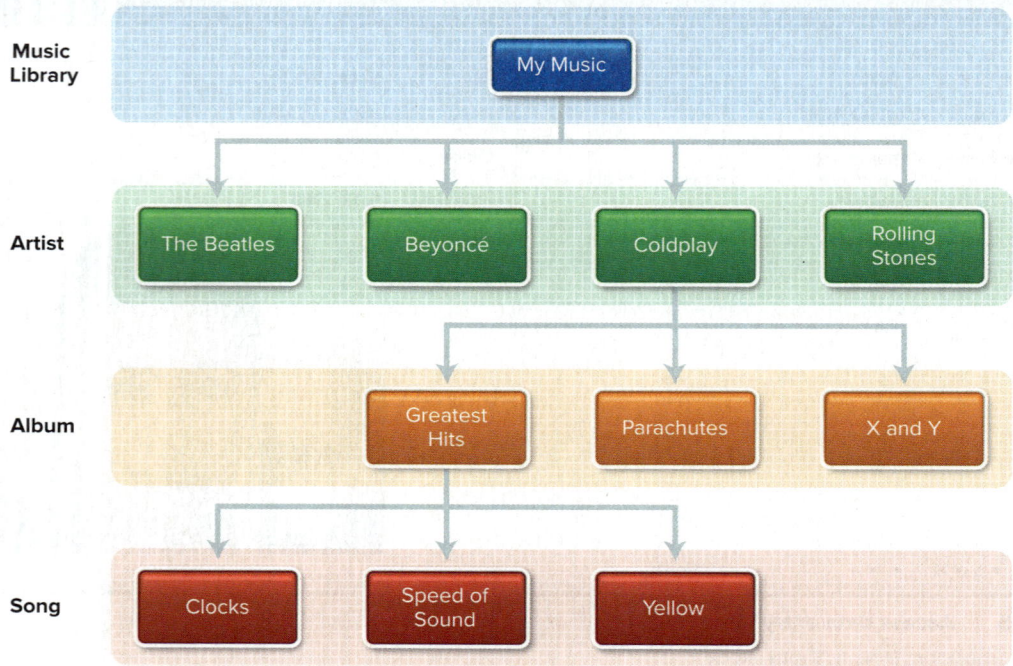

Figure 11-6 **Hierarchical database**

The problem with a hierarchical database is that if one parent node is deleted, so are all the subordinate child nodes. Moreover, a child node cannot be added unless a parent node is added first. The most significant limitation is the rigid structure: one parent only per child, and no relationships or connections between the child nodes themselves.

Network Database

Responding to the limitations of the hierarchical data model, network models were developed. A **network database** also has a hierarchical arrangement of nodes. However, each child node may have more than one parent node. This is sometimes described as a **many-to-many relationship**. There are additional connections—called **pointers**—between parent nodes and child nodes. Thus, a node may be reached through more than one path. It may be traced down through different branches.

For example, a university could use this type of organization to record students taking classes. (See **Figure 11-7**.) If you trace through the logic of this organization, you can see that each student can have more than one teacher. Each teacher also can teach more than one course. Students may take more than a single course. This demonstrates how the network arrangement is more flexible and, in many cases, more efficient than the hierarchical arrangement.

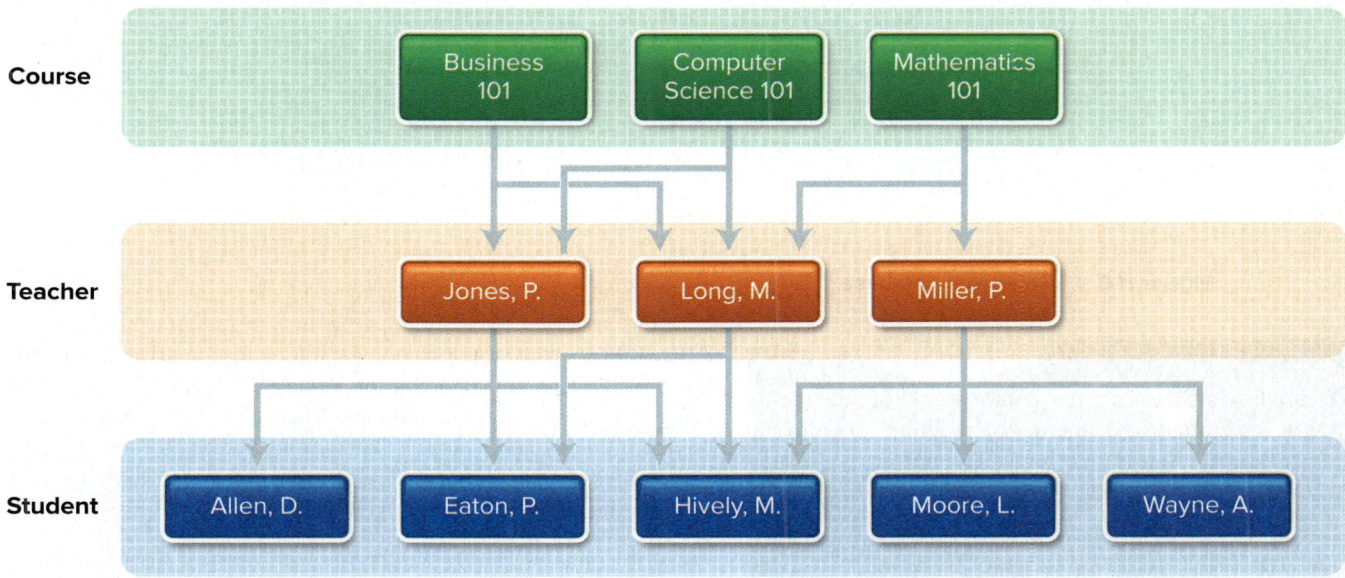

Figure 11-7 Network database

Relational Database

The most common type of organization is the **relational database**. In this structure, there are no access paths down a hierarchy. Rather, the data elements are stored in different tables, each of which consists of rows and columns. A table and its data are called a **relation**.

An example of a relational database is shown in **Figure 11-8**. The Vehicle Owner Table contains license numbers, names, and addresses for all registered drivers. Within the table, a row is a record containing information about one driver. Each column is a field. The fields are License Number, Last Name, First Name, Street, City, State, and Zip. All related tables must have a **common data item**, or shared key field, in both tables, enabling information stored in one table to be linked with information stored in another. In this case, the three tables are related by the License Number field.

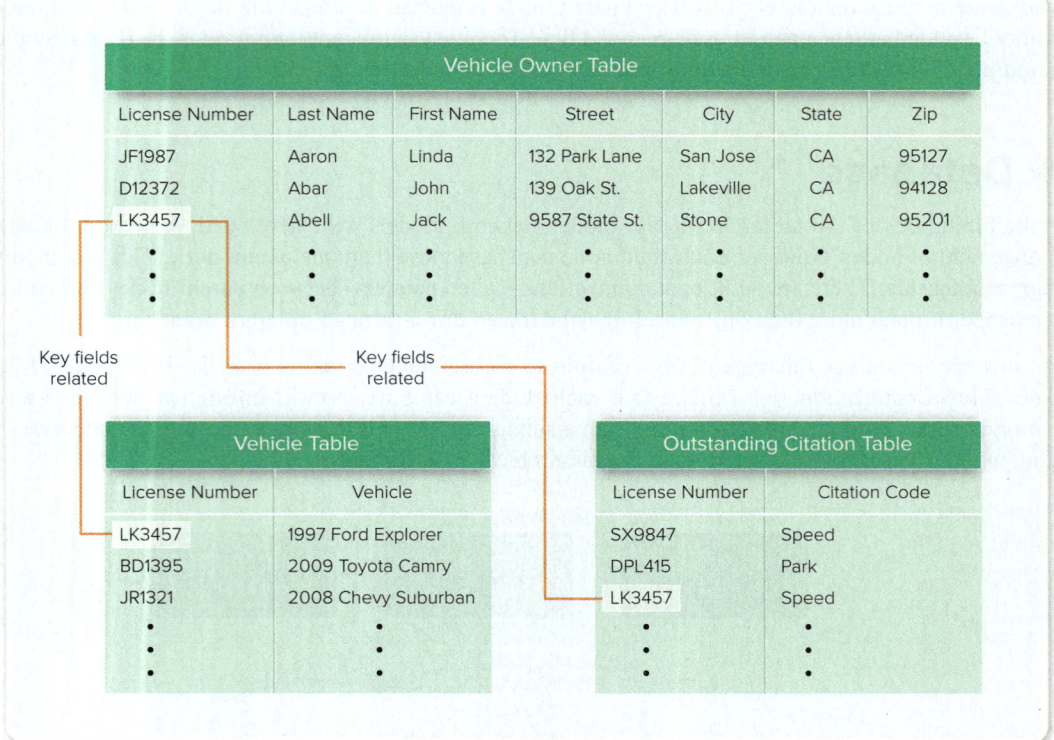

Figure 11-8 Relational database

Figure 11-9 The Department of Motor Vehicles uses a relational database
Thinkstock/Stockbyte/Getty Images

Police officers who stop a speeding car look up the driver's information in the Department of Motor Vehicles database (**Figure 11-9**) using the driver's license number. They also can check for any unpaid traffic violations in the Outstanding Citations Table. Finally, if the officers suspect that the car is stolen, they can look up what vehicles the driver owns in the Vehicle Table.

The most valuable feature of relational databases is their simplicity. Entries can be easily added, deleted, and modified. The hierarchical and network databases are more rigid. The relational organization is common for personal computer DBMSs such as Access. Relational databases are also widely used for mainframe and midrange systems.

Multidimensional Database

The multidimensional data model is a variation and an extension of the relational data model. Whereas relational databases use tables consisting of rows and columns, **multidimensional databases** extend this two-dimensional data model to include additional or multiple dimensions, sometimes called a **data cube**. Data can be viewed as a cube having three or more sides and consisting of cells. Each side of the cube is considered a dimension of the data. In this way, complex relationships between data can be represented and efficiently analyzed.

Multidimensional databases provide several advantages over relational databases. Two of the most significant advantages are

- **Conceptualization:** Multidimensional databases and data cubes provide users with an intuitive model in which complex data and relationships can be analyzed without specialized database programming knowledge.
- **Processing speed:** Analyzing and querying a large multidimensional database can be much faster. For example, a query requiring just a few seconds on a multidimensional database could take minutes or hours to perform on a relational database.

Object-Oriented Database

The other data structures are primarily designed to access and summarize structured data such as names, addresses, and pay rates. **Object-oriented databases** are more flexible and store data as well as instructions to manipulate the data. Additionally, this structure is ideally designed to provide input for object-oriented software development, which is described in **Chapter 13**.

Object-oriented databases organize data using classes, objects, attributes, and methods.

- **Classes** are general definitions.
- **Objects** are specific instances of a class that can contain both data and instructions to manipulate the data.
- **Attributes** are the data fields an object possesses.
- **Methods** are instructions for retrieving or manipulating attribute values.

For example, a health club might use an object-oriented employment database. (See **Figure 11-10**.) The database uses a class, Employee, to define employee objects that are stored in the database. This definition includes the attributes First name, Last name, Address, and Wage and the method Pay. Bob, Sarah, and Omar are objects each with specific attributes designated by their object type. For example, Bob is an employee and an accountant. Bob has attributes common to all employees, such as a name and address. He also has attributes unique to accountants, such as his certification as a public accountant. Different objects use different methods to determine some information. For example, accountants earn an annual salary, while salespeople earn an hourly wage as well as a commission.

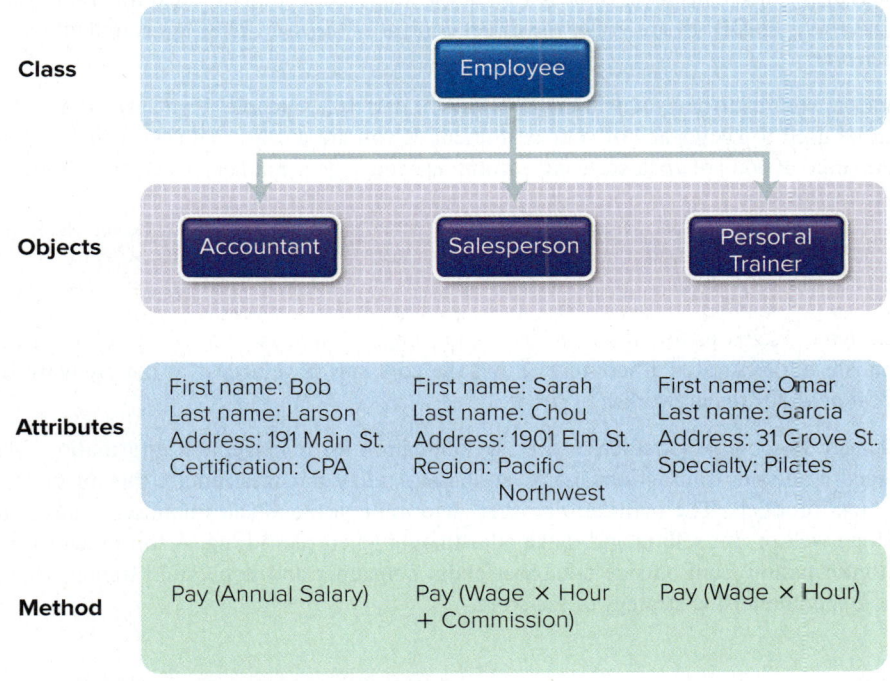

Figure 11-10 Object-oriented database

Although hierarchical and network databases are still widely used, relational and multidimensional databases are the most popular today. Object-oriented databases are becoming more widely used and are part of a new category of databases known as **NoSQL**. For a summary of DBMS organization, see **Figure 11-11**.

Organization	Description
Hierarchical	Data structured in nodes organized like an upside-down tree; each parent node can have several children; each child node can have only one parent
Network	Like hierarchical except that each child can have several parents
Relational	Data stored in tables consisting of rows and columns
Multidimensional	Data stored in data cubes with three or more dimensions
Object-oriented	Organizes data using classes, objects, attributes, and methods

Figure 11-11 Summary of DBMS organization

concept check

☐ What is a database model?
☐ List the five database models and discuss each.
☐ What is the difference between a relational database and an object-oriented database?

Types of Databases

Databases may be small or large, limited in accessibility or widely accessible. Databases may be classified into four types: *individual, company, distributed,* and *commercial*.

Individual

The **individual database** is also called a **personal computer database**. It is a collection of integrated files primarily used by just one person. Typically, the data and the DBMS are under the direct control of the user. They are stored either on the user's hard-disk drive or on a LAN file server.

There may be many times in your life when you will find this kind of database valuable. If you are in sales, for instance, a personal computer database can be used to keep track of your customers. If you are a sales manager, you can keep track of your salespeople and their performance. If you are an advertising account executive, you can keep track of your different projects and how many hours to charge each client.

Company

Companies, of course, create databases for their own use. The **company database** may be stored on a central database server and managed by a database administrator. Users throughout the company have access to the database through their personal computers linked to local or wide area networks.

As we discussed in **Chapter 10**, company databases are the foundation for management information systems. For instance, a department store can record all sales transactions in the database. A sales manager can use this information to see which salespeople are selling the most products. The manager can then determine year-end sales bonuses. Or the store's buyer can learn which products are selling well or not selling and make adjustments when reordering. A top executive might combine overall store sales trends with information from outside databases about consumer and population trends. This information could be used to change the whole merchandising strategy of the store.

Distributed

Many times the data in a company is stored not in just one location but in several locations. It is made accessible through a variety of communications networks. The database, then, is a **distributed database**. That is, not all the data in a database is physically located in one place. Typically, database servers on a client/server network provide the link between the data.

For instance, some database information can be at regional offices. Some can be at company headquarters, some down the hall from you, and some even overseas. Sales figures for a chain of department stores, then, could be located at the various stores. But executives at district offices or at the chain's headquarters could have access to all these figures.

Commercial

A **commercial database** is generally an enormous database that an organization develops to cover particular subjects. It offers access to this database to the public or selected outside individuals for a fee. Sometimes commercial databases also are called **information utilities** or **data banks**. An example is Factiva, which offers a variety of information-gathering and reporting services. (See **Figure 11-12**.)

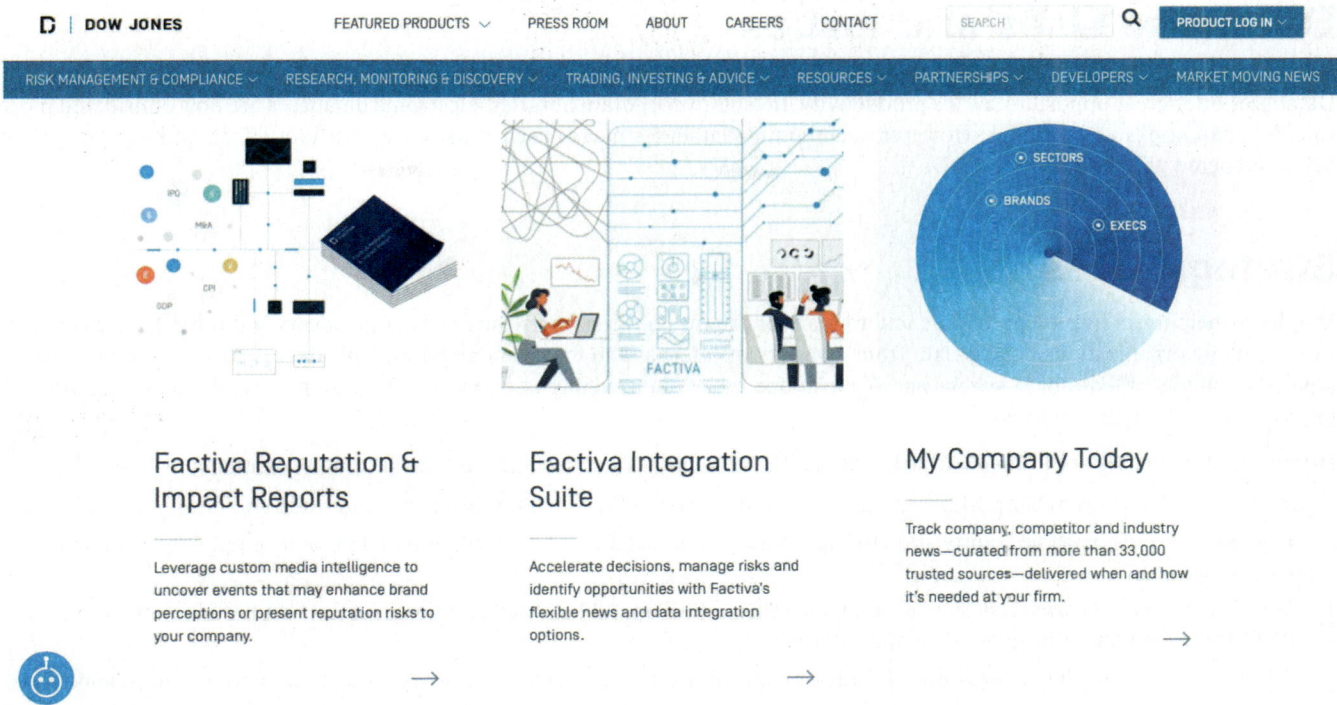

Figure 11-12   Commercial database (Dow Jones Factiva)
Dow Jones & Company, Inc.

Some of the most widely used commercial databases are

- **ProQuest Dialog**—offers business information, as well as technical and scientific information.
- **Dow Jones Factiva**—provides world news and information on business, investments, and stocks.
- **LexisNexis**—offers news and information on legal news, public records, and business issues.

Most of the commercial databases are designed for organizational as well as individual use. Organizations typically pay a membership fee plus hourly use fees. Often, individuals are able to search the database to obtain a summary of available information without charge. They pay only for those items selected for further investigation.

See **Figure 11-13** for a summary of the four types of databases.

Type	Description
Individual	Integrated files used by just one person
Company	Common operational or commonly used files shared in an organization
Distributed	Database spread geographically and accessed using database server
Commercial	Information utilities or data banks available to users on a wide range of topics

Figure 11-13   Summary of the four types of databases

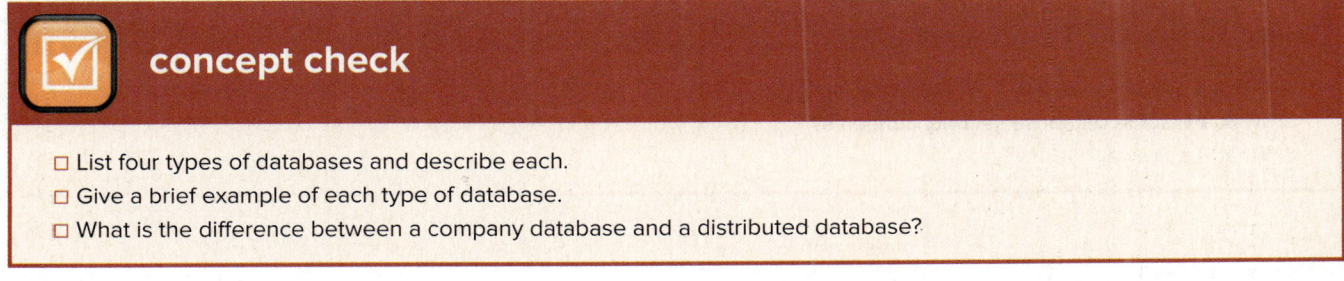

concept check

☐ List four types of databases and describe each.
☐ Give a brief example of each type of database.
☐ What is the difference between a company database and a distributed database?

Database Uses and Issues

Databases offer great opportunities for productivity. In fact, in corporate libraries, electronic databases are now considered more valuable than books and journals. However, maintaining databases means users must make constant efforts to keep them from being tampered with or misused.

Strategic Uses

Databases help users to keep up to date and to plan for the future. To support the needs of managers and other business professionals, many organizations collect data from a variety of internal and external databases. This data is then stored in a special type of database called a **data warehouse**. A technique called **data mining** is often used to search these databases to look for related information and patterns.

Hundreds of databases are available to help users with both general and specific business purposes, including

- *Business directories* providing addresses, financial and marketing information, products, and trade and brand names.
- *Demographic data,* such as county and city statistics, current estimates on population and income, employment statistics, and census data.
- *Business statistical information,* such as financial information on publicly traded companies, market potential of certain retail stores, and other business data and information.
- *Text databases* providing articles from business publications, press releases, reviews on companies and products, and so on.
- *Web databases* covering a wide range of topics, including all of those previously mentioned. As mentioned earlier, web search sites like Google maintain extensive databases of available Internet content.

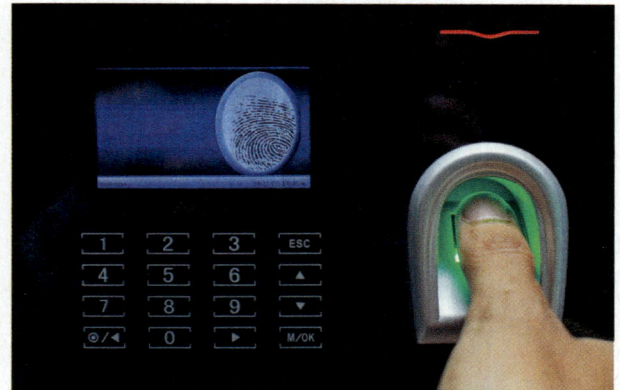

Figure 11-14 Security: Electronic fingerprint scanner
ChaikomShutterstock

Security

Precisely because databases are so valuable, their security has become a critical issue. As we discussed in **Chapter 9**, there are several database security concerns. One concern is that personal and private information about people stored in databases may be used for the wrong purposes. For instance, a person's credit history or medical records might be used to make hiring or promotion decisions. Another concern is unauthorized users gaining access to a database. For example, there have been numerous instances in which a computer virus has been launched into a database or network.

Security may require putting guards in company computer rooms and checking the identification of everyone admitted. Some security systems electronically check fingerprints. (See **Figure 11-14**.) Security is particularly important to organizations using WANs. Violations can occur without actually entering secured areas. As mentioned in previous chapters, most major corporations today use special hardware and software called **firewalls** to control access to their internal networks.

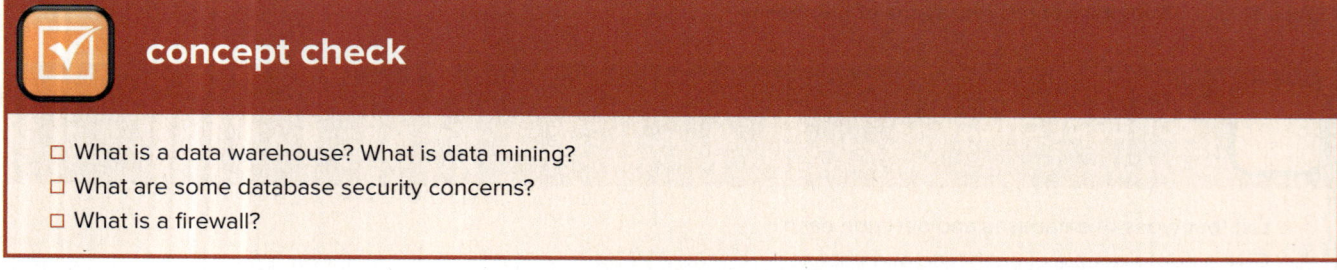

concept check

- What is a data warehouse? What is data mining?
- What are some database security concerns?
- What is a firewall?

Careers in IT

"Now that you have learned about databases, let me tell you about my career as a database administrator."

SeventyFour/Shutterstock

Database administrators use database management software to determine the most efficient ways to organize and access a company's data. Additionally, database administrators are typically responsible for maintaining database security and backing up the system. Database administration is a fast-growing industry, and substantial job growth is expected.

Database administrator positions normally require a bachelor's degree in computer science or information systems and technical experience. Internships and prior experience with the latest technology are a considerable advantage for those seeking jobs in this industry. It is possible to transfer skills learned in one industry, such as finance, to a new career in database administration. In order to accomplish this objective, many people seek additional training in computer science.

Database administrators can expect to earn an annual salary of $88,000 to $112,000. Opportunities for advancement include positions as a chief technology officer or other managerial opportunities.

A LOOK TO THE FUTURE

The Future of Crime Databases

Blue Planet Studio/Shutterstock

Have you ever imagined a world without violent crime? What would you be willing to do (or give up) if your government could guarantee that all potential criminals could be stopped before they commit their crime? Recently, we are close to making this possible through large and powerful databases, along with computer programs that can analyze data and make predictions. The tricky part is that the databases require a significant amount of personal information from everyone who lives in the country. Technology has been making better crime databases and will continue to evolve to improve our lives as we look to the future.

Currently, national crime databases in several countries, such as the United States, focus on keeping data about individuals that have committed crimes. Not only do these databases contain basic information such as name and date of birth, but they also contain fingerprints, photos, and even DNA samples. This makes it easier to figure out who committed a crime after it happens, assuming the criminal is already in the database. Although the offender will eventually be caught, it is too late for the innocent victim. For this reason, researchers are currently looking into the possibility of expanding the collection of data and then using powerful programs to figure out who is capable of committing violent crimes in the future.

Over the last few years, various research institutions have been looking into patterns that could predict criminal behavior. They analyze data ranging from childhood abuse to current employment status. Their goal is to find a combination of factors that usually leads to violent, criminal behavior. Other researchers are looking deep into human DNA, looking for any sequences that could be connected to antisocial or violent behavior. If such patterns could be found, then all we need to do is find the individuals who have these characteristics. The problem is that not all these individuals have an entry in the national crime database. Furthermore, these databases do not contain data about every aspect of a person's life. That, however, could change.

Over the years, criminal databases have been expanding. However, in the United States, each state has the ability to determine the data to be collected and from whom it will be collected. Whereas one state might take a DNA sample from only violent criminals and sex offenders, another state might collect that data from someone who committed a misdemeanor. If a future crime database is to make predictions, law enforcement will have to take DNA samples from every person living in the United States. Furthermore, the government will need access to all databases that contain information about that person—including databases from schools, businesses, insurance companies, and medical practices. Only then can these future programs be able to predict which individuals have the sort of patterns that may lead to future criminal behavior. Once those individuals are spotted, law enforcement could be authorized to monitor them closely or perhaps even intervene with psychological or medical assistance.

Supposing a crime-predicting program could be developed, there will be legal challenges to the type of data collection required. Individuals will be asked to weigh their privacy against the possibility of reducing crime. Inevitably, our trust of the government will also come into play. What do you think about this sort of future database? Would you trust the government with all this personal and biological information? Do you believe it is worth giving up privacy for the sake of having security?

VISUAL SUMMARY | Databases

DATA ORGANIZATION

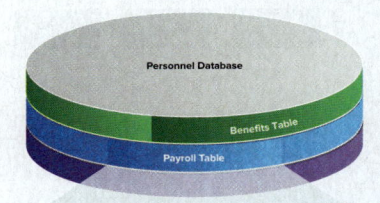

Data is organized by the following groups:

- **Character**—the most basic logical element, consisting of individual numbers, letters, and special characters.
- **Field**—next level, consisting of a set of related characters, for example, a person's last name. A data field represents an **attribute** (description or characteristic) of some **entity** (person, place, thing, or object).
- **Record**—a collection of related fields; for example, a payroll record consisting of fields of data relating to one employee.
- **Table**—a collection of related records; for example, a payroll table consisting of all the employee records.
- **Database**—an integrated collection of related tables; for example, a personnel database contains all related employee tables.

Key Field

A **key field (primary key)** is the field in a record that uniquely identifies each record.

- Tables can be related (connected) to other tables by key fields.
- Key fields in different files can be used to integrate the data in a database.
- Common key fields are employee ID numbers and driver's license numbers.

DATA ORGANIZATION

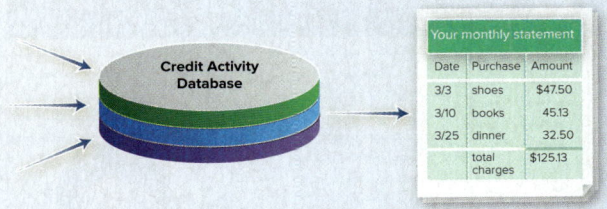

Batch versus Real-Time Processing

Traditionally, data is processed in one of two ways: batch or real-time processing.

- **Batch processing**—data is collected over time and then processed later all at one time (batched). For example, monthly credit card bills are typically created by processing credit card purchases throughout the past month.
- **Real-time processing (online processing)**—data is processed at the same time the transaction occurs; direct access storage devices make real-time processing possible. For example, a request for cash using an ATM machine initiates a verification of funds, approval or disapproval, disbursement of cash, and an update of the account balance.

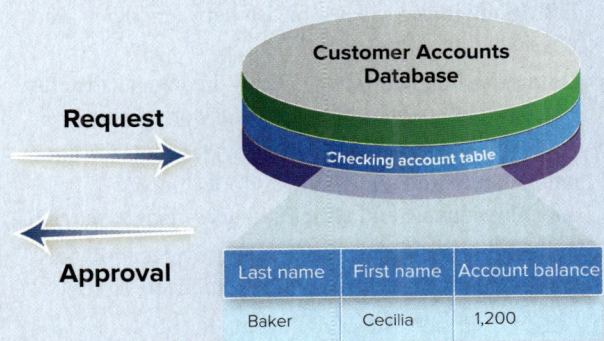

DATABASES

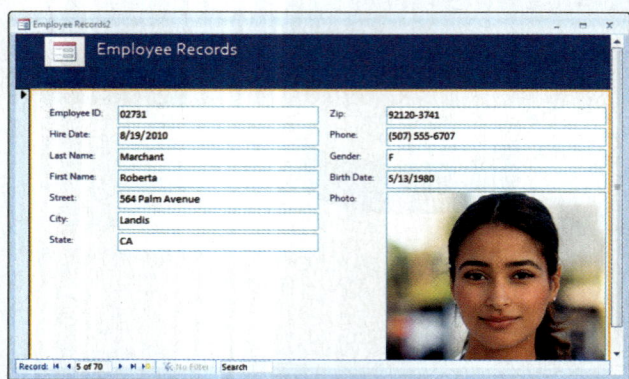

Microsoft Corporation; El Nariz/Shutterstock

A **database** is a collection of integrated data—logically related files and records.

Need for Databases

Advantages of databases are sharing data, improved security, reduced **data redundancy**, and higher **data integrity**.

Database Management

A **database management system (DBMS)** is the software for creating, modifying, and gaining access to the database. A DBMS consists of five subsystems:

- **DBMS engine** provides a bridge between logical and physical data views.
- **Data definition subsystem** defines the logical structure of a database using a **data dictionary** or **schema**.
- **Data manipulation subsystem** provides tools for **data maintenance** and data analysis; tools include **query-by-example** and **structured query language (SQL)**.
- **Application generation subsystem** provides tools for creating data entry forms with specialized programming languages.
- **Data administration subsystem** manages the database; **database administrators (DBAs)** are computer professionals who help define **processing rights**.

DBMS STRUCTURE

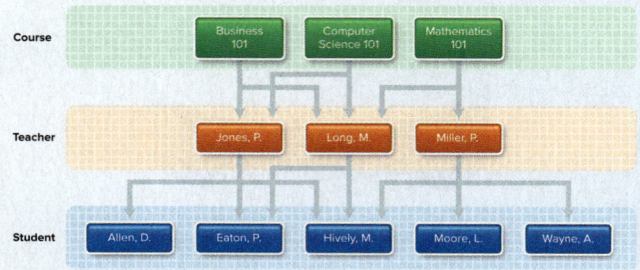

DBMS programs are designed to work with specific data structures or **database models**. These models define rules and standards for all the data in the database. Five principal database models are *hierarchical, network, relational, multidimensional,* and *object-oriented.*

Hierarchical Database

Hierarchical database uses **nodes** to link and structure fields and records; entries may have one **parent node** with several **child nodes** in a **one-to-many relationship**.

Network Database

Network database is like hierarchical except a child node may have more than one parent in a **many-to-many relationship**; additional connections are called **pointers**.

Relational Database

Relational database data is stored in tables **(relations)**; related tables must have a **common data item** (key field). A table and its data are called a **relation**.

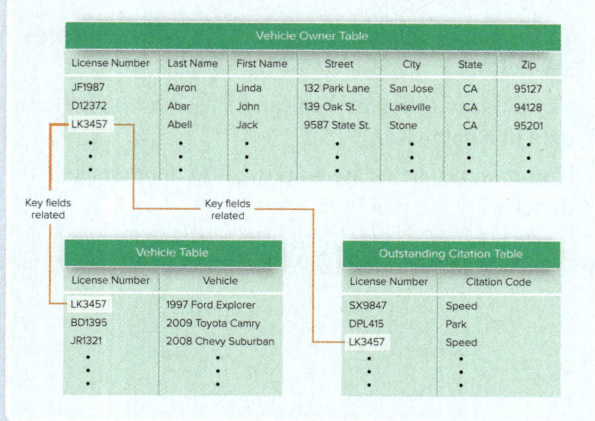

DBMS STRUCTURE

Thinkstock/Stockbyte/Getty Images

Multidimensional Database

Multidimensional databases extend two-dimensional relational tables to three or more dimensions, sometimes called a **data cube**.

Multidimensional databases tend to be more flexible and intuitive than relational databases.

Object-Oriented Database

Object-oriented databases store data, instructions, and unstructured data. Data is organized using

- **Classes** are general definitions.
- **Objects** are specific instances of a class that can contain both data and instructions to manipulate the data.
- **Attributes** are the data fields an object possesses.
- **Methods** are instructions for retrieving or manipulating attribute values.

Object-oriented databases are part of a new category of databases known as **NoSQL**.

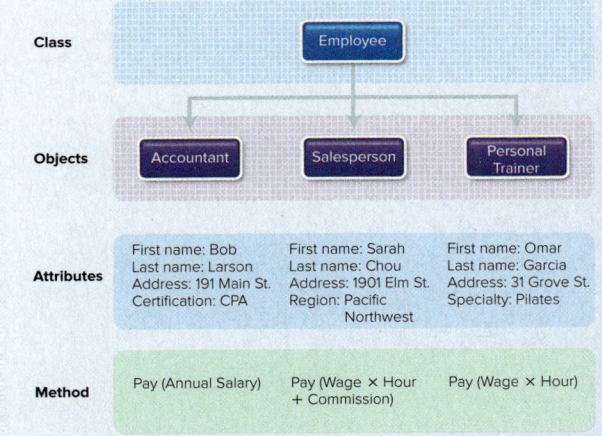

TYPES OF DATABASES

Dow Jones & Company, Inc.

There are four types of databases:

- **Individual (personal computer) database:** Used by one person.
- **Company database:** Stored on central server; accessed by multiple people.
- **Distributed database:** Spread out geographically; accessed by communications links.
- **Commercial databases (information utilities** and **data banks):** Enormous; for particular subjects.

DATABASE USES AND ISSUES

Databases offer a great opportunity for increased productivity; however, security is always a concern.

Strategic Uses

Data warehouses support data mining. **Data mining** is a technique for searching and exploring databases for related information and patterns.

Security

Two important security concerns are illegal use of data and unauthorized access. Most organizations use **firewalls** to protect their internal networks.

CAREERS in IT

Database administrators use database management software to determine the most efficient ways to organize and access a company's data. They are also responsible for database security and system backup. A bachelor's degree in computer science or information systems and technical experience are required. Expected salary range is $47,000 to $111,000.

KEY TERMS

- application generation subsystem
- attribute
- batch processing
- character
- child node
- class
- commercial database
- common data item
- company database
- data
- data administration subsystem
- data bank
- data cube
- data definition subsystem
- data dictionary
- data integrity
- data maintenance
- data manipulation subsystem
- data mining
- data redundancy
- data warehouse
- database
- database administrator (DBA)
- database management system (DBMS)
- database model
- DBMS engine
- distributed database
- entity
- field
- firewall
- hierarchical database
- individual database
- information utility
- key field
- logical view
- many-to-many relationship
- method
- multidimensional database
- network database
- node
- NoSQL
- object
- object-oriented database
- one-to-many relationship
- online processing
- parent node
- personal computer database
- physical view
- pointers
- primary key
- processing rights
- query-by-example
- real-time processing
- record
- relation
- relational database
- schema
- structured query language (SQL)
- table

MULTIPLE CHOICE

Circle the correct answer.

1. View that focuses on the actual format and location of the data.
   - a. logical
   - b. physical
   - c. field
   - d. record

2. A group of related characters is a _____.
   - a. character
   - b. record
   - c. table
   - d. field

3. Type of processing in which data is collected over several hours, days, or even weeks and then processed all at once.
   - a. real-time
   - b. batch
   - c. online
   - d. hierarchical database

4. What is the advantage of an organization using a database instead of allowing individual departments to create and maintain their own data?
   - a. sharing
   - b. less data redundancy
   - c. security
   - d. all of the above

5. Another name for a data dictionary is _____.
   - a. physical view
   - b. logical view
   - c. schema
   - d. DBMS engine

6. Type of database structure where fields or records are structured in nodes that are connected like the branches of an upside-down tree.
   - a. hierarchical
   - b. network
   - c. relational
   - d. multidimensional

7. Type of database structure where the data elements are stored in different tables.
   - a. hierarchical
   - b. network
   - c. relational
   - d. multidimensional

8. One advantage of a multidimensional database is its ability to improve processing _____.
   - a. accuracy
   - b. power
   - c. cost
   - d. speed

9. Object-oriented databases organize data by classes, objects, methods, and _____.
   - a. attributes
   - b. fields
   - c. dimensions
   - d. nodes

10. Type of database that uses communication networks to link data stored in different locations.
    - a. individual
    - b. company
    - c. distributed
    - d. commercial

MATCHING

Match each numbered item with the most closely related lettered item. Write your answers in the spaces provided.

a. many-to-many relationship
b. objects
c. pointer
d. data cube
e. database administrators
f. key field
g. security
h. character
i. DBMS engine
j. data

____ 1. Another name for a multidimensional database.
____ 2. Object-oriented databases organize data by classes, attributes, methods, and _____.
____ 3. The distinctive field(s) found in each record in a database.
____ 4. This is defined as the facts or observations about people, places, things, and events.
____ 5. The connection between parent and child nodes.
____ 6. A network database where each child node may have more than one parent node.
____ 7. This advantage of using databases is achieved by restricting database access to only the information needed by the user.
____ 8. The bridge between the logical and physical views of the data.
____ 9. This is the most basic logical data element such as a single letter, number, or special character.
____ 10. Highly trained computer specialists who interact with the data administration subsystem.

OPEN-ENDED

On a separate sheet of paper, respond to each question or statement.
1. Describe the five logical data groups or categories.
2. What is the difference between batch processing and real-time processing?
3. Identify and define the five parts of DBMS programs.
4. Describe each of the five common database models.
5. What are some of the benefits and limitations of databases? Why is security a concern?

DISCUSSION

Respond to each of the following questions.

Applying Technology: INTERNET MOVIE DATABASE

One popular commercial database is the Internet Movie Database, or IMDb. Connect to its website, explore its content, search for a few movies, and then answer the following questions: (a) What types of information does the IMDb contain? (b) What searches did you try? What were the results? (c) Based on your knowledge of databases, would you expect the IMDb to be relational or hierarchical? Justify your answer.

Writing about Technology: INFORMATION SHARING

Corporations currently collect information about the purchases you make and your personal spending habits. Sometimes corporations will share information to build a more informative profile about you. There have been proposals for legislation to regulate or halt this type of exchange. Consider how you feel about this exchange of information, and then respond to the following: (a) What ethics and privacy concerns are related to corporations sharing personal data? (b) How might the consumer benefit from this? (c) Could this harm the consumer? What could happen if your grocery store shared information about your purchases with your life insurance carrier? (d) What rights do you feel consumers should have with regard to privacy of information collected about them? How should these rights be enforced? Defend your answer.

Writing about Technology: DATABASE SECURITY

Securing the data in a database is typically as important a concern as its design. Research database security on the web, and then respond to the following: (a) Describe a few security risks that databases must be protected against. (b) Describe some steps that can be taken to ensure that a database is secured.

Design Elements: Concept Check icon: Dizzle52/Getty Images

chapter 12
Systems Analysis and Design

Funtap/Shutterstock

Why should I read this chapter?

PopTika/Shutterstock

When an organization designs and implements a new system, jobs are on the line. A well-designed system can make a career; a poor one can destroy a company. Systems analysis and design details the framework for creating new systems. In the future, information systems will rely on cloud-based services, giving easier upgrades, improved security, and higher reliability.

This chapter covers the things you need to know to be prepared for this ever-changing digital world, including:

- The systems life cycle—understand the phases of information systems development and avoid confusion, missteps, and inefficiency.
- Prototyping and rapid development—learn the newest alternatives to the systems life cycle to respond quickly and effectively to unexpected systems design challenges.

Learning Objectives

After you have read this chapter, you should be able to:

1. Describe the six phases of the systems life cycle.
2. Identify information needs and formulate possible solutions.
3. Analyze existing information systems and evaluate the feasibility of alternative systems.
4. Identify, acquire, and test new system software and hardware.
5. Switch from an existing information system to a new one with minimal risk.
6. Explain system audits and periodic evaluations.
7. Describe prototyping and rapid applications development.

Introduction

"Hi, I'm Marie, and I'm a systems analyst. I'd like to talk with you about analyzing and designing information systems for organizations."

Kateryna Onyshchuk/Shutterstock

Most people in an organization are involved with an information system of some kind. For an organization to create and effectively use a system requires considerable thought and effort. Fortunately, there is a six-step process for accomplishing this. It is known as systems analysis and design.

Big organizations can make big mistakes. For example, a large automobile manufacturer once spent $40 billion putting in factory robots and other high technology in its automaking plants. Unfortunately, the manufacturer could never make these new changes work and removed much of the equipment and reinstalled its original production systems. Why did the high-tech production systems fail? The probable reason was that not enough energy was devoted to training its workforce in using the new systems.

The government also can make big mistakes. In one year, the Internal Revenue Service computer system was so overwhelmed it could not deliver tax refunds on time. How did this happen? Despite extensive testing of much of the system, not all testing was completed. Thus, when the new system was phased in, the IRS found it could not process tax returns as quickly as it had hoped.

Both of these examples show the necessity for thorough planning—especially when an organization is trying to implement a new kind of system. Systems analysis and design reduces the chances of such spectacular failures.

To efficiently and effectively use computers, you need to understand the importance of systems analysis and design. You need to be aware of the relationship of an organization's chart to its managerial structure. Additionally, you need to know the six phases of the systems development life cycle: preliminary investigation, systems analysis, systems design, systems development, systems implementation, and systems maintenance.

Systems Analysis and Design

A **system** is a collection of activities and elements organized to accomplish a goal. As we saw in **Chapter 10**, an *information system* is a collection of hardware, software, people, procedures, data, and the Internet. These work together to provide information essential to running an organization. This information helps produce a product or service and, for profit-oriented businesses, derive a profit.

Information about orders received, products shipped, money owed, and so on flows into an organization from the outside. Information about what supplies have been received, which customers have paid their bills, and so on also flows within the organization. To avoid confusion, the flow of information must follow a route that is defined by a set of rules and procedures. However, from time to time, organizations need to change their information systems. Reasons include organizational growth, mergers and acquisitions, new marketing opportunities, revisions in governmental regulations, and availability of new technology.

Systems analysis and design is a six-phase problem-solving procedure for examining and improving an information system. The six phases make up the **systems life cycle**. (See **Figure 12-1**.) The phases are as follows:

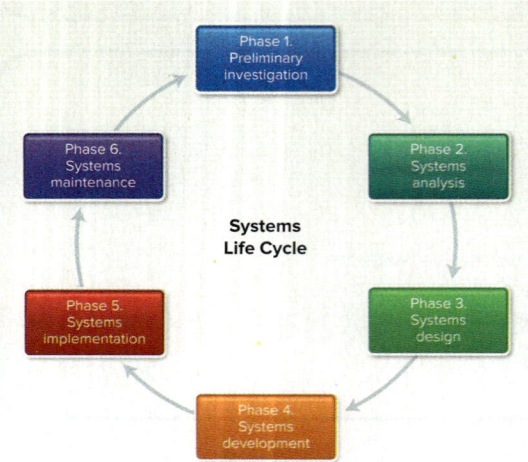

Figure 12-1 The six-phase systems life cycle

1. **Preliminary investigation:** The organization's problems or needs are identified and summarized in a short report.
2. **Systems analysis:** The present system is studied in depth. New requirements are specified and documented.
3. **Systems design:** A new or alternative information system is designed and a design report created.
4. **Systems development:** New hardware and software are acquired, developed, and tested.
5. **Systems implementation:** The new information system is installed, and people are trained to use it.
6. **Systems maintenance:** In this ongoing phase, the system is periodically evaluated and updated as needed.

In organizations, this six-phase systems life cycle is used by computer professionals known as **systems analysts**. These people study an organization's systems to determine what actions to take and how to use computer technology to assist them.

As an end user, working alone or with a systems analyst, it is important that you understand how the systems life cycle works. In fact, you may *have* to use the procedure. Every career is affected by information systems, and understanding how they are created and used will allow you to work smarter and avoid common mistakes.

There are many misconceptions about the life cycle of an information system. Those unfamiliar with the life cycle model are likely to consider the design, development, and implementation phases as the most important. However, many information systems fail because of a lack of investigation and analysis. Further, many information systems go over budget or become obsolete too quickly because of a failure to recognize the importance of system maintenance. It can be a thankless task, insisting that design and development be delayed to ensure that investigation and analysis be done thoroughly. Further, inexperienced developers, in an effort to save money, will look to reduce or underestimate maintenance expenses. However, the long-term success of an information system is dependent on understanding the value and necessity of each of the phases.

Learning the six steps described in this chapter will raise your computer efficiency and effectiveness. It also will give you skills to solve a wide range of problems. These skills can make you more valuable to an organization.

concept check

- What is a system?
- Name the six phases of the systems life cycle.
- What do systems analysts do?

Phase 1: Preliminary Investigation

The first phase of the systems life cycle is a **preliminary investigation** of a proposed project to determine the need for a new information system. This usually is requested by an end user or a manager who wants something done that is not presently being done. For example, suppose you work for Advantage Advertising, a fast-growing advertising agency. Advantage Advertising produces a variety of different ads for a wide range of different clients. The agency employs both regular staff workers and on-call freelancers. One of your responsibilities is keeping track of the work performed for each client and the employees who performed the work. In addition, you are responsible for tabulating the final bill for each project.

How do you figure out how to charge which clients for which work done by which employees? This kind of problem is common to many service organizations (such as lawyers' and contractors' offices). Indeed, it is a problem in any organization where people charge for their time and clients need proof of hours worked.

In Phase 1, the systems analyst—or the end user—is concerned with three tasks: (1) briefly defining the problem, (2) suggesting alternative solutions, and (3) preparing a short report. (See **Figure 12-2**.) This report will help management decide whether to pursue the project further. (If you are an end user, you may not produce a written report. Rather, you might report your findings directly to your supervisor.)

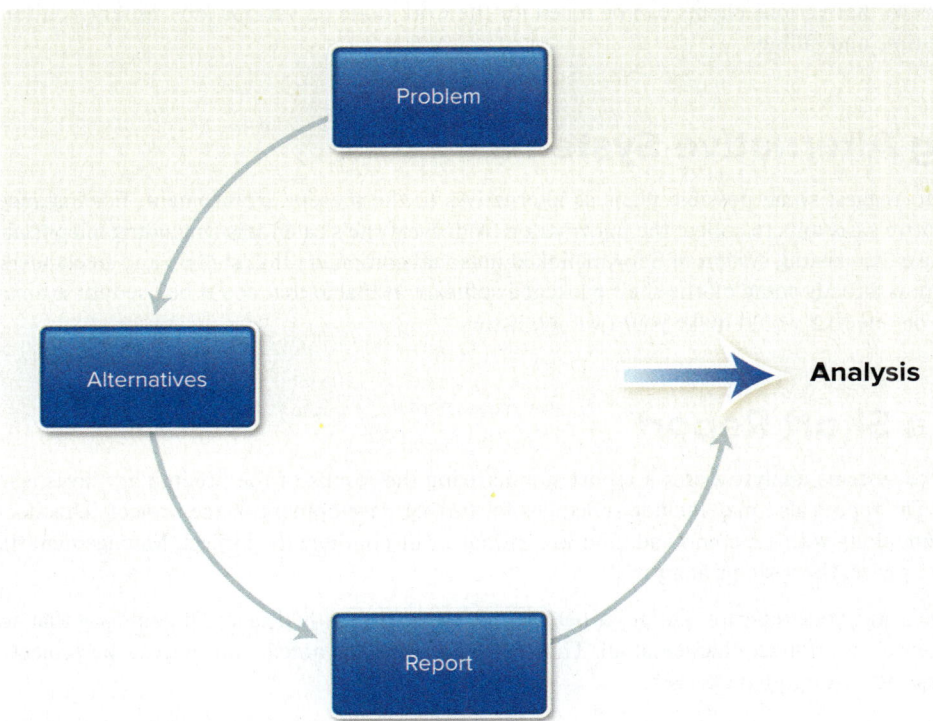

Figure 12-2   Phase 1: Preliminary investigation

Defining the Problem

Defining the problem means examining whatever current information system is in use. Determining what information is needed, by whom, when, and why is accomplished by interviewing and making observations. If the information system is large, this survey is done by a systems analyst. If the system is small, the survey can be done by the end user.

Figure 12-3 One step in defining problems with the current system is to interview executives

fizkes/Shutterstock

For example, suppose Advantage Advertising account executives, copywriters, and graphic artists currently record the time spent on different jobs on their desk calendars. (Examples might be "Client A, telephone conference, 15 minutes"; "Client B, design layout, 2 hours.") After interviewing several account executives and listening to their frustrations, it becomes clear that the approach is somewhat disorganized. (See **Figure 12-3**.) Written calendar entries are too unprofessional to be shown to clients. Moreover, a large job often has many people working on it. It is difficult to pull together all their notations to make up a bill for the client. Some freelancers work at home, and their time slips are not readily available. These matters constitute a statement of the problem: The company has a manual time-and-billing system that is slow and difficult to use.

As an end user, you might experience difficulties with this system yourself. You're in someone else's office, and a telephone call comes in for you from a client. Your desk calendar is back in your own office. You have two choices. You can always carry your calendar with you, or you can remember to note the time you spent on various tasks when you return to your office. The administrative assistant reporting to the account executive is continually after you (and everyone else at Advantage) to provide photocopies of your calendar. This is so that various clients can be billed for the work done on various jobs. Surely, you think, there must be a better way to handle time and billing.

Suggesting Alternative Systems

This step is simply to suggest some possible plans as alternatives to the present arrangement. For instance, Advantage could hire more administrative assistants to collect the information from everyone's calendars (including telephoning those working at home). Or it could use the existing system of network-linked personal computers that staffers and freelancers presently use. Perhaps, you think, there is already some off-the-shelf packaged software available that could be used for a time-and-billing system. At least there might be one that would make your own job easier.

Preparing a Short Report

For large projects, the systems analyst writes a report summarizing the results of the preliminary investigation and suggesting alternative systems. The report also may include schedules for further development of the project. This document is presented to higher management, along with a recommendation to continue or discontinue the project. Management then decides whether to finance the second phase, the systems analysis.

For Advantage Advertising, your report might point out that billing is frequently delayed. It could say that some tasks may even "slip through the cracks" and not get charged at all. Thus, as the analyst has noted, you suggest the project might pay for itself merely by eliminating lost or forgotten charges.

concept check

- What is the purpose of the preliminary investigation phase?
- What are the three tasks the systems analyst is concerned with during this phase?
- Who determines whether to finance the second phase?

Phase 2: Systems Analysis

In Phase 2, **systems analysis**, data is collected about the present system. This data is then analyzed, and new requirements are determined. We are not concerned with a new design here, only with determining the *requirements* for a new system. Systems analysis is concerned with gathering and analyzing the data. This usually is completed by documenting the analysis in a report. (See **Figure 12-4**.)

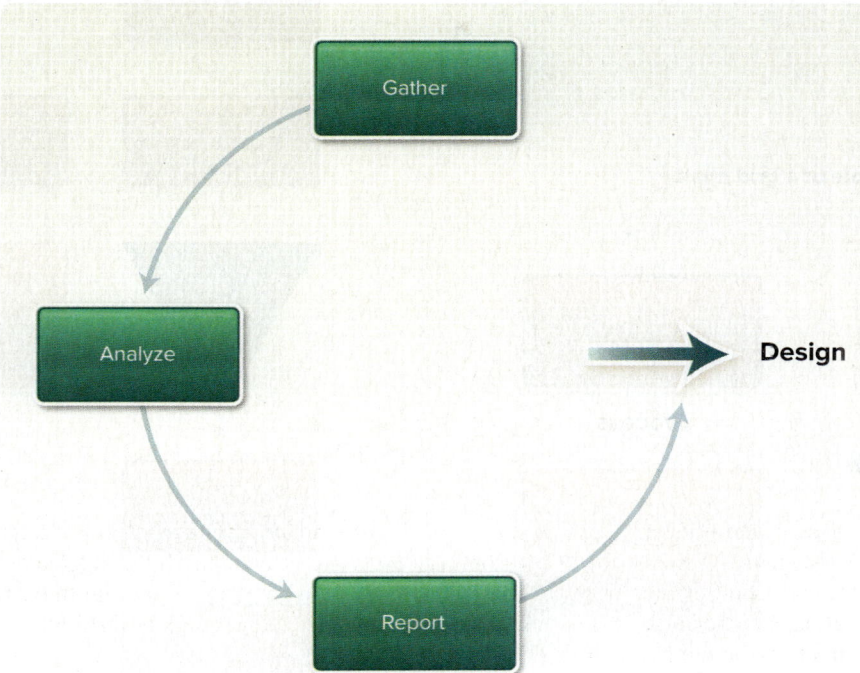

Figure 12-4 Phase 2: Systems analysis

Gathering Data

When gathering data, the systems analyst—or the end user doing systems analysis—expands on the data gathered during Phase 1. He or she adds details about how the current system works. Data is obtained from observation and interviews. In addition, data may be obtained from questionnaires given to people using the system. Data also is obtained from studying documents that describe the formal lines of authority and standard operating procedures. One such document is the **organization chart**, which shows levels of management as well as formal lines of authority.

Analyzing the Data

In the data analysis step, the idea is to learn how information currently flows and to pinpoint why it isn't flowing appropriately. The whole point of this step is to apply logic to the existing arrangement to see how workable it is. Many times, the current system is not operating correctly because prescribed procedures are not being followed. That is, the system may not really need to be redesigned. Rather, the people in it may need to be shown how to follow correct procedures.

Many different tools are available to assist systems analysts and end users in the analysis phase. Some of the most important are the top-down analysis method, grid charts, system flowcharts, data flow diagrams, and automated design tools.

- **Top-down analysis method:** The **top-down analysis method** is used to identify the top-level components of a complex system. Each component is then broken down into smaller and smaller components. This approach makes each component easier to analyze and deal with.

  For instance, the systems analyst might look at the present kind of bill submitted to a client for a complex advertising campaign. The analyst might note the categories of costs—employee salaries, telephone and mailing charges, travel, supplies, and so on.

Figure 12-5 Example of a grid chart

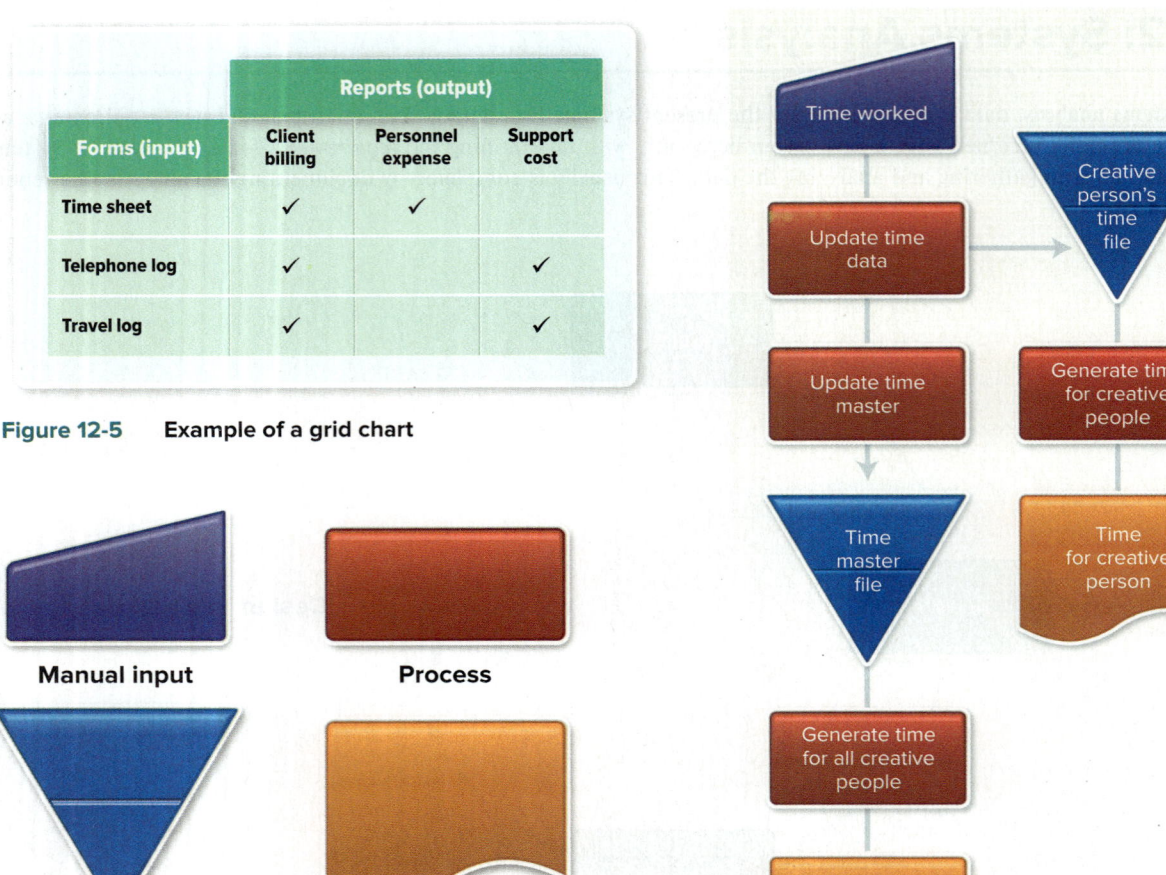

Figure 12-7 System flowchart symbols

Figure 12-6 Example of a system flowchart

- **Grid charts:** A **grid chart** shows the relationship between input and output documents. An example is shown in **Figure 12-5** that indicates the relationship between the data input and the outputs.

  For instance, a time sheet is one of many inputs that produces a particular report, such as a client's bill. Other inputs might be forms having to do with telephone conferences and travel expenses. On a grid sheet, rows represent inputs, such as time sheet forms. Columns represent output documents, such as different clients' bills. A check mark at the intersection of a row and column means that the input document is used to create the output document.

- **System flowcharts: System flowcharts** show the flow of input data to processing and finally to output, or distribution of information. An example of a system flowchart keeping track of time for advertising "creative people" is shown in **Figure 12-6**. The explanation of the symbols used appears in **Figure 12-7**. Note that this describes the present manual, or noncomputerized, system. (A system flowchart is not the same as a program flowchart, which is very detailed. Program flowcharts are discussed in **Chapter 13**.)

- **Data flow diagrams: Data flow diagrams** show the data or information flow within an information system. The data is traced from its origin through processing, storage, and output. An example of a data flow diagram is shown in **Figure 12-8**. The explanation of the symbols used appears in **Figure 12-9**.

- **Automated design tools: Automated design tools** are software packages that evaluate hardware and software alternatives according to requirements given by the systems analyst. They are also called **computer-aided software engineering (CASE) tools**. These tools are not limited to systems analysis. They are used in systems design and development as well. CASE tools relieve the systems analysts of many repetitive tasks, develop clear documentation, and, for larger projects, coordinate team member activities.

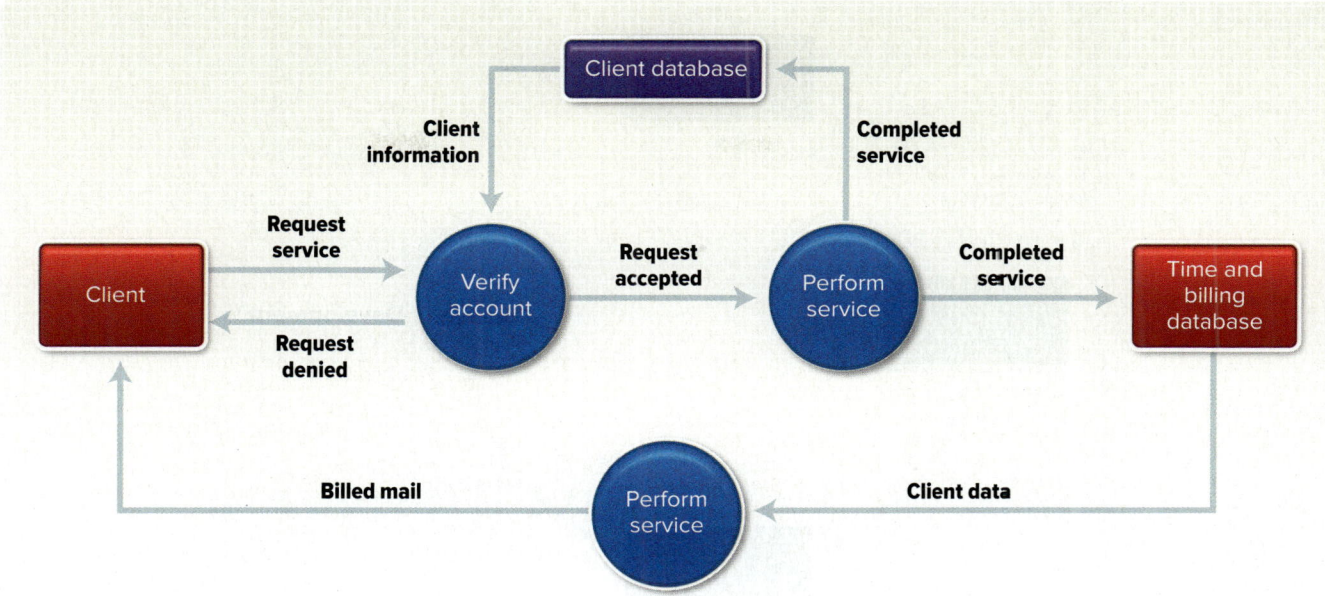

Figure 12-8 Example of a data flow diagram

Documenting Systems Analysis

In larger organizations, the systems analysis stage is typically documented in a report for higher management. The **systems analysis report** describes the current information system, the requirements for a new system, and a possible development schedule. For example, at Advantage Advertising, the system flowcharts show the present flow of information in a manual time-and-billing system. Some boxes in the system flowchart might be replaced with symbols showing where a computerized information system could work better.

Management studies the report and decides whether to continue with the project. Let us assume your boss and higher management have decided to continue. You now move on to Phase 3, systems design.

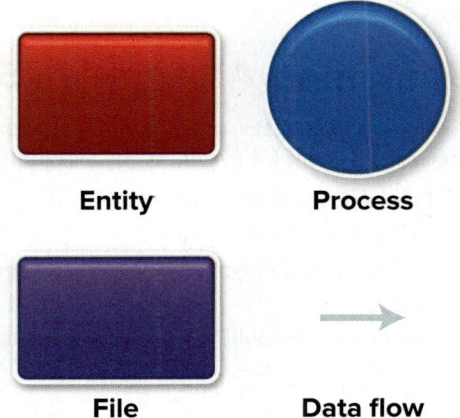

Figure 12-9 Data flow diagram symbols

- ☐ What is the purpose of the analysis phase?
- ☐ List and describe five important analysis tools.
- ☐ What is a systems analysis report?

Phase 3: Systems Design

Phase 3 is **systems design**. It consists of three tasks: (1) designing alternative systems, (2) selecting the best system, and (3) writing a systems design report. (See **Figure 12-10**.)

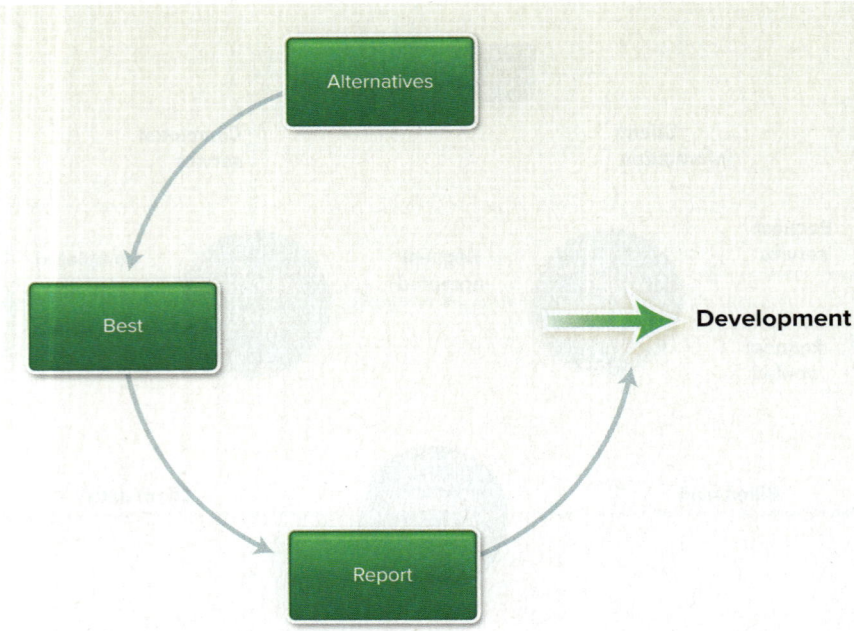

Figure 12-10   Phase 3: Systems design

Designing Alternative Systems

In almost all instances, more than one design can be developed to meet the information needs. Systems designers evaluate each alternative system for feasibility. By feasibility we mean three things:

- **Economic feasibility:** Will the costs of the new system be justified by the benefits it promises? How long will it take for the new system to pay for itself?
- **Technical feasibility:** Are reliable hardware, software, and training available to make the system work? If not, can they be obtained?
- **Operational feasibility:** Can the system actually be made to operate in the organization, or will people—employees, managers, clients—resist it?

Selecting the Best System

When choosing the best design, managers must consider these four questions: (1) Will the system work with the organization's overall information system? (2) Will the system be flexible enough so it can be modified in the future? (3) Can it be made secure against unauthorized use? (4) Are the benefits worth the costs?

For example, one aspect you have to consider at Advantage Advertising is security. Should freelancers and outside vendors enter data directly into a computerized time-and-billing system, or should they keep submitting time sheets manually? In allowing these outside people to input information directly, are you also allowing them access to files they should not see? Do these files contain confidential information, perhaps information of value to rival advertising agencies?

Writing the Systems Design Report

The **systems design report** is prepared for higher management and describes the alternative designs. It presents the costs versus the benefits and outlines the effect of alternative designs on the organization. It usually concludes by recommending one of the alternatives.

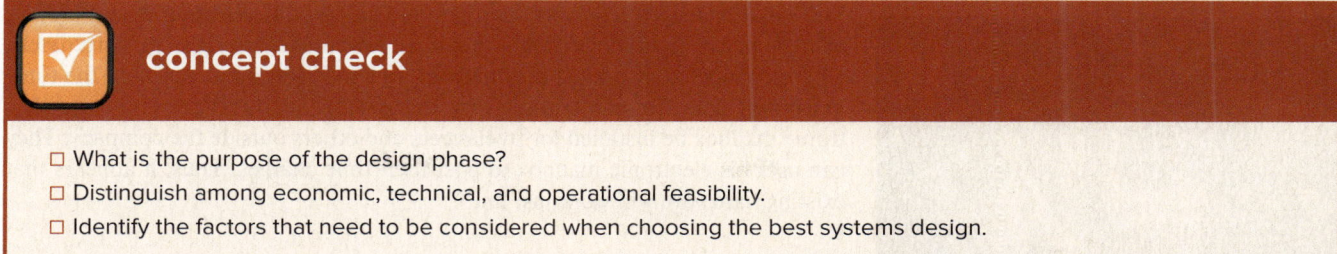

- What is the purpose of the design phase?
- Distinguish among economic, technical, and operational feasibility.
- Identify the factors that need to be considered when choosing the best systems design.

Phase 4: Systems Development

Phase 4 is **systems development**. It has three steps: (1) acquiring software, (2) acquiring hardware, and (3) testing the new system. (See **Figure 12-11**.)

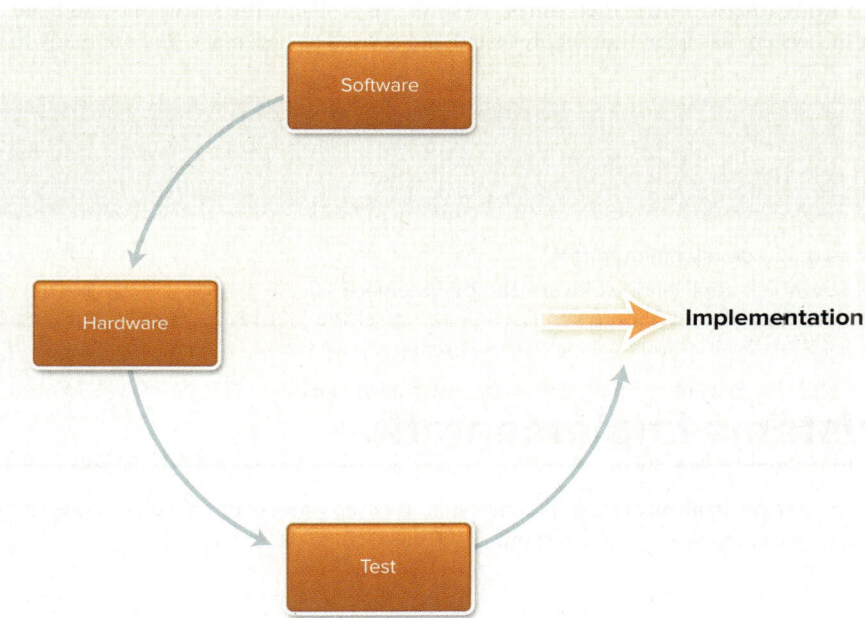

Figure 12-11 Phase 4: Systems development

Acquiring Software

Application software for the new information system can be obtained in two ways. It can be purchased as off-the-shelf packaged software and possibly modified, or it can be custom-designed. Custom software is expensive but precisely meets the needs of the organization. This must be weighed against the option of off-the-shelf software, which is less expensive and less precise. If any of the software is being specially created, the programming steps we will outline in **Chapter 13** should be followed.

With the systems analyst's help, you have looked at time-and-billing packaged software designed for service organizations. Unfortunately, you find that none of the packaged software will do. Most of the packages seem to work well for one person (you). However, none seems to be designed for many people working together. It appears, then, that software will have to be custom-designed. (We discuss the process of developing software in **Chapter 13**, on programming.)

Acquiring Hardware

Some new systems may not require new computer equipment, but others will. The equipment needed and the places where they are to be installed must be determined. This is a very critical area. Switching or upgrading equipment can be a tremendously expensive proposition. Will a personal computer system be sufficient as a company grows? Are networks expandable? Will people have to undergo costly training?

Figure 12-12   To test a system, sample data is entered and problems are resolved
PeopleImages.com - Yuri A/Shutterstock

The systems analyst tells you that there are several different makes and models of personal computers currently in use at Advantage Advertising. Fortunately, all are connected by a local area network to a file server that can hold the time-and-billing data. To maintain security, the systems analyst suggests that an electronic mailbox be installed for freelancers and others outside the company. They can use this electronic mailbox to post their time charges. Thus, it appears that existing hardware will work just fine.

Testing the New System

After the software and equipment have been installed, the system should be tested. Sample data is fed into the system. The processed information is then evaluated to see whether the results are correct. Testing may take several months if the new system is complex.

For this step, you ask some people in Creative Services to test the system. (See **Figure 12-12**.) You observe that some of the people have problems knowing where to enter their times. To solve the problem, the software is modified to display an improved user entry screen. After the system has been thoroughly tested and revised as necessary, you are ready to put it into use.

 concept check

☐ What is the purpose of the development phase?
☐ What are the ways by which application software can be obtained?

Phase 5: Systems Implementation

Another name for Phase 5, **systems implementation**, is **conversion**. It is the process of changing—converting—from the old system to the new one and training people to use the new system.

Types of Conversion

There are four approaches to conversion: *direct, parallel, pilot,* and *phased.*

- In the **direct approach**, the conversion is done simply by abandoning the old and starting up the new. This can be risky. If anything is still wrong with the new system, the old system is no longer available to fall back on.
  The direct approach is not recommended precisely because it is so risky. Problems, big or small, invariably crop up in a new system. In a large system, a problem might just mean catastrophe.
- In the **parallel approach**, old and new systems are operated side by side until the new one proves to be reliable.
  This approach is low risk. If the new system fails, the organization can just switch to the old system to keep going. However, keeping enough equipment and people active to manage two systems at the same time can be very expensive. Thus, the parallel approach is used only in cases in which the cost of failure or of interrupted operation is great.
- In the **pilot approach**, the new system is tried out in only one part of the organization. Once the system is working smoothly in that part, it is implemented throughout the rest of the organization.
  The pilot approach is certainly less expensive than the parallel approach. It also is somewhat riskier. However, the risks can be controlled because problems will be confined to only certain areas of the organization. Difficulties will not affect the entire organization.
- In the **phased approach**, the new system is implemented gradually over a period of time. The entire implementation process is broken down into parts or phases. Implementation begins with the first phase, and once it is successfully implemented, the second phase begins. This process continues until all phases are operating smoothly. Typically, this is an expensive proposition because the implementation is done slowly. However, it is certainly one of the least risky approaches.

In general, the pilot and phased approaches are the favored methods. Pilot is preferred when there are many people in an organization performing similar operations—for instance, all salesclerks in a department store. Phased is more appropriate for organizations in which people are performing different operations.

You and the systems analyst, with top management support, have decided on a pilot implementation. This approach was selected in part based on cost and the availability of a representative group of users. The Creative Services department previously tested the system and has expressed enthusiastic support for it. A group from this department will pilot the implementation of the time-and-billing system.

Training

Training people is important, of course. Unfortunately, it is one of the most commonly overlooked activities. Some people may begin training early, even before the equipment is delivered, so that they can adjust more easily. In some cases, a professional software trainer may be brought in to show people how to operate the system. However, at Advantage Advertising, the time-and-billing software is simple enough that the systems analyst can act as the trainer.

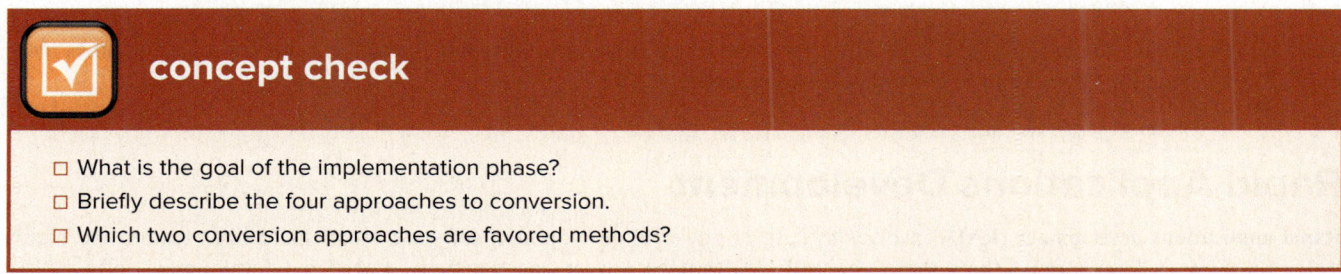

- What is the goal of the implementation phase?
- Briefly describe the four approaches to conversion.
- Which two conversion approaches are favored methods?

Phase 6: Systems Maintenance

After implementation comes **systems maintenance**, the last step in the systems life cycle. This phase is a very important, ongoing activity. Most organizations spend more time and money on this phase than on any of the others. Maintenance has two parts: a *systems audit* and a *periodic evaluation*.

In the **systems audit**, the system's performance is compared to the original design specifications. This is to determine whether the new procedures are actually furthering productivity. If they are not, some redesign may be necessary.

After the systems audit, the new information system is further modified, if necessary. All systems should be evaluated from time to time to determine whether they are meeting the goals and providing the service they are supposed to.

The six-step systems life cycle is summarized in **Figure 12-13**.

Phase	Activity
1. Preliminary investigation	Define problem, suggest alternatives, prepare short report
2. Systems analysis	Gather data, analyze data, document
3. Systems design	Design alternatives, select best alternative, write report
4. Systems development	Develop software, acquire hardware, test system
5. Systems implementation	Convert, train
6. Systems maintenance	Perform systems audit, evaluate periodically

Figure 12-13 Summary of systems life cycle

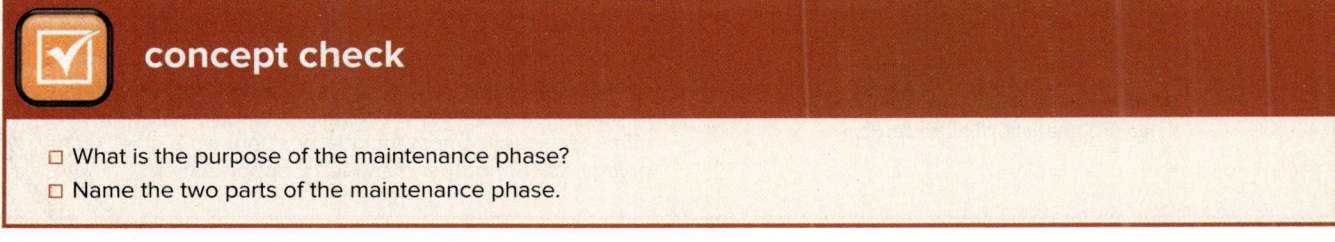

- What is the purpose of the maintenance phase?
- Name the two parts of the maintenance phase.

Prototyping and Rapid Applications Development

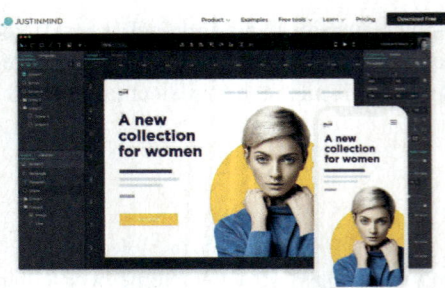

Figure 12-14 Micro Focus offers prototyping software
Justinmind

Is it necessary to follow every phase of the systems life cycle? It may be desirable, but often there is no time to do so. For instance, hardware may change so fast that there is no opportunity for evaluation, design, and testing as just described. Two alternative approaches that require much less time are *prototyping* and *rapid applications development*.

Prototyping

Prototyping means to build a *model* or *prototype* that can be modified before the actual system is installed. For instance, the systems analyst for Advantage Advertising might develop a proposed or prototype menu as a possible screen display for the time-and-billing system. Users would try it out and provide feedback to the systems analyst. The systems analyst would revise the prototype until the users felt it was ready to put into place. Typically, the development time for prototyping is shorter; however, it is sometimes more difficult to manage the project and to control costs. (See **Figure 12-14**.)

Rapid Applications Development

Rapid applications development (RAD) involves the use of powerful development software, small specialized teams, and highly trained personnel. For example, the systems analyst for Advantage Advertising would use specialized development software like CASE, form small teams consisting of select users and managers, and obtain assistance from other highly qualified analysts. Although the resulting time-and-billing system would likely cost more, the development time would be shorter and the quality of the completed system would be better.

concept check

- What is prototyping?
- What is RAD?
- What is the advantage of these two approaches over the systems life cycle approach?

Careers in IT

"Now that you have learned about systems analysis and design, let me tell you about my career as a systems analyst."

Kateryna Onyshchuk/Shutterstock

A **systems analyst** follows the steps described in the systems life cycle. Analysts plan and design new systems or reorganize a company's computer resources to best utilize them. Analysts follow the systems life cycle through all of its steps: preliminary investigation, analysis, design, development, implementation, and maintenance.

Systems analyst positions normally require either an advanced associate's degree or a bachelor's degree in computer science or information systems and technical experience. Internships and prior experience with the latest technology are a considerable advantage for those seeking jobs in this industry. Systems analysts can expect to earn an annual salary of $48,000 to $95,000. Opportunities for advancement include positions as a chief technology officer or other managerial opportunities.

A LOOK TO THE FUTURE

The Challenge of Keeping Pace

Blue Planet Studio/Shutterstock

Have you noticed the speed with which new (or competing) products and services are being released? Does your favorite website change often to keep up with its competitors? Most observers firmly believe that the pace of business is accelerating. The time to develop a product and bring it to market, in many cases, is now months rather than years. Internet technologies, in particular, have provided tools to support the rapid introduction of new products and services. Technology has been making better business tools and will continue to evolve to improve our lives as we look to the future.

To stay competitive, corporations must integrate these new technologies into their existing ways of doing business. In many cases, the traditional systems life cycle approach takes too long—sometimes years—to develop a system. Many organizations are responding by aggressively implementing prototyping and RAD. Others are enlisting the services of outside consulting groups that specialize in systems development. However, many experts believe that the future of life cycle management lies in relying on the cloud—businesses will turn to companies that offer both processing and software as a service, rather than hosting these systems on their own.

In the future, many companies will no longer have large servers and database systems under their own roof. They will instead pay a monthly fee to a company, such as Amazon, that has large data centers that are accessible via the Internet. These data centers offer security and reliability, and they can grow, or scale, based on the needs of the business. The systems analyst of the future will not have to worry about the hardware requirements of a new piece of software or database management system. The implementation of new systems will be much easier, for both the business and its customers. All hardware upgrades will now be managed by the company offering the cloud service, and the software will be hosted there. Of course, all this requires a good communications infrastructure—one that is being improved each year by telecommunications companies.

What do you think about moving so many aspects of a system to the cloud? Is there a danger in trusting another company with your business's data? Do you think cloud computing will enable a business to release reliable products more quickly and at a lower cost?

VISUAL SUMMARY | Systems Analysis and Design

SYSTEMS ANALYSIS AND DESIGN

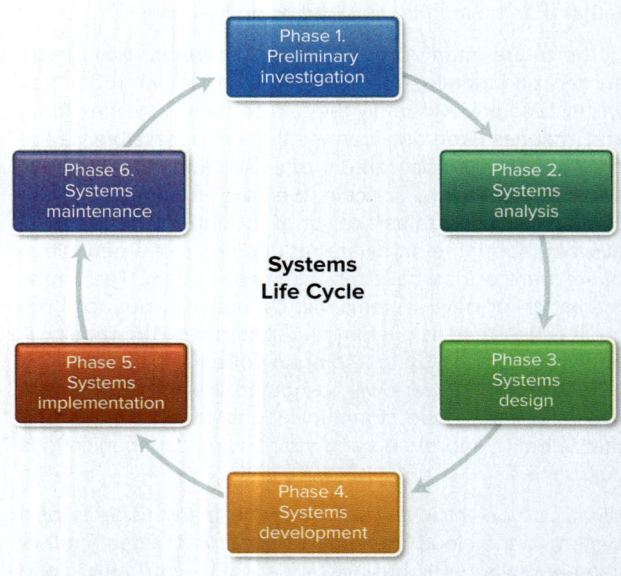

A **system** is a collection of activities and elements organized to accomplish a goal. **Systems analysis and design** is a six-phase problem-solving procedure that makes up the **systems life cycle**. The phases are

- Preliminary investigation—identifying organization's problems or needs and summarizing in short report.
- Systems analysis—studying present system in depth, specifying new requirements, and documenting findings.
- Systems design—designing new or alternative system to meet new requirements and creating a design report.
- Systems development—acquiring, developing, and testing needed hardware and software.
- Systems implementation—installing new system and training people.
- Systems maintenance—periodically evaluating and updating system as needed.

Systems analysts are the computer professionals who typically follow the six-phase systems life cycle.

PHASE 1: PRELIMINARY INVESTIGATION

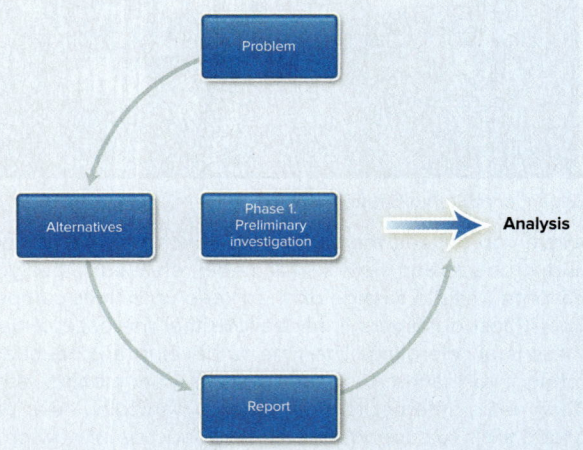

The **preliminary investigation** determines the need for a new information system. It is typically requested by an end user or a manager. Three tasks of this phase are defining the problem, suggesting alternative systems, and preparing a short report.

Defining the Problem

The current information system is examined to determine who needs what information, when the information is needed, and why it is needed.

If the existing information system is large, then a *systems analyst* conducts the survey. Otherwise, the end user conducts the survey.

Suggesting Alternative Systems

Some possible alternative systems are suggested. Based on interviews and observations made in defining the problem, alternative information systems are identified.

Preparing a Short Report

To document and communicate the findings of Phase 1, preliminary investigation, a short report is prepared and presented to management.

To efficiently and effectively use computers, you need to understand the importance of systems analysis and design. You need to know the six phases of the systems development life cycle: preliminary investigation, analysis, design, development, implementation, and maintenance. Additionally, you need to understand prototyping and RAD.

PHASE 2: SYSTEMS ANALYSIS

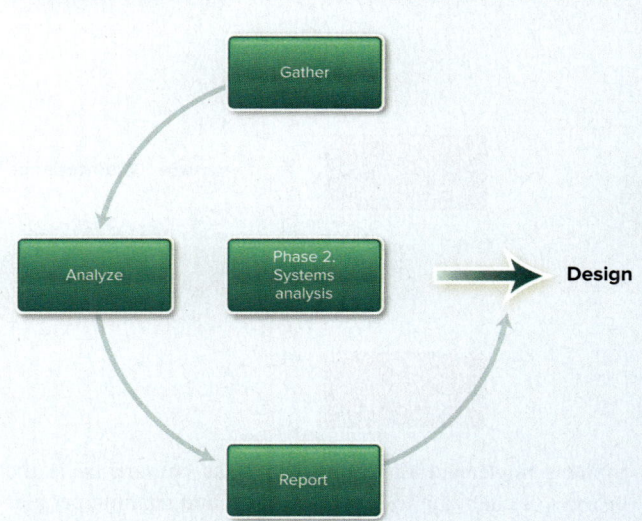

In **systems analysis**, data is collected about the present system. The focus is on determining the requirements for a new system. Three tasks of this phase are gathering data, analyzing the data, and documenting the analysis.

Gathering Data

Data is gathered by observation, interviews, questionnaires, and looking at documents. One helpful document is the **organization chart**, which shows a company's functions and levels of management.

Analyzing the Data

There are several tools for the analysis of data, including **top-down analysis**, **grid charts**, and **system flowcharts**.

Documenting Systems Analysis

To document and communicate the findings of Phase 2, a **systems analysis report** is prepared for higher management.

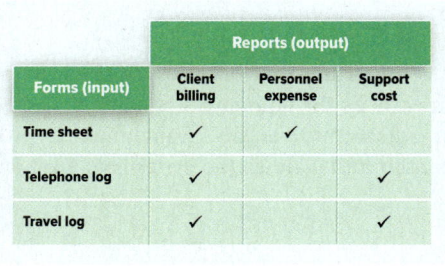

PHASE 3: SYSTEMS DESIGN

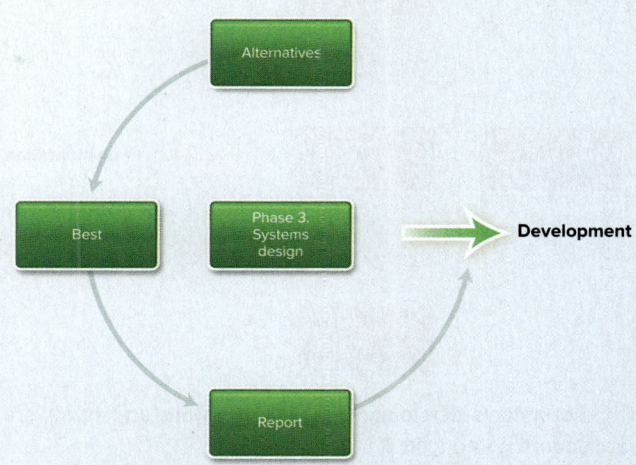

In the **systems design** phase, a new or alternative information system is designed. This phase consists of three tasks:

Designing Alternative Systems

Alternative information systems are designed. Each alternative is evaluated for

- **Economic feasibility**—cost versus benefits; time for the system to pay for itself.
- **Technical feasibility**—hardware and software reliability; available training.
- **Operational feasibility**—will the system work within the organization?

Selecting the Best System

Four questions should be considered when selecting the best system:

- Will the system fit into an overall information system?
- Will the system be flexible enough to be modified as needed in the future?
- Will it be secure against unauthorized use?
- Will the system's benefits exceed its costs?

Writing the Systems Design Report

To document and communicate the findings of Phase 3, a **systems design report** is prepared for higher management.

PHASE 4: SYSTEMS DEVELOPMENT

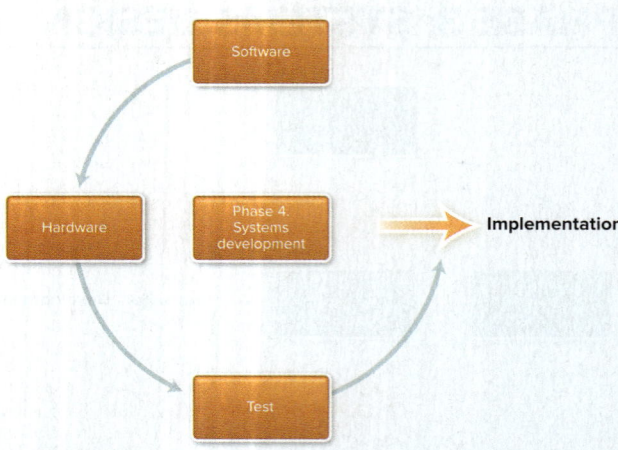

In the **systems development** phase, software and hardware are acquired and tested.

Acquiring Software

Two ways to acquire software are purchasing off-the-shelf packaged software and designing custom programs.

Acquiring Hardware

Acquiring hardware involves consideration for future company growth, existing networks, communication capabilities, and training.

Testing the New System

Using sample data, the new system is tested. This step can take several months for a complex system.

PeopleImages.com - Yuri A/Shutterstock

PHASE 5: SYSTEMS IMPLEMENTATION

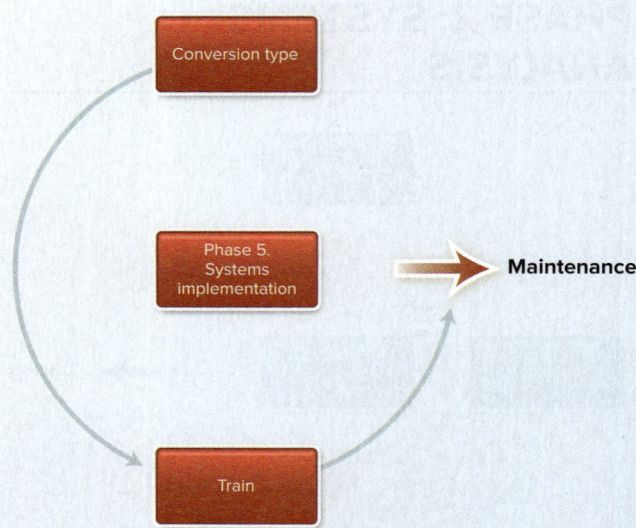

Systems implementation, also known as **conversion**, is the process of changing to the new system and training people.

Types of Conversion

Four ways to convert are direct, parallel, pilot, and phased approaches.

- **Direct approach**—abandoning the old system and starting up the new system; can be very risky and not recommended.
- **Parallel approach**—running the old and new side by side until the new system proves its worth; very low risk; however, very expensive; not generally recommended.
- **Pilot approach**—converting only one part of the organization to the new system until the new system proves its worth; less expensive but riskier than parallel conversion; recommended for situations with many people performing similar operations.
- **Phased approach**—gradually implementing the new system to the entire organization; less risky but more expensive than parallel conversion; recommended for situations with many people performing different operations.

Training

Training is important, but often overlooked. Some people may train early as the equipment is being delivered so that they can adjust more easily. Sometimes a professional trainer is used; other times the systems analyst acts as the trainer.

PHASE 6: SYSTEMS MAINTENANCE

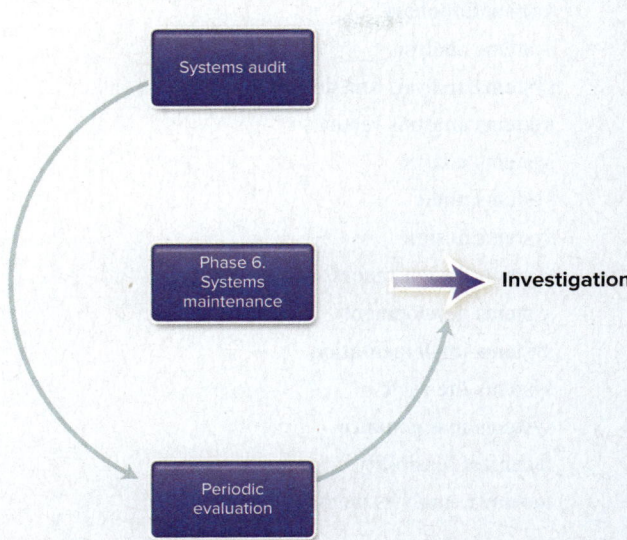

Systems maintenance consists of a systems audit followed by periodic evaluation.

Systems Audit

Once the system is operational, the systems analyst performs a **systems audit** by comparing the new system to its original design specifications.

Periodic Evaluation

The new system is periodically evaluated to ensure that it is operating efficiently.

Phase	Activity
1. Preliminary investigation	Define problem, suggest alternatives, prepare short report
2. Systems analysis	Gather data, analyze data, document
3. Systems design	Design alternatives, select best alternative, write report
4. Systems development	Develop software, acquire hardware, test system
5. Systems implementation	Convert, train
6. Systems maintenance	Perform systems audit, evaluate periodically

PROTOTYPING AND RAD

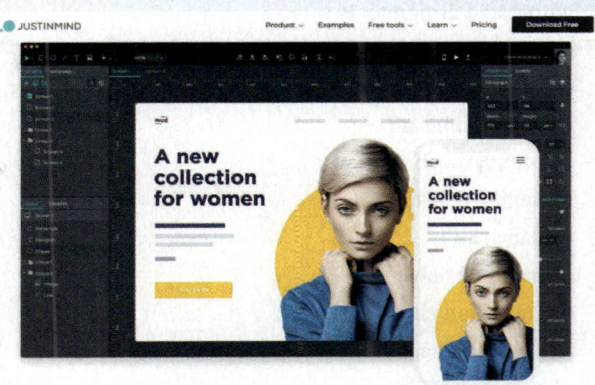

Justinmind

Due to time pressures, it is not always feasible to follow every phase of the systems life cycle. Two alternatives that require less time are *prototyping* and *RAD*.

Prototyping

Prototyping means to build a model or prototype that can be modified before the actual system is installed. Typically, the development time for prototyping is shorter; however, it can be more difficult to manage the project and to control costs.

Rapid Applications Development

Rapid applications development (RAD) uses powerful development software, small specialized teams, and highly trained personnel. Typically, the development costs more. However, the time is much less and the quality is often better.

CAREERS in IT

Systems analysts plan and design new systems or reorganize a company's computer resources to better utilize them. They follow the systems life cycle through all its steps. Either an advanced associate's degree or a bachelor's degree in computer science or information systems and technical experience are required. Expected salary range is $48,000 to $95,000.

KEY TERMS

- automated design tools
- computer-aided software engineering (CASE) tools
- conversion
- data flow diagram
- direct approach
- economic feasibility
- grid chart
- operational feasibility
- organization chart
- parallel approach
- phased approach
- pilot approach
- preliminary investigation
- prototyping
- rapid applications development (RAD)
- system
- system flowchart
- systems analysis
- systems analysis and design
- systems analysis report
- systems analyst
- systems audit
- systems design
- systems design report
- systems development
- systems implementation
- systems life cycle
- systems maintenance
- technical feasibility
- top-down analysis method

MULTIPLE CHOICE

Circle the correct answer.
1. This is examined and improved through systems analysis.
   - a. information technologies
   - b. software analysis
   - c. software implementation
   - d. information systems
2. Systems life cycle phase that studies the present system in depth.
   - a. analysis
   - b. design
   - c. implementation
   - d. conversion
3. Systems analysis involves suggesting alternative _____.
   - a. implementations
   - b. facts
   - c. solutions
   - d. software
4. The last and ongoing phase of the systems life cycle is systems _____.
   - a. maintenance
   - b. implementation
   - c. analysis
   - d. information
5. The document that shows the levels of management and formal lines of authority is a(n) _____.
   - a. systems tree
   - b. organization chart
   - c. authority hierarchy
   - d. direct conversion graph
6. This phase begins with designing alternative systems.
   - a. systems design
   - b. implementation
   - c. maintenance
   - d. investigation
7. Another name for systems implementation.
   - a. reveal
   - b. conversion
   - c. pre-maintenance
   - d. piloting
8. The phase in which the old system is replaced and training begins.
   - a. maintenance
   - b. implementation
   - c. analysis
   - d. information
9. The four approaches to conversion are parallel, pilot, phased, and _____.
   - a. implementation
   - b. preliminary
   - c. direct
   - d. designed
10. The approach in which the new system is implemented gradually over a period of time.
    - a. parallel
    - b. phased
    - c. pilot
    - d. designed

MATCHING

Match each numbered item with the most closely related lettered item. Write your answers in the spaces provided.

a. CASE
b. data
c. grid chart
d. operational
e. pilot
f. preliminary investigation
g. RAD
h. systems implementation
i. systems analysis
j. systems design

_____ 1. This systems analysis tool shows the relationship between input and output documents.
_____ 2. This phase involves installing the new system and training people.
_____ 3. This approach to conversion begins by trying out a new system in only one part of an organization.
_____ 4. These tools relieve the systems analysts of many repetitive tasks; develop clear documentation; and, for larger projects, coordinate team member activities.
_____ 5. This phase is concerned with economic, technical, and operational feasibility.
_____ 6. An information system is a collection of hardware, software, people, procedures, the Internet, and _____.
_____ 7. The first phase in the systems life cycle.
_____ 8. An alternative to the systems life cycle approach using powerful development software, small specialized teams, and highly trained personnel.
_____ 9. This phase is concerned about determining system requirements, not with design.
_____ 10. This type of feasibility evaluates whether the people within the organization will embrace or resist a new system.

OPEN-ENDED

On a separate sheet of paper, respond to each question or statement.

1. What is a system? What are the six phases of the systems life cycle? Why do corporations undergo this process?
2. What are the tools used in the analysis phase? What is top-down analysis? How is it used?
3. Describe each type of system conversion. Which is the most commonly used?
4. What is systems maintenance? When does it occur?
5. Explain prototyping and RAD. When might they be used by corporations?

DISCUSSION

Respond to each of the following questions.

 Applying Technology: SYSTEMS ANALYSIS SOFTWARE

Several companies specialize in systems analysis support software. Using the Internet, search for and connect to one of these companies. Then answer the following: (a) Describe the products designed to enhance systems analysis. (b) For each product you described, list the phase or phases of the systems life cycle it applies to. (c) Select the product that you would prefer to use and justify your selection.

 Writing about Technology: MANAGING CHOICES

Consider the following scenario, and then respond to the following: You're a manager who comes up with a new system that will make your company more efficient. However, implementing this system would make several tasks obsolete and cost many of your co-workers their jobs. (a) What is your ethical obligation to your company in this situation? (b) What is your ethical obligation to your co-workers? (c) What would you do in this situation? Defend your answer.

Design Elements: Concept Check icon: Dizzle52/Getty Images

chapter 13
Programming and Languages

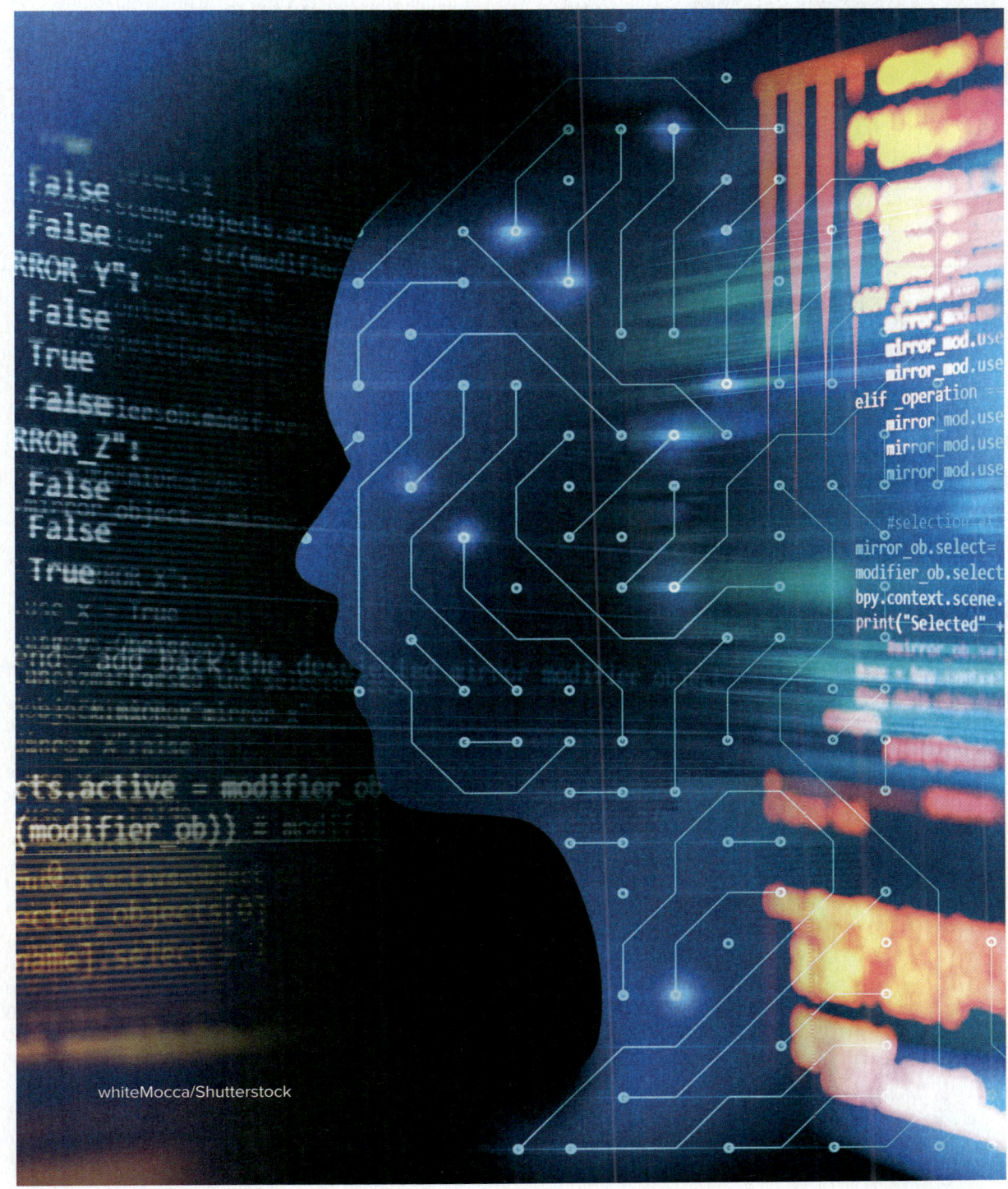

whiteMocca/Shutterstock

Why should I read this chapter?

TippaPatt/Shutterstock

Poorly designed software has destroyed companies in the blink of an eye. From stock purchases to pacemakers, our lives are increasingly dependent on digital devices and the programs that run them. In the future, robots with sophisticated artificial intelligence will handle everyday chores, and you will program these robots using conversational English instruction.

This chapter covers the things you need to know to be prepared for this ever-changing digital world, including:

- The software development life cycle—understand the steps of software development to be prepared to assist or manage software development projects.
- Programming languages—understand the differences among assembly, procedural, and natural languages to choose the best language for your needs.

Learning Objectives

After you have read this chapter, you should be able to:

1. Define programming and describe the six steps of programming.
2. Compare design tools, including top-down design, pseudocode, flowcharts, and logic structures.
3. Describe program testing and the tools for finding and removing errors.
4. Describe CASE tools and object-oriented software development.
5. Explain the five generations of programming languages.

Introduction

"Hi, I'm Alice, and I'm a computer programmer. I'd like to talk with you about programming and programming languages."

Chay_Tee/Shutterstock

In the previous chapter, we discussed systems analysis and design. We discussed the six-phase systems life cycle approach for examining and improving an information system. One of the phases is systems development, or the acquisition of new hardware and software. This chapter relates to this phase, systems development. More specifically, this chapter focuses on developing new software or programming. We will describe programming in two parts: (1) the steps in the programming process and (2) some of the programming languages available.

Why should you need to know anything about programming? The answer is simple. You might need to deal with programmers in the course of your work. You also may be required to do some programming yourself in the future. A growing trend is toward end-user software development. This means that end users, like you, are developing their own application programs.

To efficiently and effectively use computers, you need to understand the relationship between systems development and programming. Additionally, you need to know the six steps of programming, including program specification, program design, program code, program test, program documentation, and program maintenance.

Programs and Programming

What exactly is programming? Many people think of it as simply typing words into a computer. That may be part of it, but that is certainly not all of it. Programming, as we've hinted before, is actually a *problem-solving procedure*.

What Is a Program?

To see how programming works, think about what a program is. A **program** is a list of instructions for the computer to follow to accomplish the task of processing data into information. The instructions are made up of statements used in a programming language, such as C++, Java, or Python.

You are already familiar with some types of programs. As we discussed in **Chapters 1** and **3**, application programs are widely used to accomplish a variety of different types of tasks. For example, we use word processors to create documents and spreadsheets to analyze data. These can be purchased and are referred to as prewritten or packaged programs. Programs also can be created or custom-made. In **Chapter 12**, we saw that the systems analyst looked into the availability of time-and-billing software for Advantage Advertising. Will off-the-shelf software do the job, or should it be custom-written? This is one of the first things that needs to be decided in programming.

What Is Programming?

A program is a list of instructions for the computer to follow to process data. **Programming**, also known as **software development**, typically follows a six-step process known as the **software development life cycle (SDLC)**. (See **Figure 13-1**.)

The six steps are as follows:

1. **Program specification:** The program's objectives, outputs, inputs, and processing requirements are determined.
2. **Program design:** A solution is created using programming techniques such as top-down program design, pseudocode, flowcharts, and logic structures.
3. **Program code:** The program is written or coded using a programming language.
4. **Program test:** The program is tested or debugged by looking for syntax and logic errors.
5. **Program documentation:** Documentation is an ongoing process throughout the programming process. This phase focuses on formalizing the written description and processes used in the program.
6. **Program maintenance:** Completed programs are periodically reviewed to evaluate their accuracy, efficiency, standardization, and ease of use. Changes are made to the program's code as needed.

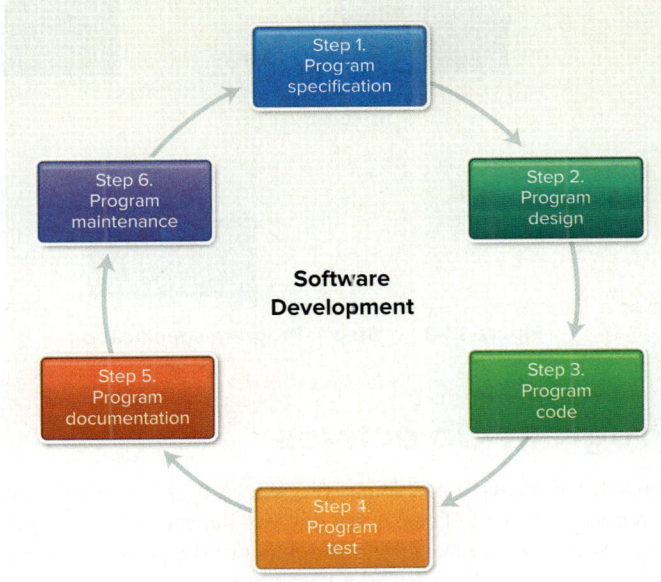

Figure 13-1 Software development

In organizations, computer professionals known as **software engineers** or **programmers** use this six-step procedure. In a recent survey by *Money* magazine, software engineers were ranked near the top of over 100 widely held jobs based on salary, prestige, and security.

You may well find yourself working directly with a programmer or indirectly through a systems analyst. Or you may actually do the programming for a system that you develop. Whatever the case, it's important that you understand the six-step programming procedure.

- ☐ What is a program?
- ☐ What are the six programming steps?

Step 1: Program Specification

Program specification is also called **program definition** or **program analysis**. It requires that the programmer—or you, the end user, if you are following this procedure—specify five items: (1) the program's objectives, (2) the desired output, (3) the input data required, (4) the processing requirements, and (5) the documentation. (See **Figure 13-2**.)

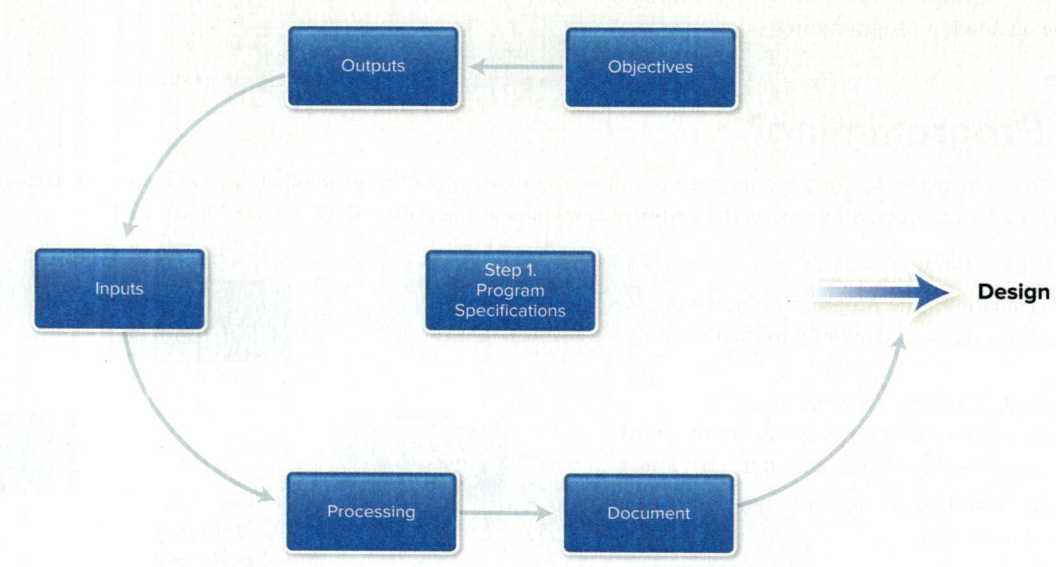

Figure 13-2   **Step 1: Program specification**

Program Objectives

You solve all kinds of problems every day. A problem might be deciding how to commute to school or work or which homework or report to do first. Thus, every day you determine your **objectives**—the problems you are trying to solve. Programming is the same. You need to make a clear statement of the problem you are trying to solve. An example would be "I want a time-and-billing system to record the time I spend on different jobs for different clients of Advantage Advertising."

Desired Output

It is best always to specify outputs before inputs. That is, you need to list what you want to *get out* of the computer system. Then you should determine what will *go into it*. The best way to do this is to draw a picture. You—the end user, not the programmer—should sketch or write how you want the output to look when it's done, including the vocabulary commonly used in your industry. It might be printed out or displayed on the monitor.

For example, if you want a time-and-billing report, you might write or draw something like **Figure 13-3**. Another form of output from the program might be bills to clients.

Client name: Allen Realty				Month and Year: Jan. '21
Date	Worker	Regular Hours & Rate	Overtime Hours & Rate	Bill
1/2	M. Jones	5 @ $10	1 @ $15	$65.00
	K. Williams	4 @ $30	2 @ $45	$210.00

Figure 13-3 End user's sketch of desired output

Input Data

Once you know the output you want, you can determine the input data and the source of this data. For example, for a time-and-billing report, you can specify that one source of data to be processed should be time cards. These are usually logs or statements of hours worked submitted either electronically or on paper forms. The log shown in **Figure 13-4** is an example of the kind of input data used in Advantage Advertising's manual system. Note that military time is used. For example, instead of writing "5:45 P.M.," people would write "17:45."

Daily Log

Worker:
Date:

Client	Job	Time in	Time out
A	TV commercial	800	915
B	Billboard ad	935	1200
C	Brochure	1315	1545
D	Magazine ad	1600	1745

Figure 13-4 Example of input data for hours worked, expressed in military time

Processing Requirements

Here you define the processing tasks that must happen for input data to be processed into output. For Advantage, one of the tasks for the program will be to add the hours worked for different jobs for different clients.

Program Specifications Document

As in the systems life cycle, ongoing documentation is essential. You should record program objectives, desired outputs, needed inputs, and required processing. This leads to the next step, program design.

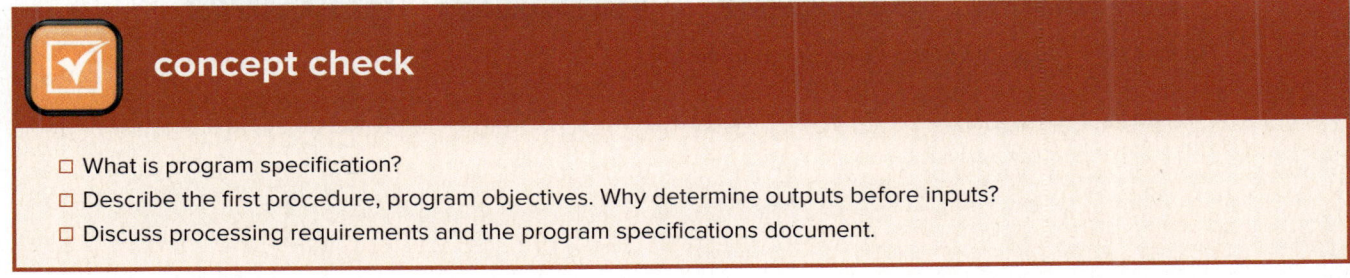

concept check

- What is program specification?
- Describe the first procedure, program objectives. Why determine outputs before inputs?
- Discuss processing requirements and the program specifications document.

Step 2: Program Design

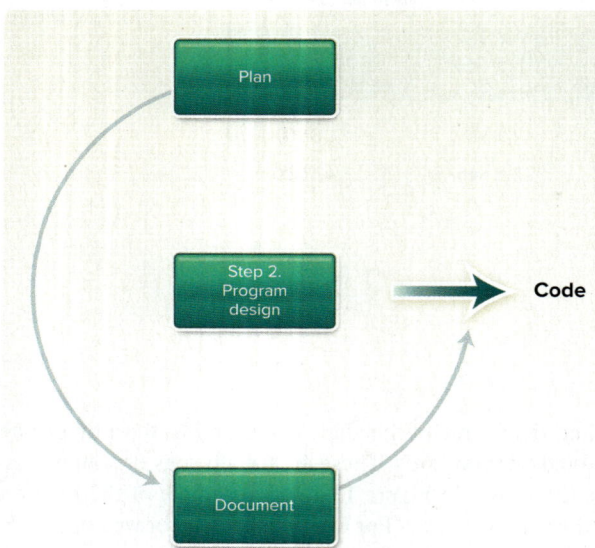

Figure 13-5 Step 2: Program design

After program specification, you begin **program design**. (See **Figure 13-5**.) Here you plan a solution, preferably using **structured programming techniques**. These techniques consist of the following: (1) top-down program design, (2) pseudocode, (3) flowcharts, and (4) logic structures.

Top-Down Program Design

First determine the outputs and inputs for the program. Then use **top-down program design** to identify the program's processing steps. Such steps are called **program modules** (or just **modules**). Each module is made up of logically related program statements.

An example of a top-down program design for a time-and-billing report is shown in **Figure 13-6**. Each of the boxes shown is a module. Under the rules of top-down design, each module should have a single function. The program must pass in sequence from one module to the next until all modules have been processed by the computer. Three of the boxes—"Obtain input," "Compute hours for billing," and "Produce output"—correspond to the three principal computer system operations: *input, process,* and *output.*

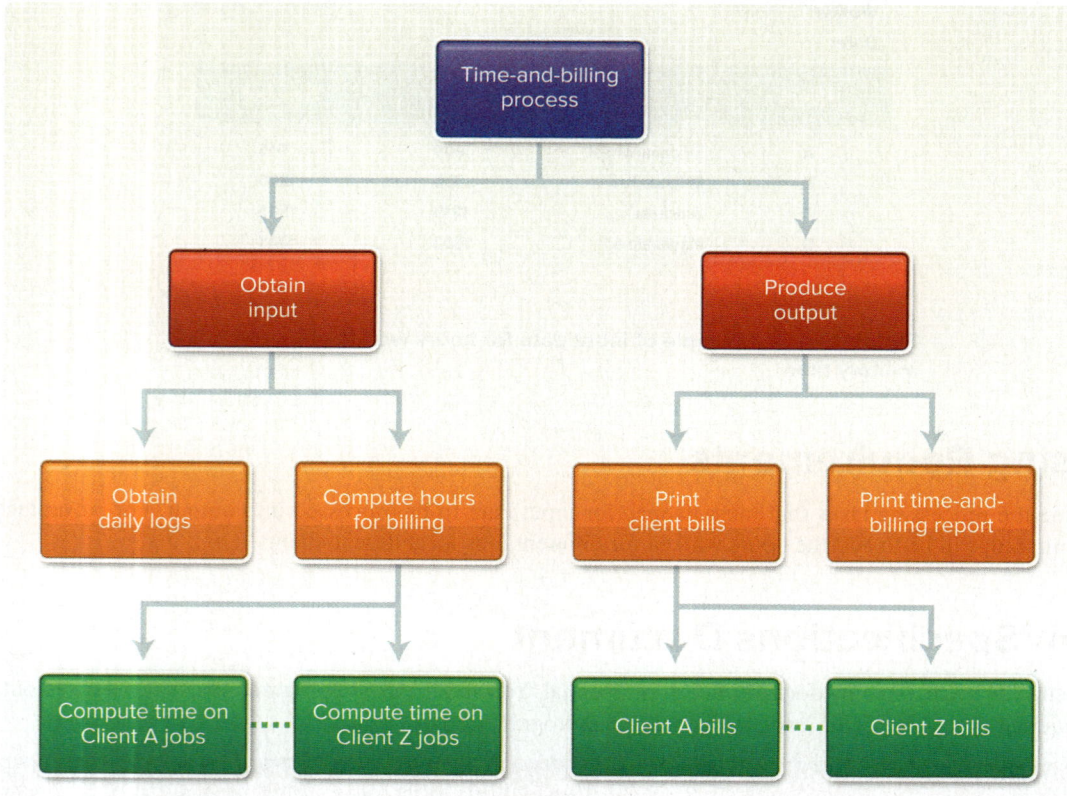

Figure 13-6 Example of top-down program design

Pseudocode

Pseudocode (pronounced "soo-doh-code") is an outline of the logic of the program you will write. It is like doing a summary of the program before it is written. **Figure 13-7** shows the pseudocode you might write for one module in the time-and-billing program. This shows the reasoning behind determining hours—including overtime hours—worked for different jobs for one client, Client A. Again, note this expresses the *logic* of what you want the program to do.

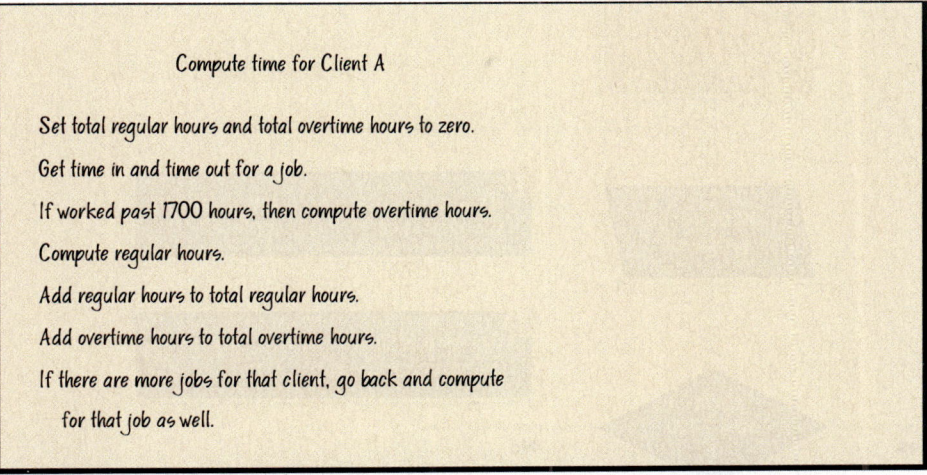

Figure 13-7   **Example of pseudocode**

Flowcharts

We mentioned system flowcharts in the previous chapter. Here we are concerned with **program flowcharts**. These graphically present the detailed sequence of steps needed to solve a programming problem. **Figure 13-8** presents several of the standard flowcharting symbols. An example of a program flowchart is presented in **Figure 13-9**. This flowchart expresses all the logic for just one module—"Compute time on Client A jobs"—in the top-down program design.

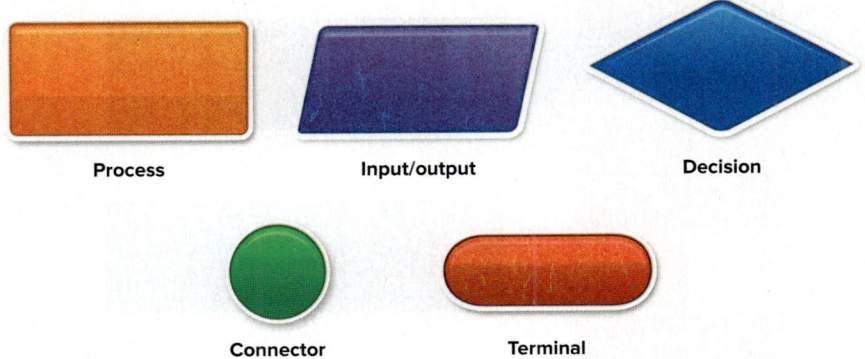

Figure 13-8   **Flowchart symbols**

Perhaps you can see from this flowchart why a computer is a computer, and not just an adding machine. One of the most powerful capabilities of computers is their ability to make logical comparisons. For example, a computer can compare two items to determine if one is less than, greater than, or equal to the other item.

But have we skipped something? How do we know which kinds of twists and turns to put in a flowchart so that it will work logically? The answer is based on the use of logic structures, as we will explain.

Logic Structures

How do you link the various parts of the flowchart? The best way is a combination of three **logic structures** called *sequential, selection,* and *repetition.* Using these arrangements enables you to write structured programs, which take much of the guesswork out of programming. Let us look at the logic structures.

CHAPTER 13: Programming and Languages

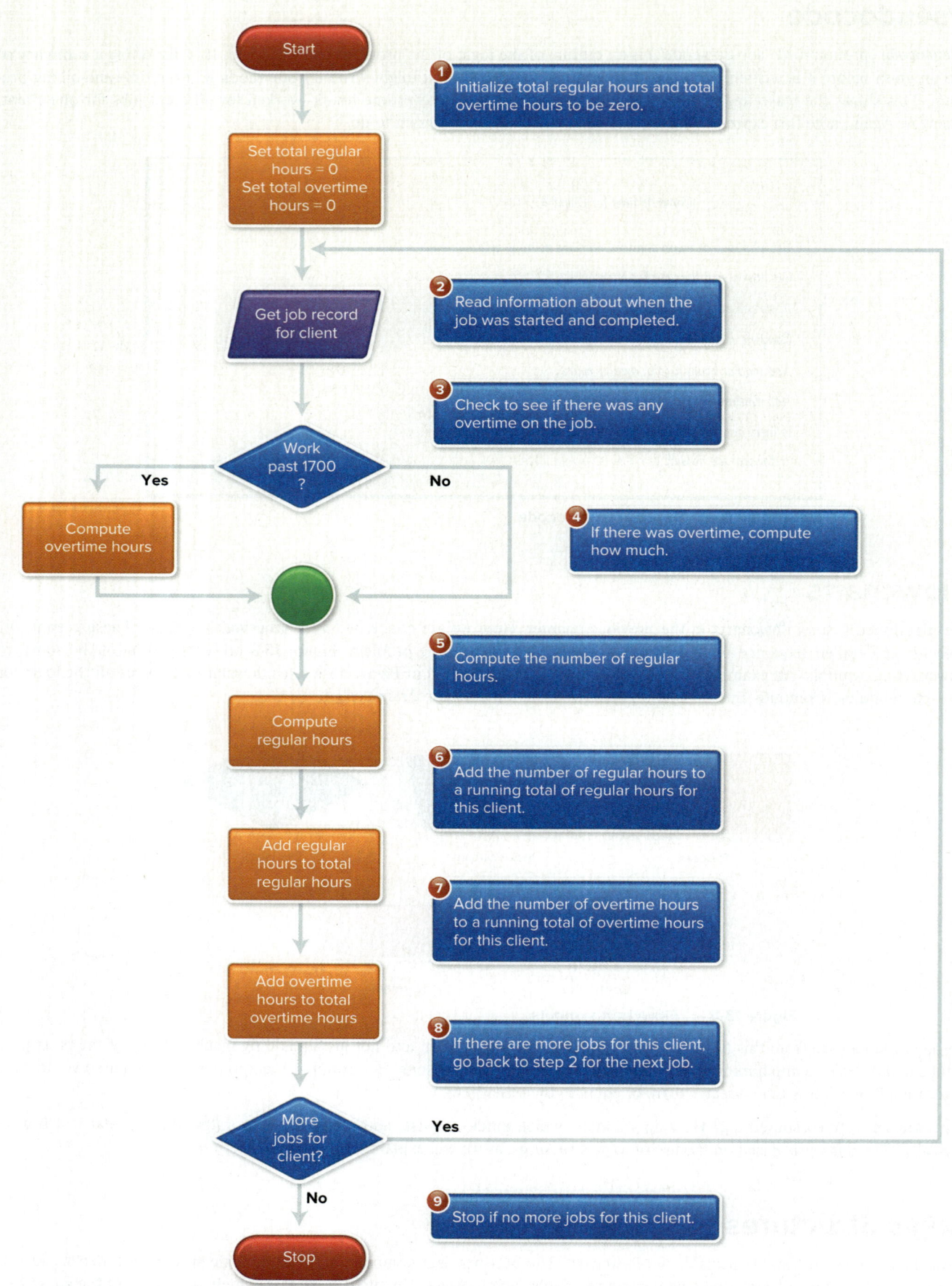

Figure 13-9 Flowchart for "Compute time on Client A jobs"

- In the **sequential structure**, one program statement follows another. Consider, for example, the "compute time" flowchart. (Refer back to **Figure 13-9**.) *Add regular hours to total regular hours* and *Add overtime hours to total overtime hours* to form a sequential structure. They logically follow each other. There is no question of "yes" or "no" or a decision suggesting other consequences. (See **Figure 13-10**.)
- The **selection structure** occurs when a decision must be made. The outcome of the decision determines which of two paths to follow. (See **Figure 13-11**.) This structure is also known as an **IF-THEN-ELSE structure** because that is how you can formulate the decision. Consider, for example, the selection structure in the "compute time" flowchart, which is concerned about computing overtime hours (*Work past 1700?*). (Refer back to **Figure 13-9**.) It might be expressed in detail as follows:
  IF hour finished for this job is later than 1700 hours (5:00 P.M.), THEN overtime hours equal the number of hours past 1700 hours, ELSE overtime hours equal zero.
  (See **Figure 13-11**.)
- The **repetition** or **loop structure** describes a process that may be repeated as long as a certain condition remains true. The structure is called a "loop" or "iteration" because the program loops around (iterates or repeats) again and again. Consider the loop structure in the "compute time" flowchart, which is concerned with testing if there are more jobs (*More jobs for clients?*). It might be expressed in detail as follows:
  DO read in job information WHILE (i.e., as long as) there are more jobs.
  (See **Figure 13-12**.)

Figure 13-10 Sequential logic structure

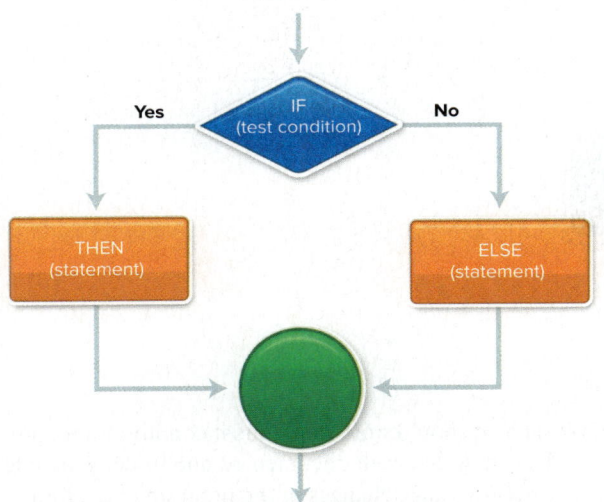

Figure 13-11 Selection (IF-THEN-ELSE) logic structure

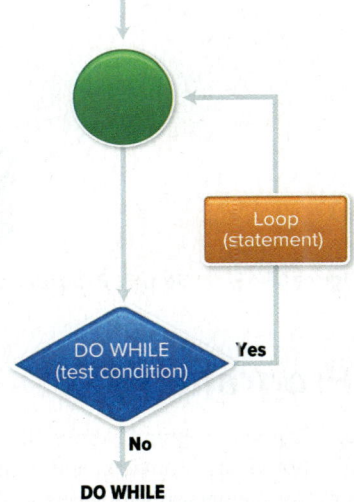

Figure 13-12 Repetition logic structure

The last thing to do before leaving the program design step is to document the logic of the design. This report typically includes pseudocode, flowcharts, and logic structures. Now you are ready for the next step, program code.

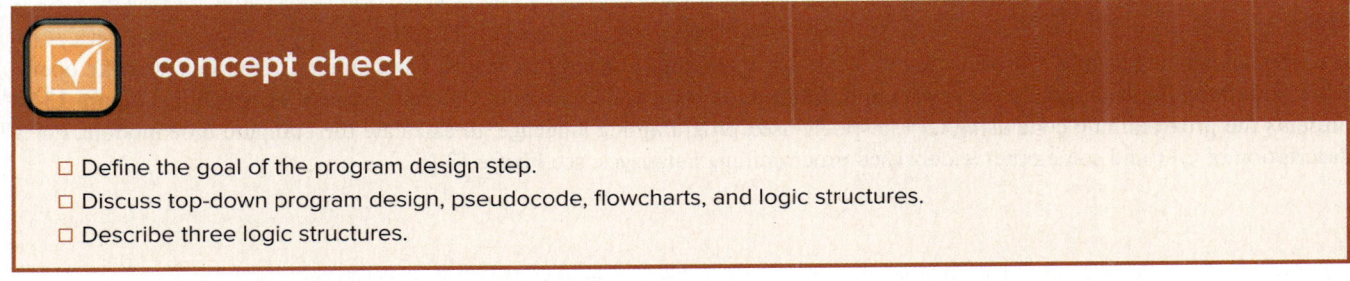

concept check

☐ Define the goal of the program design step.
☐ Discuss top-down program design, pseudocode, flowcharts, and logic structures.
☐ Describe three logic structures.

Step 3: Program Code

Writing the program is called **coding**. Here you use the logic you developed in the program design step to actually write the program. (See **Figure 13-13**.) This is the "program code" that instructs the computer what to do. Coding is what many people think of when they think of programming. As we've pointed out, however, it is only one of the six steps in the programming process.

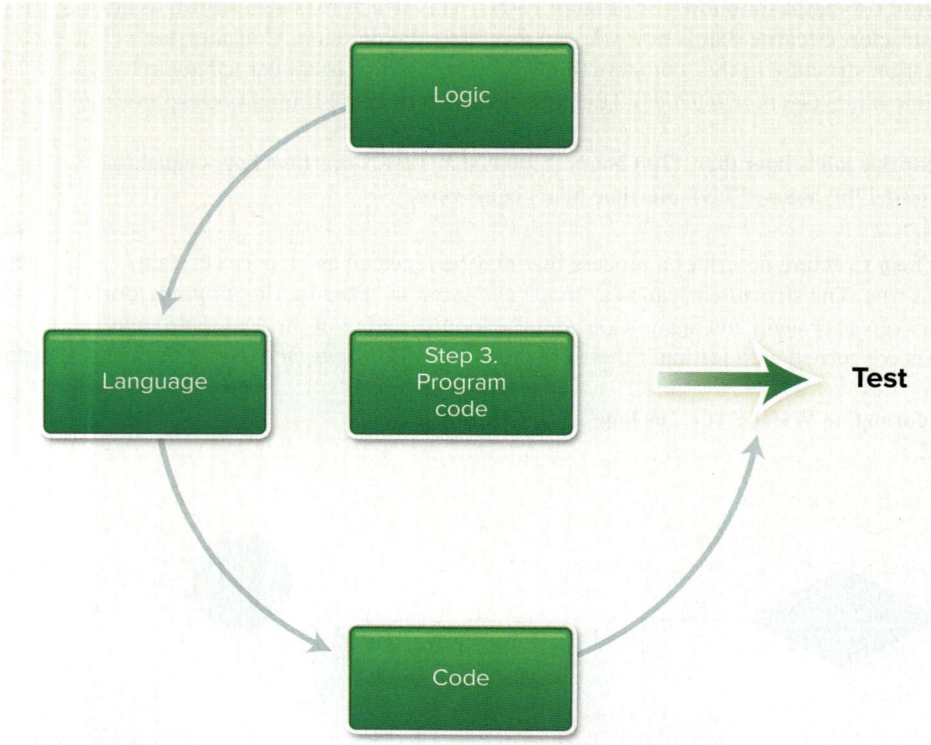

Figure 13-13   **Step 3: Program code**

The Good Program

What are the qualities of a good program? Above all, it should be reliable—that is, it should work under most conditions and produce correct output. It should catch obvious and common input errors. It also should be well documented and understandable by programmers other than the person who wrote it. After all, someone may need to make changes in the program in the future. One of the best ways to code effective programs is to write so-called **structured programs**, using the logic structures described in Step 2: Program Design.

Coding

After the program logic has been formulated, the next step is to **code**, or write the program using the appropriate computer language.

A **programming language** uses a collection of symbols, words, and phrases that instruct a computer to perform specific operations. Programming languages process data and information for a wide variety of different types of applications. **Figure 13-14** presents the programming code using C++, a widely used programming language, to calculate the compute time module. For a description of C++ and some other widely used programming languages, see **Figure 13-15**.

```
#include <fstream.h>

void main (void)
{
    ifstream input_file;

    float total_regular, total_overtime, regular, overtime;
    int hour_in, minute_in, hour_out, minute_out;
    input_file.open("time.txt",ios::in);

    total_regular = 0;
    total_overtime = 0;

    while (input_file != NULL)
    {
        input_file >> hour_in >> minute_in >> hour_out >> minute_out;

        if (hour_out > 17)
            overtime = (hour_out-17) +(minute_out/(float)60);
        else
            overtime = 0;
            regular = ((hour_out - hour_in) +(minute_out
                        - minute_in)/(float)60)   - overtime;
        total_regular += regular;
        total_overtime += overtime;
    }

    cout <<"Regular: " << total_regular <<endl;
    cout <<"Overtime " << total_overtime <<endl;
}
```

Figure 13-14 C++ code for computing regular and overtime hours

isocpp.org

Language	Description
C++	Extends C to use objects or program modules that can be reused and interchanged between programs
C#	A programming language designed by Microsoft to extend C++ for developing applications in the Windows environment
Java	Primarily used for Internet applications; similar to C++; runs with a variety of operating systems
JavaScript	Embedded into web pages to provide dynamic and interactive content
Python	General-purpose programming language that is simple and easy to learn. Frequently used in introductory programming courses
Swift	Uses graphical user interface and special code for touchscreen interfaces to create apps for Apple iOS devices

Figure 13-15 Widely used programming languages

Once the program has been coded, the next step is testing, or debugging, the program.

concept check

☐ What is coding?
☐ What makes a good program?
☐ What is a programming language?

Step 4: Program Test

Debugging refers to the process of testing and then eliminating errors ("getting the bugs out"). (See **Figure 13-16**.) It means running the program on a computer and then fixing the parts that do not work. Programming errors are of two types: *syntax errors* and *logic errors*.

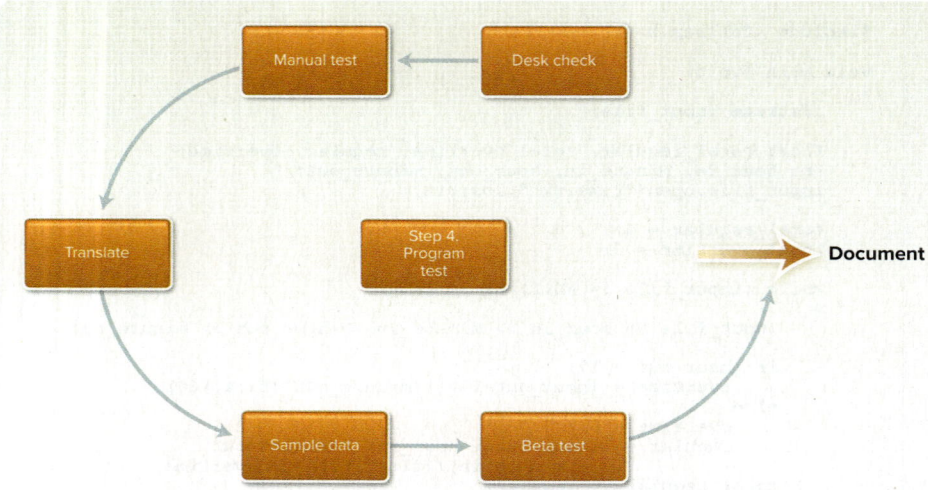

Figure 13-16   Step 4: Program test

Syntax Errors

A **syntax error** is a violation of the rules of the programming language. For example, in C++, each statement must end with a semicolon (;). If the semicolon is omitted, then the program will not run or execute due to a syntax error. For example, **Figure 13-17** shows testing of the compute time module in which a syntax error was identified.

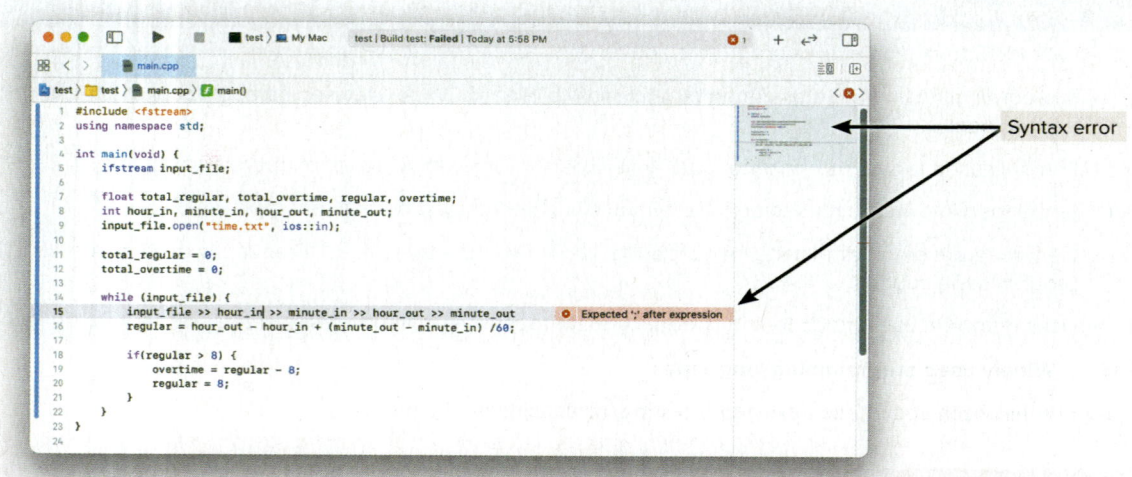

Figure 13-17   Syntax error identified
Apple Inc.

Logic Errors

A **logic error** occurs when the programmer uses an incorrect calculation or leaves out a programming procedure. For example, a payroll program that did not compute overtime hours would have a logic error.

Testing Process

Several methods have been devised for finding and removing both types of errors, including desk checking, manually testing, translating, running, and beta testing.

- **Desk checking:** In **desk checking** or **code review**, a programmer sitting at a desk checks (proofreads) a printout of the program. The programmer goes through the listing line by line carefully looking for syntax errors and logic errors.
- **Manually testing with sample data:** Using a calculator and sample data, a programmer follows each program statement and performs every calculation. Looking for programming logic errors, the programmer compares the manually calculated values to those calculated by the programs.
- **Attempt at translation:** The program is run through a computer, using a translator program. The translator attempts to translate the written program from the programming language (such as C++) into the machine language. Before the program will run, it must be free of syntax errors. Such errors will be identified by the translating program. (See **Figure 13-17**.)
- **Testing sample data on the computer:** After all syntax errors have been corrected, the program is tested for logic errors. Sample data is used to test the correct execution of each program statement.
- **Testing by a select group of potential users:** This is sometimes called **beta testing**. It is usually the final step in testing a program. Potential users try out the program and provide feedback.

For a summary of Step 4: Program test, see **Figure 13-18**.

Task	Description
1	Desk check for syntax and logic errors.
2	Manually test with sample data.
3	Translate program to identify syntax errors.
4	Run program with sample data.
5	Beta test with potential users.

Figure 13-18 Step 4: Program testing process

concept check

- What is debugging?
- What is the difference between syntax errors and logic errors?
- Briefly describe the testing process.

Step 5: Program Documentation

Documentation consists of written descriptions and procedures about a program and how to use it. (See **Figure 13-19**.) It is not something done just at the end of the programming process. **Program documentation** is carried on throughout all the programming steps. In this step, all the prior documentation is reviewed, finalized, and distributed. Documentation is important for people who may be involved with the program in the future. These people may include the following:

- **Users.** Users need to know how to use the software. Some organizations may offer training courses to guide users through the program. However, other organizations may expect users to be able to learn a package just from the written documentation. Two examples of this sort of documentation are printed manuals and the help option within most applications.
- **Operators.** Documentation must be provided for computer **operators**. If the program sends them error messages, for instance, they need to know what to do about them.
- **Programmers.** As time passes, even the creator of the original program may not remember much about it. Other **programmers** wishing to update and modify it—that is, perform program maintenance—may find themselves frustrated without adequate documentation. This kind of documentation should include text and program flowcharts, program listings, and sample output. It also might include system flowcharts to show how the particular program relates to other programs within an information system.

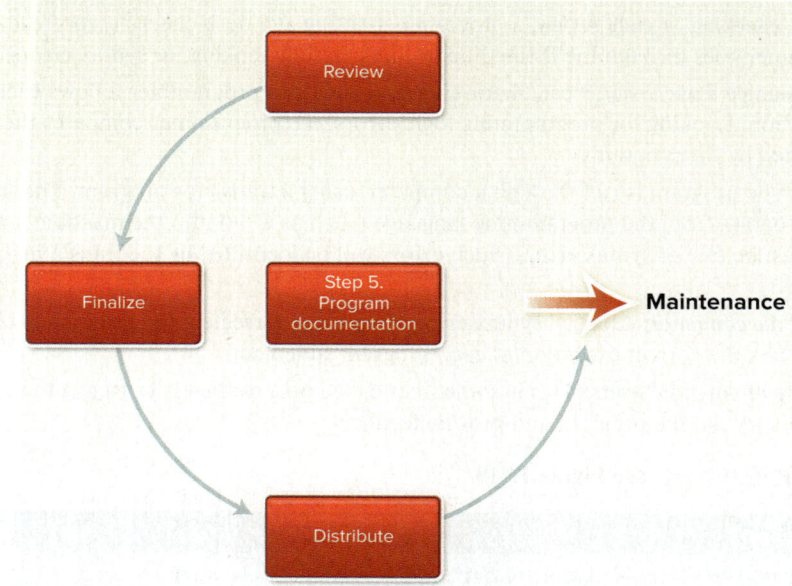

Figure 13-19 Step 5: Program documentation

- What is documentation?
- When does program documentation occur?
- Who is affected by documentation?

Step 6: Program Maintenance

The final step is **program maintenance**. (See **Figure 13-20**.) As much as 75 percent of the total lifetime cost for an application program is for maintenance. This activity is so commonplace that a special job title, **maintenance programmer**, exists.

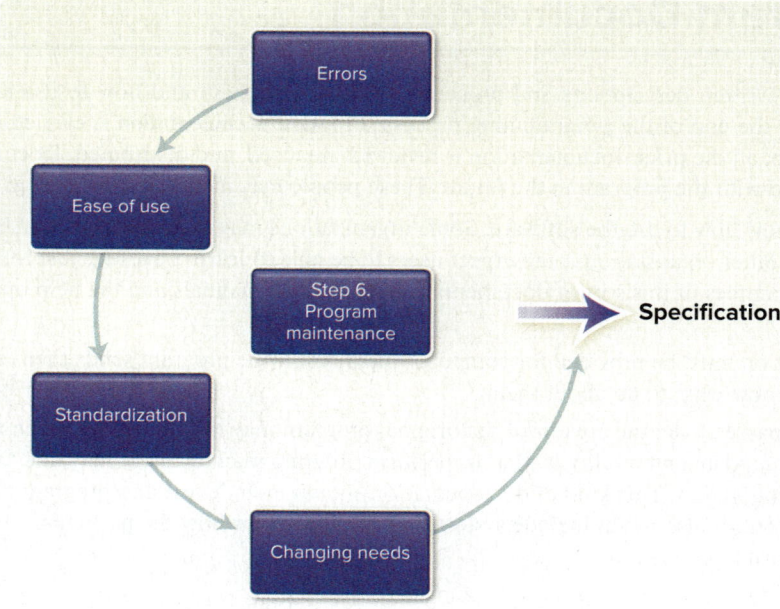

Figure 13-20 Step 6: Program maintenance

The purpose of program maintenance is to ensure that current programs are operating error-free, efficiently, and effectively. Activities in this area fall into two categories: operations and changing needs.

Operations

Operations activities concern locating and correcting operational errors, making programs easier to use, and standardizing software using structured programming techniques. For properly designed programs, these activities should be minimal.

Programming modifications or corrections are often referred to as **patches**. For software that is acquired, it is common for the software manufacturer to periodically send patches or updates for its software. If the patches are significant, they are known as **software updates**.

Changing Needs

All organizations change over time, and their programs must change with them. Programs need to be adjusted for a variety of reasons, including new tax laws, new information needs, and new company policies. Significant revisions may require that the entire programming process begin again with program specification.

Ideally, a software project sequentially follows the six steps of software development. However, some projects start before all requirements are known. In these cases, the SDLC becomes a more cyclical process, repeated several times throughout the development of the software. For example, **agile development**, a popular development methodology, starts by getting core functionality of a program working, and then expands on it until the customer is satisfied with the results. All six steps are repeated over and over as quickly as possible to create incrementally more functional versions of the application.

Figure 13-21 summarizes the six steps of the programming process.

Step	Primary Activity
1. Program specification	Determine program objectives, desired output, required input, and processing requirements.
2. Program design	Use structured programming techniques.
3. Program code	Select programming language; write the program.
4. Program test	Perform desk check (code review) and manual checks; attempt translation; test using sample data; beta test with potential users.
5. Program documentation	Write procedure for users, operators, and programmers.
6. Program maintenance	Adjust for errors, inefficient or ineffective operations, nonstandard code, and changes over time.

Figure 13-21 Summary of six steps in programming

concept check

- What is the purpose of program maintenance?
- Discuss operations activities. What are patches? Software updates?
- What are changing needs, and how do they affect programs?

CASE and OOP

You hear about efficiency and productivity everywhere. They are particularly important for software development. Two resources that promise to help are *CASE tools* and *object-oriented software development*.

CASE Tools

Professional programmers are constantly looking for ways to make their work easier, faster, and more reliable. One tool we mentioned in **Chapter 12**, CASE, is meeting this need. **Computer-aided software engineering (CASE) tools** provide some automation and assistance in program design, coding, and testing. (See **Figure 13-22**.)

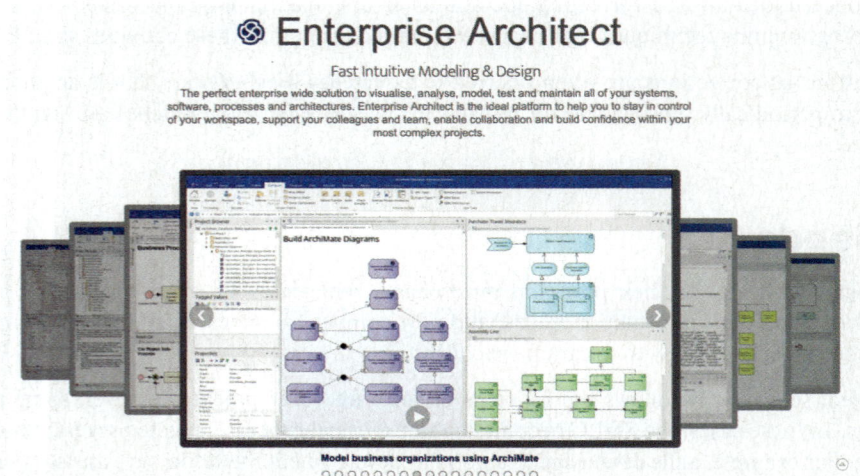

Figure 13-22   **Sparx Systems offers the CASE tool Enterprise Architect**
Smartdraw Software, LLC.

Object-Oriented Software Development

Traditional systems development is a careful, step-by-step approach focusing on the procedures needed to complete a certain objective. **Object-oriented software development** focuses less on the procedures and more on defining the relationships between previously defined procedures or "objects." **Object-oriented programming (OOP)** is a process by which a program is organized into objects. Each **object** contains both the data and processing operations necessary to perform a task. Let's explain what this means.

In the past, programs were developed as giant entities, from the first line of code to the last. This has been compared to building a car from scratch. Object-oriented programming is like building a car from prefabricated parts—carburetor, alternator, fenders, and so on. Object-oriented programs use objects that are reusable, self-contained components. Programs built with these objects assume that certain functions are the same. For example, many programs, from spreadsheets to database managers, have an instruction that will sort lists of names in alphabetical order. A programmer might use this object for alphabetizing in many other programs. There is no need to invent this activity anew every time. C++ is one of the most widely used object-oriented programming languages.

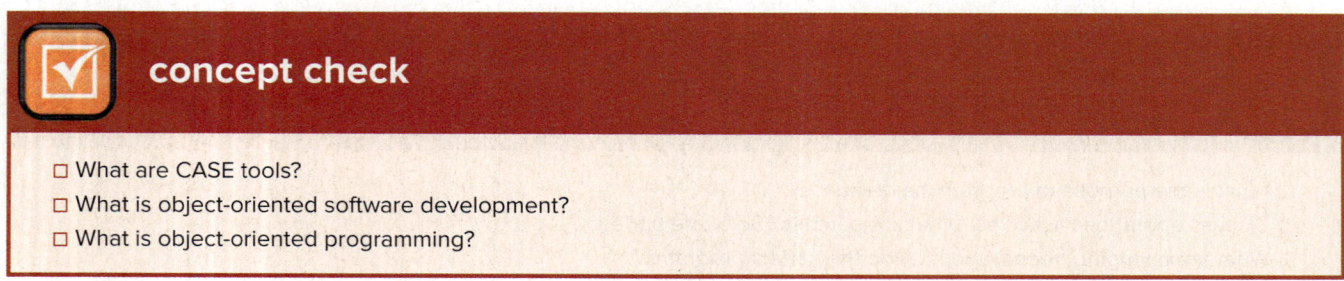

- What are CASE tools?
- What is object-oriented software development?
- What is object-oriented programming?

Generations of Programming Languages

Computer professionals talk about **levels** or **generations** of programming languages, ranging from "low" to "high." Programming languages are called **lower level** when they are closer to the language the computer itself uses. The computer understands the 0s and 1s that make up bits and bytes. Programming languages are called **higher level** when they are closer to the language humans use—that is, for English speakers, more like English.

There are five generations of programming languages: (1) machine languages, (2) assembly languages, (3) procedural languages, (4) task-oriented languages, and (5) problem and constraint languages.

Machine Languages: The First Generation

We mentioned in **Chapter 5** that a byte is made up of bits, consisting of 1s and 0s. These 1s and 0s may correspond to electricity being on or off in the computer. They also may correspond to a magnetic charge being present or absent on storage media such as disc or tape. From this two-state system, coding schemes have been developed that allow us to construct letters, numbers, punctuation marks, and other special characters. Examples of these coding schemes, as we saw, are ASCII, EBCDIC, and Unicode.

Data represented in 1s and 0s is said to be written in **machine language**. To see how hard this is to understand, imagine if you had to code this:

> 11110010011100111101001000010000011100000010 1011

Machine languages also vary according to make of computer—another characteristic that makes them hard to work with.

Assembly Languages: The Second Generation

Before a computer can process or run any program, the program must be converted or translated into machine language. **Assembly languages** use abbreviations or mnemonics such as ADD that are automatically converted to the appropriate sequence of 1s and 0s. Compared to machine languages, assembly languages are much easier for humans to understand and to use. The machine language code we gave above could be expressed in assembly language as

> ADD 210(8,13),02B(4,7)

This is still pretty obscure, of course, and so assembly language is also considered low level.

Assembly languages also vary from computer to computer. With the third generation, we advance to high-level languages, many of which are considered **portable languages**. That is, they can be run on more than one kind of computer—they are "portable" from one machine to another.

High-Level Procedural Languages: The Third Generation

People are able to understand languages that are more like their own (e.g., English) than machine languages or assembly languages. These more English-like programming languages are called "high-level" languages. However, most people still require some training to use higher-level languages. This is particularly true of procedural languages.

Procedural languages, also known as **3GLs (third-generation languages)**, are designed to express the logic—the procedures—that can solve general problems. Procedural languages, then, are intended to solve general problems and are the most widely used languages to create software applications. C++ is a procedural language widely used by today's programmers. For example, C++ was used in Advantage Advertising's time-and-billing report. (See **Figure 13-14** again for the compute time module of this program.)

Consider the following C++ statement from a program that assigns letter grades based on the score of an exam:

> if (score > = 90) grade = 'A';

This statement tests whether the score is greater than or equal to 90. If it is, then the letter grade of A is assigned.

Like assembly languages, procedural languages must be translated into machine language so that the computer processes them. Depending on the language, this translation is performed by either a *compiler* or an *interpreter*.

- A **compiler** converts the programmer's procedural language program, called the **source code**, into a machine language code, called the **object code**. This object code can then be saved and run later. The standard version of C++ is a procedural language that uses a compiler.
- An **interpreter** converts the procedural language one statement at a time into machine code just before it is to be executed. No object code is saved. An example of a procedural language using an interpreter is the standard version of BASIC.

What is the difference between using a compiler and using an interpreter? When a program is run, the compiler requires two steps. The first step is to convert the entire program's source code to object code. The second step is to run the object code. The interpreter, in contrast, converts and runs the program one line at a time. The advantage of a compiler language is that once the object code has been obtained, the program executes faster. The advantage of an interpreter language is that programs are easier to develop.

Task-Oriented Languages: The Fourth Generation

Third-generation languages are valuable, but they require training in programming. Task-oriented languages, also known as **4GLs (fourth-generation languages)** and **very high level languages**, require little special training on the part of the user.

Unlike general-purpose languages, **task-oriented languages** are designed to solve specific problems. While 3GLs focus on procedures and how logic can be combined to solve a variety of problems, 4GLs are nonprocedural and focus on specifying the specific tasks the program is to accomplish. 4GLs are more English-like, easier to program, and widely used by nonprogrammers. Some of these fourth-generation languages are used for very specific applications. For example, **IFPS (interactive financial planning system)** is used to develop financial models. Many 4GLs are part of a database management system. 4GLs include query languages and application generators:

- **Query languages: Query languages** enable nonprogrammers to use certain easily understood commands to search and generate reports from a database. One of the most widely used query languages is SQL (Structured Query Language). For example, let's assume that Advantage Advertising has a database containing all customer calls for service and that its management would like a listing of all clients who incurred overtime charges. The SQL command to create this list is

> **SELECT client FROM dailyLog WHERE serviceEnd >17**

This SQL statement selects or identifies all clients (a field name from the dailyLog table) that required service after 17 (military time for 5:00 P.M.). Microsoft Access can generate SQL commands like this one by using its Query wizard.

- **Application generators:** An **application generator** or a **program coder** is a program that provides modules of prewritten code. When using an application generator, a programmer can quickly create a program by referencing the module(s) that performs certain tasks. This greatly reduces the time to create an application. For example, Access has a report generation application and a Report wizard for creating a variety of different types of reports using database information.

Problem and Constraint Languages: The Fifth Generation

As they have evolved through the generations, computer languages have become more humanlike. Clearly, the fourth-generation query languages using commands that include words like SELECT, FROM, and WHERE are much more humanlike than the 0s and 1s of machine language. However, 4GLs are still a long way from the natural languages such as English and Spanish that people use.

The next step in programming languages will be the **fifth-generation language (5GL)**, or computer languages that incorporate the concepts of artificial intelligence to allow a person to provide a system with a problem and some constraints and then request a solution. Additionally, these languages would enable a computer to *learn* and to *apply* new information as people do. Rather than coding by keying in specific commands, we would communicate more directly to a computer using **natural languages**.

Consider the following natural language statement that might appear in a 5GL program for recommending medical treatment:

> **Get patientDiagnosis from patientSymptoms "sneezing", "coughing", "aching"**

When will fifth-generation languages become a reality? That's difficult to say; however, researchers are actively working on the development of 5GL languages and have demonstrated some success.

See **Figure 13-23** for a summary of the generations of programming languages.

Generation	Sample Statement
First: Machine	111100100111001111010010000100000111000000101011
Second: Assembly	ADD 210(8,13),02B(4,7)
Third: Procedural	if (score > = 90) grade = 'A';
Fourth: Task	SELECT client FROM dailyLog WHERE serviceEnd > 17
Fifth: Problems and Constraints	Get patientDiagnosis from patientSymptoms "sneezing", "coughing", "aching"

Figure 13-23 Summary of five programming generations

concept check

☐ What distinguishes a lower-level language from a higher-level language?
☐ What is the difference between machine and assembly languages?
☐ What is the difference between procedure and task-oriented languages?
☐ Define problem and constraint languages.

Careers in IT

"Now that you have learned about programming and programming languages, let me tell you about my career as a programmer."

Chay_Tee/Shutterstock

Computer **programmers** create, test, and troubleshoot programs used by computers. Programmers also may update and repair existing programs. Most computer programmers are employed by companies that create and sell software, but programmers also may be employed in various other businesses. Many computer programmers work on a project basis as consultants, meaning they are hired by a company only to complete a specific program. As technology has developed, the need for programmers to work on the most basic computer functions has decreased. However, demand for computer programmers with specializations in advanced programs continues.

Jobs in programming typically require a bachelor's degree in computer science or information systems. However, positions are available in the field for those with a two-year degree. Employers looking for programmers typically put an emphasis on previous experience. Programmers who have patience, think logically, and pay attention to detail are continually in demand. Additionally, programmers who can communicate technical information to nontechnical people are preferred.

Computer programmers can expect to earn an annual salary in the range of $65,760 to $112,120. Advancement opportunities for talented programmers include a lead programmer position or supervisory positions. Programmers with specializations and experience also may have an opportunity to consult.

A LOOK TO THE FUTURE

Your Own Programmable Robot

TippaPatt/Shutterstock

Have you ever dreamed of having your own robot that could help you with all your chores? Wouldn't it be nice if that robot understood every word you said and required no complex programming from you? Such a robot will be possible in the future as the field of robotics keeps advancing. Currently, robots are used in many manufacturing roles, making everything from cars to frozen pancakes. Recently, there are already several companies that are mass-producing programmable robots for individuals and educational institutions. It is just a matter of time before these robots can understand human instructions instead of complex programming languages. Technology has been making better programming tools and will continue to evolve to improve our lives as we look to the future.

One of the earliest robots that was made available to consumers was the Roomba from iRobot, which is essentially an automated, intelligent vacuum cleaner. Since then, the same company has released robots that wash floors, clean pools, and clear gutters. The programming is handled by the robot's developers, with the end user doing very little except turning the robot on. As well as these robots perform, their function is limited to their programmed task.

A company named Aldebaran Robotics has taken a different approach, creating small, humanoid robots, called Nao, which the end user can program. Although the Nao robots are being mass-produced, they are a bit too expensive for the average home. Currently, they are being marketed toward schools and research institutions. Using a GUI, students can create programs that the robot will follow. Alternatively, programmers can use one of several supported languages to write their own custom scripts for Nao.

In the future, it will not be necessary for someone to use software or know a programming language to communicate with a robot. Developers will use sophisticated programming to give the robot the artificial intelligence necessary to understand natural language. This software will be embedded in a chip within the robot. When you purchase a robot, all you will have to do is speak the commands in normal, conversational English. If you want the robot to help you clean the pool or lift a heavy box, you can tell the robot in the same manner you would tell another person.

The hardware components needed to make robots are becoming cheaper. However, the software remains a challenge. Human languages and conversations remain very difficult for a computer to fully understand. Speech recognition continues to improve, and we are seeing this technology embedded in the latest cell phones. Nevertheless, many improvements are necessary before a humanoid robot will be able to converse with us.

Do you think you will be able to own a humanoid robot in your lifetime? Do you believe that programmers will be able to make these robots intelligent enough to carry on conversations and perform a wide variety of tasks?

VISUAL SUMMARY | Programming and Languages

PROGRAMS AND PROGRAMMING

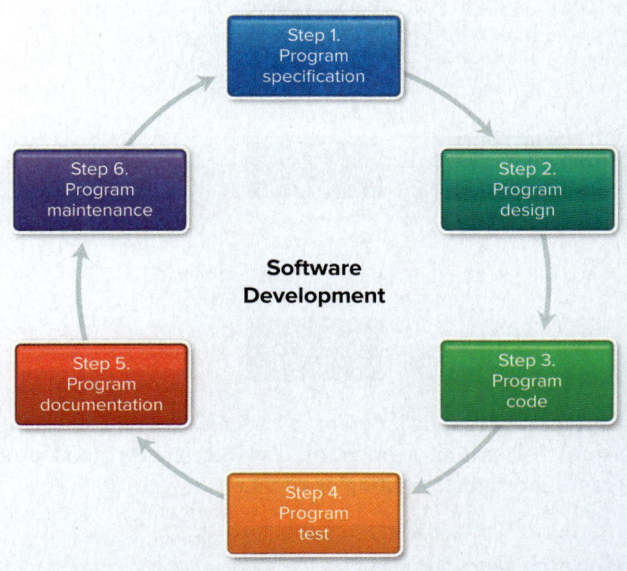

A **program** is a list of instructions for a computer to follow. **Programming (software development)** is a six-step procedure for creating programs.

The steps are

- Program specification—defining objectives, inputs, outputs, and processing requirements.
- Program design—creating a solution using structured programming tools and techniques such as **top-down program design**, **pseudocode**, **program flowcharts**, and **logic structures**.
- Program code—writing or coding the program using a **programming language**.
- Program test—testing or debugging the program by looking for **syntax** and **logic errors**.
- Program documentation—ongoing process throughout the programming process.
- Program maintenance—periodically evaluating programs for accuracy, efficiency, standardization, and ease of use and modifying program code as needed.

STEP 1: PROGRAM SPECIFICATION

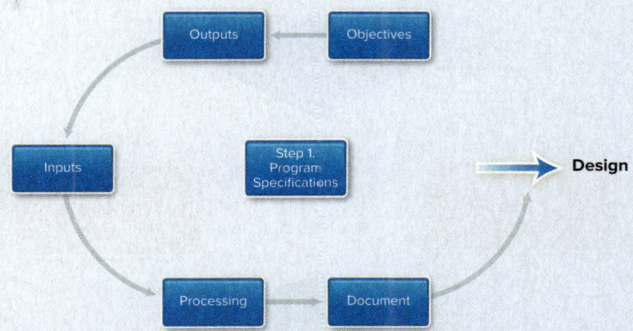

Program specification, also called **program definition** or **program analysis**, consists of specifying five tasks related to objectives, outputs, inputs, requirements, and documentation.

Program Objectives

The first task is to clearly define the problem to solve in the form of program **objectives**.

Desired Output

Next, focus on the desired output before considering the required inputs.

Input Data

Once outputs are defined, determine the necessary input data and the source of the data.

Processing Requirements

Next, determine the steps necessary (processing requirements) to use input to produce output.

Program Specifications Document

The final task is to create a specifications document to record this step's program objectives, outputs, inputs, and processing requirements.

To efficiently and effectively use computers, you need to understand the six steps of programming: program specification, program design, program coding, program test, program documentation, and program maintenance. Additionally, you need to be aware of CASE, OOP, and the generations of programming languages.

STEP 2: PROGRAM DESIGN

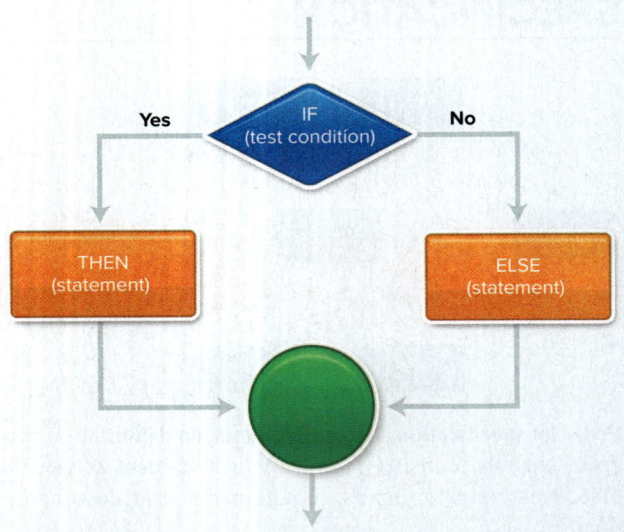

In **program design**, a solution is designed using, preferably, **structured programming techniques**, including the following.

Top-Down Program Design

In **top-down program design**, major processing steps, called **program modules (or modules),** are identified.

Pseudocode

Pseudocode is an outline of the logic of the program you will write.

Flowcharts

Program flowcharts are graphic representations of the steps necessary to solve a programming problem.

Logic Structures

Logic structures are arrangements of programming statements. Three types are

- **Sequential**—one program statement followed by another.
- **Selection (IF-THEN-ELSE)**—when a decision must be made.
- **Repetition (loop)**—when a process is repeated until the condition is true.

STEP 3: PROGRAM CODE

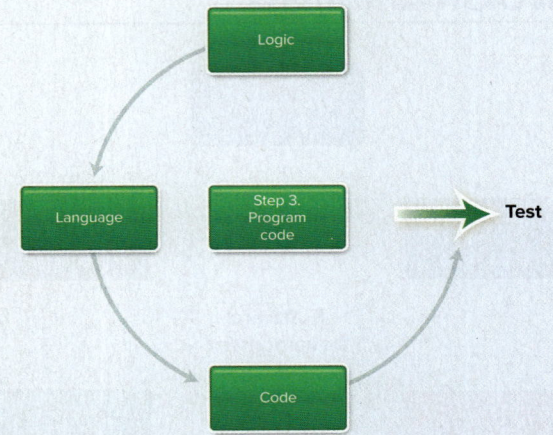

Coding is writing a program. There are several important aspects of writing a program. Two are writing good programs and actually writing or coding.

Good Programs

Good programs are reliable, detect obvious and common errors, and are well documented. The best way to create good programs is to write **structured programs** using the three basic logic structures presented in Step 2.

Coding

There are hundreds of different programming languages. **Programming languages** instruct a computer to perform specific operations. C++ is a widely used programming language.

```
#include <fstream.h>
void main (void)
{
    ifstream input_file;

    float total_regular, total_overtime, regular, overtime;
    int hour_in, minute_in, hour_out, minute_out;
    input_file.open("time.txt",ios::in);

    total_regular = 0;
    total_overtime = 0;

    while (input_file != NULL)
    {
        input_file >> hour_in >> minute_in >> hour_out >> minute_out;

        if (hour_out > 17)
            overtime = (hour_out-17) +(minute_out/(float)60);
        else
            overtime = 0;
        regular = ((hour_out - hour_in) +(minute_out
                    - minute_in)/(float)60)   - overtime;
        total_regular += regular;
        total_overtime += overtime;
    }

    cout <<"Regular: " << total_regular <<endl;
    cout <<"Overtime " << total_overtime <<endl;
}
```

isocpp.org

STEP 4: PROGRAM TEST

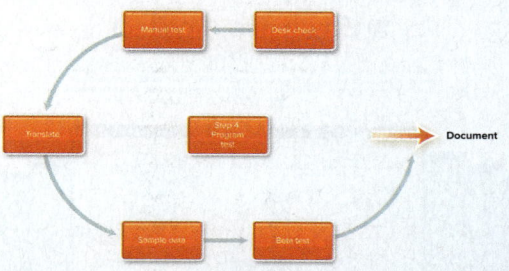

Debugging is a process of testing and eliminating errors in a program. Syntax and logic are two types of programming errors.

Syntax Errors

Syntax errors are violations of the rules of a programming language. For example, omitting a semicolon at the end of a C++ statement is a syntax error.

Logic Errors

Logic errors are incorrect calculations or procedures. For example, failure to include calculation of overtime hours in a payroll program is a logic error.

Testing Process

Five methods for testing for syntax and logic errors are

- **Desk checking (code review)**—careful reading of a printout of the program.
- **Manual testing**—using a calculator and sample data to test for correct programming logic.
- **Attempt at translation**—running the program using a translator program to identify syntax errors.
- **Testing sample data**—running the program and testing the program for logic errors using sample data.
- **Testing by users (beta testing)**—final step in which potential users try the program and provide feedback.

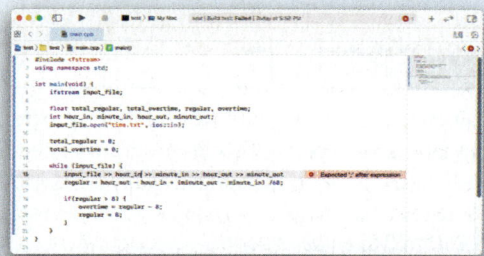

Apple Inc.

STEP 5: PROGRAM DOCUMENTATION

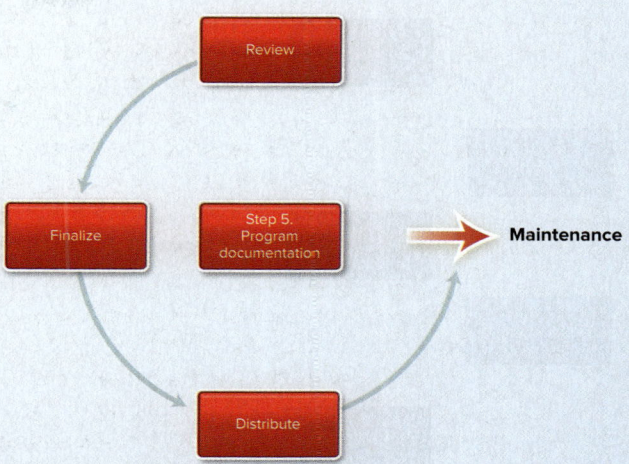

Program documentation consists of a written description of the program and the procedures for running it. People who use documentation include

- **Users**, who need to know how to use the program. Some organizations offer training courses; others expect users to learn from written documentation.
- **Operators**, who need to know how to execute the program and how to recognize and correct errors.
- **Programmers**, who may need to update and maintain the program in the future. Documentation could include text and program flowcharts, program listings, and sample outputs.

STEP 6: PROGRAM MAINTENANCE

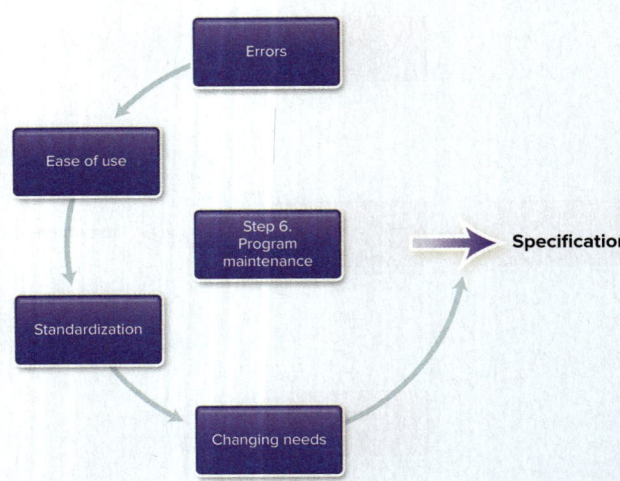

Program maintenance is designed to ensure that the program operates correctly, efficiently, and effectively. Two categories of maintenance activities are the following.

Operations

Operations activities include locating and correcting errors, improving usability, and standardizing software. Software updates are known as **patches**. Significant patches are called **software updates**.

Changing Needs

Organizations change over time, and their programs must change with them. **Agile development** starts with core program functionality and expands until the customer is satisfied with the results.

insta_photos/Shutterstock

CASE AND OOP

Smartdraw Software, LLC.

CASE

Computer-aided software engineering (CASE) tools provide automation and assistance in program design, coding, and testing.

OOP

Traditional systems development focuses on procedures to complete a specific objective.

Object-oriented software development focuses less on procedures and more on defining relationships between previously defined procedures or objects. **Object-oriented programming (OOP)** is a process by which a program is divided into modules called **objects**. Each object contains both the data and processing operations necessary to perform a task.

PROGRAMMING LANGUAGE GENERATIONS

Programming languages have **levels** or **generations** ranging from low to high. **Lower-level** languages are closer to the 0s and 1s language of computers. **Higher-level** languages are closer to the languages of humans.

CAREERS in IT

Programmers create, test, and troubleshoot programs. They also update and repair existing programs. Requirements include a bachelor's or specialized two-year degree in computer science or information systems. Expected salary range is $65,760 to $112,120.

KEY TERMS

- agile development
- application generator
- assembly language
- beta testing
- code
- code review
- coding
- compiler
- computer-aided software engineering (CASE) tools
- debugging
- desk checking
- documentation
- fifth-generation language (5GL)
- fourth-generation language (4GL)
- generation
- higher level
- IF-THEN-ELSE structure
- IFPS (interactive financial planning system)
- interpreter
- level
- logic error
- logic structure
- loop structure
- lower level
- machine language
- maintenance programmer
- module
- natural language
- object
- object code
- object-oriented programming (OOP)
- object-oriented software development
- objective
- operator
- patch
- portable language
- procedural language
- program
- program analysis
- program coder
- program definition
- program design
- program documentation
- program flowchart
- program maintenance
- program module
- program specification
- programmer
- programming
- programming language
- pseudocode
- query language
- repetition structure
- selection structure
- sequential structure
- software development
- software development life cycle (SDLC)
- software engineer
- software update
- source code
- structured program
- structured programming technique
- syntax error
- task-oriented language
- third-generation language (3GL)
- top-down program design
- user
- very high level language

MULTIPLE CHOICE

Circle the correct answer.

1. Another name for programming.
   - **a.** design
   - **b.** compiling
   - **c.** software development
   - **d.** documentation

2. An outline of the logic of the program to be written.
   - **a.** CASE
   - **b.** logic structure
   - **c.** software development life cycle
   - **d.** pseudocode

3. Logic structure, also known as IF-THEN-ELSE, that controls program flow based on a decision.
   - **a.** sequential
   - **b.** syntax
   - **c.** selection
   - **d.** IFPS

4. Programming languages that are closer to the language of humans.
   - **a.** high level
   - **b.** very high level
   - **c.** higher level
   - **d.** assembly

5. The process of testing and then eliminating program errors.
   - **a.** debugging
   - **b.** compiling
   - **c.** documenting
   - **d.** designing

6. Program step that involves creating descriptions and procedures about a program and how to use it.
   - **a.** debugging
   - **b.** documentation
   - **c.** compiling
   - **d.** designing

7. The first-generation language consisting of 1s and 0s.
   - **a.** machine
   - **b.** assembly
   - **c.** natural
   - **d.** high level

8. Converts a procedural language one statement at a time into machine code just before it is to be executed.
   - **a.** compiler
   - **b.** software developer
   - **c.** sequential
   - **d.** interpreter

9. Generation of computer languages that allows a person to provide a system with a problem and some constraints, and then request a solution.
   - **a.** 3GL
   - **b.** 4GL
   - **c.** 5GL
   - **d.** 6GL

10. 5GL that allows more direct human communication with a program.
    - **a.** assembly
    - **b.** natural language
    - **c.** task-oriented
    - **d.** portable

MATCHING

Match each numbered item with the most closely related lettered item. Write your answers in the spaces provided.

a. data
b. high-level language
c. modules
d. object code
e. object-oriented
f. program maintenance
g. program test
h. query
i. sequential
j. structured programs

____ 1. The programming logic structure in which one program statement follows another.
____ 2. A compiler converts the programmer's procedural language program, called the source code, into a machine language code, called the _____.
____ 3. These 4GL languages enable nonprogrammers to use certain easily understood commands to search and generate reports from a database.
____ 4. This step in the six-step programming procedure is the final step.
____ 5. What a program processes.
____ 6. The major processing steps identified in a top-down program design.
____ 7. Unlike traditional systems development, this software development approach focuses less on the procedures and more on defining the relationships between previously defined procedures.
____ 8. Natural languages are considered to be a(n) _____.
____ 9. One of the best ways to code effective programs is to use the three basic logic structures to create _____.
____ 10. This step in the six-step programming procedure involves desk checking and searching for syntax and logic errors.

OPEN-ENDED

On a separate sheet of paper, respond to each question or statement.

1. Identify and discuss each of the six steps of programming.
2. Describe CASE tools and OOP. How does CASE assist programmers?
3. What is meant by "generation" in reference to programming languages? What is the difference between low-level and high-level languages?
4. What is the difference between a compiler and an interpreter?
5. What are logic structures? Describe the differences between the three types.

DISCUSSION

Respond to each of the following questions.

 ### Expanding Your Knowledge: SOURCE CODE GENERATORS

Generally, the human resources that are devoted to a successful software project are its greatest single expense. Programming and testing applications are time-consuming tasks. Recently, source code generators have become popular for handling some of the more routine programming tasks. Research source code generators on the web, and answer the following questions: (a) What are source code generators? (b) How do source code generators work? (c) What programming tasks are source code generators best for? Why? (d) What programming tasks are beyond what source code generators can accomplish? Why?

 ### Writing about Technology: BUGS

Several years ago, two people died and a third was maimed after receiving excessive radiation from a medical machine. It was only after the second incident that the problem was located—a bug in the software that controlled the machine. Consider the possible consequences of software failure in situations where a life is at stake, and then respond to the following: (a) Are there situations when software bugs are unethical? Explain your answer. (b) No program of any significant complexity can reasonably be fully tested. When is it ethical to say that software is "tested enough"? (c) What responsibility does a programmer have in situations where a program fails in the field? What about the software company he or she works for? Does the consumer share any responsibility? Justify your answers.

 ### Writing about Technology: SECURITY AND PRIVACY

Security and privacy are important concerns in the development of any information system. Respond to the following: (a) In the development process, who would you expect to have the responsibility of identifying security and privacy concerns? (b) In what phase of the software development life cycle would security and privacy concerns be identified?

Design Elements: Concept Check icon: Dizzle52/Getty Images

The Evolution of the Computer Age

Many of you probably can't remember a world without computers, but for some of us, computers were virtually unknown when we were born and have rapidly come of age during our lifetime.

Although there are many predecessors to what we think of as the modern computer—reaching as far back as the 18th century, when Joseph Marie Jacquard created a loom programmed to weave cloth and Charles Babbage created the first fully modern computer design (which he could never get to work)—the computer age did not really begin until the first computer was made available to the public in 1951.

The modern age of computers thus spans slightly more than 65 years (so far), which is typically broken down into five generations. Each generation has been marked by a significant advance in technology.

- **First Generation (1951–57):** During the first generation, computers were built with vacuum tubes—electronic tubes that were made of glass and were about the size of lightbulbs.

Mathew Spolin/500px Prime/Getty Images

- **Second Generation (1958–63):** This generation began with the first computers built with transistors—small devices that transfer electronic signals across a resistor. Because transistors are much smaller, use less power, and create less heat than vacuum tubes, the new computers were faster, smaller, and more reliable than the first-generation machines.

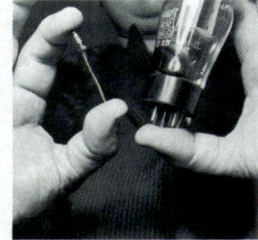

Keystone-France/Gamma-Keystone/Getty Images

- **Third Generation (1964–69):** In 1964, computer manufacturers began replacing transistors with integrated circuits. An integrated circuit (IC) is a complete electronic circuit on a small chip made of silicon (one of the most abundant elements in the earth's crust). These computers were more reliable and compact than computers made with transistors, and they cost less to manufacture.

- **Fourth Generation (1970–90):** Many key advances were made during this generation, the most significant being the microprocessor—a specialized chip developed for computer memory and logic. Use of a single chip to create a smaller "personal" computer (as well as digital watches, pocket calculators, copy machines, and so on) revolutionized the computer industry.

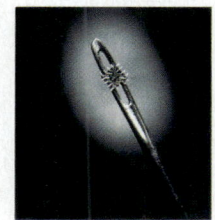

Keystone/Stringer/Getty Images

- **Fifth Generation (1991–2023 and beyond):** Our current generation has been referred to as the "Connected Generation" because of the industry's massive effort to increase the connectivity of computers. The rapidly expanding Internet, World Wide Web, and intranets have created an information superhighway that has enabled both computer professionals and home computer users to communicate with others across the globe.

John Foxx/Stockbyte/Getty Images

This appendix provides you with a timeline that describes in more detail some of the most significant events in each generation of the computer age.

First Generation: The Vacuum Tube Age

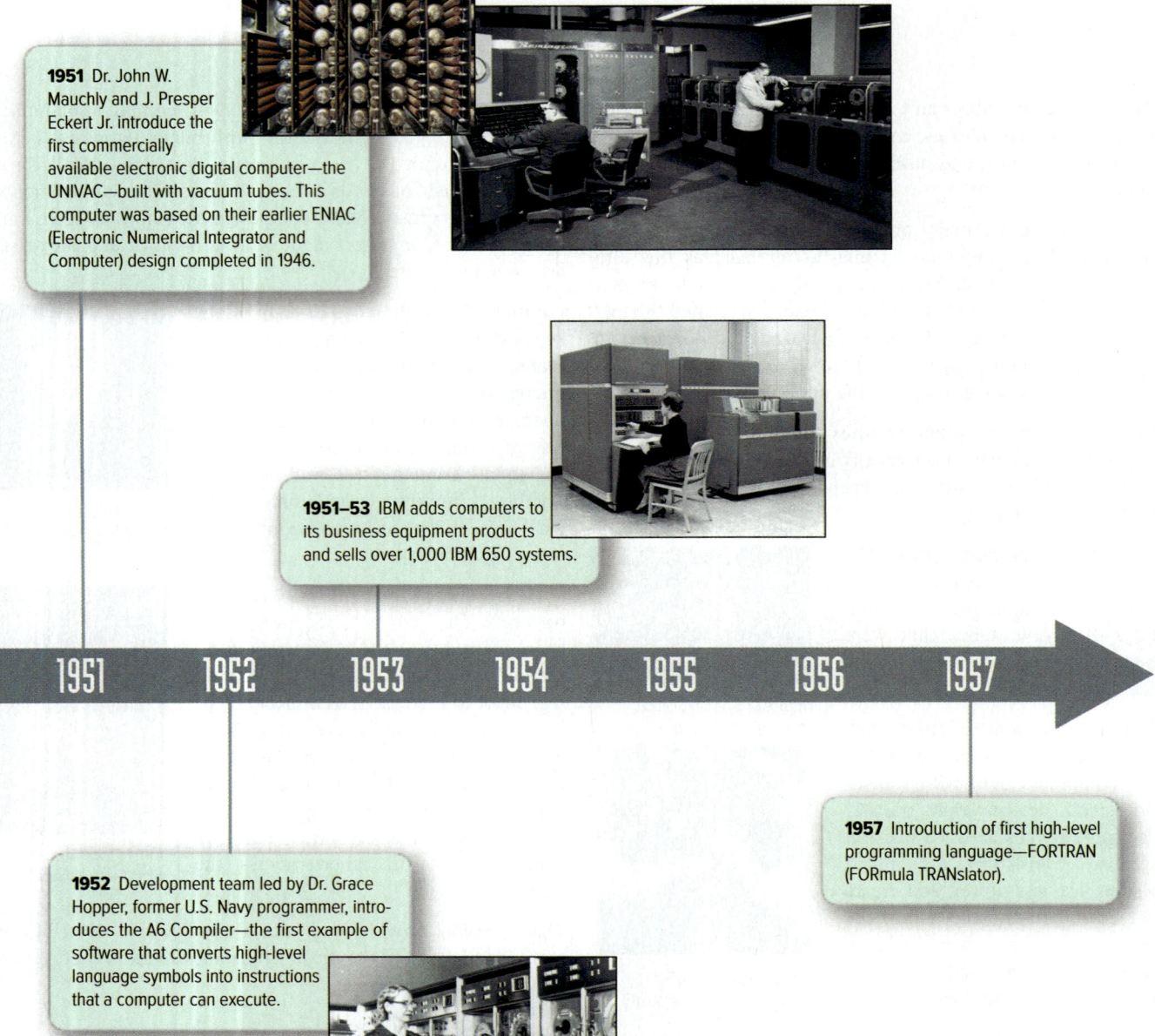

1951 Dr. John W. Mauchly and J. Presper Eckert Jr. introduce the first commercially available electronic digital computer—the UNIVAC—built with vacuum tubes. This computer was based on their earlier ENIAC (Electronic Numerical Integrator and Computer) design completed in 1946.

1951–53 IBM adds computers to its business equipment products and sells over 1,000 IBM 650 systems.

1952 Development team led by Dr. Grace Hopper, former U.S. Navy programmer, introduces the A6 Compiler—the first example of software that converts high-level language symbols into instructions that a computer can execute.

1957 Introduction of first high-level programming language—FORTRAN (FORmula TRANslator).

(1951) **left:** Mathew Spolin/500px Prime/Getty Images; **right:** Bettman/Getty Images; (1951– 53) Underwood Archives, Inc/Alamy Stock Photo; (1952) Photo Researchers/Science History Images/Alamy Stock Photo

Second Generation: The Transistor Age

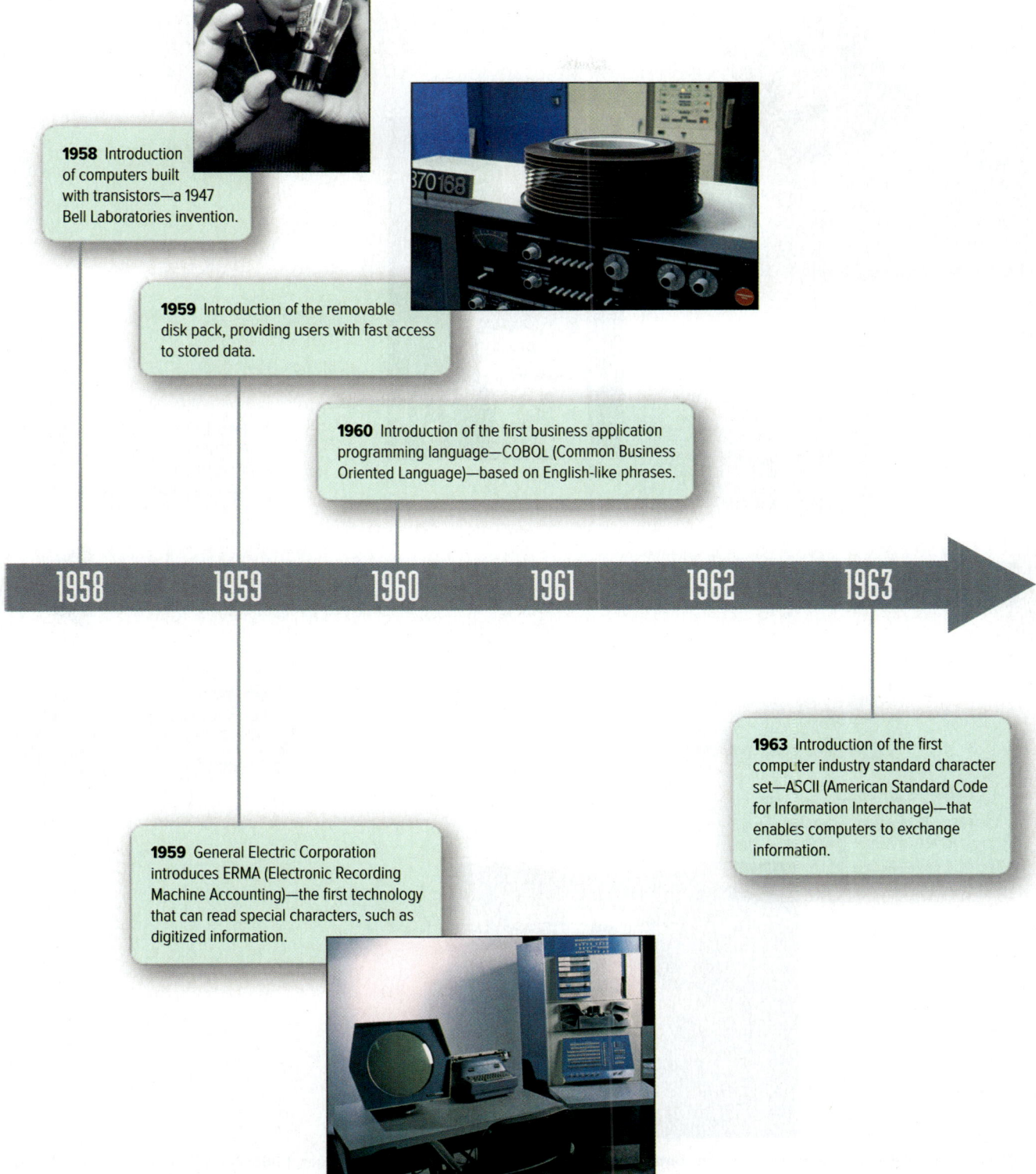

1958 Introduction of computers built with transistors—a 1947 Bell Laboratories invention.

1959 Introduction of the removable disk pack, providing users with fast access to stored data.

1960 Introduction of the first business application programming language—COBOL (Common Business Oriented Language)—based on English-like phrases.

1959 General Electric Corporation introduces ERMA (Electronic Recording Machine Accounting)—the first technology that can read special characters, such as digitized information.

1963 Introduction of the first computer industry standard character set—ASCII (American Standard Code for Information Interchange)—that enables computers to exchange information.

(1958) Keystone-France/Gamma-Keystone/Getty Images; (1959) **top:** Joseph Nettis/Science Source; **bottom:** VOLKER STEGER/Science Source

Third Generation: The Integrated Circuit Age

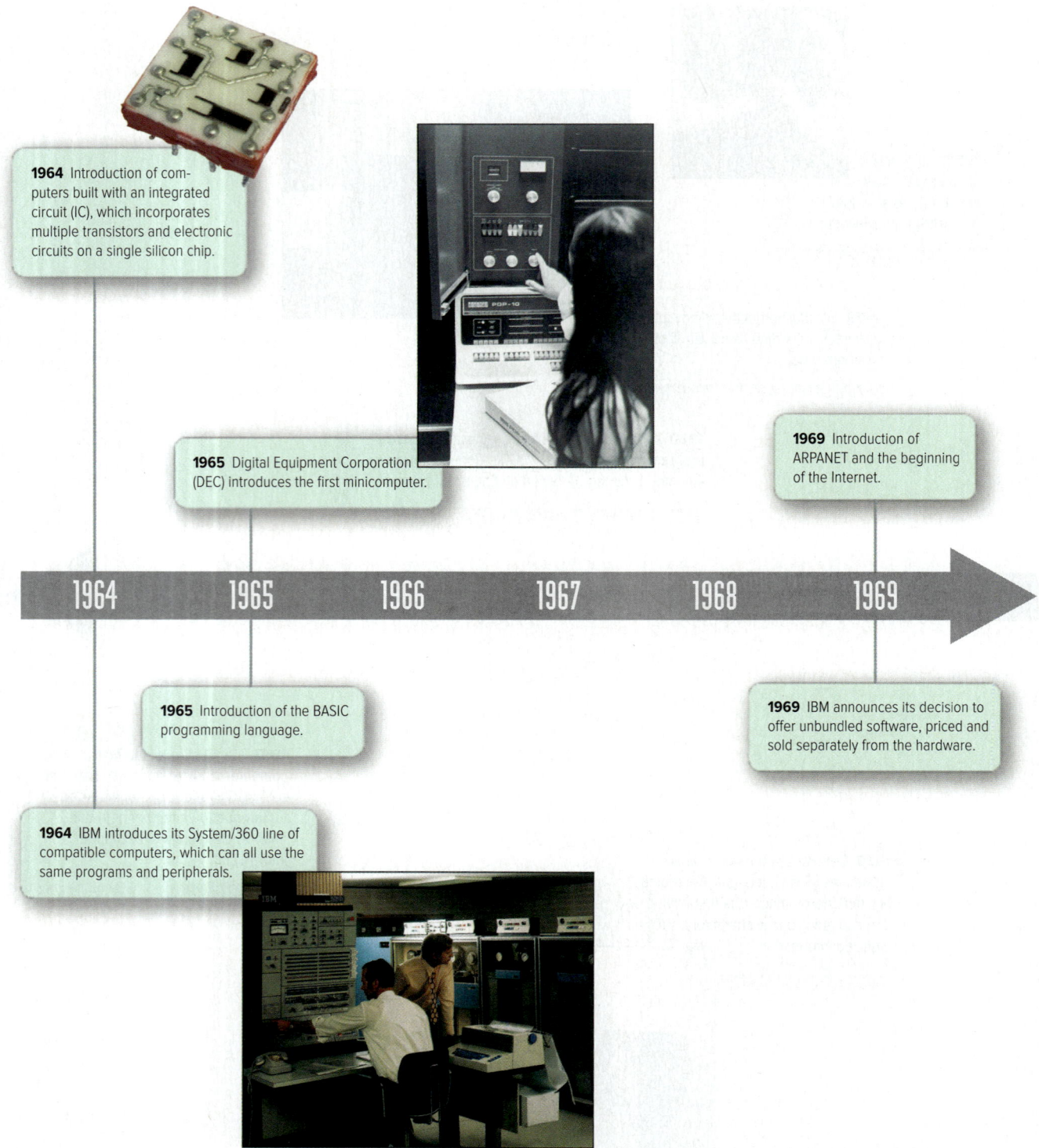

1964 Introduction of computers built with an integrated circuit (IC), which incorporates multiple transistors and electronic circuits on a single silicon chip.

1965 Digital Equipment Corporation (DEC) introduces the first minicomputer.

1969 Introduction of ARPANET and the beginning of the Internet.

1965 Introduction of the BASIC programming language.

1969 IBM announces its decision to offer unbundled software, priced and sold separately from the hardware.

1964 IBM introduces its System/360 line of compatible computers, which can all use the same programs and peripherals.

(1964) **top:** Jerry Marshall/pictureresearching.com; **bottom:** Photo by Underwood Archives/Getty Images; (1965) INTERFOTO/Alamy Stock Photo

Fourth Generation: The Microprocessor Age

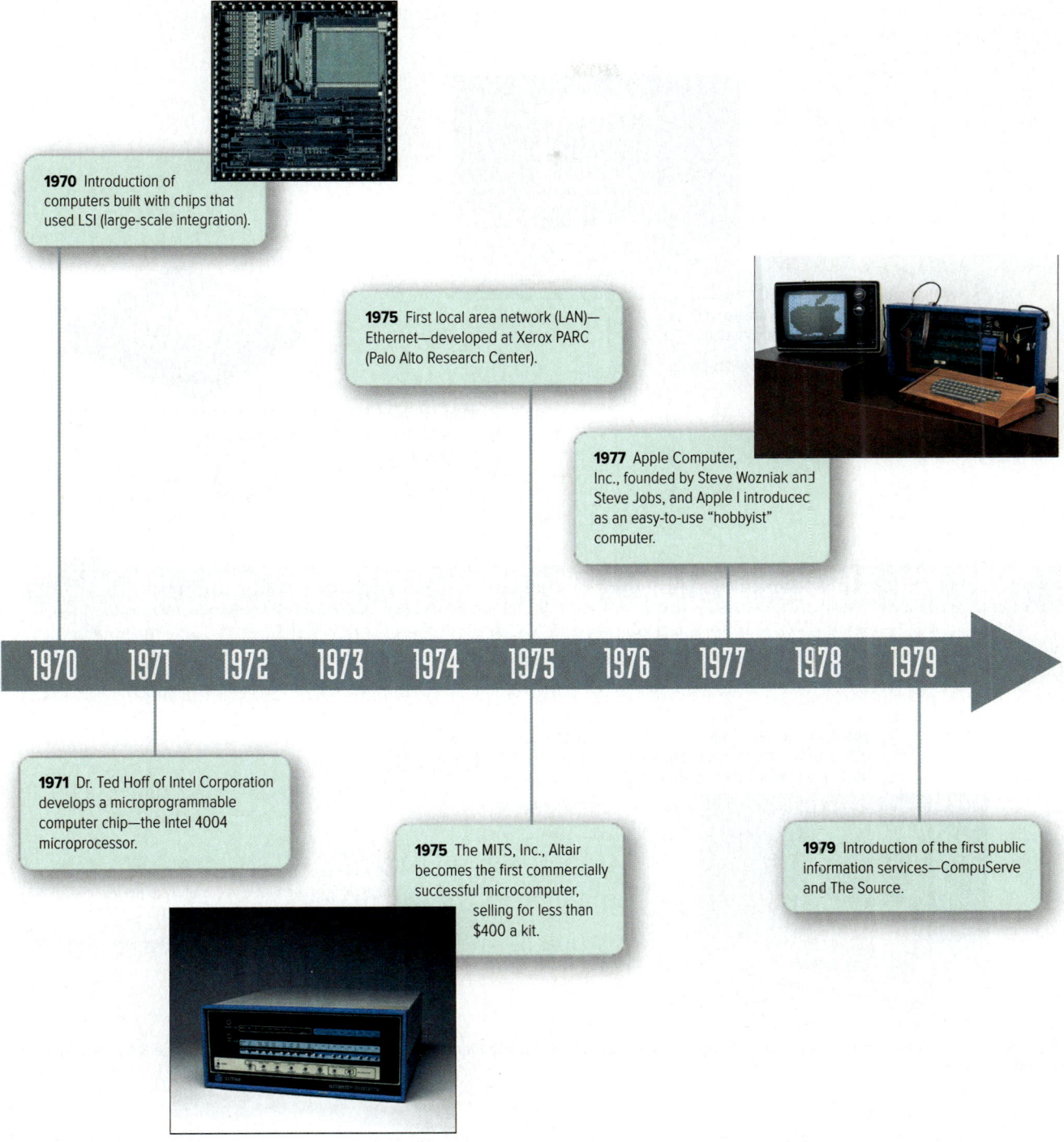

1970 Introduction of computers built with chips that used LSI (large-scale integration).

1975 First local area network (LAN)—Ethernet—developed at Xerox PARC (Palo Alto Research Center).

1977 Apple Computer, Inc., founded by Steve Wozniak and Steve Jobs, and Apple I introduced as an easy-to-use "hobbyist" computer.

1971 Dr. Ted Hoff of Intel Corporation develops a microprogrammable computer chip—the Intel 4004 microprocessor.

1975 The MITS, Inc., Altair becomes the first commercially successful microcomputer, selling for less than $400 a kit.

1979 Introduction of the first public information services—CompuServe and The Source.

(1970) RGB Ventures/SuperStock/Alamy Stock Photo; (1975) Timothy A. Clary/AFP/Getty Images; (1977) Science & Society Picture Library/Getty Images

The Evolution of the Computer Age

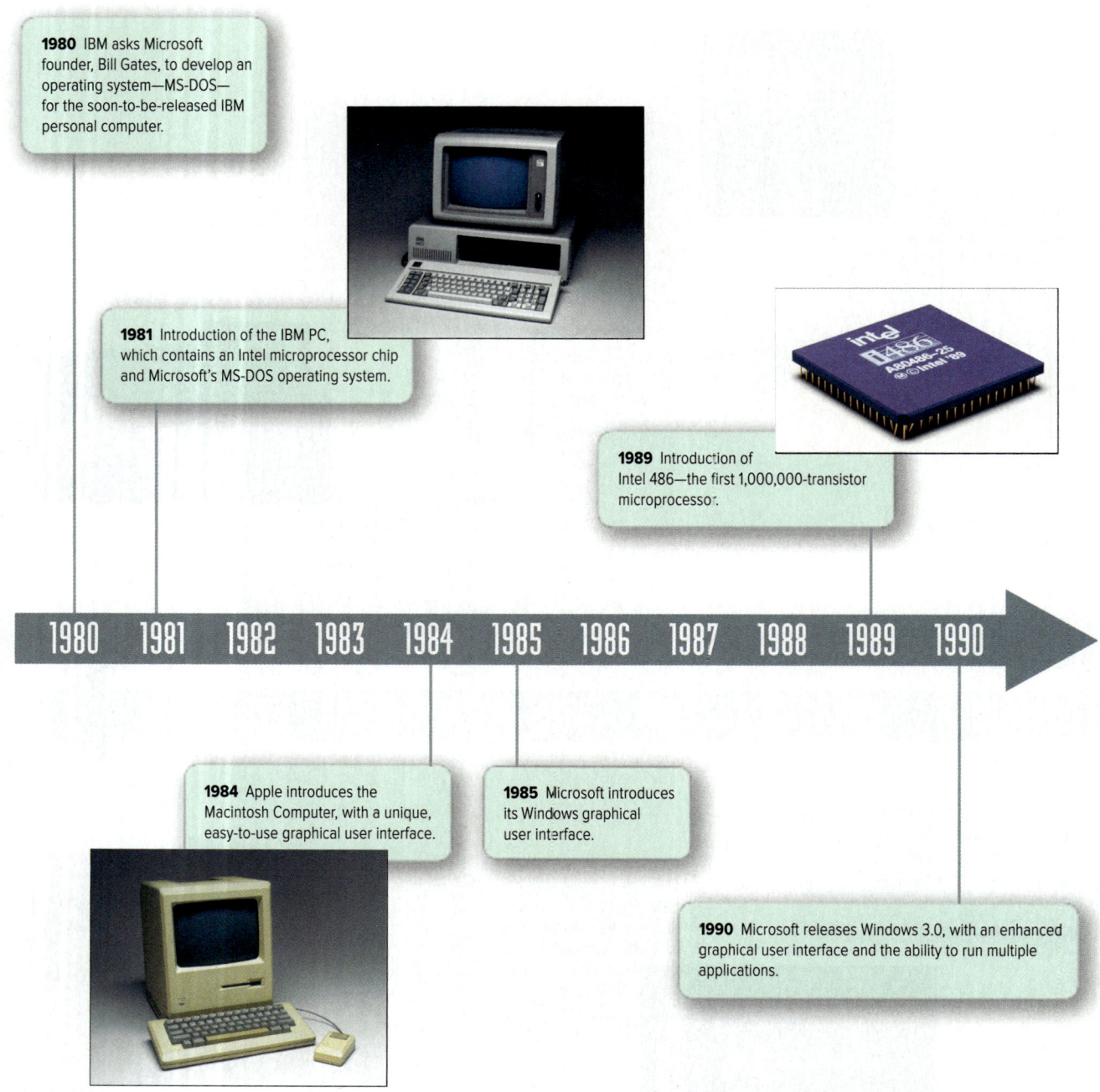

1980 IBM asks Microsoft founder, Bill Gates, to develop an operating system—MS-DOS—for the soon-to-be-released IBM personal computer.

1981 Introduction of the IBM PC, which contains an Intel microprocessor chip and Microsoft's MS-DOS operating system.

1989 Introduction of Intel 486—the first 1,000,000-transistor microprocessor.

1984 Apple introduces the Macintosh Computer, with a unique, easy-to-use graphical user interface.

1985 Microsoft introduces its Windows graphical user interface.

1990 Microsoft releases Windows 3.0, with an enhanced graphical user interface and the ability to run multiple applications.

(1981) Science & Society Picture Library/Getty Images; (1984) Science & Society Picture Library/Getty Images; (1989) Science & Society Picture Library/Getty Images

Fifth Generation: The Age of Connectivity

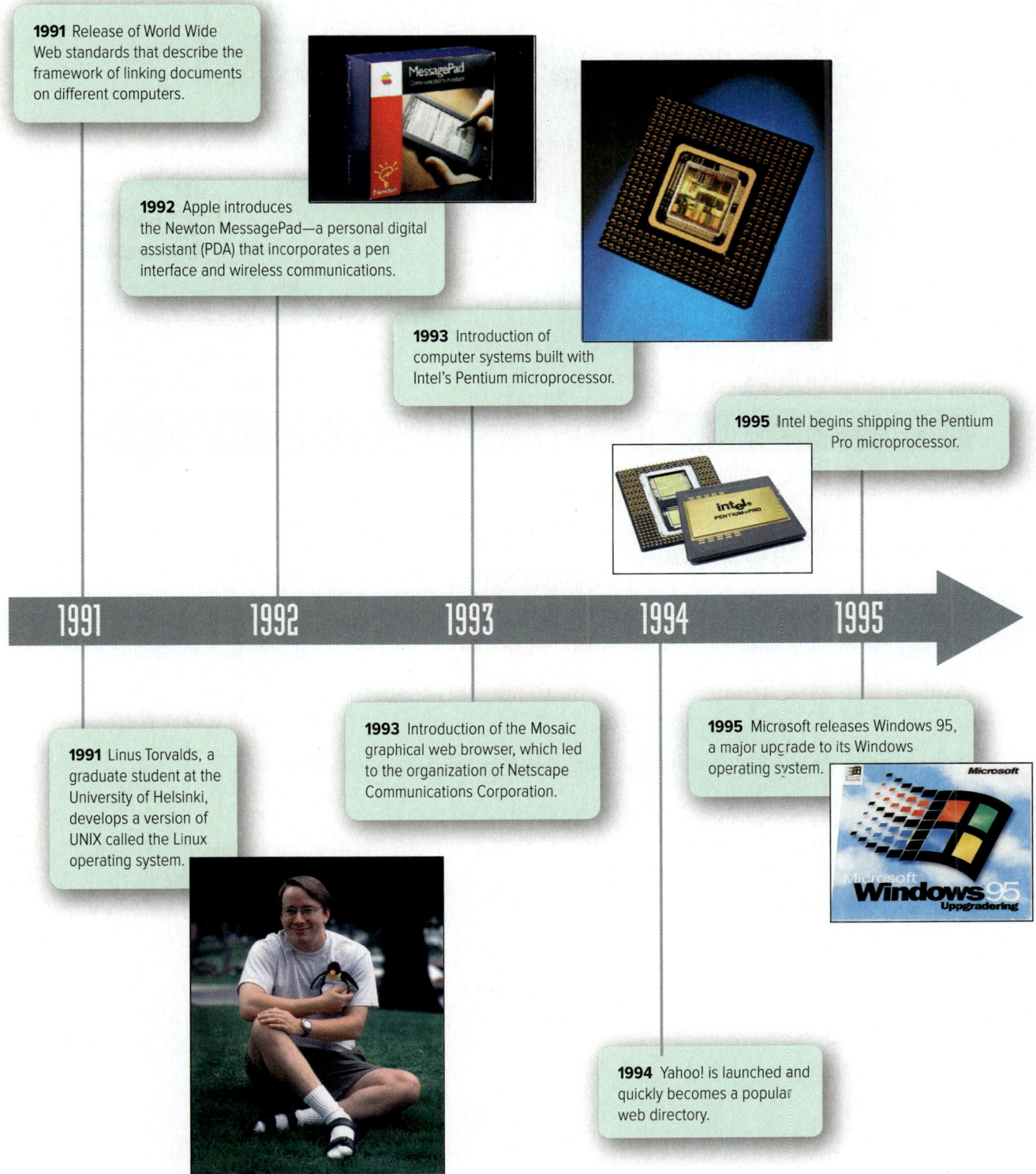

1991 Release of World Wide Web standards that describe the framework of linking documents on different computers.

1992 Apple introduces the Newton MessagePad—a personal digital assistant (PDA) that incorporates a pen interface and wireless communications.

1993 Introduction of computer systems built with Intel's Pentium microprocessor.

1995 Intel begins shipping the Pentium Pro microprocessor.

1991 Linus Torvalds, a graduate student at the University of Helsinki, develops a version of UNIX called the Linux operating system.

1993 Introduction of the Mosaic graphical web browser, which led to the organization of Netscape Communications Corporation.

1995 Microsoft releases Windows 95, a major upgrade to its Windows operating system.

1994 Yahoo! is launched and quickly becomes a popular web directory.

(1991) Chuck Nacke/Alamy Stock Photo, (1992) Science & Society Picture Library/Getty Images, (1993) Science & Society Picture Library/Getty Images, (1995) (microprocessor) Science & Society Picture Library/Getty Images, (microprocessor) Jerry Marshall/pictureresearching.com, (Windows) Roland Magnusson/Shutterstock

The Evolution of the Computer Age

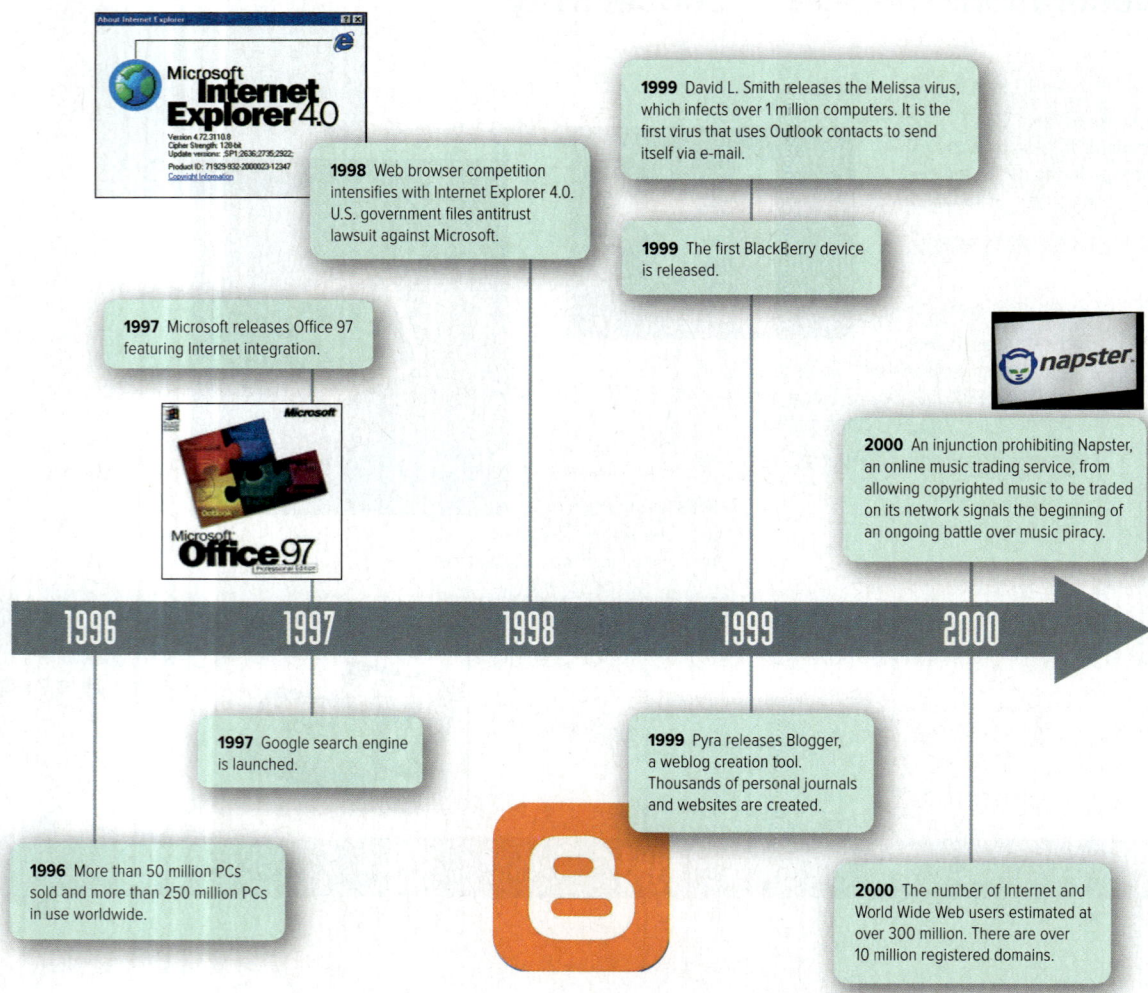

1998 Web browser competition intensifies with Internet Explorer 4.0. U.S. government files antitrust lawsuit against Microsoft.

1999 David L. Smith releases the Melissa virus, which infects over 1 million computers. It is the first virus that uses Outlook contacts to send itself via e-mail.

1999 The first BlackBerry device is released.

1997 Microsoft releases Office 97 featuring Internet integration.

2000 An injunction prohibiting Napster, an online music trading service, from allowing copyrighted music to be traded on its network signals the beginning of an ongoing battle over music piracy.

1996 1997 1998 1999 2000

1997 Google search engine is launched.

1999 Pyra releases Blogger, a weblog creation tool. Thousands of personal journals and websites are created.

1996 More than 50 million PCs sold and more than 250 million PCs in use worldwide.

2000 The number of Internet and World Wide Web users estimated at over 300 million. There are over 10 million registered domains.

(1997) Microsoft Corporation, (1998) Microsoft Corporation, (1999) Blogger.com, 2000 360b/Shutterstock

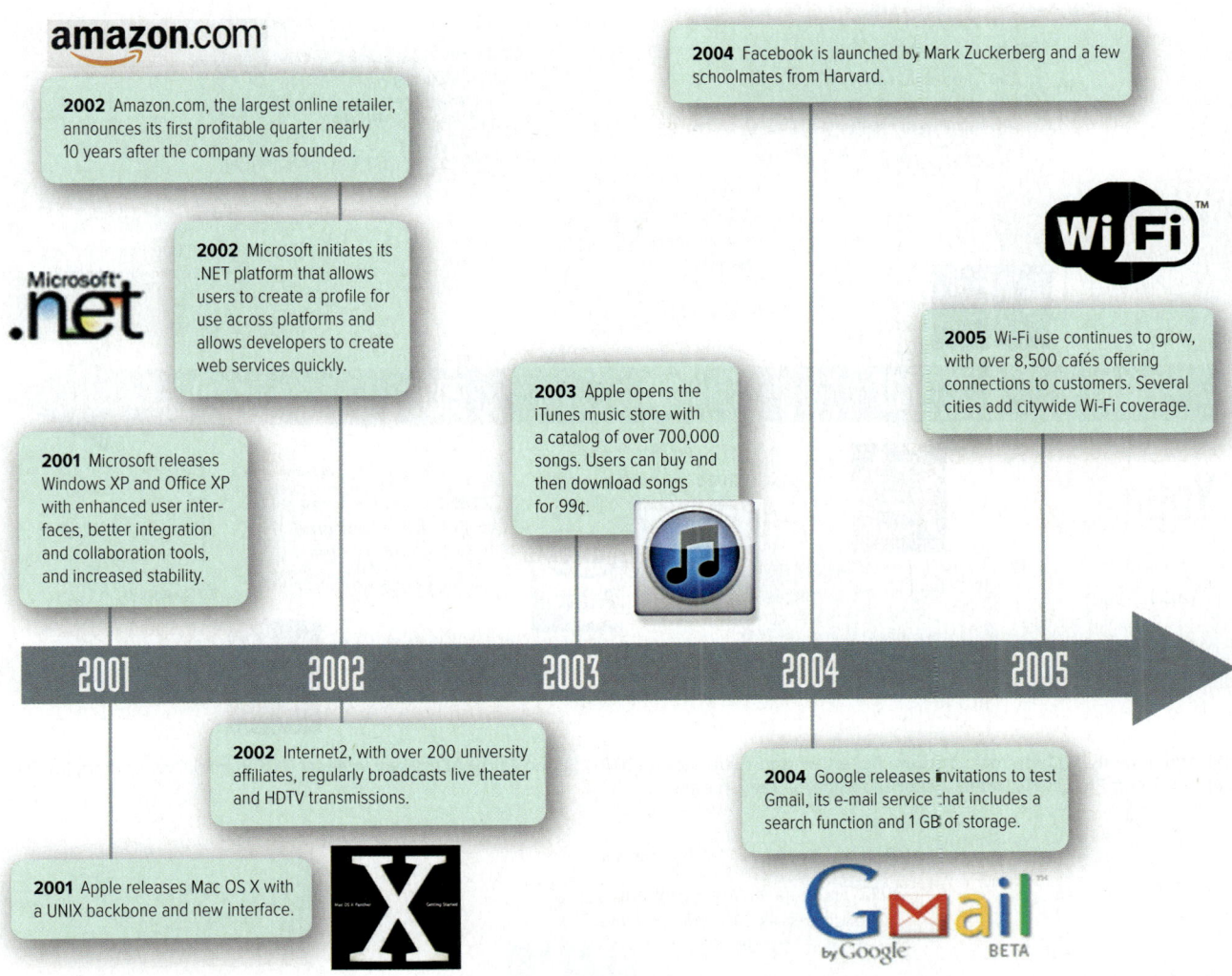

The Evolution of the Computer Age

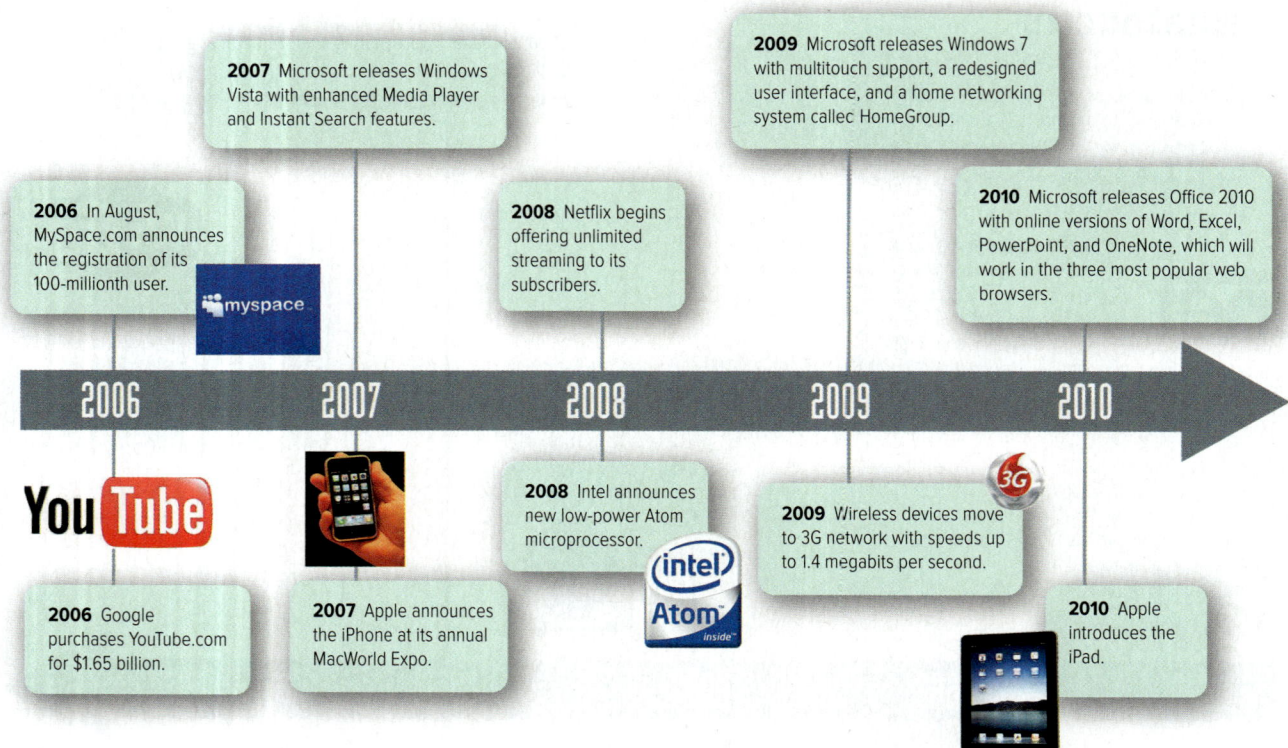

(2006) **top:** thelefty/Shutterstock; **bottom:** Rose Carson/Shutterstock; (2007) Shaun Curry/AFP/Getty Images; (2008) Intel Corporation; (2009) Vodafone Group Plc; (2010) Daniel Acker/Bloomberg/Getty Images

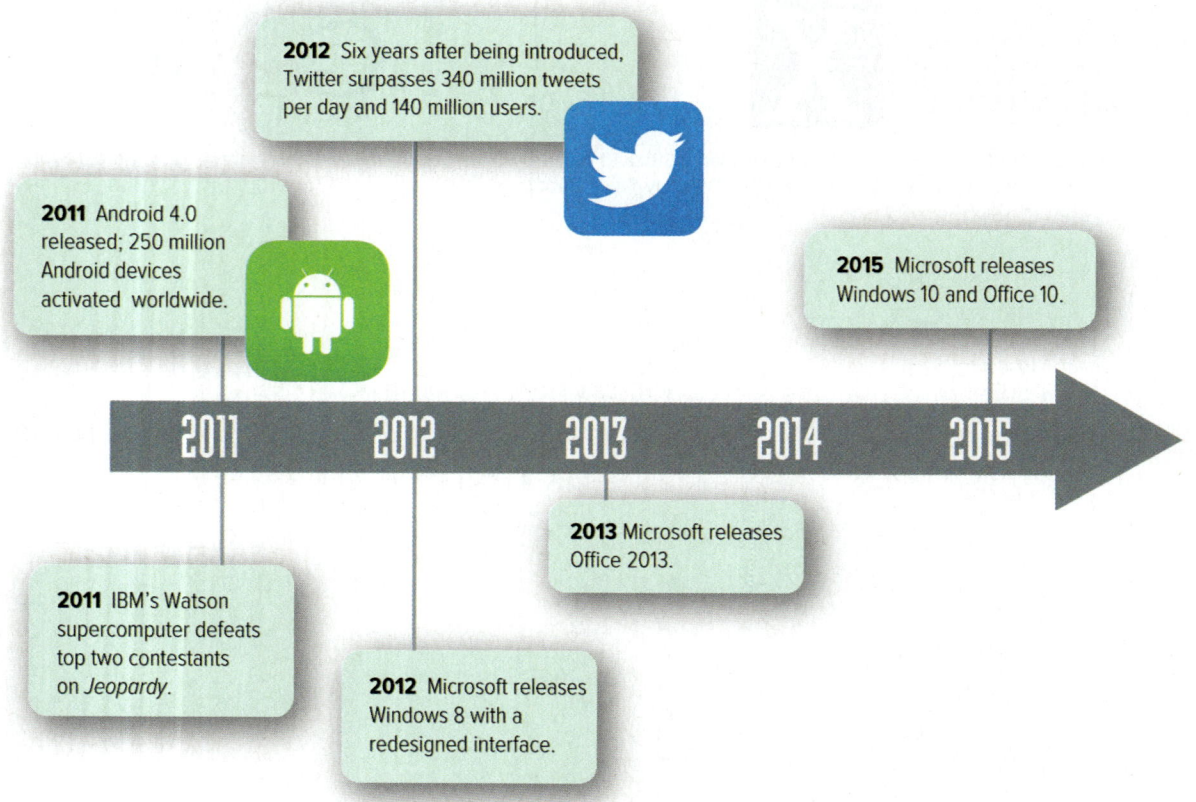

(2011) Rose Carson/Shutterstock; (2015) solomon7/Shutterstock

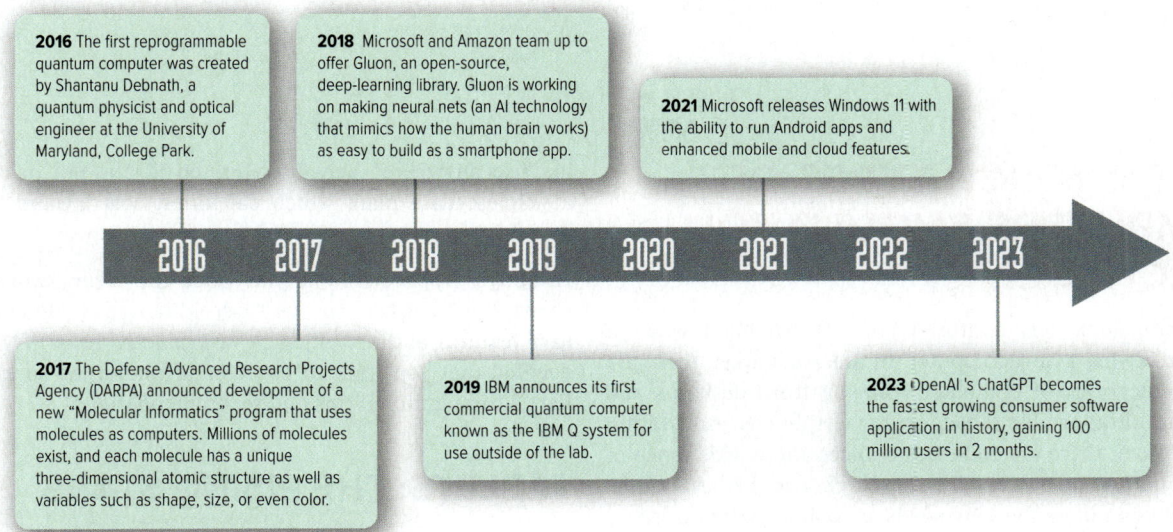

The Computer Buyer's Guide

PART 1: WHICH TYPES OF COMPUTERS SHOULD YOU PURCHASE?

Many individuals feel confused and overwhelmed when it comes to buying a new computer. In today's market, there are various categories of computers ranging from desktops and laptops to ultraportable tablets and smartphones. And within each category, there are countless choices and a wide range of prices. This buying guide will help you choose the best combination of computers for your needs, as well as point out some key specifications that they should have.

LAPTOP COMPUTERS: TODAY'S STANDARD

Typical Recommendation: Buy a $700 to $1,000 laptop with a 13-inch screen.

The laptop computer has become the must-have device that can do everything you need, wherever you need it. They are powerful enough to be your primary home computer (replacing desktops) while portable enough to take to school, work, the local coffee shop, and even on your next trip. Many students are bringing them to classes to take notes and perform research. If you can own only one computer, make it a laptop. Affordable ones start at under $700.

If you have decided that you want to buy a laptop computer, be sure to read Part 2 for additional tips and recommendations.

SMARTPHONES: COMPUTERS IN YOUR POCKET

Typical Recommendation: Obtain a free or low-cost smartphone by signing a two-year wireless contract with a data plan.

For many individuals, the smartphone has become much like a wallet, purse, or key—you don't leave home without it. This single pocket-sized device fills many roles. It is a mobile phone, digital camera, video recorder, gaming device, and personal digital assistant. Most individuals purchase a smartphone because they need to access information quickly and easily, regardless of where they are.

With many smartphones now available for free with two-year contracts, it is easier than ever to replace your older cell phone. The biggest drawback to owning a smartphone is that its "use anywhere" benefit is limited if it is not paired with a wireless data plan, which can often cost around $30 per month.

Although Apple's iPhone was once the clear leader in the smartphone market, various Android devices include competitive features and are enjoying favorable reviews. If you have decided that you want to buy a smartphone, be sure to read Part 2 for additional tips and recommendations.

TABLETS: THE RISING STARS

Typical Recommendation: Buy a 10-inch tablet with 64 GB of storage.

The release of the iPad has revived this category of computers. With an attractive, 10-inch touch-screen surface, it has quickly become a popular device for watching videos, playing games, reading e-books, and browsing the web. Tablets are light enough to hold comfortably for many hours, yet powerful enough to use a variety of apps. They generally have a much longer battery life than laptop computers.

If you are a typical computer user, then you may not benefit much from purchasing a tablet computer. Most of your computing time will be spent on your laptop computer (e.g., at home and at school) or on your smartphone. Although tablets are very popular right now, you must remember that they cost several hundred dollars. You should have a very clear need for this sort of device before you decide to spend the money on it.

Many students consider buying tablets because the low weight makes a tablet easy to carry around. Although this is true, there are lightweight laptop computers that weigh only one or two pounds more and are much more versatile than tablets. Such versatility is crucial for various types of software that are not available as apps and require installation on a Windows or Mac computer. Before you decide between a tablet and lightweight laptop, consider the type of software you will need for your schoolwork, business, or personal use. If everything you need is available as a mobile app, then a tablet (paired with a wireless keyboard) will be more affordable than a lightweight laptop computer.

In conclusion, purchase a tablet computer only if you see yourself using it often. The touch-screen interface, low weight, and long battery life do indeed make it easy to carry around and use almost anywhere you go. However, the price tag does not make the decision an obvious one if you already own a laptop computer and smartphone. If you have decided that you want to buy a tablet, be sure to read Part 2 for additional tips and recommendations.

DESKTOPS

Typical Recommendation: Buy a tower system with a 20- to 24-inch monitor for approximately $900.

Desktop (or tower) computers have been around for a long time. You will still find them in many offices and homes. However, they are steadily losing popularity because laptop computers have become very powerful and affordable. Why have a computer that remains stuck in your home or office when you can have a laptop that can be taken almost anywhere?

Although laptop computers seem to be the better choice, there are several reasons why you might want to purchase a desktop computer. First, the use of a laptop computer as your primary computer can be uncomfortable. After prolonged use, it can place stress on your neck, back, and wrists. Desktops typically come with large monitors, and various types of ergonomic keyboards are available. However, desktop critics will point out that laptop computers can be connected to external monitors and keyboards as well. The only downside is that a laptop requires the purchase of extra equipment.

The other reason to get a desktop is if you have a specific need that cannot be addressed by a laptop computer. For example, some families have a media center computer that holds all the videos, photos, and music for the entire home network. Another example involves gamers, who often seek to build or customize extremely powerful computers. They often choose desktops for this endeavor.

Most users will not need to purchase a desktop computer. In fact, they are currently the least popular of the four categories in this buying guide. However, if you have decided that you want to buy a desktop, be sure to read Part 2 for additional tips and recommendations.

PART 2: PERSONALIZED BUYING GUIDE

Now that you have decided which types of computers you need, it is time to explore computer specifications. The following areas explore the decision-making process from the perspective of various buyers.

LAPTOPS

There are two basic categories of laptops: the traditional laptop and the ultrabook. Although these two categories of laptop computers have the same general appearance, they vary greatly in power, storage capacity, weight, and battery life. The following section helps you find the device that best meets your needs.

User #1: *I am a power user. I need a portable computer that can handle the latest video games or process-intensive operations such as video editing, engineering, and design.*

Response #1: Purchase a traditional laptop computer that includes the following minimum specs:

- **The fastest categories of processors with large number of cores and high GHz count**
- **A graphics processor (GPU) outside of the main CPU**
- **16-GB RAM**
- **1-TB hard drive**
- **17-inch screen**

Expect to pay approximately $1,500, perhaps more. For games, many individuals choose Windows-based PCs. The video and design industries usually use Macs.

User #2: *I am a regular user. I need a desktop replacement and portable computer. I typically run office software, use the Internet, and listen to music.*

Response #2: Purchase an affordable traditional laptop computer that includes the following specs:

- **Middle-tier processors—not the fastest but not the slowest either**
- **8-GB RAM**
- **500-GB hard drive**
- **15-inch screen**

Expect to pay approximately $800 to $1,000. For maximum savings, as well as compatibility with most software, many buyers choose Windows-based PCs. Be sure to purchase an external monitor, keyboard, and mouse if you plan on heavy use while at home.

User #3: *I want a small, lightweight computer that I can carry anywhere. I would like long battery life for extended use.*

Response #3: Purchase an ultrabook with

- **11- to 13-inch screen**
- **Solid-state hard drive**
- **8-GB RAM**
- **Weight under 4 pounds**

Expect to pay $700 to $1,000. Many ultrabooks will not include a DVD drive. Windows-based ultrabooks tend to be more affordable. The MacBook Air is slightly more expensive, but it has always been considered a leader in the lightweight laptop field.

SMARTPHONES

Shopping for a smartphone involves three separate processes: (1) choosing an operating system, either iOS or Android; (2) choosing a device; and (3) choosing a wireless carrier. Although the following section does not review different wireless companies or data plans, it presents a list of smartphone features that you should always consider before making your choice. In addition, it presents each smartphone operating system from the perspective of typical users.

Features to Consider

- *Screen and device size*: Consider a size (and weight) that is comfortable for you to use and carry around. Four-inch screens are now typical.
- *Screen resolution*: Some devices deliver HD quality for sharp photos and videos.
- *Storage*: Having 64 GB is enough for most users. Choose 128 GB or more if you plan to store large quantities of music, photos, and videos. Some devices allow you to increase storage by using memory cards.
- *Battery life*: About 7 to 8 hours is typical.
- *Cameras*: Many include front- and rear-facing cameras. Compare megapixels and photo quality.
- *App market*: Consider the number of apps, and ensure that any apps you need are available for this operating system.

Operating Systems

- *iOS*: **The iPhone is considered by many to be the standard against which all smartphones are measured. Buyers typically choose the iPhone if they are Mac owners or prefer a tightly controlled "ecosystem" from Apple where stability and ease of use are favored over heavy customization.**
- *Android*: **There are many Android devices available from various manufacturers, some free with a wireless contract. Buyers who enjoy customizing their interface typically choose Android. In addition, this operating system is tightly integrated with many of Google's products and services.**

Remember that smartphones can cost very little when you sign a two-year contract with a wireless company. Consider whether you want to make that sort of commitment.

TABLET COMPUTERS

Because of their increased popularity, tablets are available in several sizes from many different companies. The following section helps you find the device that best meets your needs.

User #1: I want to watch videos, play games, create notes, and browse websites.

Response #1: Purchase a 10-inch tablet. Expect to pay $400 to $500. Most weigh about 1.5 pounds. Apple's iPad was very popular when first released, and it continues to be a popular choice among buyers today. Other things to consider:

- **For greater customization, consider an Android-based tablet. Several models are comparable to the iPad, which is considered by many reviewers to be the standard bearer.**
- **Make sure the apps you plan on using are available for the tablet's operating system.**
- **Having 16 GB of storage is typical. Some tablets let you expand this by using flash memory cards.**

User #2: I want to read e-books, browse websites, and have a lightweight device that I can hold with one hand.

Response #2: Purchase a 7 to 8 inch tablet. Expect to pay around $200. Most weigh less than 1 pound. Apple's iPad mini and Amazon's Fire 7 are popular Android-based tablets.

DESKTOP COMPUTERS

Desktops remain popular for offices, both at home and in many companies. In addition, many power users require them for graphics-intensive tasks. The following section helps you find the device that best meets your needs.

User #1: I need a powerful computer that can handle the latest video games or process-intensive operations such as video editing, engineering, and design.

Response #1: Purchase a tower-based computer with the following minimum specs:

- **The fastest categories of processors with large number of cores and high GHz count**
- **High-performance video card**
- **8- to 16-GB RAM**
- **1-TB SSD**
- **24-inch monitor**
- **Graphics card**

Expect to pay at least $1,500 for these powerful computers. For games, many individuals choose Windows-based PCs. The video and design industries usually use Macs.

User #2: I would like a computer that can be used by the entire family for many years.

Response #2: Purchase a tower or all-in-one computer with the following minimum specs:

- **Middle-tier processor**
- **16-GB RAM**
- **250-GB SSD (increase up to 1 TB if you plan to store a large number of videos)**
- **20- to 24-inch monitor**
- **Ergonomic keyboard**

Tower-based systems are usually offered in a package that includes the monitor and several peripherals. All-in-one systems will, of course, include the monitor, some of which are touch screens. Spending approximately $1,000 to $1,500 will help ensure that the components remain relevant for several years. For a stylish and powerful all-in-one computer, consider the slightly more expensive iMac from Apple.

Glossary

1G (first-generation mobile telecommunications) Started in the 1980s using analog signals to provide voice transmission service.

2G (second-generation mobile telecommunications) Started in the 1990s using digital radio signals.

3D printer A printer that creates objects by adding layers of material onto one another. Also known as additive manufacturing.

3D scanner Scanner that uses lasers, cameras, or robotic arms to record the shape of an object.

3G (third-generation mobile telecommunications) Started in the 2000s and provided services capable of effective connectivity to the Internet, marking the beginning of smartphones.

3GLs (third-generation languages) High-level procedural language. *See also* Procedural language.

4G (fourth-generation mobile telecommunications) Replacing 3G networks in some areas with providers using WiMax and LTE connections to provide faster transmission speeds.

4GLs (fourth-generation languages) Very high-level or problem-oriented languages. *See also* Task-oriented language.

5G (fifth-generation mobile telecommunications) A cellular network technology being developed to replace 4G networks with speeds that rival home Internet connections.

5GLs (fifth-generation languages) *See* Fifth-generation language.

a

AC adapter Notebook computers use AC adapters that are typically outside the system unit. They plug into a standard wall outlet, convert AC to DC, provide power to drive the system components, and can recharge batteries.

Access Refers to the responsibility of those who have data to control who is able to use that data.

Access speed Measures the amount of time required by the storage device to retrieve data and programs.

Accounting The organizational department that records all financial activity from billing customers to paying employees.

Accounts payable The activity that shows the money a company owes to its suppliers for the materials and services it has received.

Accounts receivable The activity that shows what money has been received or is owed by customers.

Accuracy Relates to the responsibility of those who collect data to ensure that the data is correct.

Active display area The diagonal length of a monitor's viewing area.

Activity tracker A wearable computer that typically monitors daily exercise and sleep patterns.

Additive manufacturing *See* 3D printer.

Address Located in the header of an e-mail message; the e-mail address of the persons sending, receiving, and, optionally, anyone else who is to receive copies.

Advanced Research Project Agency Network (ARPANET) A national computer network from which the Internet developed.

Agile development A development methodology that starts by getting core functionality of a program working, then expands on it until the customer is satisfied with the results.

All-in-one desktop A desktop computer that has the monitor and system unit housed together in the same case (e.g., Apple's iMac).

Analog Continuous signals that vary to represent different tones, pitches, and volume.

Analog signal Signal that represents a range of frequencies, such as the human voice. It is a continuous electronic wave signal as opposed to a digital signal that is either on or off.

Android Mobile operating system originally developed by Android Inc., and later purchased by Google.

Antispyware *See* Spy removal program.

Antivirus program A utility program that guards a computer system from viruses or other damaging programs.

App *See* Application software.

App store A website that provides access to specific mobile apps that can be downloaded either for a nominal fee or free of charge.

Application generation subsystem Provides tools to create data entry forms and specialized programming languages that interface or work with common languages, such as C or Visual Basic.

Application generator Also called program coder; provides modules of prewritten code to accomplish various tasks, such as calculation of overtime pay.

Application software Also referred to as apps. Software that can perform useful work, such as word processing, cost estimating, or accounting tasks. The user primarily interacts with application software.

Arithmetic operation Fundamental math operations: addition, subtraction, multiplication, and division.

Arithmetic-logic unit (ALU) The part of the CPU that performs arithmetic and logical operations.

ASCII (American Standard Code for Information Interchange) Binary coding scheme widely used on all computers, including personal computers. Eight bits form each byte, and each byte represents one character.

Aspect ratio The width of a monitor divided by its height. Common aspect ratios for monitors are 4:3 (standard) and 16:9 (wide screen).

Assembly language A step up from machine language, using names instead of numbers. These languages use abbreviations or mnemonics, such as ADD, that are automatically converted to the appropriate sequence of 1s and 0s.

Asymmetric digital subscriber line (ADSL) One of the most widely used types of telephone high-speed connections (DSL).

Attachment A file, such as a document or worksheet, that is attached to an e-mail message.

Attribute A data field represents an attribute (description or characteristic) of some entity (person, place, thing, or object). For example, an employee is an entity with many attributes, including his or her last name, address, phone, etc.

Authentication The process of ensuring the identity of a user.

Automated design tool Software package that evaluates hardware and software alternatives according to requirements given by the systems analyst. Also called computer-aided software engineering (CASE) tools.

b

Backbone *See* Bus.

Backend The software and hardware of a computing system that the user does not interact with.

Background Other programs running simultaneously with the program being used in an operating system. *See also* Foreground.

Backup A Windows utility program. *See* Backup program.

Backup program A utility program that helps protect you from the effects of a disk failure by making a copy of selected or all files that have been saved onto a disk.

Balance sheet Lists the overall financial condition of an organization.

Bandwidth Bandwidth determines how much information can be transmitted at one time. It is a measurement of the communication channel's capacity. There are three bandwidths: voice band, medium band, and broadband.

Bar code Code consisting of vertical zebra-striped marks printed on product containers, read with a bar code reader.

Bar code reader Photoelectric scanner that reads bar codes for processing.

Bar code scanner *See* Bar code reader.

Base station *See* Wireless access point.

Baseband Bandwidth used to connect individual computers that are located close to one another. Though it supports high-speed transmission, it can only carry a single signal at a time.

Batch processing Processing performed all at once on data that has been collected over time.

Beta testing Testing by a select group of potential users in the final stage of testing a program.

Big data Term given to describe the ever-growing volume of data currently being collected.

Binary system Numbering system in which all numbers consist of only two digits: 0 and 1.

Biometric scanning Devices that check fingerprints or retinal scans.

BIOS (basic input/output system) Information including the specifics concerning the amount of RAM and the type of keyboard, mouse, and secondary storage devices connected to the system unit.

Bit (binary digit) Each 1 or 0 is a bit; short for binary digit.

Bitcoin A form of digital cash. Bitcoin currency exists only on the Internet.

Bitmap image Graphic file in which an image is made up of thousands of dots (pixels).

BitTorrent A peer-to-peer file-sharing protocol used for distributing large amounts of data over the Internet.

Blog A type of personal website where articles are regularly posted.

Blu-ray disc (BD) A type of high-definition disc with a capacity of 25 to 50 gigabytes.

Bluetooth A wireless technology that allows nearby devices to communicate without the connection of cables or telephone systems.

Booting Starting or restarting your computer.

Botnet A collection of zombie computers.

Broadband Bandwidth that includes microwave, satellite, coaxial cable, and fiber-optic channels. It is used for very-high-speed computers.

Browser Special Internet software connecting you to remote computers; opens and transfers files, displays text and images, and provides an uncomplicated interface to the Internet and web documents. Examples of browsers are Internet Explorer, Mozilla Firefox, and Google Chrome.

Browser cache A collection of temporary Internet files that contain web page content and instructions for displaying this content.

Bus All communication travels along a common connecting cable called a bus or a backbone. As information passes along the bus, it is examined by each device on the system board to see if the information is intended for that device. *See also* Bus line *and* Ethernet.

Bus line Electronic data roadway along which bits travel; connects the parts of the CPU to each other and links the CPU with other important hardware. The common connecting cable in a bus network.

Bus network Each device is connected to a common cable called a bus or backbone, and all communications travel along this bus.

Bus width The number of bits traveling simultaneously down a bus is the bus width.

Business-to-business (B2B) A type of electronic commerce that involves the sale of a product or service from one business to another. This is typically a manufacturer-supplier relationship.

Business-to-consumer (B2C) A type of electronic commerce that involves the sale of a product or service to the general public or end users.

Button A special area you can click to make links that "navigate" through a presentation.

Byte Unit consisting of eight bits. There are 256 possible bit combinations in a byte, and each byte represents one character.

C

Cable Cords used to connect input and output devices to the system unit.

Cable modem Type of modem that uses coaxial cable to create high-speed computer connections.

Cable service Service provided by cable television companies using existing television cables.

Cache memory Area of random-access memory (RAM) set aside to store the most frequently accessed information. Cache memory improves processing by acting as a temporary high-speed holding area between memory and the CPU, allowing the computer to detect which information in RAM is most frequently used.

California Consumer Privacy Act (CCPA) State level privacy regulation that allows consumers to see all data a company has collected on them and any third parties that data has been shared with.

Capacity Capacity is how much data a particular storage medium can hold and is another characteristic of secondary storage.

Card reader A device that interprets the encoded information contained on credit, debit, access, and identification cards.

Carpal tunnel syndrome A repetitive strain injury consisting of damage to the nerves and tendons in the hands.

Cascading style sheets (CSS) Files inserted into an HTML document that control the appearance of web pages, including layout, colors, and fonts.

CD *See* Compact disc.

Cell The space created by the intersection of a vertical column and a horizontal row within a worksheet in a program like Microsoft Excel. A cell can contain text or numeric entries.

Cell phone Mobile device that uses cellular system to connect without a physical connection.

Cell tower Antennae that support cellular communication.

Cellular Type of wireless connection that uses multiple antennae (cell towers) to send and receive data within relatively small geographic regions (cells).

Cellular service provider Supports voice and data transmission to wireless devices.

Central processing unit (CPU) The part of the computer that holds data and program instructions for processing the data. The CPU consists of the control unit and the arithmetic-logic unit. In a personal computer, the CPU is on a single electronic component called a microprocessor chip.

Character A single letter, number, or special character, such as a punctuation mark or $.

Character encoding standards Assign unique sequence of bits to each character.

Child node A node one level below the node being considered in a hierarchical database or network. *See also* Parent node.

Chip A tiny circuit board etched on a small square of sand-like material called silicon. A chip is also called a silicon chip, semiconductor, or integrated circuit.

Chip card Type of credit card that contains an embedded microchip to provide added security.

Chip carrier The device onto which chips are mounted and plugged into the system board.

Chrome OS An operating system designed by Google for netbook computers and Internet connectivity through cloud computing.

Clarity Indicated by the resolution, or number of pixels, on a monitor. The greater the resolution, the better the clarity.

Class In an object-oriented database, classes are similar objects grouped together.

Client A node that requests and uses resources available from other nodes. Typically, a client is a user's personal computer.

Client-based e-mail system A system that requires a special program known as an e-mail client to be installed on your computer.

Client/server network Network in which one powerful computer coordinates and supplies services to all other nodes on the network. Server nodes coordinate and supply specialized services, and client nodes request the services.

Clock speed Also called clock rate. It is measured in gigahertz, or billions of beats per second. The faster the clock speed, the faster the computer can process information and execute instructions.

Cloud computing Data stored at a server on the Internet and available anywhere the Internet can be accessed.

Cloud storage Also known as online storage. An Internet-based space for storing data and files.

Cloud suite Suite stored at a server on the Internet and available anywhere from the Internet.

Coaxial cable High-frequency transmission cable that replaces the multiple wires of telephone lines with a single solid-copper core. It is used to deliver television signals as well as to connect computers in a network.

Code Writing a program using the appropriate computer language.

Code review *See* Desk checking.

Coding Actual writing of a computer program, using a programming language.

Cold boot Starting the computer after it has been turned off.

Combination key Keys such as the Ctrl key that perform an action when held down in combination with another key.

Commercial database Enormous database an organization develops to cover certain particular subjects. Access to this type of database is usually offered for a fee or subscription. Also known as data bank and informational utility.

Common data item In a relational database, all related tables must have a common data item or key field.

Communication channel The actual connecting medium that carries the message between sending and receiving devices. This medium can be a physical wire, cable, or wireless connection.

Communication device Computer systems that communicate with other computer systems using modems. For example, it modifies computer output into a form that can be transmitted across standard telephone lines.

Communication system Electronic system that transmits data over communication lines from one location to another.

Compact disc (CD) Widely used optical disc format. It holds 650 MB (megabytes) to 1 GB (gigabyte) of data on one side of the CD.

Company database Also called shared database. Stored on a mainframe, users throughout the company have access to the database through their personal computers linked by a network.

Compiler Software that converts the programmer's procedural language program (source code) into machine language (object code). This object code can then be saved and run later.

Computer crime *See* Cybercrime.

Computer ethics Guidelines for the morally acceptable use of computers in our society.

Computer Fraud and Abuse Act Law allowing prosecution of unauthorized access to computers and databases.

Computer monitoring software The most invasive and dangerous type of spyware. These programs record every activity made on your computer, including credit card numbers, bank account numbers, and e-mail messages.

Computer network Communications system connecting two or more computers and their peripheral devices to exchange information and share resources.

Computer support specialist Specialists include technical writers, computer trainers, computer technicians, and help-desk specialists who provide technical support to customers and other users.

Computer technician Specialist who installs hardware and software and troubleshoots problems for users.

Computer-aided design/computer-aided manufacturing (CAD/CAM) system Knowledge work systems that run programs to integrate the design and manufacturing activities. CAD/CAM is widely used in manufacturing automobiles.

Computer-aided software engineering (CASE) tool A type of software development tool that helps provide some automation and assistance in program design, coding, and testing. *See also* Automated design tool.

Connectivity Capability of the personal computer to use information from the world beyond one's desk. Data and information can be sent over telephone or cable lines and through the air so that computers can talk to each other and share information.

Consumer-to-consumer (C2C) A type of electronic commerce that involves individuals selling to individuals.

Contextual tab A type of tab found in Microsoft Word that appears only when needed and anticipates the next operations to be performed by the user.

Contrast ratio Indicates a monitor's ability to display colors. It compares the light intensity of the brightest white to the darkest black.

Control unit Section of the CPU that tells the rest of the computer how to carry out program instructions.

Conversion Also known as systems implementation; four approaches to conversion: direct, parallel, pilot, and phased. *See also* Systems implementation.

Cookies Small data files that are deposited on your hard disk from websites you have visited.

Coprocessor Specialized processing chip designed to improve specific computer operations, such as the graphics coprocessor.

Copyright A legal concept that gives content creators the right to control use and distribution of their work.

Cordless mouse A battery-powered mouse that typically uses radio waves or infrared light waves to communicate with the system unit. Also known as wireless mouse.

Cortana A virtual assistant that accepts commands through text or speech. Introduced in 2015 as a part of Windows 10.

Cracker One who gains unauthorized access to a computer system for malicious purposes.

Cryptocurrency A digital currency with no traditional cash equivalent. Cryptocurrencies use an encrypted public ledger, known as a blockchain, to ensure accuracy and security.

Cyberbullying The use of the Internet, a cell phone, or other device to deliver content intended to hurt or embarrass another person.

Cybercrime Any criminal offense that involves a computer and a network. Criminals may be employees, outside users, hackers and crackers, or organized crime members.

Cyberterrorism A politically motivated cybercrime.

Cylinder Hard disks store and organize files using tracks, sectors, and cylinders. A cylinder runs through each track of a stack of platters. Cylinders differentiate files stored on the same track and sector of different platters.

d

Dark web A part of the deep web consisting of sites that use special software to hide a user's IP address and makes it nearly impossible to identify who is using the site.

Data Raw, unprocessed facts that are input to a computer system that will give compiled information when the computer processes those facts. Data is also defined as facts or observations about people, places, things, and events.

Data administration subsystem Helps manage the overall database, including maintaining security, providing disaster recovery support, and monitoring the overall performance of database operations.

Data bank *See* Commercial database.

Data cube A multidimensional data model. *See also* Multidimensional database.

Data definition subsystem This system defines the logical structure of the database by using a data dictionary.

Data dictionary Dictionary containing a description of the structure of data in a database.

Data flow diagram Diagram showing data or information flow within an information system.

Data integrity Database characteristics relating to the consistency and accuracy of data.

Data maintenance Maintaining data includes adding new data, deleting old data, and editing existing data.

Data manipulation The unauthorized access of a computer network and copying files to or from the server.

Data manipulation subsystem Provides tools to maintain and analyze data.

Data mining Technique of searching data warehouses for related information and patterns.

Data processing system (DPS) Transaction processing system that keeps track of routine operations and records these events in a database. Also called transaction processing system (TPS).

Data redundancy A common database problem in which data is duplicated and stored in different files.

Data security Protection of software and data from unauthorized tampering or damage.

Data warehouse Data collected from a variety of internal and external databases and stored in a database called a data warehouse. Data mining is then used to search these databases.

Data worker Person involved with the distribution and communication of information, such as administrative assistants and clerks.

Database A collection of related information, like employee names, addresses, and phone numbers. It is organized so that a computer program can quickly select the desired pieces of information and display them for you.

Database administrator (DBA) Uses database management software to determine the most efficient way to organize and access data.

Database file File containing highly structured and organized data created by database management programs.

Database management system (DBMS) To organize, manage, and retrieve data. DBMS programs have five subsystems: DBMS engine, data definition, data manipulation, applications generation, and data administration. An example of a database management system is Microsoft Access. *See also* Database manager.

Database manager Software package used to set up, or structure, a database such as an inventory list of supplies. It also provides tools to edit, enter, and retrieve data from the database.

Database model Defines rules and standards for all data in a database. There are five database models: hierarchical, network, relational, multidimensional, and object-oriented. For example, Access uses the relational data model.

DBMS engine Provides a bridge between the logical view of data and the physical view of data.

Debugging Programmer's word for testing and then eliminating errors in a program. Programming errors are of two types: syntax and logic errors.

Decision model The decision model gives the decision support system its analytical capabilities. There are three types of models included in the decision model: tactical, operational, and strategic.

Decision support system (DSS) Flexible analysis tool that helps managers make decisions about unstructured problems, such as effects of events and trends outside the organization.

Deep fake Altering an individual's appearance or message in a way designed to mislead others.

Deep web Comprised of websites designed to be hidden from standard search engines to allow communication in a secure and anonymous manner.

Demand report A demand report is produced on request. An example is a report on the numbers and types of jobs held by women and minorities done at the request of the government.

Demodulation Process performed by a modem in converting analog signals to digital signals.

Denial of service (DoS) attack A variant virus in which websites are overwhelmed with data and users are unable to access the website. Unlike a worm that self-replicates, a DoS attack floods a computer or network with requests for information and data.

Density Refers to how tightly the bits (electromagnetic charges) can be packed next to one another on a disk.

Desk checking Process of checking out a computer program by studying a printout of the program line by line, looking for syntax and logic errors.

Desktop (1) The screen that is displayed on the monitor when the computer starts up. All items and icons on the screen are considered to be on your desktop and are used to interact with the computer. (2) A system unit that typically contains the system's electronic components and selected secondary storage devices. Input and output devices, such as the mouse, keyboard, and monitor, are located outside the system unit.

Desktop browser Browser designed for laptop and desktop computer.

Desktop computer Computer small enough to fit on top of or along the side of a desk and yet too big to carry around.

Desktop operating system *See* Stand-alone operating system.

Desktop publishing program Program that allows you to mix text and graphics to create publications of professional quality.

Device driver Every device that is connected to the computer has a special program associated with it called a device driver that allows communication between the operating system and the device.

Diagnostic program *See* Troubleshooting program.

Dial-up service Antiquated method of connecting to the Internet using telephones and telephone modems, which has been replaced with higher-speed connection services.

Dialog box Provides additional information and requests user input.

Dictionary attack Uses software to try thousands of common words sequentially in an attempt to gain unauthorized access to a user's account.

Digital Computers are digital machines because they can only understand 1s and 0s. For example, a digital watch states the exact time on the face, whereas an analog watch has the second hand moving in constant motion as it tells the time.

Digital camera Similar to a traditional camera except that images are recorded digitally in the camera's memory rather than on film.

Digital currency Currency for Internet purchases. Buyers purchase digital cash from a third party (a bank that specializes in electronic currency) by transferring funds from their banks.

Digital footprint A collection of data that can be searched to reveal a highly detailed account of an individual's life.

Digital Millennium Copyright Act Law that makes it legal for a program owner to make only his or her own backup copies of a software program. However, it is illegal for those copies to be resold or given away.

Digital projector A type of monitor that projects images from a traditional monitor onto a screen or wall.

Digital rights management (DRM) Encompasses various technologies that control access to electronic media and files.

Digital signal Computers can only understand digital signals. Before processing can occur within the system unit, a conversion must occur from what we understand (analog) to what the system unit can electronically process (digital). *See also* Analog signal.

Digital subscriber line (DSL) Provides high-speed connection using existing telephone lines.

Digital subscriber line (DSL) service Service provided by telephone companies using existing telephone lines to provide high-speed connections.

Digital versatile disc (DVD) A type of optical disc similar to CD-ROMs except that more data can be packed into the same amount of space. *See also* DVD (digital versatile disc).

Digital video disc *See* DVD (digital versatile disc).

Digital whiteboard A specialized device with a large display connected to a computer or projector.

DIMM (dual in-line memory module) An expansion module used to add memory to the system board.

Direct approach Approach for systems implementation whereby the old system is simply abandoned for the new system.

Directory server A specialized server that manages resources such as user accounts for an entire network.

Disaster recovery plan Plan used by large organizations describing ways to continue operations following a disaster until normal computer operations can be restored.

Disaster recovery specialist Worker responsible for recovering systems and data after a disaster strikes an organization.

Disk caching Method of improving hard-disk performance by anticipating data needs. Frequently used data is read from the hard disk into memory (cache). When needed, data is then accessed directly from memory, which has a much faster transfer rate than from the hard disk. Increases performance by as much as 30 percent.

Display Output device like a television screen that displays data processed by the computer.

Display screen *See* Display.

DisplayPort (DP) A port that an audiovisual device typically uses to connect to large monitors.

Distributed database Database that can be made accessible through a variety of communications networks, which allow portions of the database to be located in different places.

Distributed denial of service (DDoS) attack A denial of service (DoS) attack that coordinates several computers making repeated requests for service to overwhelm and shut down an ISP or website.

Document Any kind of text material.

Document file File created by a word processor to save documents such as letters, research papers, and memos.

Document scanner Similar to a flatbed scanner except that it can quickly scan multipage documents. It automatically feeds one page of a document at a time through a scanning surface.

Documentation Written descriptions and procedures about a program and how to use it. *See also* Program documentation.

Domain name The second part of the URL; it is the name of the server where the resource is located. For example, www.mtv.com.

Domain name server (DNS) Internet addressing method that assigns names and numbers to people and computers. Because the numeric IP addresses are difficult to remember, the DNS server was developed to automatically convert text-based addresses to numeric IP addresses.

Dot pitch Distance between each pixel. The lower the dot pitch, the shorter the distance between pixels and the higher the clarity of images produced.

Dots per inch (dpi) Printer resolution is measured in dpi. The higher the dpi, the better the quality of images produced.

Downlink To receive data from a satellite.

Downloading Process of transferring information from a remote computer to the computer one is using.

Drawing program Program used to help create artwork for publications. *See also* Illustration program.

Drives Storage devices that read data and programs from some type of storage media.

Drone Unmanned aerial vehicles (UAV) that acts as an output device, sending back video and sound to the user.

DS3 Provides support for very high speed, all-digital transmission for large corporations.

DSL *See* Digital subscriber line.

Duplex printing Allows automatic printing on both sides of a sheet of paper.

DVD (digital versatile disc or digital video disc) Similar to CD-ROMs except that more data can be packed into the same amount of space. DVD drives can store a typical capacity of 4.7 GB on one side.

DVI (Digital Video Interface) port A type of port that provides a connection to a digital monitor.

e

E-book *See* E-book reader.

E-book reader Handheld, book-sized device that displays text and graphics. Using content downloaded from the web or special cartridges, these devices are used to read newspapers, magazines, and books.

E-commerce Buying and selling goods over the Internet.

E-ink A black-and-white output from some e-book readers producing images that reflect light like ordinary paper.

E-learning A web application that allows one to take educational courses online.

E-mail Allows you to communicate with anyone in the world who has an Internet address or e-mail account with a system connected to the Internet. You can include a text message, graphics, photos, and file attachments.

E-mail client A special program that communicates with the e-mail service provider and must be installed on the computer first.

E-reader Dedicated mobile devices for storing and displaying e-books and other electronic media including electronic newspapers and magazines.

EBCDIC (Extended Binary Coded Decimal Interchange Code) Binary coding scheme that is a standard for minicomputers and mainframe computers.

Economic feasibility Comparing the costs of a new system to the benefits it promises.

Electronic book Traditional printed books in electronic format.

Electronic commerce (e-commerce) Buying and selling goods over the Internet.

Electronic mail Transmission of electronic messages over the Internet. Also known as e-mail.

Embedded operating system An operating system that is completely stored within the ROM (read-only memory) of the device that it is in; used for handheld computers and smaller devices like PDAs. Also known as a real-time operating system (RTOS).

Employee-monitoring software Programs that record virtually every activity on a computer system.

Encryption Coding information so that only the user can read or otherwise use it.

Encryption key A number used to gain access to encrypted information.

End user Person who uses personal computers or has access to larger computers.

Enterprise storage system Using mass storage devices, a strategy designed for organizations to promote efficient and safe use of data across the networks within their organizations.

Entity In an object-oriented database, a person, place, thing, or event that is to be described.

Ergonomics The study of human factors related to things people use.

eSIMs Unique data on a specialized microchip that allows a subscriber to access cellular networks. A digital replacement of SIM cards.

Ethernet Otherwise known as Ethernet bus or Ethernet LAN. The Ethernet bus is the pathway or arterial to which all nodes (PCs, file servers, print servers, web servers, etc.) are connected. All of this is connected to a local area network (LAN) or a wide area network (WAN). *See also* Bus network.

Ethernet cable Twisted-pair cable commonly used in networks and to connect a variety of components to the system unit.

Ethernet port A high-speed networking port that allows multiple computers to be connected for sharing files or for high-speed Internet access.

Ethics Standards of moral conduct.

Exception report Report that calls attention to unusual events.

Executive information system (EIS) Sophisticated software that can draw together data from an organization's databases in meaningful patterns and highly summarized forms.

Executive support system (ESS) *See* Executive information system.

Expansion buses Connects the CPU to slots on the system board. There are different types of expansion buses such as industry standard architecture (ISA), peripheral component interconnect (PCI), accelerated graphics port (AGP), universal serial bus (USB), and FireWire buses. *See also* System bus.

Expansion card Optional device that plugs into a slot inside the system unit to expand the computer's abilities. Ports on the system board allow cables to be connected from the expansion board to devices outside the system unit.

Expansion slot Openings on a system board. Users can insert optional devices, known as expansion cards, into these slots, allowing users to expand their systems. *See also* Expansion card.

Expert system Computer program that provides advice to decision makers who would otherwise rely on human experts. It's a type of artificial intelligence that uses a database to provide assistance to users.

External data Data gathered from outside an organization. Examples are data provided by market research firms.

External storage External storage is storage that can be removed from the system unit.

Extranet Private network that connects more than one organization.

f

Facebook One of the best-known social networking sites.

Facial recognition Software that uses specialized cameras to identify users and automatically log them in.

Fake news Information that is inaccurate or biased.

Family Educational Rights and Privacy Act (FERPA) A federal law that restricts disclosure of educational records.

Fiber-optic cable Special transmission cable made of glass tubes that are immune to electronic interference. Data is transmitted through fiber-optic cables in the form of pulses of light.

Fiber-optic service (FiOS) An Internet connection service that is not yet widely available. Current providers of FiOS include Google and Verizon with speeds faster than cable or DSL connections.

Field Each column of information within a record is called a field. A field contains related information on a specific item like employee names within a company department.

Fifth-generation language (5GL) Computer language that incorporates the concept of artificial intelligence to allow direct human communication.

File A collection of related records that can store data and programs. For example, the payroll file would include payroll information (records) for all of the employees (entities).

File compression Process of reducing the storage requirements for a file.

File decompression Process of expanding a compressed file.

File server Dedicated computer with large storage capacity providing users access to shared folders or fast storage and retrieval of information used in that business.

File transfer protocol (FTP) Internet service for uploading and downloading files.

Filter (1) A filter blocks access to selected websites. (2) A filter will locate or display records from a table that fit a set of conditions or criteria when using programs like Excel.

Firewall Security hardware and software. All communications into and out of an organization pass through a special security computer, called a proxy server, to protect all systems against external threats.

FireWire Operates much like USB buses on the system board but at higher speeds.

FireWire bus Operates much like USB buses on the system board but at higher speeds.

FireWire port Used to connect high-speed printers, and even video cameras, to system unit.

First-party cookie A cookie that is generated and then read only by the website you are currently visiting.

Flash drive *See* Universal serial bus (USB) drive.

Flash memory RAM chips that retain data even when power is disrupted. Flash memory is an example of solid-state storage and is typically used to store digitized images and record MP3 files.

Flash memory card A solid-state storage device widely used in notebook computers. Flash memory also is used in a variety of specialized input devices to capture and transfer data to desktop computers.

Flat-panel monitor Or liquid crystal display (LCD) monitor. These monitors are much thinner than CRTs and can be used for desktop systems as well.

Flatbed scanner An input device similar to a copying machine.

Flexible screen Monitor that allows digital devices to display images on nonflat surfaces.

Folder A named area on a disk that is used to store related subfolders and files.

Foreground The current program when multitasking or running multiple programs at once.

Free-to-play A business model for app sales where the initial app is free but advanced features require purchases.

Freedom of Information Act of 1970 Law giving citizens the right to examine data about them in federal government files, except for information restricted for national security reasons.

Friend An individual on a list of contacts for an instant messaging server.

Frontend The hardware and software that the user interacts with in a computing system.

g

Galleries Feature of Microsoft Office that simplifies the process of making selections from a list of alternatives by replacing dialog boxes with visual presentations of results.

Game controller A device that provides input to computer games.

Gamepad An input device designed to be held by two hands and provide a wide array of inputs, including motion, turning, stopping, and firing.

Gaming laptop A laptop with high-end graphics hardware and a very fast processor specifically designed to play computer games.

Gaming mice A game controller that provides greater precision, faster response, programming buttons, and better ergonomics than a traditional mouse.

General Data Protection Regulation (GDPR) Regulates the processing of personal data for all organizations that collect or process data about EU citizens.

General ledger Activity that produces income statements and balance sheets based on all transactions of a company.

General-purpose application Application used for doing common tasks, such as browsers and word processors, spreadsheets, databases, management systems, and presentation software. Also known as productivity applications.

Generations (of programming languages) The five generations are machine languages, assembly languages, procedural languages, problem-oriented languages, and natural languages. *See also* Levels.

Gesture control The ability to control operations with finger movements, such as swiping, sliding, and pinching.

Global positioning system (GPS) Devices that use location information to determine the geographic location of your car, for example.

Google Cloud Print A Google service that supports cloud printing.

GPU (graphics processing unit) *See* Graphics coprocessor.

Gramm-Leach-Bliley Act A law that protects personal financial information.

Graphical user interface (GUI) Special screen that allows software commands to be issued through the use of graphic symbols (icons) or pull-down menus.

Graphics card Device that provides high-quality 3D graphics and animation for games and simulations.

Graphics coprocessor Designed to handle requirements related to displaying and manipulating 2D and 3D graphic images.

Graphics program Program for making and altering digital images and video. This category of specialized applications includes video and image editors, illustration programs, and desktop publishing programs.

Grayscale The most common black ink selection in which images are displayed using many shades of gray.

Grid chart Chart that shows the relationship between input and output documents.

Group decision support system (GDSS) System used to support the collective work of a team addressing large problems.

Groups In Microsoft Word, each tab is organized into groups that contain related items. In social media, communities of individuals who share a common interest.

Guest operating system Operating system that operates on virtual machines.

h

Handwriting recognition software Translates handwritten notes into a form that the system unit can process.

Hard disk Enclosed disk drive containing one or more metallic disks. Hard disks use magnetic charges to record data and have large storage capacities and fast retrieval times.

Hard-disk drive *See* hard disk.

Hardware Equipment that includes a keyboard, monitor, printer, the computer itself, and other devices that are controlled by software programming.

Header A typical e-mail has three elements: header, message, and signature. The header appears first and includes addresses, subject, and attachments.

Headphones Typically worn around the head providing audio output.

Headsets Audio-output devices connected to a sound card in the system unit. The sound card is used to capture as well as play back recorded sound.

Health Insurance Portability and Accountability Act (HIPAA) A federal law that protects medical records.

Help A feature in most application software providing options that typically include an index, a glossary, and a search feature to locate reference information about specific commands.

Hexadecimal system (hex) Uses 16 digits to represent binary numbers.

Hi-def (high definition) The next generation of optical disc, with increased storage capacity. *See also* Blu-ray disc.

Hierarchical database Database in which fields or records are structured in nodes. Organized in the shape of a pyramid, and each node is linked directly to the nodes beneath it. Also called one-to-many relationship.

Hierarchical network *See* Tree network.

High-Definition Multimedia Interface (HDMI) port Port that provides high-definition video and audio, making it possible to use a computer as a video jukebox or an HD video recorder.

Higher level Programming languages that are closer to the language humans use.

History file Created by the browser to store information on websites visited by your computer system.

Home network LAN network for homes allowing different computers to share resources, including a common Internet connection.

Host Also called a server or provider, is a large centralized computer.

Host operating system Operating system that runs on the physical machine.

Hotspot Wireless access points that provide Internet access and are often available in public places such as coffee shops, libraries, bookstores, colleges, and universities.

HTML *See* Hypertext Markup Language.

HTML editor *See* Web authoring program.

https *See* Hypertext transfer protocol secure.

Hub The center or central node for other nodes. This device can be a server or a connection point for cables from other nodes.

Human resources The organizational department that focuses on the hiring, training, and promoting of people, as well as any number of human-centered activities within the organization.

Hybrid drive Storage drives that contain both solid-state storage and hard disk storage to provide the speed and power benefits of solid-state storage devices while still having the low cost and large capacity of hard drives.

Hyperlink Connection or link to other documents or web pages that contain related information.

Hypertext Markup Language (HTML) Programming language that creates document files used to display web pages.

Hypertext transfer protocol secure (https) A widely used protocol for web traffic and to protect the transfer of sensitive information.

I

Icon Graphic objects on the desktop used to represent programs and other files.

Identity theft The illegal assumption of someone's identity for the purpose of economic gain.

IF-THEN-ELSE structure Logical selection structure whereby one of two paths is followed according to IF, THEN, and ELSE statements in a program. *See also* Selection structure.

IFPS (interactive financial planning system) A 4GL language used for developing financial models.

Illusion of anonymity The misconception that being selective about disclosing personal information on the Internet can prevent an invasion of personal privacy.

Illustration program Also known as drawing programs; used to create digital illustrations and modify vector images and thus create line art, 3D models, and virtual reality.

Image editor An application for modifying bitmap images.

Immersive experience Allows the user to walk into a virtual reality room or view simulations on a virtual reality wall.

Incognito Mode A privacy mode available from the Google Chrome browser.

Income statement A statement that shows a company's financial performance, income, expenses, and the difference between them for a specific time period.

Individual database Collection of integrated records used mainly by just one person. Also called personal computer database.

Infected USB flash drive Flash drive that, when connected to a computer, infects that computer with viruses.

Information Data that has been processed by a computer system.

Information broker *See* Information reseller.

Information reseller Also known as information broker. It gathers personal data on people and sells it to direct marketers, fund-raisers, and others, usually for a fee.

Information system Collection of hardware, software, people, data, and procedures that work together to provide information essential to running an organization.

Information systems manager Oversees the work of programmers, computer specialists, systems analysts, and other computer professionals.

Information technology (IT) Computer and communication technologies, such as communication links to the Internet, that provide help and understanding to the end user.

Information utility *See* Commercial database.

Information worker Employee who creates, distributes, and communicates information.

Infrared Uses infrared light waves to communicate over short distances. Sometimes referred to as line-of-sight communication because light waves can only travel in a straight line.

Inkjet printer Printer that sprays small droplets of ink at high speed onto the surface of the paper, producing letter-quality images, and can print in color.

Input Any data or instructions used by a computer.

Input device Piece of equipment that translates data into a form a computer can process. The most common input devices are the keyboard and the mouse.

Instagram Microblogging site designed to share images and videos posts, with little to no written content.

Instant messaging (IM) A program allowing communication and collaboration for direct, "live" connections over the Internet between two or more people.

Integrated circuit *See* Silicon chip.

Interactive whiteboard *See* Digital whiteboard.

Internal data Data from within an organization consisting principally of transactions from the transaction processing system.

Internet A huge computer network available to everyone with a personal computer and a means to connect to it. It is the actual physical network made up of wires, cables, and satellites as opposed to the web, which is the multimedia interface to resources available on the Internet.

Internet of Things (IoT) Continuing development of the Internet that allows everyday objects embedded with electronic devices to send and receive data over the Internet.

Internet scam Using the Internet, a fraudulent act or operation designed to trick individuals into spending their time and money for little or no return.

Internet security suite Collection of utility programs designed to make using the Internet easier and safer.

Internet service provider (ISP) Provides access to the Internet.

Interpreter Software that converts a procedural language one statement at a time into machine language just before the statement is executed. No object code is saved.

Intranet Like the Internet, it typically provides e-mail, mailing lists, newsgroups, and FTP services, but it is accessible only to those within the organization. Organizations use intranets to provide information to their employees.

Intrusion detection system (IDS) Using sophisticated statistical techniques to analyze all incoming and outgoing network traffic, this system works with firewalls to protect an organization's network.

Inventory Material or products that a company has in stock.

Inventory control system A system that keeps records of the number of each kind of part or finished good in the warehouse.

iOS Previously known as iPhone OS, mobile operating system developed for Apple's iPhone, iPod Touch, and iPad.

IP address (Internet Protocol address) The unique numeric address of a computer on the Internet that facilitates the delivery of e-mail.

IT security analyst Person responsible for maintaining the security of a company's network, systems, and data. Employers look for candidates with a bachelor's or associate's degree in information systems or computer science and network experience.

j

JavaScript A scripting language that adds basic interactivity to web pages.

Joystick Popular input device for computer games. You control game actions by varying the pressure, speed, and direction of the joystick.

k

Key Another term for encryption key.

Key field The common field by which tables in a database are related to each other. This field uniquely identifies the record. For example, in university databases, a key field is the Social Security number. Also known as primary key.

Keyboard Input device that looks like a typewriter keyboard but has additional keys.

Keylogger Also known as computer monitoring software and sniffer programs. They can be loaded onto your computer without your knowledge.

Knowledge base A system that uses a database containing specific facts, rules to relate these facts, and user input to formulate recommendations and decisions.

Knowledge work system (KWS) Specialized information system used to create information in a specific area of expertise.

Knowledge worker Person involved in the creation of information, such as an engineer or a scientist.

Knowledge-based systems Programs duplicating human knowledge. A type of artificial intelligence that uses a database to provide assistance to users.

l

Land See Lands and pits.

Lands and pits The flat and bumpy areas on an optical disc surface. that represent 0s and 1s.

Language translator Converts programming instructions into a machine language that can be processed by a computer.

Laptop A small, portable system unit that contains electronic components, selected secondary storage devices, and input devices.

Laptop computer Portable computer, also known as a notebook computer, weighing between 4 and 10 pounds.

Laptop keyboard Almost all laptops have attached keyboards and screens.

Laser printer Printer that creates dotlike images on a drum, using a laser beam light source.

Levels Generations or levels of programming languages ranging from "low" to "high." See also Generations (of programming languages).

Light-emitting diode (LED) A technology for flat-panel monitors that have a more advanced backlighting technology. They produce better-quality images, are slimmer, and are more environmentally friendly as they require less power and use fewer toxic chemicals to manufacture. Most new monitors are LED.

Link A connection to related information.

LinkedIn The premier business-oriented social networking site.

Linux Type of UNIX operating system initially developed by Linus Torvalds, it is one of the most popular and powerful alternatives to the Windows operating system.

Liquid crystal display (LCD) A technology used for flat-panel monitors.

Local area network (LAN) Network consisting of computers and other devices that are physically near each other, such as within the same building.

Location For browsers to connect to resources, locations or addresses must be specified. Also known as uniform resource locators or URLs.

Logic error Error that occurs when a programmer has used an incorrect calculation or left out a programming procedure.

Logic structure Programming statements or structures called sequence, selection, or loop that control the logical sequence in which computer program instructions are executed.

Logical operation Comparing two pieces of data to see whether one is equal to (=), less than (<), or greater than (>) the other.

Logical view Focuses on the meaning and content of the data. End users and computer professionals are concerned with this view as opposed to the physical view, with which only specialized computer professionals are concerned.

Loop structure Logic structure in which a process may be repeated as long as a certain condition remains true. This structure is called a "loop" because the program loops around or repeats again and again.

Low bandwidth See Voiceband.

Lower level Programming language closer to the language the computer itself uses. The computer understands the 0s and 1s that make up bits and bytes.

LTE (Long Term Evolution) A wireless standard, comparable to WiMax.

m

Machine language Language in which data is represented in 1s and 0s. Most languages have to be translated into machine language for the computer to process the data. Either a compiler or an interpreter performs this translation.

macOS Operating system designed for Macintosh computers.

macOS 12 Sonoma Mac operating system announced in 2021 that provides new interoperability between Mac devices, increased customization of notifications, and improved video conferencing tools.

Magnetic card reader A card reader that reads encoded information from a magnetic strip on the back of a card.

Magnetic-ink character recognition (MICR) Direct-entry scanning devices used in banks. This technology is used to automatically read the numbers on the bottom of checks.

Mainboard *See* Motherboard or System board.

Mainframe computer This computer can process several million program instructions per second. Sizable organizations rely on these room-size systems to handle large programs and a great deal of data.

Maintenance programmer Programmer who maintains software by updating programs to protect them from errors, improve usability, standardize, and adjust to organizational changes.

Malware Short for malicious software.

Management information system (MIS) Computer-based information system that produces standardized reports in a summarized and structured form. Generally used to support middle managers.

Many-to-many relationship In a network database, each child node may have more than one parent node, and vice versa.

Marketing The organizational department that plans, prices, promotes, sells, and distributes an organization's goods and services.

Mass storage Refers to the tremendous amount of secondary storage required by large organizations.

Mass storage device Devices such as file servers, RAID systems, tape library, optical jukebox, and more.

MaxiCode A code widely used by the United Parcel Service (UPS) and others to automate the process of routing packages, tracking in-transit packages, and locating lost packages.

Media Media are the actual physical material that holds the data, such as a hard disk, which is one of the important characteristics of secondary storage. Singular of media is medium.

Medium band Bandwidth of special leased lines, used mainly with minicomputers and mainframe computers.

Megabits per second (Mbps) The transfer rate of millions of bits per second.

Memory Memory is contained on chips connected to the system board and is a holding area for data instructions and information (processed data waiting to be output to secondary storage). RAM, ROM, and CMOS are three types of memory chips.

Menu List of commands.

Menu bar Menus are displayed in a menu bar at the top of the screen.

Mesh network A topology requiring each node to have more than one connection to the other nodes so that if a path between two nodes is disrupted, data can be automatically rerouted around the failure using another path.

Message The content portion of e-mail correspondence.

Methods Instructions for retrieving or manipulating attribute values in an object-oriented database.

Metropolitan area network (MAN) These networks are used as links between office buildings in a city.

Microblog Publishes short sentences that only take a few seconds to write, rather than long stories or posts like a traditional blog.

Microprocessor The central processing unit (CPU) of a personal computer controls and manipulates data to produce information. The microprocessor is contained on a single integrated circuit chip and is the brain of the system. Also known as a processor.

Microwave Communication using high-frequency radio waves that travel in straight lines through the air.

Middle-level manager Middle-level managers deal with control and planning. They implement the long-term goals of the organization.

Midrange computer Refrigerator-sized machines falling in between personal computers and mainframes in processing speed and data-storing capacity. Medium-sized companies or departments of large companies use midrange computers.

Mini notebook A type of very portable laptop. They are lighter and thinner and have a longer battery life than other laptops. Also known as ultrabooks or ultraportables.

Mini tablet A type of tablet with a smaller screen.

Mistaken identity Occurs when the electronic profile of one person is switched with another.

Mobile apps (application) Add-on feature for a variety of mobile devices, including smartphones, netbooks, and tablets.

Mobile browser Special browsers designed to run on portable devices.

Mobile operating system Embedded operating system that controls a smartphone.

Mobile OS *See* Mobile operating system.

Modem Short for modulator-demodulator. It is a communication device that translates the electronic signals from a computer into electronic signals that can travel over telephone lines.

Modulation Process of converting digital signals to analog signals.

Module *See* Program module.

Monitor *See* Display.

Motherboard Also called a system board; the communications medium for the entire system.

Motion-sensing device An input device that controls games with user movements.

Mouse Device that typically is moved across the desktop and directs the cursor on the display screen.

Mouse pointer Typically in the shape of an arrow.

Multicore processor A new type of chip that provides two or more separate and independent CPUs.

Multidimensional database Data can be viewed as a cube with three or more sides consisting of cells. Each side of the cube is considered a dimension of the data; thus, complex relationships between data can be represented and efficiently analyzed. Sometimes called a data cube and designed for analyzing large groups of records.

Multifunctional devices (MFD) Devices that typically combines the capabilities of a scanner, printer, fax, and copying machine.

Multimedia messaging service (MMS) Supports the sending of images, video, and sound using text messaging.

Multitasking Operating system that allows a single user to run several application programs at the same time.

Multitouch screen Can be touched with more than one finger, which allows for interactions such as rotating graphical objects on the screen with your hand or zooming in and out by pinching and stretching your fingers.

n

Natural language A fifth-generation computer language (5GL) that allows a person to describe a problem and some constraints and then request a solution using a common language like English.

Network The arrangement in which various communications channels are connected through two or more computers. The largest network in the world is the Internet.

Network adapter card Connects the system unit to a cable that connects to other devices on the network.

Network administrator Also known as network manager. Computer professional who ensures that existing information and communication systems are operating effectively and that new ones are implemented as needed. Also responsible for meeting security and privacy requirements.

Network architecture Describes how networks are configured and how the resources are shared.

Network attached storage (NAS) Similar to a file server except simpler and less expensive. Widely used for home and small business storage needs.

Network database Database with a hierarchical arrangement of nodes, except that each child node may have more than one parent node. Also called many-to-many relationship.

Network drive Storage device attached to a network that can be accessed by other nodes of the network.

Network gateway Connection by which a local area network may be linked to other local area networks or to larger networks.

Network interface card (NIC) Also known as a network adapter card. Used to connect a computer to one or more computers forming a communication network whereby users can share data, programs, and hardware.

Network operating system (NOS) Interactive software between applications and computers coordinating and directing activities between computers on a network. This operating system is located on one of the connected computers' hard disks, making that system the network server.

Network server This computer coordinates all communication between the other computers. Popular network operating systems include NetWare and Windows NT Server. *See also* Network operating system.

News feed The first page you see after logging into a social networking site. It typically consists of a collection of recent posts from friends, trending topics on the site, people's responses to your posts, and advertisements.

Node Any device connected to a network. For example, a node is a computer, printer, or data storage device and each device has its own address on the network. Also, within hierarchical databases, fields or records are structured in nodes.

NoSQL A new category of object-oriented databases.

Notebook computer *See* Laptop *and* Laptop computer.

o

Object An element, such as a text box, that can be added to a workbook and can be selected, sized, and moved. For example, if a chart (object) in an Excel workbook file (source file) is linked to a Word document (destination file), the chart appears in the Word document. In this manner, the object contains both data and instructions to manipulate the data.

Object code Machine language code converted by a compiler from source code. Object code can be saved and run later.

Object-oriented database A more flexible type of database that stores data as well as instructions to manipulate data and is able to handle unstructured data such as photographs, audio, and video. Object-oriented databases organize data using objects, classes, entities, attributes, and methods.

Object-oriented programming (OOP) Methodology in which a program is organized into self-contained, reusable modules called objects. Each object contains both the data and processing operations necessary to perform a task.

Object-oriented software development Software development approach that focuses less on the tasks and more on defining the relationships between previously defined procedures or objects.

Objectives In programming, it is necessary to make clear the problems you are trying to solve to create a functional program.

Office automation system (OAS) System designed primarily to support data workers. It focuses on managing documents, communicating, and scheduling.

Office software suites *See* Productivity suites.

Office suites *See* Productivity suites.

One-to-many relationship In a hierarchical database, each entry has one parent node, and a parent may have several child nodes.

Online Being connected to the Internet is described as being online.

Online identity The information that people voluntarily post about themselves online.

Online office suite Office suite stored online and available anywhere the Internet can be accessed.

Online processing *See* Real-time processing.

Online storage Provides users with storage space that can be accessed from a website.

Open source A free and openly distributed software program intended to allow users to improve upon and further develop the program.

Operating system (OS) Software that interacts between application software and the computer, handling such details as running programs, storing and processing data, and coordinating all computer resources, including attached peripheral devices.

Operational feasibility Making sure the design of a new system will be able to function within the existing framework of an organization.

Operational model A decision model that helps lower-level managers accomplish the organization's day-to-day activities, such as evaluating and maintaining quality control.

Operator An operator handles correcting operational errors in any programs. To do that, they need documentation, which lets them understand the program, thus enabling them to fix any errors.

Optical carrier (OC) Provides support for very high speed, all-digital transmission for large corporations.

Optical disc Storage device that can hold over 17 gigabytes of data, which is the equivalent of several million typewritten pages. Lasers are used to record and read data on the disc. The three basic types of optical discs are compact discs (CDs), digital versatile or video discs (DVDs), and Blu-ray discs (BDs).

Optical disc drive A disc is read by an optical disc drive using a laser that projects a tiny beam of light. The amount of reflected light determines whether the area represents a 1 or a 0.

Optical scanner Device that identifies images or text on a page and automatically converts it to electronic signals that can be stored in a computer to copy or reproduce.

Optical-character recognition (OCR) Scanning device that uses special preprinted characters, such as those printed on utility bills, that can be read by a light source and changed into machine-readable code.

Optical-mark recognition (OMR) Device that senses the presence or absence of a mark, such as a pencil mark. As an example, an OMR device is used to score multiple-choice tests.

Organic light-emitting diode (OLED) Replaces LED monitor's backlighting technology with a thin layer of organic compound that produces light.

Organization chart Chart showing the levels of management and formal lines of authority in an organization.

Organizational cloud storage High-speed Internet connection to a dedicated remote organizational Internet drive site.

Output Processed data or information.

Output device Equipment that translates processed information from the central processing unit into a form that can be understood by humans. The most common output devices are monitors and printers.

p

Packet Before a message is sent on the Internet, it is broken down into small parts called packets. Each packet is then sent separately over the Internet. At the receiving end, the packets are reassembled into the correct order.

Page A social networking tool often used by companies to promote their business. These pages can include hours of operations, upcoming sales, and information about their products.

Page layout program *See* Desktop publishing program.

Parallel approach Systems implementation in which old and new systems are operated side by side until the new one has shown it is reliable.

Parallel processing Used by supercomputers to run large and complex programs.

Parent node Node one level above the node being considered in a hierarchical database or network. Each entry has one parent node, although a parent may have several child nodes. Also called one-to-many relationship.

Password Special sequence of numbers, letters, and characters that limits access to information, such as electronic mail.

Password manager Program that helps users create, use, and recall strong passwords.

Patches Programming modifications or corrections.

Payroll Activity concerned with calculating employee paychecks.

PC *See* Personal computer.

PCI Express (PCIe) New type of bus that is 30 times faster than PCI bus.

Peer-to-peer (P2P) network Network in which nodes can act as both servers and clients. For example, one personal computer can obtain files located on another personal computer and also can provide files to other personal computers.

People End users who use computers to make themselves more productive.

Periodic report Report for a specific time period as to the health of the company or a particular department of the company.

Peripheral External device, such as a monitor and keyboard.

Personal area network (PAN) A type of wireless network that works within a very small area—your immediate surroundings.

Personal computer Small, low-cost computer designed for individual users. These include desktop computers, laptops, and mobile devices.

Personal computer database *See* Individual database.

Personal laser printer Inexpensive laser printer widely used by single users to produce black-and-white documents.

Phased approach Systems implementation whereby a new system is implemented gradually over a period of time.

Phishing An attempt to trick Internet users into thinking a fake but official-looking website or e-mail is legitimate.

Photo editor *See* Image editor.

PHP A language often used within HTML documents to improve a website's interactivity.

Physical security Activity concerned with protecting hardware from possible human and natural disasters.

Physical view This focuses on the actual format and location of the data. *See also* Logical view.

Picture element *See* Pixel.

Picture Password A security application in Windows 10 that accepts a series of gestures over a picture of the user's choice to gain access to a user's account.

Pilot approach Systems implementation in which a new system is tried out in only one part of the organization. Later it is implemented throughout the rest of the organization.

Pit *See* Lands and pits.

Pixel (picture element) Smallest unit on the screen that can be turned on and off or made different shades. Pixels are individual dots that form images on a monitor. The greater the resolution, the more pixels and the better the clarity.

Pixel pitch The distance between each pixel on a monitor.

Plagiarism Representation of some other person's work and ideas as your own without giving credit to the original source.

Plagiarist Someone who engages in plagiarism.

Platform scanner Handheld direct-entry device used to read special characters on price tags. Also known as wand reader.

Platter Rigid metallic disk; multiple platters are stacked one on top of another within a hard-disk drive.

Plotter Special-purpose output device for producing bar charts, maps, architectural drawings, and three-dimensional illustrations.

Podcast An Internet-based medium for delivering music and movie files from the Internet to a computer.

Pointer For a monitor, a pointer is typically displayed as an arrow and controlled by a mouse. For a database, a pointer is a connection between a parent node and a child node in a hierarchical database.

Pointers Within a network database, pointers are additional connections between parent nodes and child nodes. Thus, a node may be reached through more than one path and can be traced down through different branches.

Pointing device A device that provides an intuitive interface with the system unit by accepting pointing gestures and converting them into machine-readable input.

Port Connecting socket on the outside of the system unit. Used to connect input and output devices to the system unit.

Portable language Language that can be run on more than one type of computer.

Portable scanner A handheld device that slides across an image to be scanned, making direct contact.

Post A message uploaded to a microblogging site, such as X.

Power supply unit Desktop computers have a power supply unit located within the system unit that plugs into a standard wall outlet, converting AC to DC, which becomes the power to drive all of the system unit components.

Preliminary investigation First phase of the systems life cycle. It involves defining the problem, suggesting alternative systems, and preparing a short report.

Presentation file A file created by presentation software to save presentation materials. For example, a file might contain audience handouts, speaker notes, and electronic slides.

Presentation software Software used to combine a variety of visual objects to create attractive and interesting presentations.

Primary key *See* Key field.

Primary storage Holds data and program instructions for processing data. It also holds processed information before it is output. *See also* Memory.

Printer Device that produces printed paper output.

Privacy Computer ethics issue concerning the collection and use of data about individuals.

Privacy mode A browser feature that eliminates history files and blocks most cookies.

Private Browsing A privacy mode provided by Safari. *See also* Privacy mode.

Procedural language Programming language designed to focus on procedures and how a program will accomplish a specific task. Also known as 3GL or third-generation language.

Procedure Rule or guideline to follow when using hardware, software, and data.

Processing rights Refers to which people have access to what kind of data.

Processor *See* Central processing unit.

Production The organizational department that actually creates finished goods and services using raw materials and personnel.

Productivity suites Also known as office suites; contain professional-grade application programs, including word processing, spreadsheets, and more. A good example is Microsoft Office.

Profile A social networking tool used by individuals to share information and often includes photos, personal details, and contact information.

Program Instructions for the computer to follow to process data. *See also* Software.

Program analysis *See* Program specification.

Program coder *See* Application generator.

Program definition *See* Program specification.

Program design Creating a solution using programming techniques, such as top-down program design, pseudocode, flowcharts, logic structures, object-oriented programming, and CASE tools.

Program documentation Written description of the purpose and process of a program. Documentation is written within the program itself and in printed documents. Programmers will find themselves frustrated without adequate documentation, especially when it comes time to update or modify the program.

Program flowchart Flowchart graphically presents a detailed sequence of steps needed to solve a programming problem.

Program maintenance Activity of updating software to correct errors, improve usability, standardize, and adjust to organizational changes.

Program module Each module is made up of logically related program statements. The program must pass in sequence from one module to the next until the computer has processed all modules.

Program specification Programming step in which objectives, output, input, and processing requirements are determined.

Programmer Computer professional who creates new software or revises existing software.

Programming A program is a list of instructions a computer will follow to process data. Programming, also known as software development, is a six-step procedure for creating that list of instructions. The six steps are program specification, program design, program code (or coding), program test, program documentation, and program maintenance.

Programming language A collection of symbols, words, and phrases that instruct a computer to perform a specific task.

Project manager Software that enables users to plan, schedule, and control the people, resources, and costs needed to complete a project on time.

Property Computer ethics issue relating to who owns data and rights to software.

Protocol Rules for exchanging data between computers. The protocol http:// is the most common.

Prototyping Building a model or prototype that can be modified before the actual system is installed.

Proxy server Computer that acts as a gateway or checkpoint in an organization's firewall. *See also* Firewall.

Pseudocode An outline of the logic of the program to be written. It is the steps or the summary of the program before you actually write the program for the computer. Consequently, you can see beforehand what the program is to accomplish.

Purchase order A form that shows the name of the company supplying the material or service and what is being purchased.

Purchasing Buying of raw materials and services.

q

Query language Easy-to-use language and understandable to most users. It is used to search and generate reports from a database. An example is the language used in an airline reservation system.

Query-by-example A specific tool in database management that shows a blank record and lets you specify the information needed, like the fields and values of the topic you are looking to obtain.

r

RAID system Several inexpensive hard-disk drives connected to improve performance and provide reliable storage.

RAM *See* Random-access memory.

Random-access memory (RAM) Volatile, temporary storage that holds the program and data the CPU is presently processing. It is called temporary storage because its contents will be lost if electrical power to the computer is disrupted or the computer is turned off.

Ransomware Malicious software that encrypts your computer's data and ransoms the password to the user.

Rapid applications development (RAD) Involves the use of powerful development software and specialized teams as an alternative to the systems development life cycle approach. Time for development is shorter and quality of the completed systems development is better, although cost is greater.

Raster *See* Bitmap image.

Read-only Optical disc format that does not allow a disc to be written on or erased by the user.

Read-only memory (ROM) Refers to chips that have programs built into them at the factory. The user cannot change the contents of such chips. The CPU can read or retrieve the programs on the chips but cannot write or change information. ROM stores programs that boot the computer, for example. Also called firmware.

Real-time operating system (RTOS) *See* Embedded operating system.

Real-time processing Or online processing. Occurs when data is processed at the same time a transaction occurs.

Record Each row of information in a database is a record. Each record contains fields of data about some specific item, like employee name, address, phone, and so forth. A record represents a collection of attributes describing an entity.

Recordable (R) Optical disc format that allows a disc to be written on once. Also known as write-once disc.

Redundant arrays of inexpensive disks (RAID) Groups of inexpensive hard-disk drives related or grouped together using networks and special software. They improve performance by expanding external storage.

Refresh rate The measure of how quickly new images are displayed on the monitor.

Relation A table in a relational database in which data elements are stored in rows and columns.

Relational database A widely used database structure in which data is organized into related tables. Each table is made up of rows called records and columns called fields. Each record contains fields of data about a specific item.

Repetition structure *See* Loop structure.

Repetitive strain injury (RSI) Any injury that is caused by fast, repetitive work that can generate neck, wrist, hand, and arm pain.

Research The organizational department that identifies, investigates, and develops new products and services.

Resolution A measurement reflecting the quality of an image.

Rewritable (RW) Optical disc format that allows a disc to be written to multiple times.

RFID (radio-frequency identification) reader A device used to read radio-frequency identification information.

RFID tag Microchip that contains electronically stored information and can be embedded in items such as consumer products, driver's licenses, passports, etc.

Ribbon Feature of Microsoft Office that replaces menus and toolbars by organizing commonly used commands into a set of tabs.

Ribbon GUI An interface that uses a system of ribbons, tabs, and galleries to make it easier to find and use all the features of an application.

Ring network Network in which each device is connected to two other devices, forming a ring. There is no host computer, and messages are passed around the ring until they reach the correct destination.

Robot Typically use cameras, microphones, and other sensors as inputs. Widely used in factories and other applications involving repetitive actions.

Robot network *See* Botnet.

Rogue Wi-Fi hotspot Imitation hotspot intended to capture personal information.

Role playing game (RPG) Computer game in which players assume persona of one of the game's actors.

ROM *See* Read-only memory.

Router A node that forwards or routes data packets from one network to their destination in another network.

S

Sales order processing Activity that records the demands of customers for a company's products or services.

Satellite This type of communication uses satellites orbiting about 22,000 miles above the Earth as microwave relay stations.

Satellite connection services Connection services that use satellites and the air to download or send data to users at a rate seven times faster than dial-up connections.

Scanner *See* Optical scanner.

Scanning devices Convert scanned text and images into a form that the system unit can process.

Schema *See* Data dictionary.

SD card Type of expansion card designed for laptops, tablets, and smartphones.

Search engine Specialized programs assisting in locating information on the web and the Internet.

Search program A utility that provides a quick and easy way to search or examine an entire computer system to find specific applications, data, or other files.

Search service Organization that maintains databases relating to information provided on the Internet and also provides search engines to locate information.

Secondary storage Permanent storage used to preserve programs and data that can be retained after the computer is turned off. These devices include hard disks, magnetic tape, CDs, DVDs, and more.

Secondary storage device These devices are used to save, back up, and transport files from one location or computer to another. *See also* Secondary storage.

Sector Section shaped like a pie wedge that divides the tracks on a disk.

Secure file transfer protocol (SFTP) *See* File transfer protocol.

Security The protection of information, hardware, and software.

Security suites A collection of utility programs designed to protect your privacy and security while you are on the web.

Selection structure Logic structure that determines which of two paths will be followed when a program must make a decision. Also called IF-THEN-ELSE structures. IF something is true, THEN do option one, or ELSE do option two.

Semiconductor Silicon chip through which electricity flows with some resistance.

Sequential structure Logic structure in which one program statement follows another.

Server A host computer with a connection to the Internet that stores document files used to display web pages. Depending on the resources shared, it may be called a file server, printer server, communication server, web server, or database server.

Share setting A feature on social media accounts that determines who can see your social networking data.

Shared laser printer More expensive laser printer used by a group of users to produce black-and-white documents. These printers can produce over 30 pages a minute.

Signature Provides additional information about a sender of an e-mail message, such as name, address, and telephone number.

Silicon chip Tiny circuit board etched on a small square of sandlike material called silicon. Chips are mounted on carrier packages, which then plug into sockets on the system board.

SIM (Subscriber Identity Module) Cards A small card inserted into mobile devices that allow customers to access cellular networks.

Siri A virtual assistant designed by Apple that accepts commands through text or speech. Introduced as a utility in iOS and incorporated with macOS Yosemite in 2016.

Slot Area on a system board that accepts expansion cards to expand a computer system's capabilities.

Smartphone A type of cell phone that offers a variety of advanced functionality, including Internet and e-mail.

Smartwatch A wearable computer that acts as a watch, fitness monitor, and communication device.

SMS (short messaging service) Texting or process of sending a short electronic message using a wireless network to another person.

Social engineering The practice of manipulating people to divulge private data.

Social networking Using the Internet to connect individuals.

Socket A socket provides connection points on the system board for holding electronic parts.

Software Computer program consisting of step-by-step instructions, directing the computer on each task it will perform.

Software development *See* Programming.

Software development life cycle (SDLC) A six-step procedure for software development.

Software engineer Programming professional or programmer who analyzes users' needs and creates application software.

Software environment Operating system, also known as software platform, consisting of a collection of programs to handle technical details depending on the type of operating system. For example, software designed to run on an Apple computer is compatible with the macOS environment.

Software piracy Unauthorized copying of programs for personal gain.

Software platform *See* Software environment.

Software suite Individual application programs that are sold together as a group.

Software update Patch in which modifications to the software are typically more extensive and significant.

Solid-state drive (SSD) Designed to be connected inside a microcomputer system the same way an internal hard disk would be but contains solid-state memory instead of magnetic disks to store data.

Solid-state storage A secondary storage device that has no moving parts. Data is stored and retrieved electronically directly from these devices, much as they would be from conventional computer memory.

Source code Occurs when a programmer originally writes the code for a program in a particular language. This is called source code until it is translated by a compiler for the computer to execute the program. It then becomes object code.

Spam Unwelcome and unsolicited e-mail that can carry attached viruses.

Spam blocker Also referred to as spam filter. Software that uses a variety of different approaches to identify and eliminate spam or junk mail.

Spam filter *See* Spam blocker.

Speakers Audio-output devices connected to a sound card in the system unit. The sound card is used to capture as well as play back recorded sound.

Specialized applications Program that is narrowly focused on specific disciplines and occupations. Some of the best known are multimedia, web authoring, graphics, virtual reality, and artificial intelligence.

Specialized suite Programs that focus on specialized applications such as graphics or financial planning.

Spider Special program that continually looks for new information and updates a search server's databases.

Spreadsheet Computer-produced spreadsheet based on the traditional accounting worksheet that has rows and columns used to present and analyze data.

Spy removal program Program such as Spybot and Spysweeper, designed to detect web bugs and monitor software.

Spyware Wide range of programs designed to secretly record and report an individual's activities on the Internet.

Stand-alone operating system Also called desktop operating system; a type of operating system that controls a single desktop or notebook computer.

Star network Network of computers or peripheral devices linked to a central computer through which all communications pass. Control is maintained by polling. The configuration of the computers looks like a star surrounding and connected to the central computer in the middle.

Storage area network (SAN) An architecture that links remote computer storage devices such as enterprise storage systems to computers so that the devices are available as locally attached drives.

Storage device Hardware that reads data and programs from storage media. Most also write to storage media.

Storage management program Utility program designed to solve the problem of running out of storage space by providing lists of application programs, stored videos, and other program files so that unused applications or archived large files can be moved or eliminated.

Storage media Storage media are the physical material that holds data and programs.

Strategic model A decision model that assists top managers in long-range planning, such as stating company objectives or planning plant locations.

Strategy A way of coordinating the sharing of information and resources. The most common network strategies are terminal, peer-to-peer, and client/server networks.

Structured program Program that uses logic structures according to the program design and the language in which you have chosen to write the program. Each language follows techniques like pseudocode, flowcharts, and logic structures.

Structured programming techniques Techniques consisting of top-down program design, pseudocode, flowcharts, and logic structures.

Structured Query Language (SQL) A program control language used to create sophisticated database applications for requesting information from a database.

Stylus Penlike device used with tablets and PDAs that uses pressure to draw images on a screen. A stylus interacts with the computer through handwriting recognition software.

Subject Located in the header of an e-mail message; a one-line description used to present the topic of the message.

Supercomputer Fastest calculating device ever invented, processing billions of program instructions per second. Used by very large organizations like NASA.

Supervisor Manager responsible for managing and monitoring workers. Supervisors have responsibility for operational matters.

Switch The center or central node for other nodes. This device coordinates the flow of data by sending messages directly between sender and receiver nodes.

Syntax error Violation of the rules of a language in which the computer program is written. For example, leaving out a semicolon would stop the entire program from working because it is not the exact form the computer expects for that language.

System Collection of activities and elements designed to accomplish a goal.

System board Flat board that usually contains the CPU and memory chips connecting all system components to one another.

System bus There are two categories of buses. One is the system bus that connects the CPU to the system board. The other is the expansion bus that connects the CPU to slots on the system board.

System chassis *See* System unit.

System flowchart A flowchart that shows the flow of input data to processing and finally to output, or distribution of information.

System software "Background" software that enables the application software to interact with the computer. System software consists of the operating system, utilities, device drivers, and language translators. It works with application software to handle the majority of technical details.

System unit Part of a personal computer that contains the CPU. Also known as the system cabinet or chassis, it is the container that houses most of the electronic components that make up the computer system.

Systems analysis This second phase of the systems life cycle determines the requirements for a new system. Data is collected about the present system and analyzed, and new requirements are determined.

Systems analysis and design Six phases of problem-solving procedures for examining information systems and improving them.

Systems analysis report Report prepared for higher management describing the current information system, the requirements for a new system, and a possible development schedule.

Systems analyst Plans and designs information systems.

Systems audit A systems audit compares the performance of a new system to the original design specifications to determine if the new procedures are actually improving productivity.

Systems design Phase three of the systems life cycle, consisting of designing alternative systems, selecting the best system, and writing a systems design report.

Systems design report Report prepared for higher management describing alternative designs, presenting costs versus benefits, and outlining the effects of alternative designs on the organization.

Systems development Phase four of the systems life cycle, consisting of developing software, acquiring hardware, and testing the new system.

Systems implementation Phase five of the systems life cycle is converting the old system to the new one and training people to use the new system. Also known as conversion.

Systems life cycle The six phases of systems analysis and design are called the systems life cycle. The phases are preliminary investigation, systems analysis, systems design, systems development, systems implementation, and systems maintenance.

Systems maintenance Phase six of the systems life cycle, consisting of a systems audit and periodic evaluation.

t

T1 High-speed lines that support all digital communications, provide very high capacity, and are very expensive.

T3 Copper lines combined to form higher-capacity options.

Tab Used to divide the ribbon into major activity areas, with each tab being organized into groups that contain related items.

Table (in database) The list of records in a database. Tables make up the basic structure of a database. Their columns display field data and their rows display records. *See also* Field *and* Record.

Tablet A type of personal computer that contains a thin system unit, most of which is the monitor.

Tablet computer *See* Tablet.

Tactical model A decision model that assists middle-level managers to control the work of the organization, such as financial planning and sales promotion planning.

Task-oriented language Programming language that is non-procedural and focuses on specifying what the program is to accomplish. Also known as 4GL or very high level language.

Technical feasibility Making sure hardware, software, and training will be available to facilitate the design of a new system.

Technical writer Prepares instruction manuals, technical reports, and other scientific or technical documents.

Telephone line A transmission medium for both voice and data.

Temporary Internet file File that has web page content and instructions for displaying this content.

Text messaging (texting) The process of sending a short electronic message, typically fewer than 160 characters, using a wireless network to another person who views the message on a mobile device, such as a smartphone.

Thermal printer Printer that uses heat elements to produce images on heat-sensitive paper.

Third-generation language (3GL) *See* Procedural language.

Third-party cookie A cookie generated by an advertising company that is affiliated with the website you are currently visiting. Often also referred to as a tracking cookie.

Thunderbolt 3 A specialty port for high-speed connections to up to seven separate devices connected one to another.

Toggle key These keys turn a feature on or off, like the Caps Lock key.

Toolbar Bar located typically below the menu bar containing icons or graphical representations for commonly used commands.

Top-down analysis method Method used to identify top-level components of a system, then break these components down into smaller parts for analysis.

Top-down program design Used to identify the program's processing steps, called program modules. The program must pass in sequence from one module to the next until the computer has processed all modules.

Top-level domain (TLD) Last part of an Internet address; identifies the geographic description or organizational identification. For example, using www.aol.com, the .com is the top-level domain code and indicates it is a commercial site. *See also* Domain name.

Top-level manager Top-level managers are concerned with long-range (strategic) planning. They supervise middle management.

Topology The configuration of a network. The five principal network topologies are *ring, bus, star, tree,* and *mesh.*

Touch pad Used to control the pointer by moving and tapping your finger on the surface of a pad.

Touch screen Monitor screen allowing actions or commands to be entered by the touch of a finger.

Tower computer A desktop system unit placed vertically.

Tower unit *See* Tower computer.

Track Closed, concentric ring on a disk on which data is recorded. Each track is divided into sections called sectors.

Tracking cookie *See* Third-party cookie.

Traditional keyboard Full-sized, rigid, rectangular keyboard that includes function, navigational, and numeric keys.

Transaction processing system (TPS) System that records day-to-day transactions, such as customer orders, bills, inventory levels, and production output. The TPS tracks operations and creates databases.

Transfer rate Or transfer speed, is the speed at which modems transmit data, typically measured in bits per second (bps).

Transmission control protocol/Internet protocol (TCP/IP) TCP/IP is the standard protocol for the Internet. The essential features of this protocol involve (1) identifying sending and receiving devices and (2) reformatting information for transmission across the Internet.

Tree network Also known as a hierarchical network. A topology in which each device is connected to a central node, either directly or through one or more other devices. The central node is then connected to two or more subordinate nodes that in turn are connected to other subordinate nodes, and so forth, forming a treelike structure.

Trojan horse Program that is not a virus but is a carrier of virus(es). The most common Trojan horses appear as free computer games, screen savers, or antivirus programs. Once downloaded, they locate and disable existing virus protection and then deposit the virus.

Troubleshooting program A utility program that recognizes and corrects computer-related problems before they become serious. Also called diagnostic program.

Twisted-pair cable Cable consisting of pairs of copper wire that are twisted together.

Two-factor authentication A type of authentication that requires two types (or factors) of data to verify a user's identity.

Two-in-one laptop Laptop that includes a touch screen and the ability to fold flat like a tablet computer. It offers the advantages of a laptop with the convenience of a tablet.

Two-step authentication A type of authentication that uses one type of authentication twice.

u

Ultra HD Blu-rays (UHD BD) Blu-ray discs that are able to play back 4K video content and store up to 100 GB of data.

Ultrabook A very portable laptop that is lighter and thinner with a longer battery life than other laptops.

Ultraportable *See* Ultrabook.

Unicode A 16-bit code designed to support international languages, like Chinese and Japanese.

Uniform resource locator (URL) For browsers to connect you to resources on the web, the location or address of the resources must be specified. These addresses are called URLs.

Universal product code (UPC) A bar code system that identifies the product to the computer, which has a description and the latest price for the product.

Universal serial bus (USB) Combines with a PCI bus on the system board to support several external devices without inserting cards for each device. USB buses are used to support high-speed scanners, printers, and video-capturing devices.

Universal serial bus (USB) port These ports have replaced serial and parallel ports. They are faster, and one USB port can be used to connect several devices to the system unit.

Universal serial bus—A (USB-A) The type of USB found on most laptops and desktops.

Universal serial bus—B (USB-B) The type of USB found on most peripherals.

Universal serial bus—C (USB-C) The newest USB type, expected to replace both USB types A and B.

UNIX An operating system originally developed for midrange computers. It is now important because it can run on many of the more powerful personal computers.

Unmanned aerial vehicle (UAV) *See* Drone.

Uplink To send data to a satellite.

Uploading Process of transferring information from the computer the user is operating to a remote computer.

USB flash drive The size of a key chain, these hard drives connect to a computer's USB port enabling a transfer of files.

User Any individual who uses a computer. *See also* End user.

User interface Means by which users interact with application programs and hardware. A window is displayed with information for the user to enter or choose, and that is how users communicate with the program.

Utility Performs specific tasks related to managing computer resources or files. Norton Utility for virus control and system maintenance is a good example of a utility. Also known as service programs.

Utility suite A program that combines several utilities in one package to improve system performance.

V

Vector illustration *See* Vector image.

Vector image Graphics file made up of a collection of objects such as lines, rectangles, and ovals. Vector images are more flexible than bitmaps because they are defined by mathematical equations so they can be stretched and resized. Illustration programs create and manipulate vector graphics. Also known as vector illustrations.

Very high level languages Task-oriented languages that require little special training on the part of the user.

Video editor Program for editing videos to enhance their quality and appearance.

Video game design software Program for organizing and providing guidance for the entire game design process, including character development and environmental design.

Videoconferencing system Computer system that allows people located at various geographic locations to have in-person meetings.

Virtual assistant A utility that accepts commands through text or speech to allow intuitive interaction with your computer, cell phone, or tablet and coordinates personal data across multiple applications.

Virtual Desktop Infrastructure (VDI) A collection of hardware and software that allow a user to remotely access a desktop environment over the Internet.

Virtual keyboard Displays an image of a keyboard on a touch-screen device. The screen functions as the actual input device, which is why the keyboard is considered virtual.

Virtual machine A software implementation of a computer that executes programs like a physical computer.

Virtual memory Feature of an operating system that increases the amount of memory available to run programs. With large programs, parts are stored on a secondary device like your hard disk. Then each part is read in RAM only when needed.

Virtual private network (VPN) Creates a secure private connection between a remote user and an organization's internal network. Special VPN protocols create the equivalent of a dedicated line between a user's home or laptop computer and a company server.

Virtual reality (VR) Interactive sensory equipment (head-mounted display and controller) allowing users to experience alternative realities generated in 3D by a computer, thus imitating the physical world.

Virtualization A process that allows a single physical computer to support multiple operating systems that operate independently.

Virtualization software Software that creates virtual machines.

Virus Hidden instructions that migrate through networks and operating systems and become embedded in different programs. They may be designed to destroy data or simply to display messages.

Voice assist tools Operating system program that provides a visual way to interact with application programs and computer hardware and allows users to directly issue commands.

Voice recognition system Using a microphone, sound card, and specialty software, the user can operate a computer and create documents using voice commands.

Voiceband Bandwidth of a standard telephone line. Also known as low bandwidth.

VR controller Device that provides input to a virtual reality environment about a user's hand movements.

VR head-mounted display Interactive sensory equipment that collects data from embedded earphones and three-dimensional stereoscopic screens worn on the head.

W

Wand reader Special-purpose handheld device used to read OCR characters.

Warm boot Restarting your computer while the computer is already on and the power is not turned off.

Wearable computer *See* Wearable device.

Wearable device A type of mobile computer such as Apple's Watch that contains an embedded computer chip. Also known as a smartwatch.

Web Prior to the introduction of the web in 1992, the Internet was all text. The web made it possible to provide a multimedia interface that includes graphics, animations, sound, and video.

Web 1.0 The first generation of the web, which focused on linking existing information.

Web 2.0 The second generation of the web, which evolved to support more dynamic content creation and social interaction.

Web 3.0 The third generation of the web, which focuses on computer-generated information requiring less human interaction to locate and to integrate information.

Web 4.0 The fourth generation of the web, which uses mobile devices to gather data and seamlessly inform and respond to the user's needs.

Web 5.0 The fifth generation of the web, which focuses on programs that can recognize and respond to a user's emotional state.

Web auction Similar to traditional auctions except that all transactions occur over the web; buyers and sellers seldom meet face to face.

Web authoring Creating a website.

Web authoring program Word processing program for generating web pages. Also called HTML editor or web page editor. Widely used web authoring programs include Adobe Dreamweaver and Microsoft FrontPage.

Web bug Program hidden in the HTML code for a web page or e-mail message as a graphical image. Web bugs can migrate whenever a user visits a website containing a web bug or opens infected e-mail. They collect information on the users and report back to a predefined server.

Web developer Develops and maintains websites and web resources.

Web page Browsers interpret HTML documents to display web pages.

Web page editor *See* Web authoring program.

Web suffix Identifies type of organization in a URL.

Web utilities Specialized utility programs making the Internet and the web easier and safer. Some examples are plug-ins that operate as part of a browser and filters that block access and monitor use of selected websites.

Web-based e-mail system An e-mail system that does not require an e-mail program to be installed on your computer.

Web-based file transfer services A type of file transfer service that uses a web browser to upload and download files, allowing you to copy files to and from your computer across the Internet.

Webcam Specialized digital video camera for capturing images and broadcasting to the Internet.

Webmail E-mail that uses a webmail client.

Webmail client A special program that runs on an e-mail provider's computer that supports webmail.

Wheel button Some mice have a wheel button that can be rotated to scroll through information displayed on the monitor.

Wi-Fi (wireless fidelity) Wireless standard also known as 802.11, used to connect computers to each other and to the Internet.

Wide area network (WAN) Countrywide and worldwide networks that use microwave relays and satellites to reach users over long distances.

Wiki A website that allows people to fill in missing information or correct inaccuracies on it by directly editing the pages.

Wikipedia An online encyclopedia, written and edited by anyone who wants to contribute.

Window A rectangular area containing a document or message.

Windows An operating environment extending the capability of DOS.

Windows 11 OS announced in 2021 that includes significant changes to the graphical user interface for a simplified and unified experience for tablet, desktop, and two-in-one devices. It introduced the ability to run Android apps from the Windows environment.

Wireless access point Or base station. The receiver interprets incoming radio frequencies from a wireless LAN and routes communications to the appropriate devices, which could be separate computers, a shared printer, or a modem.

Wireless charging platform Recharging device for laptops, tablets, and wearable computers that does not require a connecting cable.

Wireless communication The revolutionary way we now communicate on devices like tablets, smartphones, and wearable devices.

Wireless LAN (WLAN) Uses radio frequencies to connect computers and other devices. All communications pass through the network's centrally located wireless receiver or base station and are routed to the appropriate devices.

Wireless modem Typically a small plug-in USB or ExpressCard device that provides very portable high-speed connectivity from virtually anywhere.

Wireless mouse *See* Cordless mouse.

Wireless network card Allows computers to be connected without cables.

Wireless network encryption Restricts access to authorized users on wireless networks.

Wireless revolution A revolution that is expected to dramatically affect the way we communicate and use computer technology.

Wireless wide area network (WWAN) modem *See* Wireless modem.

Word The number of bits (such as 16, 32, or 64) that can be accessed at one time by the CPU.

Word processor The computer and the program that allow you to create, edit, save, and print documents composed of text.

Worksheet file Created by electronic spreadsheets to analyze things like budgets and to predict sales.

Worm Virus that doesn't attach itself to programs and databases but fills a computer system with self-replicating information, clogging the system so that its operations are slowed or stopped.

WPA3 (Wi-Fi Protected Access 3) A secure encryption protocol.

Write once read many (WORM) External storage discs that can be written on once but read many times.

Write-once *See* Recordable.

WWW (World Wide Web) Provides a multimedia interface to the Internet. Also known as the web.

WYSIWYG (what you see is what you get) editors Web authoring programs that build a page without requiring direct interaction with the HTML code and then preview the page described by the HTML code.

X

X Previously named Twitter, the most popular microblogging site that enables you to add new content from your browser, instant messaging application, or even a mobile phone.

Z

Zombie A computer infected by a virus, worm, or Trojan horse that allows it to be remotely controlled for malicious purposes.

Index

1G. *See* First generation
2G. *See* Second generation
3D Modeling programs, 68
3D printers, 145, 156
3D scanners, 137, 155
3G. *See* Third generation
3GLs. *See* Third-generation languages
4G. *See* Fourth generation
4GLs. *See* Fourth-generation languages
5G. *See* Fifth generation
32-bit-word computer, 114
64-bit-word computer, 114
802.11ac, 188
802.11ax, 188
802.11be, 188
802.11g, 188
802.11n, 188

A

AC. *See* Alternating current
AC adapters, 120, 127
Access, 214, 233
    data dictionary form, 266–267
    data entry form, 267
    restricting, 227
    speed, 164, 177
Accounting, 243, 255
    applications, 70
    area, 247
    TPS for, 247
Accounts payable, 248, 256
Accounts receivable, 247, 256
Accuracy, 39, 48, 214, 233
Active display area, 142, 156
Activity trackers, 109, 125
Ad-Aware, 219
Additive manufacturing. *See* 3D printers
Addresses, 29, 36, 46, 48
Adobe Creative Cloud, graphics suites, 71
Adobe Dreamweaver, 44, 69
Adobe Illustrator Draw, 44
Adobe InDesign, 68
Adobe Photoshop, 68
Adobe Premier, 67

ADSL. *See* Asymmetric digital subscriber line
Advanced-fee scam, 221
Advanced Research Project Agency Network (ARPANET), 25, 46, 336
Age of Connectivity, 339–342
Agile development, 319, 328
Aldebaran Robotics, 324
Alignment, 58
All-in-one desktops, 109, 125
Alternating current (AC), 120
ALU. *See* Arithmetic-logic unit
Amazon, 41, 55, 56, 58, 75, 171, 341
Amazon Prime, 27
Amazon's Alexa, 92, 93
American Standard Code for Information Interchange (ASCII), 122, 127, 165, 321, 335
Analog signals, 121, 127, 189
Android, 88, 100
    backup programs, 96
    search program, 94
    storage management programs, 95
Animation, 63
Anonymity
    end of, 232
    illusion of, 216, 233
Anticipating disasters, 223, 226, 234
Antispyware, 219, 233
Antitrack, 219
Antivirus programs, 91, 101
Apple, 337–339, 341–342
Apple App Store, 56, 75
Apple FileMaker, 66, 76
Apple Health App, 49
Apple iMac, 109
Apple iMovie, 67, 87
Apple iOS 15, 224
Apple iPhone, 60
Apple iTunes Music Store, 229, 235
Apple iWorks, 70, 73
Apple Keynote, 63, 76
Apple macOS, 87
Apple Mail, 37
Apple Numbers, 64, 76
Apple Pages, 60, 76
Apple Pay, 40

Apple Safari, 29
Apple Watch, 10, 86, 109
Application generation subsystem, 266, 278
Application generators, 322
Application programs, 307
Application software, 7, 17, 56–58
   app stores, 56
   common features, 58, 75
   general-purpose applications. *See* General-purpose applications
   mobile apps/applications, 7, 17, 56, 59–60
   software suites, 70
   specialized applications, 7, 17, 56, 67–70
   user interface, 57–58
Apps, 7, 17, 59–60, 75
App Stores, 56, 75
AR. *See* Augmented reality
Arithmetic-logic unit (ALU), 114, 126
Arithmetic operations, 114
ARPANET. *See* Advanced Research Project Agency Network
ASCII. *See* American Standard Code for Information Interchange
Aspect ratio, 141
Assembly languages. *See* Second generation (2G)
Asymmetric digital subscriber line (ADSL), 191, 206
AT&T, 26
Attachments, 37, 48
Attributes, 263, 271, 277, 279
Audio-input devices, 140, 155
Audio-output devices, 146, 157
Augmented reality (AR), 74, 110
Authentication, 225, 234
   two-factor, 225, 234
   two-step, 225, 234
Authority, 38, 48
Autodesk Sketchbook, 68
Automated design tools, 290
Automatic teller machines, 248
Autonomic computing systems, 98
AVG, 8
   Antitrack, 219
   TuneUp, 71

B

B2B commerce. *See* Business-to-business commerce
B2C commerce. *See* Business-to-consumer commerce
Babbage, Charles, 333
Backbone, 199, 207
Background, 85, 99
Back pain, 151
Backup programs, 91, 96, 101
Backups, 226–227
Balance sheets, 248, 256
Bandwidth, 194, 206
Bank loan/credit card scam, 221
Bar codes, 138, 155
   readers, 138, 155
Baseband, 194, 206
Base station, 198, 206
Basic input/output system (BIOS), 116
Batch processing, 264, 277
BDs. *See* Blu-ray discs (BDs)
Beta testing, 317, 327
Big data, 214–216, 233
Binary system, 121, 127
Bing, 30, 38, 48
Biometric scanning, 223, 234
   devices, 223
BIOS. *See* Basic input/output system
Bit, 121
Bitcoins, 40, 49
Bitdefender Internet Security, 32, 97
Bitmap, 68
BitTorrent, 31, 47
Blogs, 34, 47, 69, 77
Bluetooth, 146, 157, 188, 198, 205
Blu-ray discs (BDs), 11, 17, 169, 179
Booting, 85, 99
Botnet, 222, 234
Brain-computer interfaces, 124
Broadband, 194
Browser cache. *See* Temporary Internet files
Browsers, 29–30, 46
Bus. *See* Bus lines
Business directories, 274
Business statistical information, 274
Business-to-business (B2B) commerce, 40, 49
Business-to-consumer (B2C) commerce, 39, 49
Bus lines, 107, 113, 118, 125, 126
   expansion buses, 118
Bus network, 199, 207
Bus width, 118, 126
Buttons, 57, 75
Bytes, 121

C

C#, 315
C++, 315, 321
    code, 315
C2C commerce. *See* Consumer-to-consumer commerce
Cables, 26, 46, 119, 127, 206
    coaxial, 189, 190
    ethernet, 189, 190
    fiber-optic, 189
    modem, 190
    service, 191, 206
    twisted-pair, 189
Cache memory, 116, 126
CAN-SPAM, 37
Capacity, 164, 177
Card readers, 138, 155
    magnetic, 138, 155
Cards, 107
    chip, 138, 155
Carpal tunnel syndrome, 151, 157
Cascading style sheets (CSS), 30, 46, 69
Casual gaming, 110
CDs. *See* Compact discs
Cell phone. *See* Mobile
Cells, 64, 188, 205
Cell towers, 188, 205
Cellular, 188, 205
Cellular service providers, 191, 206
Censorship, 25
Central processing unit (CPU), 114, 126
Channels
    communication, 187–189
    physical connections, 189
    wireless connections, 188
Character, 139, 263, 277
    effects, 58, 61
    encoding, 122, 127
    encoding standards, 122, 127
ChatGPT, 254
Chief executive officer (CEO), 245
Child nodes, 268, 278
Chip cards, 138, 155
Chips, 113, 333, 336
    carriers, 113, 125
Circle, web filter, 31
Circle with Disney, 31

Cisco TelePresence, 204
Clarity, 141
Classes, 271, 279
Client-based e-mail systems, 37, 48
Clients, 41, 195, 206
Client/server networks, 200–201, 207
Clock speed, 114, 126
Cloud computing, 13, 18, 41–42, 49, 70, 170, 179
Cloud office suites, 5, 72–73
Cloud storage, 5, 170–171, 193
    Google Drive Docs, 170
    organizational, 174, 179
    services, 171
Cloud suites, 70, 72
Coaxial cable, 189, 205
Code review. *See* Desk checking
Coding, 314–315, 326
Cold boot, 85, 99
Color, 144
.com, 30
Combination input and output devices, 146–147, 157
    drones, 147, 157
    headsets, 146, 157
    multifunctional devices (MFD), 146, 157
    robots, 147, 157
    virtual reality, 147, 157
Combination keys, 135, 154
Comcast, 26
Commercial database, 272–273, 279
Common Business Oriented Language (COBOL), 335
Common data item, 269, 278
Communication(s), 33–37, 46–48, 186–187
    blogs, 34, 47
    channels, 187–189
    computer, 186
    connection devices, 189–191
    connectivity, 186, 205
    data transmission, 191, 194–195
    devices, 11, 17
    e-mail, 36–37
    Internet, 32, 33, 35–37
    links, 250
    messaging, 35
    microblogs, 34, 47
    mobile office, 192–193
    networks, 195–201

and networks, 185
organizational networks, 201–202
podcasts, 35, 47
social networking, 33–34, 47
systems, 186–187, 205
telepresence, 204
wikis, 35, 47
wireless revolution, 186, 205
Community, 5, 16, 33, 35, 39, 136, 165, 215
fake reviews, 58
impact of technology, 5
operating system, 85
recycling, 113
Compact discs (CDs), 11, 17, 169, 179
Company database, 272–273, 279
Compatibility, 111
Compiler, 322
Computer age, evolution of, 333–343
Computer-aided design/computer-aided manufacturing systems (CAD/CAM systems), 253, 256
Computer-aided software engineering (CASE) tools, 290, 320, 328
Computer-based information systems, 246
Computer Buyer's Guide, 344–346
desktops, 345–346
laptop computers, 344, 345
smartphones, 344–346
tablets, 344
Computer crime. See Cyberterrorism
Computer Fraud and Abuse Act, 223, 234
Computer(s), 108, 186
communications, 186
desktop, 346
monitoring software, 218, 233
networks, 195, 196
technicians, 123, 127
types of, 9–10, 17
Computer security, 220
anticipating disasters, 226
encrypting data, 226–227
measures to protecting, 223–228, 234
preventing data loss, 226–227
restricting access, 223–226
Computer support specialists, 14, 97, 101
Computer technician, 14
"Compute time" flowchart, 311
Conceptualization, 270
Connected Generation, 333

Connection
service, 190–191
technologies, 26, 46
Connection devices, 186–187, 189
connection service, 190–191
modems, 190
Connectivity, 3, 13, 18, 144, 156, 186, 205
Console gaming, 110–111
Consumer-to-consumer (C2C) commerce, 39, 49
Contactless payment, 138
Content evaluation, 38–39, 48
Contextual tabs, 57, 75
Contrast ratios, 142
Control, 244
unit, 114, 126
Cookies, 217
accepting, 217
blocking, 217
first-party, 217
third-party, 217–218
tracking, 217
Copper lines, 190
Coprocessors, 115
Copyright, 229
Cordless/wireless mouse, 136, 154
Cortana, 91, 140, 155
CPU. See Central processing unit (CPU)
Crackers, 222, 234
Credit card, 40
Credit card fraud, 40
Crime databases, 276
Crossy-road and Stardew Valley, 59
Cryptocurrency, 40
Cryptoprocessors, 115
CSS. See Cascading style sheets
Currency, 39, 48
Cyberbullying, 229, 235
Cybercrime, 220–221, 234
Cyberterrorism, 220, 234
Cylinders, 166, 178

D

Dark web, 217, 233
Data, 3, 4, 12, 16, 18, 250, 263
administration subsystem, 266, 278
analysis, 65
cube, 270, 279

definition subsystem, 266, 278
dictionary, 266
encryption, 226–227
integrity, 266
less redundancy, 266
loss prevention, 226–227
maintenance, 266, 278
mining, 274
redundancy, 266
security, 226
storage, 164
warehouse, 274
workers, 252
Data banks, 272, 279
Database, 66, 76
    creation of, 66
    files, 12, 18
Database administrators (DBAs), 266, 275, 278–279
Database management system (DBMS), 7, 66, 76, 248, 266, 278
    access data dictionary form, 266–267
    access data entry form, 266, 267
    department of motor vehicles, 270
    engine, 266, 278
    hierarchical database, 268–269, 278
    multidimensional database, 270, 279
    network database, 269, 278
    object-oriented database, 271, 279
    organization, 271
    relational database, 269–270, 278
    structure, 268–271
Database manager, 66, 76
Databases, 242, 263, 266, 277
    commercial, 272–273
    company, 272
    data organization, 263
    DBMS, 266–267
    DBMS structure, 268–271
    distributed, 272–273
    individual, 272–273
    model, 268, 278
    need for, 266, 278
    security, 274, 279
    strategic uses, 274, 279
    types of, 272–273
    uses and issues, 274
Data flow
    diagrams, 290
    diagram symbols, 290
Data manipulation, 221
    subsystem, 266, 278
Data organization, 263–264
    batch processing, 264, 277
    key field, 264, 277
    logical, 264
    real-time processing, 264, 277
Data processing systems (DPSs). *See* Transaction processing system
Data transmission, 191, 194–195
    bandwidth, 194
    protocols, 194–195
    specifications, 187
DBMS. *See* Database management system
Debnath, Shantanu, 343
Debts, 248
Debugging, 315, 327
Decision making, 244
Decision models, 250, 256
Decision support system (DSS), 246, 249–250
Deep fakes, 38, 48
Deep web, 217, 233
Demand report, 248
Demodulation, 190
Demographic data, 274
Denial of service attacks (DoS attacks), 221, 234
Density, 166, 178
Department of Defense, 188
Desired output, end user's sketch of, 308–309, 325
Desk checking, 317, 327
Desktop browsers, 29, 46
Desktop operating systems, 87–91
    Linux, 87, 101
    macOS, 88, 100
    UNIX, 87, 101
    virtualization, 90–91, 101
    Windows, 88–89, 100
Desktop(s), 86, 99, 109, 125, 345
    all-in-one, 109, 125
    computers, 10, 17, 107, 346
    publishing programs, 68, 77
Device drivers, 84, 99
Diagnostic program, 91, 101
Dialog boxes, 57, 75, 86, 99
Dial-up services, 191, 206

Dictionary attacks, 223, 234
Digital cameras, 139–140, 155
Digital cash, 40, 49
    providers, 40
Digital currency, 40, 49
Digital electronic signals, 121
Digital Equipment Corporation (DEC), 336
Digital footprint, 214, 233
Digital Millennium Copyright Act, 229, 235
Digital projectors, 143, 156
Digital rights management (DRM), 229, 235
Digital signals, 189, 206
Digital subscriber line (DSL) service, 191, 206
Digital versatile discs (DVDs), 11, 17, 169, 179
Digital video disc (DVD), 11
Digital Video Interface (DVI) ports, 119, 127
Digital whiteboards, 143, 156
Direct approach, 294, 300
Direct current (DC), 120
Directory server, 195, 206
Disaster recovery plans, 226
Disaster recovery specialists, 175, 179
Disk caching, 167, 178
Display, 11, 17
Display Ports (DP), 119, 127
Display screens. *See* Monitors
Distributed database, 272–273
Distributed denial of service (DDoS), 221, 234
DNA, 176
Documentation, 307, 317
Documents, 58
    files, 12, 18
    scanner, 137, 155
    theme, 63
Domain name, 29, 46
Domain name server (DNS), 194
    converting text-based addresses to numeric IP addresses, 195
Dot (pixel) pitch, 142
Dots per inch (dpi), 144, 156
Dow Jones Factiva, 273
Downlink, 188
Downloading, 31, 47
DP. *See* DisplayPort
Drawing programs, 68, 77
Drives, 164
Drones, 147, 157

DS3 lines. *See* T3 lines
DSL (digital subscriber line), 26, 46
DSS. *See* Decision support system
Dual in-line memory module (DIMM), 116, 126
DuckDuckGo, 38, 48
Duplex printing, 144, 156
DVDs. *See* Digital versatile discs
DVI (Digital Video Interface) ports, 119

E

eBay, 40
EBCDIC. *See* Extended Binary Coded Decimal Interchange Code
Eckert, J. Presper, Jr., 334
Economic feasibility, 292, 299
.edu, 30
Education, 26, 46
E-ink, 143
EIS. *See* Executive information system
E-learning, 26, 46
Electronic-book readers (e-readers), 143, 156
Electronic books (e-books), 143, 156
Electronic commerce (e-commerce), 39–40, 49, 186
    security, 40, 49
    types of, 39–40, 49
Electronic data and instructions, 121
    character encoding, 122, 127
    numeric representation, 121–122, 127
Electronic mail (e-mail), 36–37, 48, 186
    client-based, 37, 48
    encryption, 226
    web-based, 37
Electronic Numerical Integrator and Computer (ENIAC), 334
Electronic Recording Machine Accounting (ERMA), 335
E-mail. *See* Electronic mail
Embedded operating systems, 7, 17, 86, 99
Employee-monitoring software, 216, 233
Encryption, 226, 234
    e-mail, 226
    file, 226
    key, 226, 234
    website, 226
End users, 3, 16
Enterprise storage system, 174, 179
Entertainment, online, 26–28, 46
Entity, 263, 277

Environment, 2
Equity, 248
Ergonomic(s), 150, 157
    keyboard, 151
    physical problems, 150–151
    portable computers, 151, 157
    recommendations, 157
ESSs. *See* Executive support systems
Ethernet, 197, 206
    cables, 189, 205
    ports, 119, 127
Ethics, 6, 15, 16, 25, 187, 219, 235
    computer, 227, 235
    copyright, 229
    cyberbullying, 229
    DRM, 229
    end of anonymity, 232
    image editing software, 68
    IT security analysts, 231, 235
    voice assist tools, 85
Etsy, 40
Exception reports, 248
Executive information system (EIS), 246
Executive support systems (ESSs), 246, 250–252, 256
Expansion buses, 118, 126
Expansion cards, 117, 126
Expansion slots, 117, 126
Expert systems, 253
Extended Binary Coded Decimal Interchange Code (EBCDIC), 122, 127, 165
External data, 250
External hard drives, 127, 167
External storage, 178
Extranet, 201, 207
Eyestrain, 150

F

Facebook, 24, 28, 34, 47, 75, 215, 341
Facebook Messenger, 36
Facial recognition, 224, 234
Fake news, 38, 48
Fake reviews, 58
Family Educational Rights and Privacy Act (FERPA), 220, 233
Femtosecond, 114
Fiber-optic cable, 189, 205
Fiber-optic service (FiOS), 191, 206
Field, 263, 277
Fields, database, 66
Fifth generation (5G), 333, 339–343
    mobile telecommunications, 191, 206
Fifth-generation language (5GL), 322–323
File compression, 168, 178
File decompression, 168, 178
File(s), 86, 99
    database, 12, 18
    encryption, 226
    transfer utilities, 31, 47
    types, 12
File servers, 174
File transfer, 31
File transfer protocol (FTP), 31, 47
Filters, 30, 47
Final Fantasy, 75
Fingerprint, 223
Firewalls, 201, 207, 224, 274
FireWire buses, 118, 126
FireWire ports, 119
First generation (1G), 321, 333, 334
    mobile telecommunications, 191, 206
First-party cookie, 217
Fitbit, 49
Flash, 115
Flash drives, 165
Flash memory, 116, 126, 165, 176
    cards, 165, 177
Flatbed scanner, 137, 155
Flat-panel monitors, 142
Flexible screens, 143
Flowcharts, 289, 290, 307, 310–313, 317, 326, 327
    for "Compute time on Client A jobs,", 312
    symbols, 311
    system, 290
Flyer creation, 61
Folders, 86, 99
Foreground, 99
Forms, database, 66
Formulas, 64
FORmula TRANslator (FORTRAN), 334
Fourth generation (4G), 333, 337–338
    mobile telecommunications, 191, 206
Fourth-generation languages (4GLs), 322
Free antivirus program, 5, 8
Freedom of Information Act, 215, 233

Free-to-play, 59
Friends, 33, 35, 47, 48, 110
Froala 4.0, 69
FTP. *See* File transfer protocol
Functions, 64, 85

G

Galleries, 57, 75
Game controllers, 136–137, 154
Gamepads, 136, 137, 154
Games, 110–111
Games app, 59
Gaming, 5, 110–111
    console, 110–111
    gamepads, 136, 137, 154
    joysticks, 136
    laptops, 109, 125
    mice, 136
    mobile, 110
    motion-sensing devices, 136, 137
    PC, 111
Gates, Bill, 338
Gbps, 188
General Electric Corporation, 335
General ledger, 248, 256
General-purpose applications, 7, 17, 56, 60–66, 76
    database management system, 66
    presentation software, 76
    spreadsheets, 64–65, 76
    word processors, 60–62, 76
Generations of programming languages, 320–323
Gesture control, 86, 99
Global positioning system (GPS), 188, 205
    navigation, 188
    virtual assistants, 93
Google, 38, 40, 48, 55, 88, 341–342
Google Android, 75, 88
Google Assistant, 25, 92
Google Chrome OS, 29, 89, 218
Google Docs, 60, 72
Google Drive Apps, 42
Google Drive Docs, 170
Google Gmail, 37
Google Nest, 43
Google Play, 56, 75
Google Podcasts, 28
Google Search engine, 340

Google Sheets, 64
Google Slides, 63, 76
Google Street View, 216
Google Translate, 140
.gov, 30
GPU. *See* Graphics processing unit
Grammar checker, 58
Gramm-Leach-Bliley Act, 220, 233
Graphical user interface (GUI), 57, 75, 85, 99
    Ribbon, 57, 75
    traditional, 57
Graphics, 67–68
    cards, 117, 126
    coprocessor, 115, 126
Graphics processing unit (GPU), 115, 126
Grayscale, 144, 156
Greeting card scam, 221
Grid charts, 289, 290, 299
Group decision support systems (GDSS), 250, 256
Groups, 33, 47, 57
Guest operating system, 90, 101
GUI. *See* Graphical user interface

H

Handwriting recognition software, 135, 154
Haptic technology, 204
Hard disks, 11, 17, 164, 166–168, 178
    drives, 103
    external, 167, 178
    internal, 167, 178
    network drives, 167, 178
    performance enhancements, 167–168, 178
Hardware, 3, 4, 9–11, 16. *See also* Computer(s)
    personal computer, 10, 17
HBO Max (hbomax.com), 27
HDMI ports. *See* High Definition Multimedia Interface ports
Headache, 150
Header, 36, 48
Headphones, 148–149, 157
Headsets, 146, 157
Health Insurance Portability and Accountability Act (HIPAA), 220, 233
HealthWise, 245
    Group, 243
    monthly sales and production reports, 248
Heat-assisted magnetic recording (HAMR), 176

Help, 86, 99
Hexadecimal system (hex), 122
Hidden costs, 110
Hierarchical database, 268–269
Hierarchical network, 199
Hierarchical organization, 271
High-definition (hi-def), 169, 179
High Definition Multimedia Interface (HDMI) ports, 119, 127
Higher-level languages, 321–322, 328
High-frequency transmission cable, 189
High-level procedural languages, 321–322
High-speed Internet wireless technology, 186
History files, 217, 233
Hoff, Ted, Dr., 337
Home, 199
Home networks, 198, 206
Hopper, Grace, Dr., 334
Host, 195, 206
Host operating system, 90, 101
Hotspots, 198
HTML editors, 69, 77
HTML. *See* Hypertext Markup Language
Hub, 195
Hulu (hulu.com), 27
Human resources, 244, 255
Hybrid cloud, 42
Hybrid drives, 168, 178
Hyperlinks, 30, 46
Hypertext Markup Language (HTML), 30, 46, 69
Hypertext transfer protocol secure (https), 194, 234

I

IBM, 9, 41, 254, 334, 336, 338, 343
    Blue Gene supercomputer, 9
    Watson, 254, 342
Icons, 57, 85, 99
Identification, 194
Identity theft, 220, 221, 234
IFPS (interactive financial planning system), 322
IF-THEN-ELSE structure, 313
Illusion of anonymity, 216, 233
Illustration programs, 68, 77
IM. *See* Instant messaging
Image-capturing devices, 139–140, 155
Image editing software, 67
Image editors, 68, 77

Immersive experience, 147, 157
Incognito Mode, 218
Income statements, 248, 256
Individual database, 272–273, 279
Industrial robots assemble automobiles, 147
Indy games, 111
Infected USB flash drives, 223, 234
Information, 3, 16
    brokers, 214, 233
    flow, 245, 255
    resellers, 214, 233
    workers, 252
Information systems, 16, 252–253, 256, 286
    computer-based information systems, 246
    data, 3, 4, 12, 16
    DSS, 246, 249–250
    ESSs, 246, 250–252, 256
    expert systems, 253
    hardware, 3, 4, 9–11, 16
    IBM's Watson, 254, 357
    Internet, 3, 4, 16
    managers, 253
    MIS, 246–248, 252, 256
    organizational information flow, 243–245
    parts of, 3–4
    people, 3–6, 16
    procedures, 3, 4, 16
    software, 3, 4, 6–7, 16
    TPS, 247–248, 252, 256
Information technology (IT), 6, 16
    free antivirus program, 5, 8
    future trends, 15
    information systems. *See* Information systems
Information utilities, 272, 279
Infrared, 188, 205
Inkjet printers, 145, 156
Input, 134, 154
    audio-input devices, 140
    character and mark recognition devices, 139
    data, 309, 325
    devices, 11, 17, 109, 134, 154
    image capturing devices, 139–140
    keyboard entry, 134–135
    pointing devices, 135–137
    RFID reader, 139
    scanning devices, 137
Instagram, 34, 47, 59, 75

Instant messaging (IM), 36, 48
Integrated Circuit Age, 336
Integrated circuit (IC). *See* Chips
Intel, 41
Intel Corporation, 41, 337, 342
Interactive financial planning system (IFPS), 322
Interactive whiteboards, 143, 156
Internal data, 250
Internal hard disk, 164, 167
Internal storage, 178
International Telecommunications Satellite Consortium (Intelsat), 188
Internet, 3, 4, 11, 13, 15, 16, 18, 30, 46, 185, 216–219, 233, 263
    access, 26, 29–30
    browsers, 29–30, 46
    cloud computing, 41–42
    common uses of, 46
    communications, 33–37
    e-commerce, 39–40, 49
    IoT, 43, 49
    providers, 26
    scams, 221
    search tools, 38–39
    security suite, 32, 47
    smart homes, 45
    technologies, 201
    and web, 25–26, 46
Internet of Things (IoT), 13, 18, 43, 45, 49, 109, 153
Internet protocol address (IP address), 194, 206, 216
Internet scams, 221, 234
Internet security suite, 32, 47
Internet service providers (ISPs), 26, 46, 221
Internet technologies, 201
Interpreter, 322
Intranet, 201
Intrusion detection systems (IDS), 202, 207
Intuit QuickBooks, 70
Inventory, 247, 256
    control system, 247
iOS, 88, 100
    backup programs, 96
    search program, 94
    storage management programs, 95
IoT. *See* Internet of Things
iPhone OS (iOS), 88
Iris scanners, 223

iRobot, 324
ISP. *See* Internet service provider
IT security analysts, 231, 235

J
Jacquard, Joseph Marie, 333
Java, 315
JavaScript, 30, 46, 315
Jobs, Steve, 337
Joysticks, 136, 137, 154

K
Key, 226
Keyboard, 11, 17, 154
    attachments, 109
    entry, 134–135
    ergonomic, 151
    laptop, 134, 154
    traditional, 135, 154
    virtual, 134, 154
Key field, 264, 277
Keylogger, 218, 233
Knowledge base, 253, 256
Knowledge-based systems. *See* Expert systems
Knowledge workers, 253
Knowledge work systems (KWSs), 253, 256

L
Lands, 169, 179
Language translators, 84, 99
LANs. *See* Local area networks (LANs)
Laptop(s), 10, 17, 109, 125, 151, 345
    computers, 344
    gaming, 109, 125
    keyboards, 134, 154
    two-in-one, 108, 109, 125
Large databases, 214–216
Large-scale integration (LSI), 337
Laser printer, 145, 156
Laws on privacy, 220
LCD. *See* Liquid crystal display
LED. *See* Light-emitting diode
Less data redundancy, 266
Levels of programming languages, 320–321
LexisNexis, 273
Light-emitting diode (LED), 142, 156

Line-of-sight communication, 188
LinkedIn, 34, 47
Links, 30, 46
Linux, 87, 89–90, 200
Liquid crystal display (LCD), 142, 156
Local area networks (LANs), 197, 206, 337
    adapter, 196
Location, 29, 46
Logical operations, 114
Logical view, 263
Logic errors, 316, 325, 327
Logic structures, 311, 313, 325–326
Long-range planning, 244
Long Term Evolution (LTE), 191, 206
Loop. *See* Repetition logic structure
Lottery scam, 221
Low bandwidth. *See* Voiceband
Lower-level languages, 320–321, 328

M

Machine languages. *See* First generation (1G)
Macintosh, 338
macOS, 7, 87, 89, 100
    backup programs, 96
    search program, 94
    storage management programs, 95
macOS 12 Sonoma, 89
macOS X, 200
macOS X Server, 200
Magix's Sound Forge Pro 15, mastering suites, 71
Magnetic card reader, 138
Magnetic field, 176
Magnetic-ink character recognition (MICR), 139, 155
Mainboard. *See* System board
Mainframe computers, 9, 17
Maintenance programmer, 318–319
Malicious programs, 7
Malicious software, 222, 234
Malware, 222, 234
Management information system (MIS), 246–248, 252, 256
Management levels, 244, 255
MANs. *See* Metropolitan area networks (MANs)
Manual testing, 317, 327
Many-to-many relationship, 269, 278
Marketing, 244, 255
Mark recognition devices, 139

Mass storage, 171, 179
    devices, 171, 174–175, 179
    enterprise storage system, 174, 179
    storage area network, 175, 179
Mauchly, John W., Dr., 334
MaxiCode, 138, 155
Mbps, 190
Media, 164, 177
Medium band, 194
Megabits per second (Mbps), 190, 206
Megabytes (MB), 169
Memory, 11, 17, 107, 115, 116, 126, 144
    cache, 116, 126
    capacity, 116
    DIMM, 116, 126
    flash, 116, 126, 165
    RAM, 11, 17, 115, 126
    ROM, 116, 126
    virtual, 116, 126
Menu bar, 57, 75
Menus, 57, 75, 85, 99
Mesh network, 200, 207
Message, 37, 48
Messaging, 35–36, 48
Methods, 271, 279
Metropolitan area networks (MANs), 198–199, 206
MICR. *See* Magnetic-ink character recognition
Microblogs, 34, 47
Microprocessor, 11, 17, 107, 112, 114–115, 126
    chips, 114–115, 126
    specialty processors, 115
Microprocessor Age, 337–338
Microsecond, 114
Microsoft, 41, 48, 338–343
Microsoft Access, 66, 76
Microsoft Cortana monitors, 93
Microsoft Edge, 29
Microsoft Excel, 64
Microsoft Office, 70, 87
Microsoft PowerPoint, 63, 76
Microsoft Project, 70, 252
Microsoft Publisher, 68
Microsoft search engine, 30
Microsoft's Outlook, 37
Microsoft's Photos, 68
Microsoft Store, 56, 75
Microsoft Windows operating systems, 87

Microsoft Word, 60, 69, 76
Microsoft Xbox, 111
Microwave, 188, 205
    dish, 188
    transmissions, 188
Middle-level managers, 245, 255
Middle management, 244
Middle managers, 248
Midrange computers, 9, 17
.mil, 30

Minecraft, 75
Mini DisplayPort (MiniDP/mDP) ports, 119, 127
Mini notebooks. *See* Ultrabooks
Mini tablets, 108
MIS. *See* Management information system (MIS)
Mistaken identity, 215
MMS (Multimedia Messaging Service), 35
Mobile, 10, 17
    browsers, 29, 46
    computing, 164
    devices, 7
    gaming, 110
    hotspot device, 192
    office, 192-193
    operating systems, 88, 100
Mobile apps/applications, 7, 17, 56, 59-60, 75
Mobile Internet, 13, 18
Mobile office, 5
Modems, 11, 17, 26, 190, 206
Modulation, 190
Modules, 310
Moneytree Software's TOTAL Planning Suite, 71
Monitors, 11, 17, 141-143, 156
    e-readers, 143, 156
    features, 141-142, 156
    flat-panel, 142
    resolution, 141
Motherboard. *See* System board
Motion-sensing devices, 136, 154
Mouse, 11, 17, 136, 154
    cordless/wireless, 136
    pointer, 136, 154
Mozilla Firefox, 29
Multicore processors, 115, 126
Multidimensional database, 270, 279
Multidimensional data model, 270
Multidimensional organization, 271

Multifunctional devices (MFD), 146, 157
Multilevel cells (MLCs), 176
Multitasking, 85, 99
Multitouch screens, 136, 154
Music apps, 75

N

Nanosecond, 114
Nao robots, 324
Natural languages, 322
Neck pain, 151
.net, 30
Netflix, 30, 59, 75, 342
Net Nanny, 31
Net neutrality, 208
Network, 13, 18
    administrator, 203, 206, 207
    architecture, 199-201, 207
    communications and, 185
    computer, 196, 206
    database, 269, 278
    gateway, 197, 206
    organizational networks, 201-202
    security, 201-202, 207
    strategies, 199
    telepresence, 204
    terms, 195-196
    topology, 199
    types, 197-199, 206
Network adapter cards. *See* Network interface cards (NIC)
Network administrator, 14, 175
Network attached storage (NAS), 174, 179
Network drives, 167, 178
Network interface cards (NIC), 117, 126, 196, 206
Network operating systems (NOS), 7, 17, 87, 99, 196, 206
Network organization, 271
Network server, 87, 99
News feeds, 33, 47
Newspapers, 34, 38, 47, 143
Next-generation storage, 176
Nintendo Switch, 110
Nodes, 195, 206, 268, 278
Norton Security, 219
Norton's utility suite, 97
NOS. *See* Network operating systems
NoSQL, 271, 279
Notebook computers, 10, 17
Numeric representation, 121-122, 127

O

OASs. *See* Office automation systems (OASs)
Object code, 322
Objectivity, 39, 48
Object-oriented database, 271
Object-oriented organization, 271
Object-oriented programming (OOP), 320, 328
Object-oriented software development, 320, 328
Objects, 271, 279, 328
OCR. *See* Optical-character recognition
Office automation systems (OASs), 252, 256
Office software suites. *See* Office suites
Office suites, 70, 77
OLED. *See* Organic light-emitting diode
OMR. *See* Optical-mark recognition
One-to-many relationship, 268, 278
Online, 25
    entertainment, 27–28
    music, 27
    processing. *See* Real-time processing
    registration systems, 248
    video, 27
Online identity, 219, 233
Online office suites. *See* Cloud suites
Online storage. *See* Cloud storage
Open-source software, 101
Operating systems (OS), 7, 17, 84–91, 99
    defined, 85
    embedded/real-time, 7, 17, 86, 99
    features, 85–86, 99
    functions, 85, 99
    mobile, 88, 100
    network, 7, 17, 87, 99
    stand-alone, 7, 17, 87
    utilities, 94–96, 101
    voice assist tools, 85
Operational feasibility, 292, 299
Operational models, 250, 256
Operators, 317, 327
Optical carrier (OC) lines, 190, 206
Optical-character recognition (OCR), 139, 155
Optical discs, 11, 17, 168–170, 179
    BDs, 169, 179
    CDs, 169, 179
    drive, 169, 179
    DVDs, 169, 179
    read-only, 169, 179
    recordable (R) disc, 169
    rewritable, 169, 179
    types of, 170
    UHD BD, 169, 179
    write-once, 169, 179
Optical-mark recognition (OMR), 139, 155
Optical scanners, 137, 155
    document scanner, 137, 155
    flatbed scanner, 137, 155
    portable scanner, 137, 155
    3D scanners, 137, 155
Oracle Database Express Edition, 66
.org, 30
Organic light-emitting diode (OLED), 142, 156
Organizational cloud storage, 174, 179
Organizational information flow, 243–245
    functions, 243–244
    information flow, 245
    management levels, 244
Organization chart, 289, 299
Organizations, 15
OS. *See* Operating systems
Output, 141, 156
    audio-output devices, 146
    devices, 11, 17, 109, 141, 156
    monitors, 141–143
    printers, 144–145

P

P2P network. *See* Peer-to-peer network
Packetization, 195
Packets, 195, 206
Page layout programs, 68, 77
Pages, 33, 47
PAN. *See* Personal area network (PAN)
Pandora, 75, 229
Parallel approach, 294, 300
Parallel processing, 115, 126
Parallels (virtual machine program), 90, 101
Parent node, 268, 278
Password managers, 224, 234
Passwords, 223, 234
Patches, 319, 328
Paychecks, 248
PayPal, 40
Payroll, 248, 256
PCIe. *See* PCI Express

PCI Express (PCIe), 118, 126
PCs. *See* Personal computers
Peer-to-peer (P2P) network, 201, 207
People, 3-6, 16, 214
Performance enhancements, 167-168, 178
    techniques, 168
Periodic evaluation, 295
Periodic reports, 248
Peripherals, 119
Personal area network (PAN), 198, 206
Personal computers (PCs), 10, 17, 108, 196
    gaming, 111
    hardware, 10-11, 17
Personal consent, spreading information without, 215
Personal hotspot, 192
Personalized buying guide, 345-346
    desktop computers, 346
    laptops, 345
    smartphones, 345-346
    tablet computers, 346
Personal laser printers, 145, 156
Personal security, 220
Phased approach, 294-295, 300
Phishing, 221, 222, 234
Photo editors. *See* Image editors
PHP, 30, 44, 46
Physical connections, 189, 205
Physical security, 226
Physical view, 263
Picosecond, 114
Picture elements. *See* Pixels
Picture Password, 223, 234
Pilot approach, 294, 300
Pinterest, 34
Pits, 169, 179
Pixels, 68, 77, 141
Plagiarism, 230, 235
Plagiarists, 230, 235
Planning, 244
Platform scanners, 138, 155
Platters, 166, 178
Plotters, 145, 156
Podcasts, 28, 35, 47
Pointer, 57, 75, 85, 99, 269, 278
Pointing devices, 135-137, 154
    game controllers, 136-137, 154
    mouse, 136, 154

    touch screen, 135-136, 154
Popular microprocessors, 114
Pop-ups, 26
Portable computers, 151, 157
Portable languages, 321
Portable scanner, 137, 155
Ports, 107, 119, 127
    cables, 119, 120, 127
    cell phone, 119, 127
    DVI, 119, 127
    Ethernet, 119, 127
    FireWire, 119, 127
    HDMI, 119, 127
    specialized, 119, 127
    standard, 119, 127
    Thunderbolt, 119, 127
    USB-A, 119, 127
    USB-B, 119, 127
    USB-C, 119, 127
Power supply, 120, 121, 127
    units, 120, 121, 127
Preliminary investigation, 287-288, 298
Presentation files, 12, 18
Presentation software, 7, 63, 76
    creating presentation, 63
Pretty Good Privacy, 226
Prewritten programs, 307
Primary key, 66, 264, 277
Primary storage, 164, 177
Printers, 145-146, 156
    features, 144, 156
    inkjet, 145, 156
    laser, 145, 156
    plotters, 145, 156
    thermal, 145, 156
    3D, 145, 156
Privacy, 6, 15, 16, 33, 115, 214, 233
    end of anonymity, 232
    Google Street View, 214
    information reseller's website, 215
    Internet and web, 216-219, 233
    IT security analysts, 231, 235
    large databases, 214-215
    laws on, 220
    mobile apps, 75
    mode, 218, 233
    online identity, 219

private networks, 216, 233
Private Browsing, 218, 233
Private cloud, 42
Private networks, 216, 233
Problem and constraint languages. *See* Fifth-generation language (5GL)
Problem-solving procedure, 307
Procedural languages. *See* Third-generation languages (3GLs)
Procedures, 3, 4, 16
Processing rights, 266, 278
Processing speeds, 114, 270
Processor, 114, 126
Production, 244, 255
Productivity suites. *See* Office suites
Profiles, 33, 47
Program, 307, 325
    code, 307, 314-315, 319, 326
    definition, 308
    design, 307, 310-313, 319, 326
    documentation, 307, 317-318, 327
    flowcharts, 311
    maintenance, 307, 318-319, 328
    modules, 310
    objectives, 308, 325
    specification, 307-309, 319, 325
    test, 307, 315-317, 319, 327
Program analysis. *See* Program specification
Program code, 307, 314-315, 319, 326
Program coder, 322
Program definition. *See* Program specification
Program design, 307, 310-313, 319, 326
    flowcharts, 310-313, 325-326
    logic structures, 311, 325-326
    pseudocode, 311, 325-326
    top-down, 310, 325-326
Program documentation, 307, 317-318, 327
Programmable robot, 324
Program maintenance, 307, 318-319, 328
Programmers. *See* Software engineers
Programming, 307, 325
Programming language, 314, 319, 325, 326
    generations of, 320-323, 328
Programs. *See also* Software
Program specification, 307-309, 319, 325
Program test, 307, 315-317, 319, 327
Project managers, 252, 256

Property, 214, 233
ProQuest Dialog, 273
Protocols, 29, 46, 194-195, 206
    feature, 194
    identification, 194-195
    packetization, 195
Prototyping, 296, 301
Proxy server, 201
Pseudocode, 311, 325-326
Public cloud, 42
Public Wi-Fi, 192
Purchase order, 247
Purchasing, 247, 256
Python, 315

Q

Query-by-example, 266
Query languages, 322
Quicken Starter Edition, 70
Qustodio Parental Control, 31
QWERTY keyboard, 134

R

Radio frequencies, 198
Radio-frequency identification (RFID) tags, 139, 155
RAID. *See* Redundant arrays of inexpensive disks
Random-access memory (RAM), 11, 17, 115, 126, 169, 179
Ransomware, 221, 234
Rapid applications development (RAD), 296, 301
Raster images, 68, 77
Reader/sorter, 139
Reading, 164
Read-only memory (ROM), 115, 116, 126, 169, 179
Real-time operating systems (RTOS), 7, 17, 86, 99
Real-time processing, 264, 265, 277
Receiving devices, 186, 188
Record, 263, 277
Recordable (R) disc, 169
Records, database, 66
Recycling, 113
Redundant arrays of inexpensive disks (RAID), 168, 174, 178, 179, 226
Regional manager, 248
Relation, 269, 278
Relational database, 269-270, 278
Relational organization, 271
Repetition logic structure, 313, 326

Repetitive strain injury (RSI), 151, 157
Report creation, 62
Research, 244, 255
Resolution, 141, 144, 156
    standards, 141
Restricting access, 223-225
Rewritable discs, 169, 179
RFID reader, 139, 155. *See also* Radio-frequency identification (RFID) tags
Ribbon GUI, 57, 75
Ribbons, 57, 75
Ring network, 199, 207
Robot network, 222, 234
Robots, 147, 157
Rogue Wi-Fi hotspots, 223, 234
Role playing game (RPG), 59, 75
Router, 195, 206
RPG. *See* Role playing game
RTOS. *See* Real-time operating systems

S

Sales order processing, 247, 256
Satellite, 188, 205
Satellite connection services, 191
Scanners, 138, 155. *See also* Optical scanners
Scanning devices, 137-139, 155
    bar code readers, 138, 155
    card readers, 138, 155
    character and mark recognition devices, 139, 155
    optical scanners, 137-138, 155
    RFID readers, 139, 155
Schema, 266
SD card, 117, 126
Search engines, 38, 48
Searching, for information, 25, 46
Search programs, 91, 94, 101
Search services, 38, 48
Search tools, 38-39
Secondary storage, 11, 17, 164, 177
    cloud storage, 170-171
    devices, 164, 177
    hard disks, 11, 166-168
    mass storage devices, 171, 174-175
    next-generation storage, 176
    optical discs, 11, 168-170
    solid-state storage, 11, 165-166
Second generation (2G), 321, 333, 335

    mobile telecommunications, 191, 206
Sectors, 166, 178
Secure file transfer protocol (SFTP), 31, 47
Security, 15, 40, 49, 220, 234, 266, 274, 279
    cybercrime, 220-221, 234
    data, 226
    end of anonymity, 232
    IT security analysts, 231, 235
    malicious hardware, 222-223, 234
    malicious software, 222, 234
    measures to protecting computer security, 223-227, 234
    physical, 226
    social engineering, 222, 234
    suites, 224, 234
    and technology, 228
Selection logic structure, 313, 326
Semiconductor, 113, 125
Sending devices, 186
Sequential logic structure, 313, 325
Servers, 9, 17, 206. *See also* Midrange computers
Service providers, 42
SFTP. *See* Secure file transfer protocol
Shared laser printers, 145, 156
Share settings, 33, 47
Sharing, 266
Shopping, 25
Shopping apps, 59
Signature, 36, 48
Silicon chip, 113, 125
SIM (Subscriber Identity Module) cards, 117
Siri, 25, 91, 92, 155
Six-phase systems life cycle, 286
Slots, 113, 125
Smart homes, 45
Smartphones, 2-4, 7, 10, 17, 108, 109, 125, 344-346. *See also* Mobile bar code reader
Smartwatches, 109, 125
Smith, David L., 340
SMS (short message service), 35, 48. *See also* Text messaging
Snapchat, 34
Snapseed, 68
Social engineering, 222, 234
Social media, 28
Social networking, 33-34, 47, 59, 214
Sockets, 113, 125
Software, 3, 4, 6-7, 16, 17

application. *See* Application software
environment/platform, 87, 99
piracy, 229, 235
presentation, 7
suites, 70-73
system. *See* System software updates
Software development, 307, 325. *See also* Programming
Software development life cycle (SDLC), 307, 319
Software engineers, 14, 71, 307
Solaris, 200
Solid-state drives (SSDs), 147, 158
Solid-state storage, 11, 17, 165-166
    flash memory cards, 165, 177
    internal drive, 165
    USB drives, 165-166, 177
Source code, 322
Spam, 37, 48
Spam blockers/filters, 37, 48
Speakers, 146, 157
Specialized applications, 7, 17, 56, 67-70
    graphics programs, 67-68, 77
    video game design software, 68-69, 77
    web authoring programs, 69, 77
Specialized ports, 119
Specialized suites, 71, 77
Specialty processors, 115, 126
Speed, 144
Spell checker, 61
Spiders, 38, 48
Spotify, 27, 59, 75
Spreading inaccurate information, 215
Spreading information without personal consent, 215
Spreadsheets, 7, 64-65, 76
    flyer, creation of, 61
    report, creation of, 62
    sales forecast, creation of, 64
Sprint, 191
Spy removal programs, 219, 233
Spyware, 219, 233
SSDs. *See* Solid-state drives
Stand-alone operating systems, 7, 17, 87, 99
Standard ports, 119, 127
Storage, 164, 177
    devices, 164, 177
    next-generation, 176
Storage area network (SAN), 175, 179
Storage management programs, 91, 95, 101

Storage media, 164
Strategic models, 250
Strategic uses, 274, 279
Strategy, 200, 207
Structured programming techniques, 310
Structured programs, 314, 326
Structured query language (SQL), 266, 322
Stylus, 135, 154
Subject, 36, 48
Subscriber Identity Module (SIM) cards, 117
Subscription services, 27
Supercomputers, 9, 17
Supervisors, 244, 255
Supervisory managers, 245
Swift, 315
Switch, 195, 206
Symantec Norton Family Premier, 31
Symantec Norton Internet Security, 32
Syntax errors, 316, 325, 327
System, 286, 298
    analysts, 286, 296, 298, 301
    audit, 295, 301
    buses, 118, 126
    design, 286, 291-292, 298-299
    design report, 292, 299
    development, 286, 293-294, 300
    flowcharts, 290, 299
    flowchart symbols, 290
    implementation, 286, 294-295, 300
    life cycle, 286, 295-296, 298
    maintenance, 286, 295, 301
System board, 107, 112, 125
System chassis. *See* System unit
System conversion, 294
    types, 294-295
Systems analysis, 286, 289-291
    analyzing data, 289-291
    documenting, 291
    gathering data, 289
    one step in defining problems, 288
    preliminary investigation, 287-288, 298
    prototyping, 296, 301
    RAD, 296, 301
    report, 291, 299
    systems design, 291-292, 299
    systems development, 293-294
    systems implementation, 294-295

    systems maintenance, 295, 301
    testing system, 294
Systems maintenance, 295, 301
System software, 7, 17, 56, 84, 99, 249, 256
    device drivers, 84
    language translator, 84
    operating systems. *See* Operating systems
    utilities, 84, 91–97
System unit, 11, 17, 108–112, 125
    brain-computer interfaces, 124
    bus lines, 118, 126
    components, 112, 125
    computer technicians, 123, 127
    desktops, 109, 125
    electronic data and instructions, 121–122
    expansion cards and slots, 117, 126
    gaming, 110–111
    laptops, 109, 125
    memory, 115, 126
    microprocessor, 114, 126
    ports, 119, 127
    power supply, 120, 127
    smartphones, 108, 125
    system board, 112, 125
    tablets, 108, 125
    wearable computers, 109, 125

T

T1 lines, 190, 206
T3 lines, 190, 206
Tables, 66, 263, 277
Tablet computers. *See* Tablets
Tablet hunch, 151
Tablets, 10, 17, 108, 125, 151, 344, 346
    mini, 108, 125
Tabs, 57, 75, 86, 99
Tactical models, 250, 256
Task-oriented languages. *See* Fourth-generation languages (4GLs)
Tbps, 190
Technical feasibility, 292, 299
Technical writer, 14, 152, 157
Technology, 213
Telephone lines, 189, 205
Templates, 63
Temporary Internet files, 217, 233
Temporary storage, 11, 115

Testing process, 316–317, 327
Tethering, 192
Text databases, 274
Text entries, 64
Texting. *See* Text messaging
Text messaging, 35, 48, 186
Thermal printers, 145, 156
Third generation (3G), 333, 336
    mobile telecommunications, 191, 206
Third-generation languages (3GLs), 321–322
Third-party cookie, 217–218
Thunderbolt ports, 119, 127
Tidal (tidal.com), 27
TikTok, 28, 34, 59
TLD. *See* Top-level domain
T-Mobile, 26
Toggle keys, 135, 154
Toolbars, 57, 75
Top-down analysis method, 289, 299
Top-down program design, 310, 325–326
Top-level domain (TLD), 29, 46
Top-level executives, 250
Top-level managers, 245, 255
Top management, 244
Topologies, 199–200, 207
Torvalds, Linus, 339
Touch pad, 136, 154
Touch screen, 135–136, 154
Tower computer, 109, 125
Tower unit, 109
TPS. *See* Transaction processing system
Tracking cookies, 217
Tracks, 166, 178
Traditional graphical user interface, 57
Traditional keyboards, 135, 154
Traditional laptop, 345
Transaction processing system (TPS), 247–248, 252, 256
Transfer rate, 190, 206
Transistor Age, 335
Transmission control protocol/Internet protocol (TCP/IP), 194, 206
Tree network, 199, 207
Trojan horses, 222, 234
Troubleshooting program, 91, 101
Tumblr, 34
TuneUp, 71
Turnitin website, 230

Tweets, 232
Twisted-pair cable, 189, 205
Twitter, 25, 28, 34, 47
Two-in-one laptops, 109, 125

U

UHD 4K, 141
UHD 8K, 141
UHD BD. *See* Ultra HD Blu-rays discs
Ultrabooks, 109, 125, 345
Ultra HD Blu-rays (UHD BD) discs, 169, 179
Ultraportables. *See* Ultrabooks
Unicode, 122, 127
    binary codes, 165
Uniform resource locators (URLs), 29, 46
United Parcel Service (UPS), 138
UNIVAC, 334
Universal product codes (UPCs), 138
Universal serial bus (USB), 118, 119, 126
    flash drives, 165-166, 177
    port types, 119
UNIX, 87, 89, 101
Unmanned aerial vehicles (UAVs), 147, 157
UPCs. *See* Universal product codes
Upgrades, 111
Uplink, 188
Uploading, 31, 47
URLs. *See* Uniform resource locators
USB. *See* Universal serial bus
USB flash drives, 165-166
User interface, 57-58, 75, 85, 99
    graphical. *See* Graphical user interface
Users, 249, 256, 317, 327
Utilities, 7, 17, 84, 91-97, 99, 101
    operating system, 94-96, 101
    virtual assistant, 92-93
Utility suites, 71, 77, 97, 101

V

Vacuum Tube Age, 334
Vector illustrations/images, 68, 77
Venmo, 40
Verizon, 26
Very high level languages (4GLs), 322
Video apps, 59
Videoconferencing, 186, 193

Videoconferencing systems, 252, 256
Video editors, 67, 77
Video game design software, 68-69, 77
Virtual assistants, 91, 101
Virtual Desktop Infrastructure (VDI), 91
Virtualization, 90-91, 101
Virtualization software, 90, 101
Virtual keyboards, 134, 154
Virtual machines, 90, 101
Virtual memory, 116, 126
Virtual private networks (VPNs), 207, 226, 234
Virtual reality (VR), 74, 147, 157
    head-mounted displays and controllers, 147, 157
Viruses, 7, 37, 48, 97, 101, 222, 234
    tracking, 225
VMware (virtual machine program), 90, 101
Voice assist tools, 85, 99
Voiceband, 194
Voice recognition systems, 140, 155
Volatile storage, 115
VR controllers, 147
VR. *See* Virtual reality
VR head-mounted displays, 147

W

WAN. *See* Wide area network (WAN)
Wand readers, 138, 155
Warm boot, 85, 99
Wearable computers, 10, 17, 109, 125
Wearable devices. *See* Wearable computers
Web, 13, 15, 25-26, 46, 216-219, 233, 339
Web 1.0, 25, 46
Web 2.0, 25, 33, 34, 46
Web 3.0, 46
    application, 25, 43, 49
Web 4.0, 25, 46
Web 5.0, 25, 45, 46
Web auctions, 49
Web authoring, 69, 77
Web authoring programs, 61, 77
Web-based e-mail systems, 37, 48
Web-based file transfer services, 31, 47
Web bugs, 218
Webcams, 140, 155
Web databases, 274
Web developers, 14, 44, 49
Webmail, 37, 48

Webmail client, 37
Web page, 30, 46
Web page editors, 69, 77
Website encryption, 204
Web suffix, 29, 46
Web utilities, 30–32, 47
What-if analysis, 65
WhatsApp, 36
Wheel button, 136, 154
Wide area networks (WANs), 199, 206
Widely used programming languages, 266, 315, 326
Wi-Fi Protected Access (WPA3), 226, 234
Wi-Fi standards, 188
Wikipedia, 35, 47
Wikis, 35, 47
Windows, 56, 57, 75, 85, 88, 91, 99, 100
    backup programs, 96
    search program, 94
    storage management programs, 95
Windows Defender, 219
Windows key, 135
Windows operating systems, 57, 88, 89, 100
Windows Server, 87, 200
WinZip, 168
Wired channels, 197
Wired networks, 185
Wireless
    keyboard, 344
    modem, 26, 46, 190
    mouse, 136, 154
Wireless access point, 198, 205
Wireless adapter, 190, 198
Wireless channels, 197
Wireless charging platform, 120, 127
Wireless communications, 186
    devices, 13
Wireless connections, 188, 205
Wireless fidelity (Wi-Fi), 188, 205
    public, 192

Wireless local area network (WLAN), 198, 199, 206
Wireless modems, 26, 46
Wireless network cards, 117, 126
Wireless network encryption, 226
Wireless networks, 185
Wireless revolution, 13, 18, 186, 205
Wireless technology, 186
Wireless wide area network (WWAN) modem, 190, 206
WLAN. *See* Wireless local area network
Word, 114
Word processors, 7, 60–62, 76
Workbook, 65
Worksheet files, 12, 18
Worksheets, 64
    name, 65
World Wide Web (WWW), 25, 46. *See also* Web
Worms, 222, 234
Wozniak, Steve, 337
WQXGA, 141
Write-once discs, 169, 179
Writing, 164
WWW. *See* World Wide Web
WYSIWYG (what you see is what you get) editors, 69, 77

X

X (formerly Twitter), 34

Y

Yahoo!, 38, 48, 339
Yahoo! Mail, 37
Yoga mats, 243, 244
YouTube, 59
    music, 27, 59

Z

Zombies, 222–223, 234
Zuckerberg, Mark, 341